Communications
in Computer and Information Science   716

Commenced Publication in 2007
Founding and Former Series Editors:
Alfredo Cuzzocrea, Dominik Ślęzak, and Xiaokang Yang

More information about this series at http://www.springer.com/series/7899

Stanisław Kozielski · Dariusz Mrozek
Paweł Kasprowski · Bożena Małysiak-Mrozek
Daniel Kostrzewa (Eds.)

Beyond Databases, Architectures and Structures

Towards Efficient Solutions for Data Analysis and Knowledge Representation

13th International Conference, BDAS 2017
Ustroń, Poland, May 30 – June 2, 2017
Proceedings

Springer

Editors
Stanisław Kozielski
Institute of Informatics
Silesian University of Technology
Gliwice
Poland

Bożena Małysiak-Mrozek
Institute of Informatics
Silesian University of Technology
Gliwice
Poland

Dariusz Mrozek ⓘ
Institute of Informatics
Silesian University of Technology
Gliwice
Poland

Daniel Kostrzewa ⓘ
Institute of Informatics
Silesian University of Technology
Gliwice
Poland

Paweł Kasprowski ⓘ
Institute of Informatics
Silesian University of Technology
Gliwice
Poland

ISSN 1865-0929   ISSN 1865-0937 (electronic)
Communications in Computer and Information Science
ISBN 978-3-319-58273-3   ISBN 978-3-319-58274-0 (eBook)
DOI 10.1007/978-3-319-58274-0

Library of Congress Control Number: 2017938564

Printed on acid-free paper

This Springer imprint is published by Springer Nature
The registered company is Springer International Publishing AG
The registered company address is: Gewerbestrasse 11, 6330 Cham, Switzerland

Preface

Collecting, processing, and analyzing data have become important branches of computer science. Many areas of our existence generate a wealth of information that must be stored in a structured manner and processed appropriately in order to gain the knowledge from the inside. Databases have become a ubiquitous way of collecting and storing data. They are used to hold data describing many areas of human life and activity, and as a consequence, they are also present in almost every IT system. Today's databases have to face the problem of data proliferation and growing variety. More efficient methods for data processing are needed more than ever. New areas of interests that deliver data require innovative algorithms for data analysis.

Beyond Databases, Architectures and Structures (BDAS) is a series of conferences located in Central Europe and very important for this geographic region. The conference intends to give the state of the art of the research that satisfies the needs of modern, widely understood database systems, architectures, models, structures, and algorithms focused on processing various types of data. The aim of the conference is to reflect the most recent developments of databases and allied techniques used for solving problems in a variety of areas related to database systems, or even go one step forward — beyond the horizon of existing databases, architectures, and data structures.

The 13th International BDAS Scientific Conference (BDAS 2017), held in Ustroń, Poland, from May 30 to June 2, 2017, was a continuation of the highly successful BDAS conference series started in 2005. For many years BDAS has been attracting hundreds or even thousands of researchers and professionals working in the field of databases. Among attendees of our conference were scientists and representatives of IT companies. Several editions of BDAS were supported by our commercial, world-renowned partners, developing solutions for the database domain, such as IBM, Microsoft, Sybase, Oracle, and others. BDAS annual meetings have become an arena for exchanging information on the widely understood database systems and data-processing algorithms.

BDAS 2017 was the 13th edition of the conference, organized under the technical co-sponsorship of the IEEE Poland Section. We also continued our successful cooperation with Springer, which resulted in the publication of this book. The conference attracted more than a hundred participants from 15 countries, who made this conference a successful and memorable event. There were three keynote talks and one tutorial given by leading scientists: Prof. Jens Allmer from the Department of Molecular Biology and Genetics, Izmir Institute of Technology, Urla, Izmir, gave an excellent keynote talk entitled "Database Integration Facilitating the Merging of MicorRNA and Gene Regulatory Pathways in ALS." Prof. Dirk Labudde from the Bioinformatics group Mittweida (bigM) and Forensic Science Investigation Lab (FoSIL), University of Applied Sciences, Mittweida, Germany, honored us with a presentation entitled "3D Crime Scene and Disaster Site Reconstruction using Open Source Software." Dr. Dominik Szczerba from Future Processing, Gliwice, Poland, gave a talk on "Computational Physiology." Prof. Jean-Charles Lamirel from SYNALP team, LORIA,

Vandœuvre-lès-Nancy, France, prepared a tutorial on "Text Mining in the Big Data Context: Existing Approaches and Challenges." The keynote speeches, tutorials, and plenary sessions allowed participants to gain insight into new areas of data analysis and data processing.

BDAS is focused on all aspects of databases. It is intended to have a broad scope, including different kinds of data acquisition, processing, and storing, and this book reflects fairly well the large span of research presented at BDAS 2017. This volume consists of 44 carefully selected papers that are assigned to seven thematic groups:

- Big data and cloud computing
- Artificial intelligence, data mining, and knowledge discovery
- Architectures, structures, and algorithms for efficient data processing
- Text mining, natural language processing, ontologies, and Semantic Web
- Bioinformatics and biological data analysis
- Industrial applications
- Data mining tools, optimization, and compression

The first group, containing four papers, is devoted to big data and cloud computing. Papers in this group discuss hot topics of stream processing with MapReduce, a tensor-based approach to temporal features modeling with application in big data, querying XML documents with SparkSQL, and automatic scaling computing infrastructure of the cloud. The second group contains six papers devoted to various methods used in data mining, knowledge discovery, and knowledge representation. Papers assembled in this group show a wide spectrum of applications of various exploration techniques, including decision rules, knowledge-based systems, clustering and classification algorithms, and rough sets, to solve many real-world problems.

The third group contains nine papers devoted to various database architectures and models, data structures, and algorithms used for efficient data processing. Papers in this group discuss the effectiveness of query execution, performance, and consistency of various database systems, including relational and NoSQL databases, indexing structures, sorting algorithms, and distributed data processing. The fourth group consists of nine papers devoted to natural language processing, text mining, ontologies, and the Semantic Web. These papers discuss problems of building recommendation systems with the use ontologies, extending expressiveness of knowledge description, ontology reuse for fast prototyping of new concepts, processing natural language instructions by robots, data integration in NLP, authorship attribution for texts, plagiarism detection, and RDF validation.

The research devoted to bioinformatics and biological data analysis is presented in six papers gathered in the fifth group. The papers cover problems connected with gene expression and chromatography but also medical diagnosing as well as face and emotion recognition. The sixth group includes four papers describing various applications of data mining — especially in coal mining and automotive industries. The last group includes six papers presenting various data-mining tools, performance optimization techniques, and a compression algorithm.

We hope that the broad scope of topics related to databases covered in this proceedings volume will help the reader to understand that databases have become an important element of nearly every branch of computer science.

We would like to thank all Program Committee members and additional reviewers for their effort in reviewing the papers. Special thanks to Piotr Kuźniacki — builder and for 12 years administrator of our website bdas.polsl.pl. The conference organization would not have been possible without the technical staff: Dorota Huget and Jacek Pietraszuk.

March 2017

Stanisław Kozielski
Dariusz Mrozek
Pawel Kasprowski
Bożena Małysiak-Mrozek
Daniel Kostrzewa

Organization

Program Committee

Chair

Stanisław Kozielski   Silesian University of Technology, Poland

Honorary Member

Lotfi A. Zadeh   University of California, Berkeley, USA

Members

Alla Anohina-Naumeca	Riga Technical University, Latvia
Sansanee Auephanwiriyakul	Chiang Mai University, Thailand
Vasile Avram	Bucharest Academy of Economic Studies, Romania
Sergii Babichev	J.E. Purkyně University in Ústí nad Labem, Czech Republic
Werner Backes	Sirrix AG Security Technologies, Bochum, Germany
Susmit Bagchi	Gyeongsang National University, Jinju, South Korea
Péter Balázs	University of Szeged, Hungary
Katalin Balla	Budapest University of Technology and Economics, Hungary
Igor Bernik	Nove Mesto University, Slovenia
Bora Bimbari	University of Tirana, Albania
Marko Bohanec	University of Nova Gorica, Slovenia
Alexandru Boicea	Polytechnic University of Bucharest, Romania
Patrick Bours	Gjovik University College, Norway
Lars Braubach	University of Hamburg, Germany
Ljiljana Brkić	University of Zagreb, Croatia
Marija Brkić Bakarić	University of Rijeka, Croatia
Germanas Budnikas	Kaunas University of Technology, Lithuania
Peter Butka	Technical University of Košice, Slovakia
Rita Butkienė	Kaunas University of Technology, Lithuania
Sanja Čandrlić	University of Rijeka, Croatia
George D.C. Cavalcanti	Universidade Federal de Pernambuco, Brazil
Chantana Chantrapornchai	Kasetsart University, Bangkok, Thailand
Ming Chen	University of Bielefeld, Germany
Andrzej Chydziński	Silesian University of Technology, Poland
Armin B. Cremers	University of Bonn, Germany
Tadeusz Czachórski	IITiS, Polish Academy of Sciences, Poland
Yixiang Chen	East China Normal University, Shanghai, P.R. China
Po-Yuan Chen	China Medical University, Taichung, Taiwan, University of British Columbia, BC, Canada

Tomas Krilavičus	Vytautas Magnus University, Lithuania
Antonín Kučera	Masaryk University, Czech Republic
Bora I. Kumova	Izmir Institute of Technology, Turkey
Andrzej Kwiecień	Silesian University of Technology, Poland
Dirk Labudde	University of Applied Sciences, Mittweida, Germany
Jean-Charles Lamirel	LORIA, Nancy, France, University of Strasbourg, France
Dejan Lavbič	University of Ljubljana, Slovenia
Fotios Liarokapis	Masaryk University, Czech Republic
Sergio Lifschitz	Pontificia Universidade Catolica do Rio de Janeiro, Brazil
Antoni Ligęza	AGH University of Science and Technology, Poland
Maciej Liśkiewicz	University of Lübeck, Germany
Ivica Lukić	Josip Juraj Strossmayer University of Osijek, Croatia
Ivan Luković	University of Novi Sad, Serbia
Bożena Małysiak-Mrozek	Silesian University of Technology, Poland, IEEE member
Algirdas Maknickas	Vilnius Gediminas Technical University, Lithuania
Violeta Manevska	St. Clement of Ohrid University of Bitola, Macedonia
Saulius Maskeliūnas	Vilnius University, Lithuania
Jelena Mamčenko	Vilnius Gediminas Technical University, Lithuania
Marco Masseroli	Politecnico di Milano, Italy
Dalius Mažeika	Vilnius Gediminas Technical University, Lithuania
Zygmunt Mazur	Wroclaw University of Technology, Poland
Peter Mikulecký	University of Hradec Králové, Czech Republic
Biljana Mileva Boshkoska	Nove Mesto University, Slovenia
Guido Moerkotte	University of Mannheim, Germany
Yasser F.O. Mohammad	Assiut University, Egypt
Tadeusz Morzy	Poznan University of Technology, Poland
Mikhail Moshkov	King Abdullah University of Science and Technology, Saudi Arabia
Dariusz Mrozek	Silesian University of Technology, Poland, IEEE member
Mieczysław Muraszkiewicz	Warsaw University of Technology, Poland
Sergio Nesmachnow	Universidad de la Republica, Uruguay
Laila Niedrīte	University of Latvia, Latvia
Mladen Nikolić	University of Belgrade, Serbia
Tadeusz Pankowski	Poznan University of Technology, Poland
Martynas Patašius	Kaunas University of Technology, Lithuania
Bogdan Pătruţ	Vasile Alecsandri University of Bacău, Romania
Mile Pavlić	University of Rijeka, Croatia
Witold Pedrycz	University of Alberta, Canada, IEEE member
Adam Pelikant	Lodz University of Technology, Poland
Horia F. Pop	Babeş-Bolyai University, Romania
Václav Přenosil	Masaryk University, Czech Republic
Hugo Proenca	University of Beira Interior, Portugal

Organizing Committee

Bożena Małysiak-Mrozek
Dariusz Mrozek
Pawel Kasprowski

Daniel Kostrzewa
Piotr Kuźniacki
Dorota Huget

Additional Reviewers

Dariusz Rafał Augustyn
Małgorzata Bach
Piotr Bajerski
Robert Brzeski
Adam Duszenko
Jacek Frączek
Katarzyna Harężlak
Michał Kozielski
Marcin Michalak
Alina Momot

Thi Hoa Hue Nguyen
Agnieszka Nowak-Brzezińska
Karolina Nurzyńska
Ewa Płuciennik
Arkadiusz Poteralski
Ewa Romuk
Robert Tutajewicz
Aleksandra Werner
Łukasz Wyciślik
Hafed Zghidi

Sponsoring Institutions

Technical co-sponsorship of the IEEE Poland Section

Contents

Architectures, Structures and Algorithms for Efficient Data Processing

Text Mining, Natural Language Processing, Ontologies and Semantic Web

Bioinformatics and Biological Data Analysis

Industrial Applications

Data Mining Tools, Optimization and Compression

Big Data and Cloud Computing

Integrating Map-Reduce and Stream-Processing for Efficiency (MRSP)

Pedro Martins[✉], Maryam Abbasi, José Cecílio, and Pedro Furtado

Polytechnic Institute of Viseu, Department of Computer Sciences,
University of Coimbra (CISUC Research Group), Coimbra, Portugal
{pmom,maryam,jose,pnf}@dei.uc.pt

Abstract. Works in the field of data warehousing (DW) do not address
Stream Processing (SP) integration in order to provide results freshness (i.e. results that include information that is not yet stored into the
DW) and at the same time to relax the DW processing load. Previous
research works focus mainly on parallelization, for instance: adding more
hardware resources; parallelizing operators, queries, and storage. A very
known and studied approach is to use Map-Reduce to scale horizontally in order to achieve more storage and processing performance. In
many contexts, high-rate data needs to be processed in small time windows without storing results (e.g. for near real-time monitoring), in other
cases, the objective is to relax the data warehouse usage (e.g. keeping
results updated for web-pages reload). In both cases, stream processing
solutions can be set to work together with the data warehouse (Map-
Reduce or not) to keep results available on the fly avoiding high query
execution times, and, this way leaving the DW servers more available to
process other heavy tasks (e.g. data mining).

In this work, we propose the integration of Stream Processing and
Map-Reduce (MRSP) for better query and DW performance. This approach allows to relax the data warehouse load, and, by consequence
reducing the network usage. This mechanism integrates into Map-Reduce
scalability mechanisms and uses the Map-Reduce nodes to process
Stream queries.

Results show/compare performance gains on the DW side and the
quality of experience (QoE) when executing queries and loading data.

Keywords: Complex event processing · Stream processing · Extraction
transformation and load · Distributed system · Data warehouse · Big
data · Small data · Map-Reduce

1 Introduction

In this paper, we investigate the problem of integrating stream processing with
data warehousing systems oriented to Map-Reduce processing, with the objective to offer scalability, query and load performance, relax the data warehouse
storage and load, and at the same time dealing efficiently with high-rate data.

© Springer International Publishing AG 2017
S. Kozielski et al. (Eds.): BDAS 2017, CCIS 716, pp. 3–15, 2017.
DOI: 10.1007/978-3-319-58274-0_1

Consequently, our approach reduces the network usage, by minimizing query access to the data warehouse. Query Data Warehouse access is performed the first time a query is registered/submitted, second+ times query results are kept for update as new data arrives.

Map-Reduce approaches are designed to scale the storage, process capacity and assure fault tolerance mechanisms. However, the Map and Reduce processing architecture for high fault tolerance, data redundancy, materialization mechanisms (resulting in excessive network usage), impairs query execution performance.

Complex Event Processing (CEP) applications deal mostly with high-right data volume processing (big-data) based on a window. However, CEP engines do not store any data. Although, they are extremely efficient processing queries and keeping updated results as new data arrives.

In this paper, we propose a solution to enable CEP systems integration with Map-Reduce based approaches. At the same time, the CEP approach should be able to load-(re)balance and support automated elasticity as happens with the Map-Reduce approach. Our proposed system is built to adapt itself when overloaded nodes appear due to an increase of data or increase of simultaneous queries. The approach allows isolating the query execution from the DW.

When a new query is registered, it is assigned to a pool of nodes (which can include the Map and Reduce nodes), choosing the one with Least Work Remaining (LWR). The first result is obtained using the Map-Reduce DW system. Once new data arrives, before storing, it is sent to a layer running on top of the Map-Reduce architecture, which will replicate data across the CEP nodes and process the registered queries in order to update the results. After the queries result updated, the data is stored in the Map-Reduce DW architecture. Thus, instead of storing data and then process query results (traditional approach), with our approach, MRSP, we first process the queries and then store the data, this way relaxing the data warehouse load.

Based on the proposed approach, only for the first time, query results need to use the Map-Reduce processing architecture, after that point, queries are keep updated as new data arrives. This method improves, the data warehouse storage system performance, availability, network usage, and saves processing resources (hardware and software).

This solution shows promising results when applied to systems monitoring (e.g. fraud detection) and for keeping heavy load websites updated in real-time (i.e. avoiding constant access to the data warehouse to retrieve updated results). In the following sections, we provide the description of the approach and experimental demonstration.

2   Related Work

In this section, we analyze related works in the field of stream processing integration with Map-Reduce approaches, for query performance and DW load relaxation.

Most current distributed stream processing systems, including Yahoo!'s S4 [26], Twitter's Storm [24], and streaming databases [4–6], are based on a record-at-a-time processing model, where nodes receive each record, update internal state, and send out new records in response. This model raises several challenges in a large-scale cloud environment, especially regarding query performance due to high amounts of data to be analyzed.

Academic related work on stream processing such as Aurora, Borealis, Tele-graph, and STREAM [2, 4–6], provide a SQL interface and achieve fault recovery through replication (an active or passive standby for each node [4, 14]), and on top of all they focus on more theoretical aspects of the paradigm.

In addition, researchers have also focused on issues like load distribution [1], load-balancing [29] and fault-tolerance [4] in stream processing systems. Stream processing systems typically provide support for instantiating real-time data analysis applications that consume high-rate data streams. However, such systems are not meant to analyze large volumes of stored data and often rely on data warehouses and other ad-hoc mechanisms for analyzing stored data, and such analysis is often the source of the analytic model used by stream processing operators.

The industry activity [1] has primarily focused on commercialization and wide availability of this novel data processing paradigm, most stream process-ing frameworks, such as, S4, Storm, and Flume [13, 24, 26], focuses on recovery from failures (e.g., by keeping all state in a replicated database) and consistency across nodes. Several recent research systems have looked at on-line processing in clusters, map-reduce Online [15] is a streaming Hadoop run-time, but can-not compose multiple map-reduce steps into a query or reduce tasks. iMR [17] is an in-situ map-reduce engine for log processing, but does not support more general computation graphs and can lose data on failure. CBP [16] and Comet [12] provide "bulk incremental processing" by running map-reduce jobs on new data every few minutes to update state in a distributed file system. However, they incur the high overhead of replicated on-disk storage. Naiad [18] runs com-putations incrementally but does not yet have a cluster implementation or a discussion of fault tolerance. Percolator [22] performs incremental computations using triggers, but does not offer consistency guarantees across nodes or high-level operators like map and join.

D-Streams [31] mechanism is conceptually similar to recovery techniques in map-reduce, GFS, and RAMCloud [9, 20, 27], which all leverage (re)partitioning for small timescales stream processing. D-Streams, re-computes lost data instead of having to replicate all data, avoiding the network and storage cost of replication.

The map-reduce programming model [9] has gained significant acclaim for its ease-of-use and scalability. It offers a parallelization framework that depends on a run-time component for scheduling and managing parallel tasks [28] and a distributed file system that is responsible for providing parallel access to different blocks of the stored data [11, 28]. The success of the programming model has resulted in the development of an open source implementation of its run-time called Hadoop [28] and the Hadoop distributed file system [27]. A number of efforts are also focused on developing languages that help the specification of map-reduce jobs [19]. Recent research efforts have also focused on adapting map-reduce for a database like operations using a map-reduce-merge framework [30]. Another effort focuses on implementing well-known machine learning algorithms for the map-reduce programming model [21]. Although map-reduce has also received a lot of criticism for ignoring a lot of research on task parallelization and data modeling conducted by database researchers [10], it continues to be the new preferred platform for developing large-scale analytic on static data.

Data analysis systems have evolved from monolithic data warehouses [23] of the past to the modern day real-time data analysis systems [1]. The nature of the task has changed from offline to online and therefore new sets of challenges have been posed to the research community. Modern day data analysis systems should not only scale with the volume of stored data but also scale with the rate at which live data arrives. In addition to that, modern frameworks must be able to provide processing capabilities to deal with both structured and unstructured data. Another recent framework created for high-volumes of data and high-performance is Spark [3] which focuses on adding an in-memory layer before the fiscal data materialized storage to provide more performance. However, this approach still focuses on storing and then querying data.

3   Architecture

In this section, we describe the most relevant aspects regarding the integration of Stream-Processing and Map-Reduce. Moreover, we also discuss how MRSP scales, and (re)balances registered queries for performance.

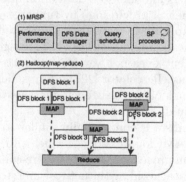

Fig. 1. Global architecture

Figure 1, shows the MRSP global proposed architecture. On top marked by (1), we have the CEP layer, part of the proposed system, (2) represents the Hadoop Map-Reduce. The MRSP (1) is composed by:

- "Performance Monitor" - this module is responsible for monitoring the performance of each query and the respective nodes. If performance is under the desired threshold, queries are (re)scheduled to other node using a least-work-remaining (LWR) algorithm. If the measure does not solve the performance problem a new node is requested to the admin. This algorithm is described in more detail bellow.
- "DFS Data Manager" - ensures the correct data insertion into (2) and data replication for all registered queries.
- "Query Scheduler" - together with the "performance monitor" module (based on its alarms), this module is responsible for (re)scheduling the registered queries across the available nodes (LWR algorithm).
- "SP process" - this module manages the running registered queries. It must assure the on-time start of each query (i.e. accordingly with its running frequency) and output results.

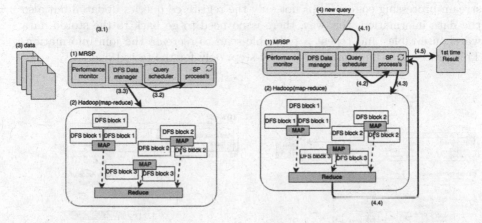

Fig. 2. Data store   **Fig. 3.** Processing new queries

Every time new data arrives to be stored, the entire MRSP and MR data distribution process are transparent to the user. Figure 2, shows in (3) the data to be stored and the steps inside the framework to replicate and store it into the (2) and at the same time the necessary replication to keep the stream processing queries results updated. In (3.1) the information is submitted, the "DFS Data manager" in (3.3) it sends the data into the Map-Reduce architecture (2) and at the same time it replicates the data into the "SP process" module (3.2) which is processing and keeping the stream queries results updated. This operation is performed automatically in all Stream Processing nodes simultaneously as data is submitted. The user does not need to deal with the information distribution across neither the MRSP (1) or the MR (2).

The first time a query is submitted the result from (2) needs to be merged with (1), all the remaining times the same query is executed the result is kept for update, thus no need to create more load in (2) - accessing it to compute

and fetch the new/same result. Figure 3, shows a new query being submitted
(4), then in (4.1) the "Query Scheduler" assigns the LWR node to register the
query and in (4.2) the query is set to run in the "SP process" module. Because
it is the first time this query is executed, there is no known result from the
DW to be updated and outputted, so the "SP process" request the query to
be executed at the DW (4.3) and waits until the result is returned back to the
"SP process" (4.4). Based on the result from the DW the query is processed and
results updated (merged with new information) if necessary to be outputted
(4.5). Note that, the first time a query is submitted the execution time depends
on the DW performance. Remaining times, second+, results performance are
assured by the MRSP proposed module.

For instance: A new tuple arrives for table T. The query we want to update is
T join S join V join ... As the query is registered in the framework and executed
for the first time all relevant information regarding the join condition is kept
on the stream-processing engine side (memory/disk depending on the available
space), in the form of column organization, as new data arrives to be stored, the
stream-processing engine keeps not only the registered queries updated but also
the data information, this way, there is no need to go back to the stored data
warehouse tables and repeat a full-table-scan to recreate the join information.
This process was implemented step-by-step, only for the tested queries.

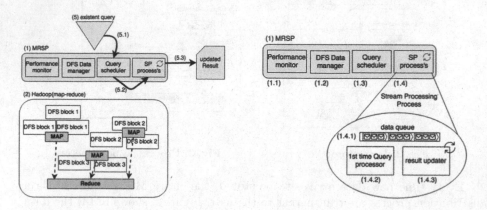

Fig. 4. Processing existent queries   **Fig. 5.** Stream processing process

Second time (and more, second+) times the same query is executed, it does
not need to access the DW (2) since the result is already known. If new data
arrives, it will be replicated to the query processor module to keep results constantly updated. Figure 4, shows the (5.2) and (5.3), where no access to the DW
(2) is required to output the query result. This way relaxing the DW access.

Each data "SP process" is formed by three main modules, Fig. 5 (1.4), the
data queue (1.4.1) used to detect overload situations by (1.1). If this queue
increases above a certain configured limit size, it means that the node is over
capacity and queries/load needs to be re-balanced. The module (1.4.2) is used

to request the first-time result of a query, and finally, the (1.4.3) keeps results updated as new data arrives.

Note that, the Performance Monitor (1.1) module monitors both the data queue size from (1:4) nodes and the executing performance time for the queries (the desired query execution time is configured by the user, and applied to all queries, future version will allow independently desired execution times for each query).

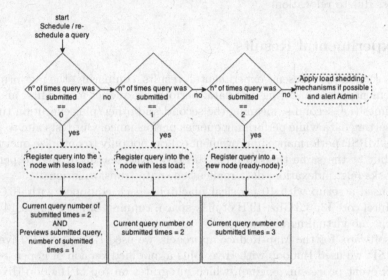

Fig. 6. Query load balance and scheduling algorithm

Figure 6 depicts the query scheduling algorithm. Each query has a counter associated, representing the number of times a query was (re)scheduled.

- If the query was (re)scheduled zero times (meaning it is a new query), then the scheduler will find the best node to register it, and the "number of scheduled times" is set to two (2 because the query was placed in the best fitting node (i.e. LWR algorithm), if it becomes overloaded (i.e. the queue starts increasing above a threshold) it will go directly into a new node ...
- ... Then the "number of scheduled times" of the previous last registered query is set to one. Because the "Query scheduler" already chooses the best node to register the query (i.e. the one with least load) and it did not become overloaded;
- Now, if the "Performance monitor", detects an overload situation in a node, it sends the last submitted query in that node (by order of registration) to the scheduler to be resubmitted into a better node, and the parameter "number of scheduled times" is increased to 2 (it was set to one after a new query was inserted, or already to 2);
- At the third ("number of scheduled times" = 2) (re)schedule of a query, it is put into a ready-node if available. If the ready node is not available or it gets overloaded, then the load shedding and admin alert algorithm will deal with the problem.

Every time a query is relocated because of an overload situation, that query is not removed from the overloaded node until the newly selected node (the one with most available resources) is providing results from an equal data window. Upon results provided, the scheduler decides in which node to leave the query running. For that, the system scheduler analyses the throughput of both nodes and removes the query from the node with less throughput. This process avoids work loss due to relocation.

4   Experimental Results

In this section we present experimental results comparing the performance improvements when inserting new data, performing queries for the first and second+ times (i.e. second+, meaning the second and other/plus execution times), and, inserting data while performing queries performance. Our tests aim to prove that the MRSP performance improvements allow not only to optimize query execution but at the same time to relax the DW load, leaving it free to perform other tasks (e.g. indexation, data replication, data transformations).

We used a setup with 10 physical machines, each equipped with 16 GB of RAM, Intel core I3, 3.3 Ghz, 1TB of disk space, running Linux Ubuntu 14 LTS. Note that, no virtual nodes were used.

As software, for the Map-Reduce approach, we used Hadoop with Hive. For the MRSP, we used Hadoop with Hive, plus a modified version of Esper [8] (i.e. complex event processing engine), which integrates on top of Hadoop-Hive, for efficient data and query information exchange.

Each test was performed 6 times, the worst and best times were discarded and remaining results were used to make an average. All performance times are shown in seconds, and load speed in MB per second.

To benchmark the different scenarios, we used TPC-H schema [7], data, and queries, with a scale factor of 1TB total. We use Hadoop with Hive [25], modified to support the integration of the proposed MRSP. For the high-rate scenario we used the same relational schema, in order to support the same results from the tested queries. The selected queries to be tested were Q2, Q3, Q5, Q8, Q21 from TPC-H. This choice was made based on their implementation complexity and integration complexity with our prototype framework.

4.1   Data Load with and Without MRSP

In order to compare loading performance, we load 1TB of information into 10 nodes. We compare Hadoop-Hive (Map-Reduce approach) with the proposed MRSP solution.

Figure 7 shows, the average performance of all nodes in MB/s for both approaches over the time. When using only the Hadoop-Hive Map-Reduce the global average load time was 35 min. When using the proposed approach it takes 43 min. Despite the difference between both systems, performance times are not very distant from each other, the MRSP is slower due to the added layer on

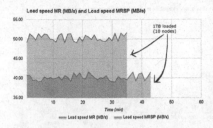

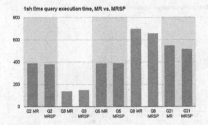

Fig. 7. Comparing data load perfor-   **Fig. 8.** 1st time query execution
mance, MR vs. MRSP

top of MR to replicate all data into the stream processing module. This module requires all data that arrives at the system to pass trough it so that it can process and keep all registered queries results updated.

We conclude that load performance involving huge amounts of data is slightly affected when using the proposed solution (MRSP). However, given the high data volume, the performance difference is not very significant.

4.2 Query Performance in MR and MRSP (First-Time Run)

Query execution time represents the most important performance part of every data storage system. In this section, we test query performance (in seconds) when executing a new query, submitted for the first time.

Figure 8 shows, a comparison between the MR and the MRSP when executing a new query for the first time. In the figure, we can see that for the first execution, both systems take almost the same time to finish. This leads us to conclude that, in MRSP the extra processing layer added on top of the MR, does not affect query performance.

4.3 Query Performance in MR and MRSP (Second+ Time Run)

Query execution in all systems runs faster second+ times. This happens because many of the queried information stays in memory (i.e. operating system, internal database engine, disk caches, and so on). In this section, we compare second+ query execution on MR, where the query is performed over all stored data, and in MRSP, where the query result is updated based only the new ingress data.

Figure 9 shows, the query performance improvement when executing for the second+ time. Both MR and MRSP improve their performance speed, with special attention to MRSP, where in just some seconds the results are outputted. We conclude that MRSP is able to improve query performance very significantly while at the same time relaxing resources of the DW to process other tasks. MRSP is able to fast process the query results because it does not need to query all 1TB of stored data, as happens in MR. Results are kept in memory or disk and updated as new data arrives (i.e. the same way as Stream Processing engines work).

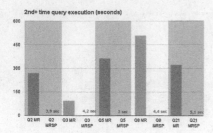

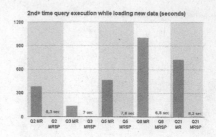

Fig. 9. 2nd time+ query execution

Fig. 10. Query performance while loading new data

4.4   Query Performance While Inserting New Data in MR and MRSP

In high-rate demanding scenarios, at the same time queries are executing, also new data is arriving to be treated and stored. This leads to an extra DW load, consequently a performance degradation. In this section, we demonstrate, for both MR and MRSP, the impact on queries execution (i.e. executing second+ times) when at the same time loading and storing new data to be queried.

Figure 10 shows, MR query performance in seconds, while at the same time loading data. When comparing MR results from Fig. 10 with 9, we notice that in Fig. 9 there is a slight increase on the query execution time. This is due to the extra tasks being performed simultaneously while loading data (i.e. load, indexation, replication of new data). The proposed solution, MRSP, every time new data arrives, the registered query results are immediately updated, and then data is stored. Although, we detect a time performance increase in MRSP while loading data. In this scenario, MRSP still shows a very big performance increase face to MR. MRSP performance decrease (i.e. execution time increase) is due to the extra necessary memory and processing to keep results updated by the stream processing engine.

With this results, we prove that MRSP allows not only to significantly optimize query results but also to relax the DW nodes to perform other data maintenance operations such as transformation, indexation, replication, and so on.

5   Conclusions and Future Work

In this work, we research a solution for almost-real-time query performance on MR systems. Our main approach consists on querying data before storing it, for that we integrate a stream processing engine on top of the MR architecture. On the other hand, traditional database engines approaches and also MR approaches, first focus on storing data, and then on queering it. Our proposed solution, MRSP, demonstrates huge performance gains, especially when executing queries for the second+ times, and on top of all, it allows to reduce the MR data warehouse nodes load.

Future work includes improvements to the MRSP prototype to support a wider diversity of queries. Another direction for the future work would be to include in MRSP query replication. In case one of the nodes computing/updating the result fails, other node takes over. More important future work, not address in this work, is the automatic query rewrite and adaptation to Hive Query Language (HQL) and at the same time to the stream processing engine.

Acknowledgements. This project was partially financed by CISUC research group from the University of Coimbra, and: This work is financed by national funds through FCT - Fundação para a Ciência e Tecnologia, I.P., under the project UID/Multi/04016/2016. Furthermore we would like to thank the Instituto Politécnico de Viseu and CI&DETS for their support.

References

1. Amini, L., Andrade, H., Bhagwan, R., Eskesen, F., King, R., Selo, P., Park, Y., Venkatramani, C.: SPC: a distributed, scalable platform for data mining. In: Proceedings of the 4th International Workshop on Data Mining Standards, Services and Platforms, pp. 27–37. ACM (2006)
2. Arasu, A., Babcock, B., Babu, S., Cieslewicz, J., Datar, M., Ito, K., Motwani, R., Srivastava, U., Widom, J.: STREAM: the stanford data stream management system. In: Garofalakis, M., Gehrke, J., Rastogi, R. (eds.) Data Stream Management. DSA, pp. 317–336. Springer, Heidelberg (2016). doi:10.1007/978-3-540-28608-0_16
3. Armbrust, M., Xin, R.S., Lian, C., Huai, Y., Liu, D., Bradley, J.K., Meng, X., Kaftan, T., Franklin, M.J., Ghodsi, A., et al.: Spark SQL: relational data processing in spark. In: Proceedings of the 2015 ACM SIGMOD International Conference on Management of Data, pp. 1383–1394. ACM (2015)
4. Balazinska, M., Balakrishnan, H., Madden, S.R., Stonebraker, M.: Fault-tolerance in the borealis distributed stream processing system. ACM Trans. Database Syst. (TODS) **33**(1), 3 (2008)
5. Chandrasekaran, S., Cooper, O., Deshpande, A., Franklin, M.J., Hellerstein, J.M., Hong, W., Krishnamurthy, S., Madden, S.R., Reiss, F., Shah, M.A.: TelegraphCQ: continuous dataflow processing. In: Proceedings of the 2003 ACM SIGMOD International Conference on Management of Data, pp. 668–668. ACM (2003)
6. Cherniack, M., Balakrishnan, H., Balazinska, M., Carney, D., Cetintemel, U., Xing, Y., Zdonik, S.B.: Scalable distributed stream processing. In: CIDR, vol. 3, pp. 257–268 (2003)
7. Council, T.P.P.: TPC-H benchmark specification (2008). http://www.tcp.org/hspec.html
8. Cugola, G., Margara, A.: Processing flows of information: from data stream to complex event processing. ACM Comput. Surv. (CSUR) **44**(3), 15 (2012)
9. Dean, J., Ghemawat, S.: MapReduce: simplified data processing on large clusters. Commun. ACM **51**(1), 107–113 (2008)
10. DeWitt, D., Stonebraker, M.: MapReduce: a major step backwards. Database Column **1**, 23 (2008)
11. Ghemawat, S., Gobioff, H., Leung, S.T.: The Google file system. In: ACM SIGOPS Operating Systems Review, vol. 37(5), pp. 29–43. ACM (2003)

12. He, B., Yang, M., Guo, Z., Chen, R., Su, B., Lin, W., Zhou, L.: Comet: batched stream processing for data intensive distributed computing. In: Proceedings of the 1st ACM Symposium on Cloud Computing, pp. 63–74. ACM (2010)
13. Hoffman, S.: Apache Flume: Distributed Log Collection for Hadoop. Packt Publishing Ltd, Birmingham (2013)
14. Krishnamurthy, S., Franklin, M.J., Davis, J., Farina, D., Golovko, P., Li, A., Thombre, N.: Continuous analytics over discontinuous streams. In: Proceedings of the 2010 ACM SIGMOD International Conference on Management of Data, pp. 1081–1092. ACM (2010)
15. Li, M., Zeng, L., Meng, S., Tan, J., Zhang, L., Butt, A.R., Fuller, N.: MRONLINE: MapReduce online performance tuning. In: Proceedings of the 23rd International Symposium on High-Performance Parallel and Distributed Computing, pp. 165–176. ACM (2014)
16. Logothetis, D., Olston, C., Reed, B., Webb, K.C., Yocum, K.: Stateful bulk processing for incremental analytics. In: Proceedings of the 1st ACM Symposium on Cloud Computing, pp. 51–62. ACM (2010)
17. Logothetis, D., Trezzo, C., Webb, K.C., Yocum, K.: In-situ MapReduce for Log processing. In: 2011 USENIX Annual Technical Conference (USENIX ATC 2011), p. 115 (2011)
18. McSherry, F., Isaacs, R., Isard, M., Murray, D.G.: Naiad: the animating spirit of rivers and streams. SOSP Poster Session (2011)
19. Olston, C., Reed, B., Srivastava, U., Kumar, R., Tomkins, A.: Pig Latin: a not-so-foreign language for data processing. In: Proceedings of the 2008 ACM SIGMOD International Conference on Management of Data, pp. 1099–1110. ACM (2008)
20. Ongaro, D., Rumble, S.M., Stutsman, R., Ousterhout, J., Rosenblum, M.: Fast crash recovery in RAMCloud. In: Proceedings of the Twenty-Third ACM Symposium on Operating Systems Principles, pp. 29–41. ACM (2011)
21. Owen, S., Anil, R., Dunning, T., Friedman, E.: Mahout in action. Manning Shelter Island (2011)
22. Peng, D., Dabek, F.: Large-scale incremental processing using distributed transactions and notifications. In: OSDI, vol. 10, pp. 1–15 (2010)
23. Rajakumar, E., Raja, R.: An overview of data warehousing and OLAP technology. Adv. Nat. Appl. Sci. 9(6 SE), 288–297 (2015)
24. Ranjan, R.: Streaming big data processing in datacenter clouds. IEEE Cloud Comput. 1, 78–83 (2014)
25. Thusoo, A., Sarma, J.S., Jain, N., Shao, Z., Chakka, P., Anthony, S., Liu, H., Wyckoff, P., Murthy, R.: Hive: a warehousing solution over a map-reduce framework. Proc. VLDB Endowment 2(2), 1626–1629 (2009)
26. Wang, C., Rayan, I.A., Schwan, K.: Faster, larger, easier: reining real-time big data processing in cloud. In: Proceedings of the Posters and Demo Track, p. 4. ACM (2012)
27. Wang, M., Li, B., Zhao, Y., Pu, G.: Formalizing Google file system. In: 2014 IEEE 20th Pacific Rim International Symposium on Dependable Computing (PRDC), pp. 190–191. IEEE (2014)
28. White, T.: Hadoop: The Definitive Guide. O'Reilly Media Inc., Sebastopol (2012)
29. Xing, Y., Zdonik, S., Hwang, J.H.: Dynamic load distribution in the borealis stream processor. In: Proceedings. 21st International Conference on Data Engineering, ICDE 2005, pp. 791–802. IEEE (2005)

30. Yang, H.c., Dasdan, A., Hsiao, R.L., Parker, D.S.: Map-Reduce-merge: simplified relational data processing on large clusters. In: Proceedings of the 2007 ACM SIGMOD International Conference on Management of Data, pp. 1029–1040. ACM (2007)
31. Zaharia, M., Das, T., Li, H., Shenker, S., Stoica, I.: Discretized streams: an efficient and fault-tolerant model for stream processing on large clusters. HotCloud **12**, 10 (2012)

Tensor-Based Modeling of Temporal Features for Big Data CTR Estimation

Andrzej Szwabe, Pawel Misiorek$^{(\boxtimes)}$, and Michal Ciesielczyk

Institute of Control and Information Engineering, Poznan University of Technology,
ul. Piotrowo 3a, 60-965 Poznan, Poland
{andrzej.szwabe,pawel.misiorek,michal.ciesielczyk}@put.poznan.pl

Abstract. In this paper we propose a simple tensor-based approach to temporal features modeling that is applicable as means for logistic regression (LR) enhancement. We evaluate experimentally the performance of an LR system based on the proposed model in the Click-Through Rate (CTR) estimation scenario involving processing of very large multi-attribute data streams. We compare our approach to the existing approaches to temporal features modeling from the perspective of the Real-Time Bidding (RTB) CTR estimation scenario. On the basis of an extensive experimental evaluation, we demonstrate that the proposed approach enables achieving an improvement of the quality of CTR estimation. We show this improvement in a Big Data application scenario of the Web user feedback prediction realized within an RTB Demand-Side Platform.

Keywords: Big data · Multidimensional data modeling · Context-aware recommendation · Data extraction · Data mining · Logistic regression · Click-through rate estimation · WWW · Real-Time Bidding

1 Introduction

Web content utility maximization is one of the main paradigms of the so-called Adaptive Web [3]. Many researchers agree that Click-Through Rate (CTR) estimation is important for maximization of Web content utility and that machine learning plays a central role in computing the expected utility of a candidate content item to a Web user. The click prediction – widely referred to as CTR estimation – is an interesting and important data mining application scenario, especially when realized on the Web scale [4,9,16]. Real-Time Bidding (RTB) belongs to the best examples of widely-used Big Data technologies [15]. As confirmed by many authors, the research on RTB algorithms involves facing many challenges that are typical for Big Data. In particular, RTB algorithms must be capable to process heterogeneous and very sparse multi-attribute data streams having the volume order of terabytes rather than gigabytes [4]. Moreover, to be applicable in a real-world environment, an RTB optimization algorithm must be able to provide its results in tens of milliseconds [15].

© Springer International Publishing AG 2017
S. Kozielski et al. (Eds.): BDAS 2017, CCIS 716, pp. 16–27, 2017.
DOI: 10.1007/978-3-319-58274-0_2

Digital advertising is a rapidly growing industry already worth billions of dollars. RTB is one of the leading sectors of the digital advertising industry. In this paper, we contribute to the intensively investigated area of research on machine learning algorithms optimizing RTB-based dynamic allocation of ads. The proposed solution faces the challenges imposed by the RTB protocol requirements and, at the same time, introduces the temporal feature engineering based on the tensor model what has not been investigated yet.

Some of the tensor-based approaches to data modeling have already been identified as addressing the Big Data challenges [5,10]. Although the area of the research on tensor-based Big Data modeling has emerged quite recently [5], the results achieved so far, indicate that, at least in online advertising application scenarios, tensor-based approaches are able to outperform many alternative ones, including those based on the matrix factorization and deep learning [16,21].

2   Related Work

As far as Big Data application scenarios are concerned, it is widely agreed that Logistic Regression (LR) is the state-of-the-art CTR estimation method [4,16, 23]. For this reason, the scope of the research presented in this paper is limited to feature modeling applicable to a data mining system based on LR.

In the context of the RTB Demand Side Platform (DSP) optimization scenario, it is important to make the CTR estimation algorithm highly contextual and capable to exploit various data augmentations [16,22]. In the relevant papers these two requirements are sometimes integrated into the single, more general requirement. Specifically, recommender systems deployed to perform CTR estimation are required to model the heterogeneous data attributes explicitly from multiple alternative and complementary 'aspects' [16]. The idea of such 'multi-aspect' data modeling is familiar to researchers working on tensor-based data representation methods [2,13]. The need for 'multi-aspect' data modeling has been recognized by the authors of tensor-based RTB CTR estimation systems [16] and by the authors of advanced classification systems theoretically-grounded on the rough set theory [12]. All these types of data mining systems perform some type of 'multi-aspect' data modeling by using combinations of multiple 'interacting' features [4,16].

There are a few approaches to building feature conjunctions that have been presented in the literature on RTB CTR estimation [4,9,16,21]. Some of the papers involve the explicit use of the cartesian product or the tensor product in the models' definitions [4,16].

It is worth recalling that the tensor space is a space formed over a cartesian product of the constituent vector spaces. In the context of an algebraic feature representation, it is a straightforward and widely-followed assumption to represent features in their vector spaces and to map the feature values to the dimensions of these spaces [8,14,18,20]. Although not all the authors of such feature conjunctions models explicitly refer to the tensor product as the means for building the algebraic representations of feature conjunctions, such a tensor-based definition is a direct consequence of the assumption that the constituent features are represented by vector spaces. On the other hand, the authors of many papers presenting tensor-based models of the data sets elements having the form of the properties' conjunctions, explicitly refer to the tensor product as the means for building algebraic multi-feature data representations [8, 14, 16, 18, 20].

To the best of our knowledge, there are no publications presenting multi-linear temporal feature models designed for CTR estimation systems based on LR. Although temporal features are included in the overall feature sets of the models proposed by the authors of the leading CTR estimation algorithms based on LR [4, 9, 16, 21], none of these models is both tensor-based and used for multi-linear representation of temporal features. In consequence, none of these models involves the use of feature conjunctions of different arity [4, 9, 16]. Moreover, the impact of an application of temporal features (with or without their conjunctions) on the quality of CTR estimation has not been presented in any of the above-recalled papers. Therefore, we believe that the research results presented herein are not only practically useful, but may also be regarded as original and interesting theoretical contribution to the field of the research on feature models for CTR estimation systems.

3   Tensor-Based Feature Modeling

Tensor-based data modeling is a broad topic – typically investigated from the perspective of various approaches to tensor-based data processing [5, 11, 20]. It has to be stressed that the scope of the tensor-based modeling that is represented by the model proposed in this paper is relatively narrow – it is limited to (i) the 'feature addressing' scheme based on the tensor product of the feature-indexing standard basis vectors and (ii) the use of a simple multi-tensor network. In such a simplified form, a tensor-based model of additional features (herein referred to as metafeatures) is equivalent to the state-of-the-art feature models defined with the use of the cartesian product that are presented in the literature on CTR estimation based on LR [4, 16]. It also has the most distinctive property of any tensor-based data representation, which is the ability to represent data in its natural form in which vector space dimensions represent feature values, rather than data/training examples [14, 20]. Thanks to this property, any combination of the features may be mapped on its dedicated tensor entry. Moreover, the use of the multi-tensor hierarchy network (presented in Sect. 3.2) provides simple means for the mapping between a conjunction features' subset and the corresponding tensor network node; the arity of the conjunction tuple maps to the level of the tensor network – the level including the tensors of the order equal to the arity of the conjunction tuple.

3.1   Tensor-Based Multidimensional Data Modeling

Let us use the notation in which $\mathcal{A}, \mathcal{B}, \ldots$ denote sets, $\mathbf{A}, \mathbf{B}, \ldots$ denote tensors and $a, b, \ldots$ denote scalars. The tensor-based feature model represents the multi-attribute data describing the given user feedback event (in the case of the application scenario presented in this paper – the event representing a user click on a given ad in result of a given impression). This data have the form of a set of logs $\mathcal{E}$, in which each of the log entries is defined as a tuple of multiple features. We model this set as an n-order tensor:

$$\mathbf{T} = [t_{i_1,\ldots,i_n}]_{m_1 \times \cdots \times m_n},$$

defined in a tensor space $\mathcal{I}_1 \otimes \cdots \otimes \mathcal{I}_n$, where each $\mathcal{I}_i$, $1 \leq i \leq n$ indicates a standard basis [14] of dimension $|\mathcal{I}_i| = m_i$ used to index elements of the domain $\mathcal{F}_i$ – the domain of feature i. The entries of tensor $t_{i_1,\ldots,i_n}$ represent the outcome of the investigated event – formally described using the function $\psi : \mathcal{F}_1 \times \cdots \times \mathcal{F}_n \to \mathbb{R}$. In the case of RTB CTR prediction task, the events' outcomes are usually described by means of the binary-valued function $\psi : \mathcal{F}_1 \times \cdots \times \mathcal{F}_n \to \{0, 1\}$ defining *non-click* and *click* events, respectively.

Since the input data form a sparse (incomplete) multidimensional structure, the tensor $\mathbf{T}$ is usually stored in the form of n-tuples, for which a given tuple γ is modeled as:

$$\gamma = (w^\gamma, f_1^\gamma, \ldots, f_n^\gamma),$$

where $f_i^\gamma \in \mathcal{F}_i$ are the feature values defining the tuple γ and $w^\gamma = \psi(f_1^\gamma, \ldots, f_n^\gamma)$ denotes its weight.

3.2 Multi-Tensor Hierarchy Network

In this paper we used the model, referred to as *Multi-Tensor Hierarchy Network* (MTHN), enabling the representation of correlations observed in any subset of the feature set. In contrast to other approaches (e.g., [16]), we do not use any heuristic method for feature grouping which is necessary when simplifying the model, e.g., to the single third-order tensor. The proposed model provides the averaging framework enabling to represent the means within a network of tensors, which is used for combinatorial exploration of all the possible subsets of the features.

Let $[n] = \{1, 2, \ldots, n\}$ denotes the set enumerating features describing the investigated event. For each subset $\mathcal{S} = \{p_1, \ldots, p_k\} \subset [n]$ we construct the tensor $\mathbf{T}(\mathcal{S})$ by averaging the data throughout all non-missing (i.e., known) values in the respective $(n - k)$-dimensional 'sub-tensor' of the input n-order tensor $\mathbf{T}$. The fibres (one-dimensional fragments of a tensor, obtained by fixing all indices but one) and slices (two-dimensional fragments of a tensor, each being obtained by fixing all indices but two) are the examples of such sub-tensors that are most commonly referenced in the literature [11,13].

Formally, for each subset $\mathcal{S} = \{p_1, \ldots, p_k\}$ of $[n]$, where $0 \leq k < n$, such that $\mathcal{R} = [n] \setminus \mathcal{S} = \{r_1, \ldots, r_{n-k}\}$, we construct tensor $\mathbf{T}(\mathcal{S}) = [t(\mathcal{S})_{j_1,\ldots,j_k}]_{m_{p_1} \times \cdots \times m_{p_k}}$ in such a way that, for a given combination of feature values in S, we have:

$$t(\mathcal{S})_{j_1,\ldots,j_k} = \frac{1}{z} \sum_{i_{r_1}=1}^{m_{r_1}} \cdots \sum_{i_{r_{n-k}}=1}^{m_{r_{n-k}}} \sum_{i_{p_1}=j_1}^{j_1} \cdots \sum_{i_{p_k}=j_k}^{j_k} t_{i_1,\ldots,i_n} \tag{1}$$

if $z > 0$, where z is a number of all known (i.e., non-missing) values in a sub-tensor of tensor $\mathbf{T}$ defined by fixing $i_{p_1}, \ldots, i_{p_k}$ to feature values indices $j_1, \ldots, j_k$. The value $t(\mathcal{S})_{j_1, \ldots, j_k}$ may be equivalently seen as the weight w^ϕ of the 'shortened' tuple $\phi = (w^\phi, f_{p_1}^\phi, \ldots, f_{p_k}^\phi)$. Note that in the case of RTB CTR modeling, value w^ϕ is just the CTR observed over all the events for which the features $p_1, \ldots, p_k$ have their values equal to $j_1, \ldots, j_k$.

Tensors $\mathbf{T}(\mathcal{S})$ form the hierarchical network of 2^n tensor structures of orders from the set $\{0, \ldots, n\}$ – called Multi-Tensor Hierarchy Network – corresponding to all possible subsets of the set of n investigated features. In particular, the model consist of:

- level 0 of MTHN – containing one node which is the tensor of order 0 $\mathbf{T}(\varnothing)$ – the scalar representing the weight value averaged over all known events (the averaged CTR in RTB CTR estimation application scenario),
- for $k \in \{1, \ldots, n-1\}$: level k of MTHN – containing $\binom{n}{k}$ k-order tensors $\mathbf{T}(\{p_1, \ldots, p_k\})$ representing the averages over events with k features with fixed values,
- level n of MTHN – containing one node – n-order tensor $\mathbf{T}([n]) = \mathbf{T}$ storing the weights of input n-tuples.

Two nodes $\mathbf{T}(\mathcal{S})$ and $\mathbf{T}(\mathcal{S}')$ of MTHN are connected if and only if they belong to neighboring levels ($\||\mathcal{S}| - |\mathcal{S}'|\| = 1$) and the set of features modeled by one of them is a subset of features modeled by the other ($|\mathcal{S} \cap \mathcal{S}'| = |\mathcal{S}|$ or $|\mathcal{S} \cap \mathcal{S}'| = |\mathcal{S}'|$) – in other words – one of them may be obtained from the other by averaging over a single tensor mode.

The example of hierarchy of tensors $\mathbf{T}(\mathcal{S})$ for set $\mathcal{S} \subset [3] = \{1, 2, 3\}$ corresponding to features 'hour', 'weekday' and 'advertiser', respectively, is illustrated in Fig. 1.

4   Modeling Temporal Features Using MTHN

Although in this paper we investigate temporal features modeling, the MTHN metamodel presented in Sect. 3 may be used to represent data tuples defined on the basis of any feature sets, not only temporal ones. To illustrate the modeling of temporal features performed in accordance with the approach proposed herein, let us present just one of many possible MTHN metamodel use cases. In Fig. 1 an order-three MTHN-based model is presented that exemplifies the case of the MTHN-based model built to represent average CTR values that reflect jointly the content of the advertisers set (five advertisers for the 2nd season of the iPinYou dataset) and two temporal features sets – the set consisted of the one-hour long nychthemeron (day and night) time slots (twenty four hours of the day) and the set of seven weekdays. This case is one of the cases that we evaluated experimentally in Sect. 6.

To build the MTHN metamodel visualization presented in Fig. 1, we used the set $\mathcal{S} = [3] = \{1, 2, 3\}$ to enumerate three features as follows: index 1 was used to represent 24 one-hour long nychthemeron time-slots, index 2 was used

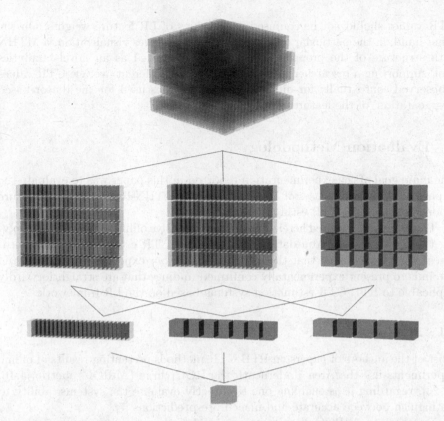

Fig. 1. Visualization of a MTHN-based representation of the average CTR values in the case of the 2nd season of the iPinYou dataset, with the advertisers set and two temporal features sets consisted of the one-hour time slots set and the weekdays set.

to represent 7 weekdays, and index 3 was used to represent 5 advertisers of the iPinYou dataset Season 2 [23]. The Fig. 1 presents MTHN consisting of 4 levels enumerated from bottom to top. In particular, level 0 is just tensor $\mathbf{T}(\varnothing)$. Level 1 contains tensors $\mathbf{T}(\{1\}), \mathbf{T}(\{2\}), \mathbf{T}(\{3\})$ (presented from left to right) representing features values corresponding to hours, weekdays and advertisers, respectively. Level 2 contains tensors $\mathbf{T}(\{1,2\}), \mathbf{T}(\{1,3\}), \mathbf{T}(\{2,3\})$ representing feature conjunctions of arity 2 of the form 'hour $\times$ weekday', 'hour $\times$ advertiser', and 'weekday $\times$ advertiser' respectively. Finally, level 3 contains the tensor $\mathbf{T}([n])$ representing the feature conjunctions of arity 3 of the form 'hour $\times$ weekday $\times$ advertiser'.

In Fig. 1 the darkness of each box – representing a given entry of the given tensor (in cases of some of the MTHN nodes, being a special case of a tensor: a vector or a scalar) – illustrates the average CTR value observed for the feature conjunction corresponding to this entry (the darker the box the higher the CTR value). It should be stressed that such a visualization is just a demonstration of an example of MTHN data structure application. In particular, the visualized

CTR values should not be confused with values of LR feature weights. On the other hand, in the particular case of average CTR values visualization, a MTHN data structure of the proposed kind may be regarded as an novel 'analytics tool' supporting a researcher in his/her analysis of different average CTR values – observed contextually for different sets of features used for the tensor-based 'segmentation' of the features.

5   Evaluation Methodology

The main goal of the experimentation reported in this paper was to evaluate the impact of the selected cases of an application of MTHN-based temporal feature models on the RTB CTR estimation quality.

The analysis presented herein is based on extensive offline experiments involving the use of the iPinYou dataset and selected CTR estimation quality measures. We have assumed that the ultimate goal of our experimentation reported herein is to present experimentally confirmed findings that are straightforwardly applicable to RTB CTR estimation systems based on the LR framework.

5.1   Measures

Most of the authors of papers on RTB CTR methods presenting results of offline experiments use the Area Underneath the ROC curve (AuROC) metric [4,16, 21,23], regarding it as enabling one to directly evaluate the systems' ability to distinguish between accurate and inaccurate predictions [7].

Nonetheless, AuROC, despite being useful for heavy-tailed recommendation or link prediction systems [6,17,19], may not provide a full insight into the RTB CTR estimation problem. Taking into account both the popularity and the limitations of AuROC, complementary to the presentation of our AuROC results, we show the Average Precision (AP) results (equivalent to the area underneath the precision-recall curve). While both metrics measure true positive rate, AP emphasizes precision while AuROC emphasizes false positive rate [6,17]. Such a difference of how the true negatives are treated is especially evident when the number of negative observations (*non-clicks*) is significantly higher than the positive ones (*clicks*).

As realized by some authors [22,23], in the context of the RTB, the quality of CTR estimation should be measured in a way that reflects the real-world requirement of the system's ability to preserve a high Key Performance Index value under the time constraints of the given campaign execution. In terms of precision and recall measures, a useful RTB optimization cannot severely reduce the bidding frequency. This means that the increase of precision should not lead to a severe reduction of recall. Under such conditions, the CTR estimation results presented as 'summarized' curves – such as AuROC and AP – are not considered as sufficiently informative for a real-world DSP.

Following the above-stated observation, in the analysis of experiments presented in Sect. 6 we additionally analyze the results from the perspective of the

CTR estimation system's ability of achieving different trade-offs between precision (i.e., CTR) and recall.

5.2 Dataset

To evaluate the proposed model we used the first publicly available large-scale RTB dataset released by iPinYou Information Technologies Co., Ltd [23]. The dataset contains impression, click, and conversion logs collected from several campaigns of different advertisers during various days and is divided into a training set and a test set. Each record contains five types of information: (i) temporal features (timestamp of the bid request), (ii) user features (iPinYou ID, browser user-agent, IP address, etc.), (iii) ad features (creative ID, advertiser ID, landing page, etc.), (iv) publisher features (domain, URL, ad slot ID, size, visibility, etc.), and (v) other features regarding the RTB auction (bid ID, bidding price, winning price, etc.).

In this paper we partition the dataset in two different ways:

(a) by timestamps, in the same manner as described in [16,23],
(b) randomly, using the same training ratios (tr) as in (a) (i.e., $tr \approx 0.7897$ for season 2 and $tr \approx 0.6667$ for season 3).

The major dataset statistics are shown in Table 1. More detailed information on the dataset may be found in [16,23].

Table 1. Dataset statistics (using the partitioning by timestamps).

Season	Dataset	Impressions	Number of feature values	Clicks	CTR (%)
2	Training set	12, 190, 438	801,890	8,838	0.073
	Test set	2, 521, 627	543,711	1,873	0.074
3	Training set	3, 147, 801	589,872	2,700	0.086
	Test set	1, 579, 071	482,208	1,135	0.072

6 Experiments

We trained the CTR estimator to predict the probability of the user click on a given ad impression using information extraction from raw user feedback data. As suggested in [23], in each of the tested variants the following pre-processing was performed:

- The timestamps were generalized into the corresponding weekday and hour value.
- The OS and browser names were extracted from the user-agent field.
- The floor prices were quantized into the buckets of 0, $[1, 10]$, $[11, 50]$, $[51, 100]$ and $[101, +\infty)$.

Table 2. CTR estimation performance in terms of AuROC and AP (%); the best results in each row are highlighted by bold font setting.

Dataset partitioning	iPinYou dataset season	day+hour		MTHN(domain)		MTHN(advertiser)	
		AuROC	AP	AuROC	AP	AuROC	AP
By timestamp	2	**90.67**	15.580	**90.67**	15.591	90.54	**16.243**
	3	75.23	0.312	**76.52**	**0.328**	76.01	0.317
Random	2	86.19	11.671	**86.52**	11.781	85.84	**11.922**
	3	78.91	0.530	**79.22**	**0.555**	78.66	0.531

We evaluated two variants based on the proposed tensor-based feature modeling metamodel:

- *MTHN(domain)* – reflecting two temporal features sets consisted of weekdays and hours, and the content of the domains set (the set of Web domains offering ad impressions),
- *MTHN(advertiser)* – reflecting two temporal features sets consisted of weekdays and hours, and the content of the advertisers set.

The state-of-the-art algorithm based on the basic temporal feature modeling proposed in [23] (referred to as *day+hour*) was chosen as a baseline.

To learn the LR model parameters, we used the Stochastic Gradient Descent (SGD) algorithm. The initial parameters (i.e., weights corresponding to the binary feature values) were set to 0. The learning rate in all the experiments was set to 0.01. The tolerance for the stopping criterion was set to 0.0001. Specifically, the learning was stopped when the logistic loss value change observed between two consecutive iterations reached the specified tolerance-defining threshold value. The training examples were randomly shuffled after each iteration so as not to introduce a bias into the optimization results [1].

The CTR estimation performance results concerning both dataset partitioning scenarios – the timestamp-based one and the random one – are presented in Tables 2 and 3. In the case of the random approach, the mean values from 10 experiments are shown for all the presented measures. Table 2 demonstrates the performance comparison provided by means of AuROC and AP measures. The standard error of each presented mean is less than 0.1% and 0.02% for AuROC and AP correspondingly.

Table 3 presents the CTR estimation system's ability of achieving different trade-offs between precision (i.e., CTR) and recall. Specifically, CTR values for recall equal to $1/8$ and $1/4$ were evaluated. Finally, Fig. 2 presents the precision vs recall curve (i.e., all CTR values) for iPinYou Dataset Season 2.

Table 3. CTR estimation performance in terms of $P(R = 1/8)$ and $P(R = 1/4)$ (%); the best results in each row are highlighted by bold font setting.

Dataset partitioning	iPinYou Dataset season	day+hour		MTHN(domain)		MTHN(advertiser)	
		$P(R = 1/8)$	$P(R = 1/4)$	$P(R = 1/8)$	$P(R = 1/4)$	$P(R = 1/8)$	$P(R = 1/4)$
By timestamp	2	33.62	29.46	**38.91**	**31.73**	34.41	32.37
	3	**0.58**	0.39	0.54	**0.44**	0.52	0.42
Random	2	33.16	23.11	**33.74**	**24.05**	31.58	23.16
	3	1.01	0.68	**1.08**	**0.73**	1.03	0.72

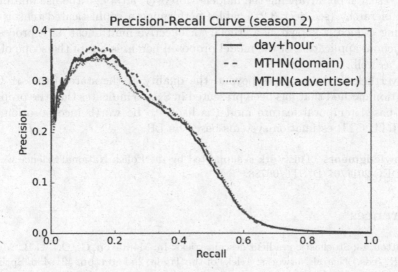

Fig. 2. PvR curve for Season 2 partitioned by timestamps.

7   Conclusions

On the basis of the extensive experimental evaluation (presented in Sect. 6), we have demonstrated that the proposed tensor-based model of temporal features enables to improve the quality of CTR estimation. We have shown this improvement in a Big Data application scenario of the Web user feedback prediction (corresponding to the task typically realized within an RTB DSP). In particular, we have shown that, in the investigated scenario, one may improve the quality of CTR estimation by using a simple, order-three MTHN-based models combining two temporal features sets – the set consisted of the one-hour long nychthemeron ('day and night') time slots (i.e., 'hours') and the set of weekdays ('days of the week') – with the set of domains and, equivalently, with the set of advertisers. The high-performance results of the approach applying the

MTHN(domain) variant indicate the potential of context-aware modeling – in this case based on the features that describe the Web publishers.

The improvement beyond the state-of-the-art algorithm based on LR (proposed in [23]) was achieved despite the referenced algorithm already involved the use of a basic (not tensor-based) temporal feature model. Additionally, our result was achieved despite the relative simplicity of the tensor-based model. The simplified form of a tensor-based feature representation model presented in this paper does not provide the properties that are widely-regarded as the key source of the practical value of tensor-based data representations. In particular, being used merely as a training data representation structure, the model itself provides no means for tensor-based feature similarity modeling nor tensor decomposition [11,13] – the techniques that naturally constitute the area of the future research on advancing the model. Moreover, although the use of a multi-tensor hierarchy (see Sect. 3.2) enables performing a sophisticated tensor data centering (which is known as a crucial for effective multilinear data processing [2]), such an application of the model proposed herein is out of the scope of this paper, as well.

Nevertheless, the progress beyond the quality of the state-of-the-art CTR estimation method that has been presented in Sect. 6 indicates that the proposed tensor-based temporal feature model is likely to be worth incorporation into many RTB CTR estimation systems based on LR.

Acknowledgments. This work is supported by the Polish National Science Centre, grant DEC-2011/01/D/ST6/06788.

References

1. Bottou, L.: Stochastic gradient descent tricks. In: Montavon, G., Orr, G.B., Müller, K.R. (eds.) Neural Networks: Tricks of the Trade, 2nd edn, pp. 421–436. Springer, Heidelberg (2012)
2. Bro, R., Smilde, A.K.: Centering and scaling in component analysis. J. Chemometr. **17**(1), 16–33 (2003)
3. Brusilovsky, P., Kobsa, A., Nejdl, W. (eds.): The Adaptive Web: Methods and Strategies of Web Personalization. Springer, Berlin (2007)
4. Chapelle, O., Manavoglu, E., Rosales, R.: Simple and scalable response prediction for display advertising. ACM Trans. Intell. Syst. Technol. **5**(4), 61:1–61:34 (2014)
5. Cichocki, A.: Era of Big Data Processing: A New Approach via Tensor Networks and Tensor Decompositions. CoRR abs/1403.2048 (2014)
6. Ciesielczyk, M., Szwabe, A., Morzy, M., Misiorek, P.: Progressive random indexing: dimensionality reduction preserving local network dependencies. ACM Trans. Internet Technol. **17**(2), 20:1–20:21 (2017). http://doi.acm.org/10.1145/2996185
7. Fawcett, T.: An introduction to ROC analysis. Pattern Recogn. Lett. **27**(8), 861–874 (2006)
8. Franz, T., Schultz, A., Sizov, S., Staab, S.: TripleRank: ranking semantic web data by tensor decomposition. In: Bernstein, A., Karger, D.R., Heath, T., Feigenbaum, L., Maynard, D., Motta, E., Thirunarayan, K. (eds.) ISWC 2009. LNCS, vol. 5823, pp. 213–228. Springer, Heidelberg (2009). doi:10.1007/978-3-642-04930-9_14

9. He, X., Pan, J., Jin, O., Xu, T., Liu, B., Xu, T., Shi, Y., Atallah, A., Herbrich, R., Bowers, S., Candela, J.Q.: Practical lessons from predicting clicks on ads at Facebook. In: Proceedings of the Eighth International Workshop on Data Mining for Online Advertising, ADKDD 2014, NY, USA, pp. 5:1–5:9. ACM, New York (2014)

10. Japkowicz, N., Stefanowski, J.: Big Data Analysis: New Algorithms for a New Society. Studies in Big Data. Springer International Publishing, Heidelberg (2015)

11. Kolda, T.G., Sun, J.: Scalable tensor decompositions for multi-aspect data mining. In: Proceedings of the 2008 Eighth IEEE International Conference on Data Mining, ICDM 2008, pp. 363–372. IEEE Computer Society, Washington, DC (2008)

12. Kruczyk, M., Baltzer, N., Mieczkowski, J., Draminski, M., Koronacki, J., Komorowski, J.: Random reducts: a Monte Carlo rough set-based method for feature selection in large datasets. Fundam. Inform. **127**(1–4), 273–288 (2013)

13. Lathauwer, L.D., Moor, B.D., Vandewalle, J.: A multilinear singular value decomposition. SIAM J. Matrix Anal. Appl. **21**, 1253–1278 (2000)

14. Nickel, M., Tresp, V.: An analysis of tensor models for learning on structured data. In: Blockeel, H., Kersting, K., Nijssen, S., Železný, F. (eds.) Machine Learning and Knowledge Discovery in Databases: European Conference, ECML PKDD 2013, Prague, Czech Republic, September 23–27, 2013, Proceedings, Part II, pp. 272–287. Springer, Heidelberg (2013)

15. Provost, F., Fawcett, T.: Data science and its relationship to big data and data-driven decision making. Big Data **1**(1), 51–59 (2013)

16. Shan, L., Lin, L., Sun, C., Wang, X.: Predicting ad click-through rates via feature-based fully coupled interaction tensor factorization. Electron. Commer. Res. Appl. **16**, 30–42 (2016)

17. Shani, G., Gunawardana, A.: Evaluating recommendation systems. In: Ricci, F., Rokach, L., Shapira, B., Kantor, P.B. (eds.) Recommender Systems Handbook, pp. 257–297. Springer US, Heidelberg (2011)

18. Sutskever, I., Tenenbaum, J.B., Salakhutdinov, R.R.: Modelling relational data using Bayesian clustered tensor factorization. In: Bengio, Y., Schuurmans, D., Lafferty, J., Williams, C., Culotta, A. (eds.) Advances in Neural Information Processing Systems 22, pp. 1821–1828. Curran Associates, Inc. (2009)

19. Szwabe, A., Ciesielczyk, M., Misiorek, P.: Long-tail recommendation based on reflective indexing. In: Wang, D., Reynolds, M. (eds.) AI 2011. LNCS (LNAI), vol. 7106, pp. 142–151. Springer, Heidelberg (2011). doi:10.1007/978-3-642-25832-9_15

20. Szwabe, A., Misiorek, P., Walkowiak, P.: Tensor-based relational learning for ontology matching. In: Advances in Knowledge-Based and Intelligent Information and Engineering Systems - 16th Annual KES Conference, San Sebastian, Spain, 10–12 September 2012, pp. 509–518 (2012)

21. Zhang, W., Du, T., Wang, J.: Deep learning over multi-field categorical data. In: Ferro, N., Crestani, F., Moens, M.-F., Mothe, J., Silvestri, F., Nunzio, G.M., Hauff, C., Silvello, G. (eds.) ECIR 2016. LNCS, vol. 9626, pp. 45–57. Springer, Cham (2016). doi:10.1007/978-3-319-30671-1_4

22. Zhang, W., Yuan, S., Wang, J.: Optimal real-time bidding for display advertising. In: Proceedings of the 20th ACM SIGKDD International Conference on Knowledge Discovery and Data Mining, KDD 2014, pp. 1077–1086. ACM, New York (2014)

23. Zhang, W., Yuan, S., Wang, J.: Real-time bidding benchmarking with iPinYou dataset. CoRR abs/1407.7, pp. 1–10 (2014)

Evaluation of XPath Queries Over XML Documents Using SparkSQL Framework

Radoslav Hricov, Adam Šenk, Petr Kroha, and Michal Valenta[(✉)]

Faculty of Information Technology, Czech Technical University in Prague,
Prague, Czech Republic
{hricorad,senkadam,krohapet,valenta}@fit.cvut.cz

Abstract. In this contribution, we present our approach to querying
XML document that is stored in a distributed system. The main goal
of this paper is to describe how to use Spark SQL framework to implement a subset of expressions from XPath query language. Five different
methods of our approach are introduced and compared, and by this, we
also demonstrate the actual state of query optimization on Spark SQL
platform. It may be taken as the next contribution of our paper. A subset of expressions from XPath query language (supported by the implemented methods) contains all XPath axes except the axes of attribute
and namespace while predicates are not implemented in our prototype.
We present our implemented system, data, measurements, tests, and
results. The evaluated results support our belief that our method significantly decreases data transfers in the distributed system that occur
during the query evaluation.

Keywords: Spark · SQL · XML · XPath · Big data

1 Introduction

Currently, XML is a very popular language for its platform independent way of
storing data. The XPath query language is one of many possibilities to formulate
queries over data stored in XML documents. Having big data stored in XML
documents, the problem is how to retrieve data efficiently, i.e., how to implement
the XPath query language for big data.

In this paper, we investigate the case in which a single unit of data, usually
a file, can be processed by in-memory processing methods. This limitation is
often met, because it is currently possible to use computers that have 128 GB
operating memory.

The main topic of this paper is to describe how to use Apache Spark SQL
framework to implement a subset of expressions from XPath query language
under conditions described above. Apache Spark is a fast evolving engine for
in-memory big data processing that powers several modules. One of the modules
is Spark SQL. It works with structured data using SQL-like query language
or domain-specific language of DataFrame. We investigate its possibilities and
potential limitations.

© Springer International Publishing AG 2017
S. Kozielski et al. (Eds.): BDAS 2017, CCIS 716, pp. 28–41, 2017.
DOI: 10.1007/978-3-319-58274-0_3

XML documents can be processed by XSL technologies including XSLT, XPath, and XQuery. We choose XPath technology, as we worked with it successfully before [7,11]. We map XML data into the relational tables, and we map XPath queries to the SQL queries.

The paper is organized as follows. In Sect. 2, we present work related to our investigations. Concretely, we start with Spark SQL in Sect. 3.1. We explain why we need to transform XML documents into relational tables in Sect. 3.2.

Section 3.3 is dedicated to the investigated methods.

Our approach to query process analysis and the architecture of our system are given in Sect. 3. In Sect. 4, we describe the data we use and measurements we made, and we evaluate the results obtained. Finally, in Sect. 5, we draw conclusions and discuss possible future work.

2 Related Work

In this section, we evaluate work related to our paper. We focus on three inherently related topics: XML-to-relation mapping, Storing XML data in NoSQL databases and distributed evaluation of XPath and XQuery queries.

Starting with strategies for mapping XML to relations, various ways have been proposed in works such as [1,3].

On the other hand, the paper [10] introduces mapping methods that preserve XML node order.

In this work, several indexing methods are described, and comparisons of creating, updating, and reading cases are given.

Paper [8] introduces mapping of XML data to quasi-relational model. The authors propose a simple, yet efficient algorithm that translates XML data into structure savable in relational columns. Stored data can be queried by SQL syntax based language - SQLxD.

The following papers describe mapping into various NoSQL database systems. Specifically, the paper [9] introduces XML format mapping into a key-value store. Three possible ways of mapping are compared, however, query evaluation was not investigated. The paper [4] describes a distributed query engine used in Amazon Cloud. Three possible index strategies are introduced, and a subset of XPath and XQuery is implemented. However, the aim of the present work is to scale queries over a big set of XML documents. It does not investigate parallel processing of a single document. The implemented subset of XPath includes fewer operators than our solution.

Finally, distributed evaluation of XPath and XQuery language over a single document is investigated in [2]. The query engine introduced in this work can compute XML queries including only a small subset of XPath axis identifiers. Some of the papers published in recent years use MR framework for efficient XML query processing. The MRQueryLanguage is introduced in [6]. The query language is designed for querying distributed data, but it brings a new syntax different from known XML query languages like XPath and XQuery. HadoopXML

[5] is a suite enabling parallel processing of XPath queries. Queries are evaluated via twig join algorithm. Unfortunately, the XPath subset (that can be evaluated in HadoopXML) includes only root-to-leaf axis identities: child, and descendant or descendant-or-self. Evaluation of all XPath axes using MR framework is described in [11]. Authors focus on mapping complete XPath axes set to a bash of MR queries. The results are satisfying, but evaluation of some axes is highly inefficient.

3   Our Approach

First, we describe the Spark technology that we used. We focus both on the Spark core and on the SparkSQL framework. Then, we introduce the architecture of our system. In this part, we show how to map the XML data in data processable by SparkSQL, how to store it, how to translate the XPath queries and evaluate them, and how to reconstruct the results back to XML.

Finally, we describe our original approach to XPath query translation. We introduce five different methods that can be used for processing XPath queries using SparkSQL. In the last part of this section, we evaluated the methods and compared them.

3.1   Used Technology - Spark

Apache Spark is a multipurpose cluster computing system for a large-scale data processing. Spark is an open source engine originally developed by UC Berkley AMPLab and later adopted by Apache Software Foundation in 2010. Spark provides fast in-memory computing, and its ecosystem consists of higher-level combinable tools including Spark Streaming, Dataframes and SQL, as well as MLlib for machine learning, and GraphX for graph processing. The core engine of Spark provides scheduling, distributing, and monitoring of applications across the computing cluster. Spark is implemented in Scala that runs on Java Virtual Machine. API of Spark and its tools are available in Scala, Java, Python and R.

The Programming of distributed operations is based on RDD (abbr. Resilient Distributed Dataset). It is a Spark's main abstraction. It is a collection of objects that can be processed in parallel.

In our work, we use the SparkSQL module. This module enables to query RDDs using SQL-like syntax, so even programmers not familiar with Spark API can query data in parallel using Spark. However, SparkSQL implements only a subset of SQL language, which brings severe limitations.

3.2   Transformation of XML Document to Data Frames

The main programming abstraction of SparkSQL is DataFrame. This is a distributed collection that is similar to the concept of relational table. The tree structure of XML document is an ordered data model based upon the order of each element within the XML document. We mainly focus on the selection of nodes, and we want to be able to reconstruct selected nodes back to the valid and ordered XML. Accordingly, we are not interested in insertion or deletion of nodes. The transformation of XML document must be able to transform data from an unordered relational model back to the XML document. Hence, we decided to follow the paper [10], since it shows that XML ordered data model can indeed be efficiently supported by a relational database system. This is accomplished by encoding the data order as a data value. There are three methods for transformation of XML documents in tables [1] global order encoding, local order encoding, and Dewey order encoding.

From the three encodings mentioned above, the Dewey order encoding will be used here, since it is the universal solution, and the information stored in Dewey path is sufficient. Compared with the global encoding, it can be a bit slower (depending upon the comparison of the paths). Dewey path implicitly contains information about the node's ancestor nodes and also about its position among the siblings. Thus, Dewey encoding is the best option for our purposes.

The process of transformation begins with a numbering of elements and text nodes. Based on the pre-order traversal of the XML tree, a Dewey path is assigned to each node.

In the second phase of the transformation, the Dewey paths are recomputed to preserve the document order information, and it also makes the paths comparable as *String*. Now, each part (parts are separated by dots) of Dewey paths has the same length. The number of zeros in its prefix depends upon the number of digits of the highest value of Dewey path part among all Dewey paths. This is also helpful in SQL ORDER BY operation. Additionally, during the first phase, i.e. during the numbering phase, the paths to the certain nodes built from the names of nodes are created and assigned to each stored node.

A file containing serialized, transformed XML document is stored on a disk. When we query the document, it has to be loaded into memory and serialized back to Data Frame. We use Spark core to create RDD of XML nodes. Each row is read and split to the Node object. Node object is then passed to RDD. DataFrame may be created directly from RDD, but it is necessary to define a schema of a table. We created a class Node, and by reflection, the schema defined through Node class was applied on the RDD. Finally, a DataFrame was created from the RDD.

We experimented with datasets shown in Table 1. See Sects. 3.3 and 4 for results.

Table 1. Size comparison of generated tables

	XML file	Text file	Number of rows
books.xml	1.1 kB	1.4 kB	60
nasa.xml	25.1 MB	45.8 MB	791 922
proteins.xml	716.9 MB	2.3 GB	37 260 927

To make it easier for the reader to follow the discussion, let us show a sample of an XML file translated into a (relational) table.

```
+---------------+------+----+------------------+
|          dewey|pathId|type|             value|
+---------------+------+----+------------------+
|          00.01|    0|   1|         bookstore|
|       00.01.01|    1|   1|              book|
|    00.01.01.01|    2|   1|             title|
|00.01.01.01.01|    3|   3| XQuery Kick Start|
|            ...|   ...| ...|               ...|
+---------------+------+----+------------------+
```

Fig. 1. Nodes' table of transformed XML

Optionally, an additional table containing the `pathId` and `Path` columns may be generated. It may be helpful for a particular query evaluation, but in principle, the table in Fig. 1 is enough. The column `type` contains the node type according to the W3C classification.

3.3 XPath Queries Evaluation

In this section, we describe our approach to processing of XPath queries. We have an XPath query and that, subsequently, must be translated into the SQL query that can be evaluated in the SQL module of Spark. According to the description given above, we developed two applications. The first one is an XML processor: it transforms an XML document into a relational table, and we introduced it in the previous section. The second one is a driver program for Spark: it processes an XPath query (using SQL query or via Spark SQL API), and it applies it on DataFrame built from the transformed XML document. The result of the driver program operation is the final table of nodes; it may be further processed.

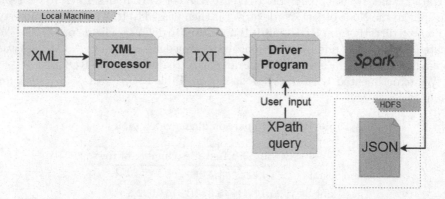

Fig. 2. Local cooperation of applications

Figure 2 shows how the applications locally cooperate to return XPath query result. The transformed XML document is stored as a text file. Both the text file and the XPath query are used as parameters of a driver program that is then running on Spark. The result of evaluation of a XPath query by the driver program is a JSON file stored in HDFS.

We implemented a simple XPath parser that parses XPath queries that were inputted as parameters of the driver program. A support for abbreviated forms of some XPath steps was added to the parser. In this prototype application, the abbreviated forms of child axis as / and descendant as // are allowed. Further, the wildcard * is an alias for any element node, which may be used in the queries that are parsed by a parser.

The whole query is split to the separated XPath steps, and the abbreviated forms are resolved. Then, all steps are evaluated step by step according to the desired axis. The step by step evaluation is implemented by an indirect recursive algorithm. It means that every next step is dependent upon the result of the previously evaluated step. This parser is working just with axes that were mentioned above and does not support predicates.

In following paragraphs, we describe five translation methods we investigated:

– Pure SQL method
– Join-based SQL method
– SQL query via DataFrame API
– Left semi join method
– Broadcasted lookup collection.

First, we introduce a native, trivial method, and then it will be improved in the next subsections. The next two methods of use apply the SQL queries to evaluate XPath queries. The other methods use a domain specific language of Spark SQL API.

Pure SQL Method. In the early familiarization with the Spark SQL module, we tried to directly translate an XPath query to the SQL query. For a faster local testing, we were working with a small table of nodes containing 60 rows. In all our tests, we performed translations of simple XPath queries that covered all XPath axes.

The generated SQL query starts with a selection of an auxiliary node that represents a parent of the root node. It is an alternative to a *document* statement doc(''xmlFile.xml'') in XPath. Then, the inputted XPath query is translated step by step. After the translation of the last step, one more selection and filtration are needed. It completes the result of query by selecting all descendant or self nodes of previously selected nodes. It is because the XPath steps traverse through the nodes, so by the last extra step, their content is appended.

Let us show a basic example to illustrate our approach and to support better understanding of following improvements. The example implements an XPath expression //book/author/:

```
SELECT p2.dewey, p2.pathId, p2.type, p2.value FROM nodes p2,
  (SELECT p1.dewey, p1.pathId, p1.type, p1.value FROM nodes p1,
```

```
(SELECT p0.dewey, p0.pathId, p0.type, p0.value FROM nodes p0,
  (SELECT '0' as dewey) n0
 WHERE p0.type=1 AND n0.dewey <= p0.dewey
                AND p0.value='book'
                AND isPrefix(n0.dewey, p0.dewey) ) n1
 WHERE p1.type=1 AND n1.dewey < p1.dewey
                AND p1.value='author'
                AND isChild(n1.dewey, p1.dewey) ) n2
WHERE n2.dewey <= p2.dewey AND isPrefix(n2.dewey, p2.dewey)
```

We start with the parent of document root, i.e., doc(''xmlFile.xml'') – label n0, followed by its descendant nodes named book, etc. Functions isChild and isPrefix are based on Dewey encoding string, and their meaning is obvious.

Using our small testing file, it was relatively fast to compute a result; however, when we started with processing larger table of nodes (containing 791922 rows), problems with performance occurred.

After we examined an execution plan, we found out that for this naive method, it was actually the Cartesian product followed by filtration that was executed. It was the bottleneck of this method.

Remember that details of the following improvements can be found in [7]. We cannot discuss the problem fully given the limited space.

Join-Based SQL. In this case, the best solution is to avoid Cartesian product and apply an SQL JOIN clause instead. Hence, the JOIN ON conditions were defined to join results of a single XPath step. The idea is to select nodes that are candidates for the next context node, then combine them with the current context node and, based upon the relation, filtrate suitable nodes from joined pairs by using user defined functions. Note that the context node is a set of nodes returned by executing one step of XPath query. We use this term in the next sections.

By analyzing Spark's execution plans, we finally decided for RIGHT JOIN (LEFT JOIN is also acceptable, but it depends upon the order in which XPaths steps are joined). Although the type of JOIN was defined, Spark has generated Cartesian product in some cases because joins conditions were not strong enough. The conditions were based on non-equality of Dewey paths, and the user defined functions were used in a filter condition. To solve this, the join condition had to be enhanced, so instead of filtering based on a user defined function, we add required UDF into the join condition. For the sake of completeness, let us add that a JOIN, whose condition is based only upon the user defined function (that requires arguments both from left and right tables), invokes the Cartesian product: all pairs must be processed by UDF.

After the changes were done, the performance was admittedly better than the performance of the Cartesian product using method.

We compare the two previously discussed methods in Table 2.

As we can see, the usage of OUTER JOIN and the proper definition of JOIN conditions have a crucial impact upon the performance. After twenty minutes

Table 2. Performance of translated queries - Cartesian product versus Right JOIN

	Table	Method	Query	Time [s]
1	Books	Cart. prod	//book/author	9.790
2	Books	RIGHT JOIN	//book/author	8.372
3	Nasa	Cart. prod	//author/suffix	1200*
4	Nasa	RIGHT JOIN	//author/suffix	232.695

of computing, we were forced to cancel the third measurement marked with asterisk. We realized that Cartesian product in Spark is really slow.

SQL Query via DataFrame API. So far, we have worked only with pure SQL queries. However, DataFrame contains its own API that may be used to obtain the same results as by using SQL queries. We rewrote the previous SQL query that used RIGHT JOIN by calling a certain combination of functions from API. Using API, we changed the order of processed axes, so instead of RIGHT JOIN, the LEFT JOIN was applied.

Since we know that SQL and DataFrame shared the same optimization pipeline, the physical plans vary in small details - depending upon the implementation - and they actually do the same work. Broadcast Nested Loop Join is realized in Spark for OUTER JOINs. It compares the sizes of tables to be joined, and it broadcasts the smaller one across the workers. As it was expected, the durations of the computations of SQL and DataFrame are almost the same, since the optimizer generates the same physical plan.

Alternative Methods Without Joins. After the previous findings, we decided to restrict the usage of JOIN in further experiments. Several alternatives - such as nested queries, SQL IN operator or SQL UNION statement - had been tested, but the results were not satisfiable.

Instead, we decided to avoid joins altogether. To simplify the previous methods, we wanted to select those nodes of some XPath step that are in desired relation with at least one node from the nodes of previously evaluated XPath step. For this purpose, the best option is to use IN operator.

We wrote a user defined function *Parent()* that cuts the last part of inputted Dewey path, and since Dewey path contains information about all ancestors, this function returns the Dewey path of its parent node. This is the valid SQL query and both - the query and the nested query - are executable, though, as it turned out, Spark does not think the same.

Unfortunately, Spark SQL is not able to execute nested SELECT following the WHERE clause. Using Spark SQL API is not applicable either: Spark evaluates it in a different way - as expected.

Left Semi Join. As mentioned above, the IN clause may be used just with a joined table. We realized that Spark SQL provides LEFT SEMI JOIN - and it turned out to be more effective than all of our previous attempts. First, let us explain how LEFT SEMI JOIN works.

"Semi" means that the result contains just rows returned from one table. In case of LEFT SEMI JOIN, just the rows from the left table are returned. LEFT SEMI JOIN is based upon the existence of records in the right table. It means that if there is a record in the right table that fulfills the JOIN ON condition, just this one record from the left table is returned.

Using this method, we implemented a translation of XPath steps for parent, child, ancestor, ancestor-and-self, descendant, and descendant-or-self axes. Other axes have to be implemented in a different manner, a one that does not use LEFT SEMI JOIN, but it uses, for example, user defined functions (UDF). It is because the implemented axes are based on prefixes.

Table 3 compares the computation times using SQL and DataFrame.

Table 3. SQL versus DataFrame

	Table	Method	Query	Time [s]
1	Books	DataFrame	//book/author	8.313
2	Books	SQL	//book/author	8.372
3	Nasa	DataFrame	//author/suffix	231.989
4	Nasa	SQL	//author/suffix	232.695
5	Nasa	DataFrame	//suffix	219.012
6	Nasa	SQL	//suffix	217.818

Broadcasted Lookup Collection. Concerning the JOIN statement, in Spark tutorials, it is recommended to set a table that is repeatedly used in joins as a broadcast variable, and then join it. This table is often considered as a lookup table.

Since when we know that it is impossible to work with two DataFrames at the same time without joining them together, we had to find a way out of this loophole.

We adapt the idea of lookup table, but since we had bad experience with the joins, we wanted to avoid them. To do that, we create a collection from the context node by applying *collect* action on the DataFrame.

First, the action *collect* creates a collection of *Strings* where each element is a Dewey path. Then, we register a user defined function, and during the registration, the broadcast variable from the collection is created. The input parameter of the user defined function is a Dewey path of a candidate for a member of the new context nodes. The candidates for a new context node are all rows whose values of column *value* fulfill the node test of XPath step. The called UDF (User Defined Function) checks whether the relationship between inputted Dewey path and the Dewey paths in the collection of the context node is as it is desired. If the UDF is evaluated as true, the currently checked node will be a member of the next lookup table. The advantage of this method is that each executor may have its own partitions of input file in memory, and just lookup collections are collected to the driver and then broadcasted among other executors.

The user defined functions used in this method are different from those that are used in the Pure SQL discussed method in Sect. 3.3. We created UDF separately for each axis. The difference is that these functions, firstly, create a broadcast variable and then, according to the axis specifier, they detect whether the examined node belongs to the desired axis that was desired. Instead of two input parameters, just one is required by UDFs in this method, given that they use the broadcast variable.

Also, in this method, the evaluation starts with a selection of a parent node of a root node, and then independent XPath steps are evaluated step by step.

By the evaluation of the last XPath step, the result nodes are obtained, but still, their content is not text nodes or other descendant elements. So, the last step of the evaluation is to get them in the required format.

Table 4 shows a comparison of the method using LEFT SEMI JOIN and the method using lookup collections. Unlike the previous measurement in Sect. 3.3, in this case, both methods do as a first step caching of partitions of processed DataFrame into the memory.

Table 4. Performance of translated queries - LEFT SEMI JOIN versus Lookup collection

	Table	Method	Query	Time [s]
1	Books	Lookup col	//book/author	2.442
2	Books	LEFT SEMI JOIN	//book/author	3.064
3	Nasa	Lookup col	//author/suffix	7.024
4	Nasa	LEFT SEMI JOIN	//author/suffix	9.492
5	Nasa	Lookup col	//suffix	5.856
6	Nasa	LEFT SEMI JOIN	//suffix	7.235
7	Protein	Lookup col	//formal	418.305
8	Protein	LEFT SEMI JOIN	//formal	423.819
9	Protein	Lookup col	//organism/formal	1088.489
10	Protein	LEFT SEMI JOIN	//organism/formal	3441.569

4 Experimental Results and Discussion

In Sect. 3, we provided several comparisons, and in this section, we summarize measured times of different methods.

Local Mode. All experiments of local performance testing were run on a virtual machine hosted on an Intel Core i3 350 M 2.27 GHz processor, with 8 GB DDR3

RAM and 100 Mbps LAN network, and with installed Windows 8.1 Pro 64-bit operating system. The virtual machine has allocated 2 CPU cores and 5 GB RAM under the operating system Ubuntu 14.04 64-bit. All experiments were run on Spark in version 1.5.2, for which 512 MB of memory has been allocated. Information about tested tables containing transformed XML documents is in Table 1. Table 5 summarizes tested queries and tested tables. All measured values are summarized in Table 6. All the measurements in Table 6 were realized locally.

Table 5. Summarizing table of tested queries

	Query	XML file	Text file	Rows count
Books 2	//book/author	1.1 kB	1.4 kB	60
Nasa 1	//suffix	25.1 MB	45.8 MB	791 922
Nasa 2	//author/suffix	25.1 MB	45.8 MB	791 922
Protein 1	//formal	716.9 MB	2.3 GB	37 260 927
Protein 2	//organism/formal	716.9 MB	2.3 GB	37 260 927

Table 6. Performance of proposed methods in seconds

Method/Tab.	Books 2	Nasa 1	Nasa 2	Protein 1	Protein 2
Cartesian	9,79	-	1200#	-	-
SQL JOIN	8,372	217,818	232,70	-	-
DF JOIN	8,313	219,012	231,99	-	-
SEMI JOIN	4,687	13,336	18,75	-	-
SEMI JOIN*	3,064	7,235	9,49	423,819	3441,569
Broadcast coll.	2,442	5,856	7,02	418,305	1088,489

Note that some values were not measured in Table 6, because the computation was very slow. The second SEMI JOIN marked with * is the caching using method. The measuring of the value marked # was so slow that it had to be interrupted. Measured values are shown graphically in Fig. 3.

It can be seen that the method using Cartesian product is really slow. SQL RIGHT JOIN and DataFrame LEFT JOIN are use the same optimization process, so given the same physical plan is generated, the computational time is almost the same. In Fig. 3, one more thing is noteworthy: a positive impact of caching. According to Table 6, the evaluation of XPath query `//author/suffix` over the table Nasa is currently (with broadcast lookup collection method) more than 150 times faster in comparison with the Cartesian product. Since our methods evaluate an XPath query step by step, the impact of the number of evaluated steps can be seen in Fig. 3. This implies that more XPath steps mean longer computation duration since there is no optimization used, so all steps are evaluated.

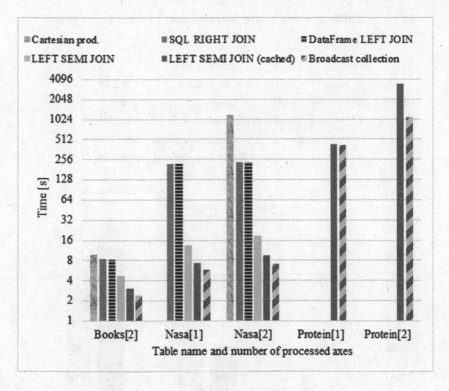

Fig. 3. Performance of proposed methods

Cluster Mode. The experiment ran on a cluster of 4 virtual machines hosted on four processors Intel Xeon 3.4 GHz (each 2 physical cores and 4 logical cores, with enabled Hyper Threading), with 32 GB DDR2 RAM and 2×1 Gbps LAN. Each virtual machine has allocated 6 GB RAM (of which 4,8 GB were used by Spark) and 2 CPU cores. Virtual machines were connected via 10 Gbps VMXNet3 LAN and the installed operating system was Ubuntu 14.04.

Using a small input file, it is impossible to see the benefits of cluster computation: in some cases, the communication load took more time then the actual computation. The cluster computation forced us to use the Hadoop Distributed File System to make our text files visible for workers. On our cluster, we continued our test attempts with bigger files, since a sufficient amount of memory was available for the worker nodes.

Comparison of Performance in Local and Cluster Mode. Comparison of computation in local mode and in cluster mode brought expected results. Admittedly, the computation on the cluster with enabled cluster mode was faster in some cases. Table 7 shows time comparison of local and cluster mode. In these experiments, it turned out that the fastest method is the one that uses nested lookup collection.

Table 7. Performance of cluster and local mode

	Table	Mode	Query	Time [s]
1	Nasa	Local	//author/suffix	7.024
2	Nasa	Cluster	//author/suffix	38.415
3	Protein	Local	//formal	418.305
4	Protein	Cluster	//formal	398.701
5	Protein	Local	//organism/formal	1088.489
6	Protein	Cluster	//organism/formal	957.729

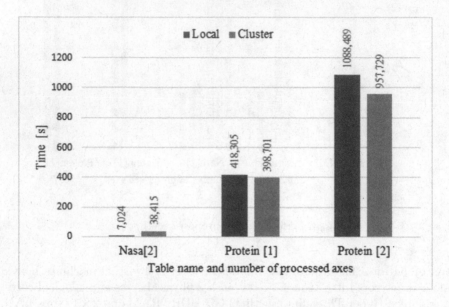

Fig. 4. Performance of cluster and local mode

As we can see in Table 7, a processing of a smaller table can be faster when it is done locally. The reason behind it are cluster overhead expenses such as serialization and transporting data among other workers.

In Fig. 4, we can see the measured times from Table 7.

5   Conclusions and Future Work

We analyzed possibilities of applying XPath queries on XML-documents by using framework Spark SQL. Additionally, we implemented, tested, and measured our initial statements concerning querying process of XML documents using the Spark SQL system - its advantages and disadvantages. We designed multiple methods and compared their efficiency. Bases upon our experiments, we conclude that the efficiency of some of the tested methods is limited. So, not all of the proposed methods can be used to query large data. We compared the method efficiency in both Spark modes, i.e. in the local mode and in the cluster mode.

One of the biggest advantages of Spark is the broadcast variables method. As our measurements show, the broadcast variables method is the fastest (and by far, the best) method from all the methods we investigated.

In the future, we plan to experiment with larger files and with more powerful clusters to evaluate scalability of our methods. We want focus on utilization using various Spark tuning possibilities to optimize computational cost and time. Finally, we plan to compare parallel XPath queries evaluation (using Spark) to other framework for distributed computing.

References

1. Amer-Yahia, S., Du, F., Freire, J.: A comprehensive solution to the XML-to-relational mapping problem. In: Proceedings of the 6th Annual ACM International Workshop on Web Information and Data Management, pp. 31–38 (2004)
2. Bidoit, N., Colazzo, D., Malla, N., Sartiani, C.: Partitioning XML documents for iterative queries. In: Proceedings of the 16th International Database Engineering & Applications Symposium, pp. 51–60. ACM (2012)
3. Bourret, R., Bornhövd, C., Buchmann, A.: A generic load/extract utility for data transfer between XML documents and relational databases. In: Advanced Issues of E-Commerce and Web-Based Information Systems, WECWIS 2000, pp. 134–143 (2000)
4. Camacho-Rodríguez, J., Colazzo, D., Manolescu, I.: Building large XML stores in the Amazon cloud. In: 2012 IEEE 28th International Conference on Data Engineering Workshops (ICDEW), pp. 151–158. IEEE (2012)
5. Choi, H., Lee, K.H., Kim, S.H., Lee, Y.J., Moon, B.: HadoopXML: a suite for parallel processing of massive XML data with multiple twig pattern queries. In: Proceedings of the 21st ACM International Conference on Information and Knowledge Management, pp. 2737–2739. ACM (2012)
6. Fegaras, L., Li, C., Gupta, U., Philip, J.: XML query optimization in Map-Reduce. In: WebDB (2011)
7. Hricov, R.: Evaluation of XPath queries over XML documents using SparkSQL framework - Master thesis. FIT CTU - Master thesis (2016)
8. Marcjan, R., Siwik, L.: The concept of transformation of XML documents into quasi-relational model. In: Kozielski, S., Mrozek, D., Kasprowski, P., Małysiak-Mrozek, B., Kostrzewa, D. (eds.) BDAS 2014. CCIS, vol. 424, pp. 569–580. Springer, Cham (2014). doi:10.1007/978-3-319-06932-6_55
9. Strnad, P., Macek, O., Jira, P.: Mapping XML to key-value database. In: The Fifth International Conference on Advances in Databases, Knowledge, and Data Applications, DBKDA 2013, pp. 121–127 (2013)
10. Tatarinov, I., Viglas, S.D., Beyer, K., Shanmugasundaram, J., Shekita, E., Zhang, C.: Storing and querying ordered XML using a relational database system. In: Proceedings of the 2002 ACM SIGMOD International Conference on Management of Data, pp. 204–215. ACM (2002)
11. Šenk, A., Valenta, M., Benn, W.: Distributed evaluation of XPath axes queries over large XML documents stored in MapReduce clusters. In: Proceedings of the 2014 International Semiconductor Laser Conference, ISLC 2014, pp. 253–257. IEEE Computer Society, Washington, DC (2014). http://dx.doi.org/10.1109/DEXA.2014.59

Metrics-Based Auto Scaling Module for Amazon Web Services Cloud Platform

Dariusz Rafal Augustyn[(✉)] and Lukasz Warchal

Institute of Informatics, Silesian University of Technology,
16 Akademicka St., 44-100 Gliwice, Poland
{draugustyn,lukasz.warchal}@polsl.pl

Abstract. One of the key benefits of moving an application to the cloud is the ability to easy scale horizontally when the workload increases. Many cloud providers offer a mechanism of auto scaling which dynamically adjusts the number of virtual server instances, on which given system is running, according to some basic resource-based metrics like CPU utilization. In this work, we propose a model of auto scaling which is based on timing statistics: a high order quantile and a mean value, which are calculated from custom metrics, like execution time of a user request, gathered on application level. Inputs to the model are user defined values of those custom metrics. We developed software module that controls a number of virtual server instances according to both auto scaling models and conducted experiments that show our model based on custom metrics can perform better, while it uses less instances and still maintains assumed time constraints.

Keywords: Cloud computing · Scalability · Auto scaling · Custom metrics · Load balancing

1 Introduction

Modern information systems should support possibility of running in a cloud environment. Thanks to capabilities offered by a cloud providers the desired efficiency level of such system may be achieved by using, among others, infrastructure elements like: load balancers, virtual server instances and data storage services. System workload can be then distributed among many computing instances and their total number may vary over time, according to current needs.

Many cloud providers offer auto scaling mechanisms which allow to optimize resource utilization with respect to substantial load. Most often the standard auto scaling policy uses a statistics based on resource metrics, for example an average CPU utilization of given virtual server. A user is obligated to arbitrary define a lower and upper limit of CPU utilization, which indicates system underload and overload, respectively. This might be inconvenient, because user does not necessarily know optimal values as they are not directly related to observed system behavior. Moreover, when system workload changes between high and

© Springer International Publishing AG 2017
S. Kozielski et al. (Eds.): BDAS 2017, CCIS 716, pp. 42–52, 2017.
DOI: 10.1007/978-3-319-58274-0_4

low level in small amount of time, it could cause many instances to launch and stop after a short while. This generates additional costs, because user often pays for compute capacity by the hour.

In our approach, user defines custom auto scaling policy by providing intuitive timing metrics, as we assume it is easier to him to obtain proper values than in case of CPU utilization. We suggest to use an execution time of the most frequently used application features as a timing metric.

Utilizing this metric values we can calculate a high order quantile statistic to detect that system is overloaded (what should result in scaling in) and a mean value statistic to detect underload, respectively (what should result in scaling out). A user may set an upper limit for the quintile and a lower limit for the mean of execution time values.

In our work, we propose also an experimental method of obtaining limits for CPU utilization in standard auto scaling policy, which are approximately equivalent to those defined explicitly by a user in our approach. This allows us to combine both auto scaling models.

The proposed solution should be better than the standard one by preserving near-optimal number of running instances even for very unstable system load. To be able to compare both models, we define a mean cost function as follows:

$$MeanCost = \frac{1}{Time} \int_0^{Time} Number\_of\_instances(t)dt \qquad (1)$$

where $Time$ denotes a total period of an experiment and $Number\_of\_instances(t)$ is a function describing number of running instances over time. The appropriate software module was designed and developed for managing instances in Amazon Web Services public cloud according to both auto scaling models. Using load testing tool we conducted experiments using real-life application which show that our solution is better than the standard one by giving lower values of the mean cost defined in Eq. 1.

The problem of auto scaling was already considered in many approaches [10]. Some of them are proactive and utilize a prediction model e.g. [8] based on the queue theory or [3,11] ARMA/ARIMA models. The other ones are reactive e.g. [5]. All of them use a high-level (gathered on application level) or/and resources level metrics. The method proposed in this paper is reactive and uses both types of metrics, however resource level ones are used implicitly. In the proposed method a user – in very simple comparing to the other methods and intuitive manner – has to only set boundaries of application level metrics according to Quality of Service (QoS) requirements.

The article is organized as follows. Sections 2 and 3 briefly describes the application we used in our research. Section 4 provides some details about standard auto scaling model in AWS cloud. In Sect. 5 the method of obtaining limits for standard auto scaling model based on resource metrics utilizing our approach was introduced. Next, in Sect. 6 a deep insight into our auto scaling model was presented. In Sect. 7 we outline the workload generation process and we present results of conducted experiments. Section 8 contains conclusions and summary.

2 System Under Test

To check the auto scaling module described in this article we used measurements from experiments conducted using existing production-ready system called RepoEDM (Repository of Electronic Medical Documents). Its main goal is to index, store, search and retrieve medical documents created according to HL7 CDA [7] specification. RepoEDM was designed with Service Oriented Architecture (SOA) and Domain Driven Design (DDD) [6] principles in mind. Therefore it is composed of several modules, each representing separated Bounded Context [6] identified during system analysis. The system as a whole exposes its services to clients via SOAP Web Services [4]. The data gathered by RepoEDM is stored in relational database (Oracle or PostgreSQL).

Although RepoEDM consists of many modules it is deployed as a monolith on application server.

3 Migrating to Cloud

Migrating existing applications to cloud infrastructure is currently a widely observed trend. Increasing number of companies decide to move their systems to external data centers in order to reduce the cost of ensuring system availability and security. However it is a non-trivial task to move an existing application to the cloud, therefore there were developed several migration strategies [1] which may help. In case of RepoEDM moving to the cloud – Amazon Web Services[1] in particular – required several changes in system architecture. In a data layer PostgreSQL instance managed by Amazon RDS service[2] was used. RepoEDM modules were transformed into microservices, thus whole application became a distributed system. Obviously, this required adding several supporting components (also microservices) like service discovery, edge server etc. [2]. Each microservice was containerized using Docker[3] technology in order to simplify deployment tasks [9].

Redesigned RepoEDM application was then deployed in AWS infrastructure using Amazon Elastic Compute Cloud[4] (EC2) instances (Fig. 1). Clients can reach web services endpoints by the Elastic Load Balancer[5] (ELB). It is a managed load balancer service provided by AWS. ELB distributes workload between instances that are in service. The instances may return a valid response on so-called health-check endpoint.

To enable monitoring system performance, RepoEDM measures execution times of most common used services (web service calls) and reports them to

---

[1] Amazon Web Services (2016) https://aws.amazon.com.
[2] Amazon Relational Database Service (RDS) (2016) https://aws.amazon.com/rds.
[3] Build, ship, run – Docker is the world's leading software containerization platform (2016) https://www.docker.com.
[4] Amazon EC2 – Virtual Server Hosting (2016) https://aws.amazon.com/ec2.
[5] Elastic Load Balancing (2016) https://aws.amazon.com/elasticloadbalancing.

Graphite[6] running on dedicated EC2 instance. This tool can then visualize values of average or 0.9^{th} quantile in real-time.

AWS provides many types of EC2 instances[7] that differ in a number of CPUs amount of memory and pricing. Machine denoted as $m4.xlarge$ with 4 CPUs and 16 GB memory was used as a baseline for RepoEDM, because it has similar hardware specification to servers used in typical on-premise installations. Machine with 2 CPUs and 8 GB of memory ($m4.large$) was used for running Graphite.

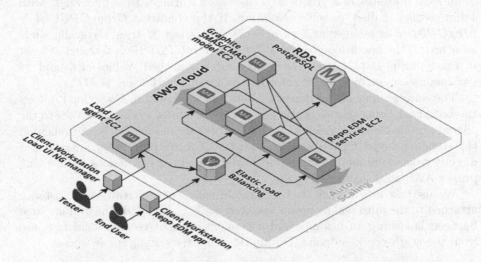

Fig. 1. The architecture of the auto scaling infrastructure (Elastic Load Balancer + four RepoEDM EC2 instances + Amazon RDS service) with the additional EC2 instance for Graphite service and SMAS/CMAS controller and the another one for a Load UI agent.

4   Standard Auto Scaling Model

AWS public cloud offers auto scaling mechanism[8] as a fully managed service. Auto scaling policy controls collections of EC2 instances labeled as auto scaling group.

Before we can define auto scaling group, first we create so-called lunch configuration[9] in which we choose AMI (Amazon Machine Image) and a machine type (e.g. $m4.xlarge$). AMI is a disk image that will be used to create and run new EC2 instance during the scaling in process.

---

[6] Graphite (2016) https://graphiteapp.org.

[7] Amazon EC2 Instance Types (2016) https://aws.amazon.com/ec2/instance-types.

[8] Auto Scaling (2016) https://aws.amazon.com/autoscaling.

[9] Launch Configurations (2016) http://docs.aws.amazon.com/autoscaling/latest/ userguide/LaunchConfiguration.htm.

To create the auto scaling group we:

- use the defined launch configuration,
- select the existing ELB,
- set size parameters of an auto scaling group (min, max and desired number of running EC2 instances).

Cloud infrastructure can make scaling decisions by means of some built–in metrics. For instance we can use $GroupCPUUtil$ – an average of CPU utilization of all EC2 instances in a group. It is calculated during a time interval T_i with 1 min as the smallest possible resolution. If the condition $GroupCPUUtil > MaxCPUUtil$ is satisfied at least m times during last M tries, the scaling in is launched. If the condition $GroupCPUUtil < MinCPUUtil$ is satisfied at least m times during last M tries, the scaling out is launched. Values of m and M parameters are set by policy creator but commonly we have $m > M/2$.

The auto scaling model described above is available in AWS cloud out-of-the-box. That being said, it is also possible to extend or modify it thanks to the offered by this cloud provider APIs. In our software module we re-implemented this standard model. CPU utilization statistics of auto scaling group are retrieved directly from cloud monitoring service[10] and scale in or out is done by invoking proper AWS auto scaling API calls.

We also take into account the time of stabilizing a newly created EC2 instance attached to the auto scaling group (denoted by $T_{unstable}$). It is the average time between launching an instance and a moment when it returns valid response from the heath-check endpoint. During this time the scaling is suspended.

5   The Method of Obtaining Limits for Scaling Policy Applied in Standard Resource-Based Auto Scaling Method

In standard auto scaling model a user is obligated to setup lower and upper limit of some resource-based metric like CPU utilization. This could be difficult task but our approach gives possibility to assign those limits indirectly by defining some timing-based constraints that are approximately equivalent to the aforementioned resource-based ones. Moreover the latter can be then obtained experimentally.

An experienced user, who works with any system on the day-to-day basis can easily say when it performs well in terms of response time. Let us introduce $T_{q\,acc}$ as the highest accepted by the user value of T_q, where T_q is the qth high order quantile of execution time of some critical business operation. At the same time let us introduce MV_{acc} as the lowest accepted value of mean execution time (MV) of the same operation. Obviously condition $MV_{acc} < T_{q\,acc}$ must be satisfied. The above user defined boundaries of timing statistics $T_{q\,acc}$ and MV_{acc} are used to obtain parameters for standard auto scaling method.

---

[10] Amazon Cloud Watch (2016) http://docs.aws.amazon.com/AmazonCloudWatch/latest/APIReference.

We start with one-instance system and experimentally find an upper limit of CPU utilization when the system becomes overloaded. This happen when the following condition is satisfied:

$$|T_q - T_{q\ acc}| \leq rT_{q\ acc} \tag{2}$$

where r is a value of the relative error. In a result we obtain $MaxCPUUtil$ as an upper limit for $GroupCPUUtil$ that when exceeded launches scaling out.

Next, also with one-instance system we experimentally find a lower limit of CPU utilization when the system becomes underloaded. This happen when the following condition is satisfied:

$$|MV - MV_{acc}| \leq rMV_{acc}. \tag{3}$$

Hence we obtain $MinCPUUtil$ as an lower limit for $GroupCPUUtil$ that when exceeded launches scaling in.

6   Auto Scaling Based on Custom Metrics

In this chapter we introduce a model of auto scaling based on custom metrics. It is labeled as CMAS (Custom Metrics Auto Scaling) to distinguish it from SMAS model (Standard Metrics Auto Scaling), which is based on a group CPU utilization.

CMAS utilizes statistics defined on T – values of execution times of the selected business operation. Values of T_q (the qth order quantile) are used to decide whether to scale out and MV (mean value) to scale in, respectively. $\widehat{T_q}, \widehat{MV}$ the estimators of T, MV are reset after each instance add/remove.

$\widehat{T_q}, \widehat{MV}$ work on measurements gathered during T_i interval. For both estimators, confidence intervals are calculated for some confidence probability level ($p = 0.9$). To obtain a confidence interval for T_q we use Maritz–Jarrett method [12]. For MV we use the well-known method based on Central Limit Theorem. This allows to use/ignore CMAS model when there is enough narrow/too wide confidence interval for T_q or MV. It also allows to verify that estimators' values are significantly different from the constants $T_{q\ acc}$ or MV_{acc} at the assumed confidence level.

Sequences of statistical confident values of $\widehat{T_q}, \widehat{MV}$ are obtained in subsequent moments determined by T_i. If the condition $\widehat{T_q} > T_{q\ acc}$ is satisfied (distr. no 2 in Fig. 2) at least in m moments during last M moments, scaling out should be launched. If the condition $\widehat{MV} < MV_{acc}$ is satisfied (distr. no 1 in Fig. 2) at least m times during last M tries, scaling in should be launched.

Both CMAS and SMAS models were implemented in a dedicated software module called CMAS/SMAS controller. It is running alongside with Graphite monitoring tool on the same EC2 instance (Fig. 1).

CMAS model supplements SMAS one: the latter is used when the former cannot work i.e. when the assumed confidence conditions for estimators $\widehat{T_q}, \widehat{MV}$ are not satisfied, because there is a lack of elementary data about values of T gathered in at least T_i intervals.

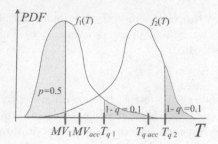

Fig. 2. Two distributions of executing times (T) for a not enough loaded system (f_1) and an overloaded one (f_2). For the distribution no 1 the condition $MV_1 < MV_{acc}$ (and $T_{q\,1} < T_{q\,acc}$) may causes scaling out. For the distribution no 2 the condition $T_{q\,2} > T_{q\,acc}$ may causes scaling in.

7 Experimental Results

In conducted experiments RepoEDM (described in Sect. 2) was put under load by running prepared test scenario that represents one of the key system capabilities. It was developed and executed using a set of tools offered by SmartBear company, which are dedicated to such tests. The load profile was fixed, i.e. number of Virtual Users (VUs) was constant during whole experiment and each VU was running test scenario repeatedly with 1 s break between each two runs. To eliminate negative impact of network latency in final results, agent generating load was in the same network inside AWS cloud (Fig. 1).

In scaling policies we used $m = 3$ an $M = 5$ either for CMAS or SMAS.

7.1 Obtaining Parameters for CMAS Model

Described above fixed load profile was used for testing a one-instance RepoEDM system in order to experimentally find an upper and lower limit of CPU utilization.

Let us assume that user accepts $T_{q\,acc} \approx 4000$ ms with $q = 0.9$. By varying the number of VUs and assuming $r = 5\%$ it is possible to find the value of CPU utilization of the instance which satisfies the condition $|T_q - 4000\,\text{ms}| \leq 5\% \cdot 4000\,\text{ms}$ (see Eq. 2). In this case it is $82 \pm 8\%$, hence $MaxCPUUtil = 82\%$.

Analogously, let us assume that user expects $MV_{acc} \approx 700$ ms. By varying a number of VUs and assuming $r = 5\%$ it is possible to find the value of CPU utilization of the instance which satisfies the condition $|MV - 700\,\text{ms}| \leq 5\% \cdot 700\,\text{ms}$ (see Eq. 3). In this case it is $14 \pm 4\%$, hence $MinCPUUtil = 14\%$.

7.2 CMAS/SMAS Differences in Scaling Behavior

To reveal differences in CMAS/SMAS scaling behavior it is sufficient to analyze results from a short 30-minute experiment with the above-mentioned fixed load profile with 30 VUs.

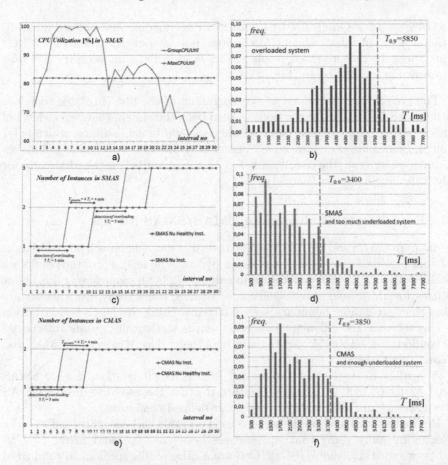

Fig. 3. A sample 30-minute test results for the system under fixed load: (a) group CPU utilization in SMAS model which influences on a course of Number of Instances in Fig. 3c (b) distribution of execution times based on T values gathered in $1 \div 6$ intervals (dashed line shows too high value of quantile: $T_q = 5850 > 4000 = T_{q\ acc}$) (c) Number of All Instances or Healthy ones handled by SMAS (d) distribution of execution times for CMAS based on T values gathered in $20 \div 26$ intervals (dashed line shows very low value of quantile: $T_q = 3400 < 4000 = T_{q\ acc}$) (e) Number of All Instances or Healthy ones handled by CMAS (f) distribution of execution times for CMAS based on T values gathered in $10 \div 16$ intervals (dashed line shows enough low value of quantile: $T_q = 3850 < 4000 = T_{q\ acc}$).

The most characteristic distinctions that can be easily spot are shown in Fig. 3:

– CMAS model can detect an overloading a little bit earlier than SMAS (commonly we noticed a $1 \div 2$ min delays in enabling *GroupCPUUtil* by AWS to SMAS). In the presented results CMAS launches an EC2 instance after interval no 6 (Fig. 3e) and SMAS later i.e. after interval no 7 (Fig. 3c),

– SMAS by using GroupCPUUtil detects an overloading and creates an additional EC2 instance while CMAS does not change the number of EC2 instances. In the presented result SMAS launches the unnecessary 3rd EC2 instance in interval no 16 (Fig. 3c).

The system was overloaded at the beginning of the test (Fig. 3b), therefore both CMAS and SMAS take actions to turn back the system to a proper level of efficiency (Fig. 3d, f). CMAS launches only one additional instance, which is far enough. Launching yet another EC2 instance by SMAS is redundant and makes an unnecessary cost (the condition based on the quantile is satisfied already for only 2 running instances (Fig. 3d)).

7.3 Experimental Evaluation of CMAS/SMAS Models

To compare CMAS and SMAS we run 20 tests in a row, where each one consists of a 30 min period of fixed load and a 20 min period of no load. From the results we obtained $MeanT_q$, which denotes mean value of $\widehat{T_q}$ in all sequences and is about 3710 ms for SMAS and about 3960 ms for CMAS, respectively. Both values are lower than acceptable boundary set to 4000 ms. Using Student's t-distribution we checked that the difference between them is statistically significant (i.e. at 0.9 confidence level we rejected the null hypothesis that $MeanT_q$ for SMAS and $MeanT_q$ for CMAS are equal).

According to Eq. 1 we also obtained $MeanCost$. It is about 2.3 for SMAS and 1.42 for CMAS. This allows to state that for the used test profile CMAS gives lower mean cost maintaining the user-defined conditions.

Beside the fixed load profile we also used a profile with variable load in order to experimentally verify that equivalences between $T_{q\ acc}$ and $MaxCPUUtil$ and between MV_{acc} and $MinCPUUtil$ are a little profile specific. In conducted experiments we used profile which has base number of VUs = 5 running test scenario for 10 s. Then number of VUs burst up to $5 \div 40$ during 30 s. After that time it comes back to 5 for next 10 s and so on. Results obtained using both profiles, when compared, do not differ much (eg. $MinCPUUtil$ value differ in less than 7%, $MaxCPUUtil$ value differ in less than 9%). But this weighs in favor of the use of CMAS over SMAS, while CMAS is directly aligned to a user expectation based on timing statistics and does not depend on a load profile.

8 Conclusions

In the paper we proposed the method of auto scaling based on times of executing selected domain critical service. A user by assuming limit values for a qth quantile and a mean value of execution time values may assign conditions for an accepted level of system efficiency.

For the used load test profile we experimentally shown that usage of a model based on such custom metrics gives lower cost than usage of a model based only on standard resource metrics.

We propose the method of obtaining parameters for standard auto scaling model based of resource metrics using parameters of our model based on timing metrics.

In future we plan to extend a set of load tests for more detail verification of advantages of the proposed model and its implementation.

The future work may also concentrate on operating on a deeper granularity of auto scaling. According to appearance of Application Load Balancer offered by AWS we may scale not only at level of instances but also at level of microservices that are run inside an instance. The architecture of RepoEDM system allows applying such approach.

Acknowledgements. This work was supported by NCBiR of Poland (No. INNOTECH-K3/IN3/46/229379/NCBR/14).

References

1. Andrikopoulos, V., Binz, T., Leymann, F., Strauch, S.: How to adapt applications for the cloud environment. Computing, 1–43 (2012). http://www2.informatik.uni-stuttgart.de/cgi-bin/NCSTRL/NCSTRL_view.pl?id=ART-2012-15&engl=1
2. Balalaie, A., Heydarnoori, A., Jamshidi, P.: Migrating to cloud-native architectures using microservices: an experience report. In: Celesti, A., Leitner, P. (eds.) ESOCC 2015. Communications in Computer and Information Science, vol. 567, pp. 201–215. Springer International Publishing, Cham (2016). doi:10.1007/978-3-319-33313-7_15
3. Calheiros, R.N., Masoumi, E., Ranjan, R., Buyya, R.: Workload prediction using ARIMA model and its impact on cloud applications' QoS. IEEE Trans. Cloud Comput. **3**(4), 449–458 (2015). http://dx.doi.org/10.1109/TCC.2014.2350475
4. Del Ra III, W.: Service design patterns: fundamental design solutions for SOAP/WSDL and RESTful web services by robert daigneau. ACM SIGSOFT Softw. Eng. Notes **37**(4), 40 (2012). http://dblp.uni-trier.de/db/journals/sigsoft/sigsoft37.html#Ra12c
5. Dias De Assuncao, M., Cardonha, C., Netto, M., Cunha, R.: Impact of user patience on auto-scaling resource capacity for cloud services. Future Gener. Comput. Syst. 1–10 (2015). https://hal.inria.fr/hal-01199207
6. Evans, E.: Domain-Driven Design: Tackling Complexity in the Heart of Software. Addison-Wesley, Boston (2004). http://www.worldcat.org/search?qt=worldcat_org_all&q=0321125215
7. Health Level Seven®: HL7CDA (2013). http://www.hl7.org/implement/standards/product_brief.cfm?product_id=7
8. Jiang, J., Lu, J., Zhang, G., Long, G.: Optimal cloud resource auto-scaling for web applications. In: 13th IEEE/ACM International Symposium on Cluster, Cloud, and Grid Computing, CCGrid 2013, Delft, Netherlands, 13–16 May 2013, pp. 58–65 (2013). http://doi.ieeecomputersociety.org/10.1109/CCGrid.2013.73
9. Pahl, C.: Containerization and the PaaS cloud. IEEE Cloud Comput. **2**(03), 24–31 (2015)
10. Qu, C., Calheiros, R.N., Buyya, R.: Auto-scaling web applications in clouds: a taxonomy and survey. CoRR abs/1609.09224 (2016). http://arxiv.org/abs/1609.09224

11. Roy, N., Dubey, A., Gokhale, A.: Efficient autoscaling in the cloud using predictive models for workload forecasting. In: Proceedings of the 2011 IEEE 4th International Conference on Cloud Computing, Cloud 2011, pp. 500–507 (2011). http://dx.doi.org/10.1109/CLOUD.2011.42

12. Wilcox, R.: Chapter 3 - estimating measures of location and scale. In: Wilcox, R. (ed.) Introduction to Robust Estimation and Hypothesis Testing. Statistical Modeling and Decision Science, pp. 43–101. Academic Press, Boston (2012). http://www.sciencedirect.com/science/article/pii/B9780123869838000032

Artificial Intelligence, Data Mining and Knowledge Discovery

Comparison of Two Versions of Formalization Method for Text Expressed Knowledge

Martina Asenbrener Katic[✉], Sanja Candrlic, and Mile Pavlic

Department of Informatics, University of Rijeka, R. Matejcic 2, Rijeka, Croatia
{masenbrener,sanjac}@inf.uniri.hr, mile.pavlic@ris.hr
http://www.inf.uniri.hr

Abstract. The Node of Knowledge (NOK) method is a method for knowledge representation. It is used as a basis for development of formalism for textual knowledge representation (FNOK). Two versions of formalization methods and, respectively, two Question Answering (QA) systems are developed. The first system uses grammars; it is written and implemented in Python. The second, improved system is based on storing text in relational databases without losing semantics and it is implemented in Oracle.

This paper presents the results of comparison of the two QA systems. The first system was tested using 42 sentences. It received 88 questions from users and provided answers. After improving the formalization method, the second system was tested with the same set of sentences and questions. The paper presents the results of the testing, the comparison of answers received from both systems and the analysis of correctness of the answers received.

Keywords: NOK · Node of Knowledge · Relational database · Question answering systems · Knowledge · Knowledge-based systems

1 Introduction

Classic information systems store data in relational databases. Unfortunately, the use of relational databases is limited: they cannot store sentences expressed in natural language; only data structured in a certain way (as requested by the database scheme) can be stored. In addition, it is not possible to query relational databases using questions expressed in natural language, but SQL queries should be used for data access and extraction. Another problem is related to performing changes within a relational database with the goal to add new knowledge. Relational databases can store sentences only by using textual data types, but knowledge stored that way cannot be used as a source for querying. Relational databases allow data insert and data store only if they comply to strictly defined rules. This leads toward the idea to store the sentence in a relational database in parts, i.e. phrases or words.

The aim of our broad research is to develop a new system that will improve the possibilities of relational database through two main capabilities: (1) to

© Springer International Publishing AG 2017
S. Kozielski et al. (Eds.): BDAS 2017, CCIS 716, pp. 55–66, 2017.
DOI: 10.1007/978-3-319-58274-0_5

enable storing of sentences expressed in natural language (unstructured or semi-structured text) and (2) to enable querying of relational databases using questions expressed in natural language and finding the answer within relational databases.

Currently, business intelligence tools and additional systems such as OLAP and data warehousing tools can improve reasoning on data used in an information system, but the problem with semantics of natural language still exists. There are many natural language processing methods and text processing tools developed, but they do not offer a solution for storing text into relational databases without loss of semantics. An example of real life application of an expert system using queries submitted by the user using natural language is explained in [3].

For example, [6,9] describe researches that focus on design of queries in natural language i.e. they translate the request for data from a query into SQL and the relational database can provide the required response, but the loss of semantics still occurs.

Our research and development of a system that enables integration of texts, dictionaries and relational databases is based on the NOK method. The Node of Knowledge (NOK) method is one of the methods for knowledge representation. It is simpler than other methods for graphic knowledge representation (it uses less elements), more expressive (it enables knowledge representation on different abstraction levels) and simpler to read (it enables reading of knowledge starting from any node using links' roles. These characteristics of the NOK method ensure its simple implementation and broad applicability, and these were the main reasons to choose NOK. Implementation of the NOK method into existing information systems would enable development of a question answering (QA) system that could communicate with users simply by questions and sentences expressed in natural language.

The research of QA systems started in the 1960-ies [8]. Recent results in the field have been reported since 1999 in a series of QA evaluations as part of the Text Retrieval Conference (TREC) [2] with various success in the evaluation of answers. Since most data of every company is usually stored in relational databases, there is an increasing demand for automated reasoning of relational databases in order to use information systems as decision support systems [7].

This paper presents a research of two Question Answering (QA) systems based on two versions of formalization methods, i.e. two research ideas. The first system is grammar-based, while the second system is based on relational databases. Motivation for this research was to determine which system should be used as a basis for further development of a QA system. The goal of the research is to obtain objective evaluation result. During the evaluation process, both systems were tested using 42 sentences and 88 questions expressed in natural language. The paper presents the results of the testing, the comparison of answers received from both systems and the analysis of correctness of the answers received.

2   Background

The Node of Knowledge method enables notation of knowledge through identification of semantic relationships between words and phrases within the sentence. The role of each word in the sentence is interpreted with the help of wh-questions. In addition to NOK, several formalisms have been developed: Diagram Node of Knowledge (DNOK) - formalism for graphical representation [5,13,14], Formalized Node of Knowledge (FNOK) - formalism for textual knowledge representation [4,12] and QFNOK - formalism for question representation.

Focus of the NOK is on research and analysis of natural human language. NOK extracts words from a sentence and creates models of knowledge expressed in the sentence. This way NOK enables an alternative storing of knowledge in a way different from the language and script, i.e. the human mind. It defines the structure of a network of knowledge contained in a textual form in any natural language [14]. Basic concepts of the NOK method [13] are node, process node, link and role. To explain these concepts, a simple example is used: *Julia swims fast on the pool* (Fig. 1).

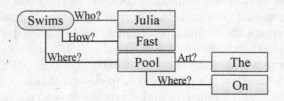

Fig. 1. DNOK for the sentence *Julia swims fast on the pool*

DNOK for this sentence consists of five regular nodes: *Julia, Fast, Pool, On* and *The* and the process node *Swims*. Nodes *Julia, Fast* and *Pool* are connected to the process node *Swims*. In addition, the node *Pool* is connected to the node *The* and node *On*. Questions that define the role are listed on each link between nodes, and they are very important for storing and querying knowledge. For example, the node *Julia* and the process node *Swims* are connected by a link role *Who?*. If we ask *Who swims?*, the answer is *Julia*. FNOK for this sentence is:

swims(*"who?" Julia, "how?" fast, "where?" pool("where?" on, "art?" the))*

Process node takes the primary position in the FNOK record, i.e. it is on the top of the hierarchy. The rest of the nodes follow. In front of every node is a question that marks semantic connection of the node and its ancestor node. For example, the node *pool* is bellow process node *swims*, they are connected by the role *where?* The node *pool* is above node *the*, their connection is defined by the role *art?* (which marks article) and node *on* (which belongs to the preposition word type and therefore uses the same role as its ancestor node).

2.1 Grammar-Based System (GBS)

The basic Question Answering system based on the NOK framework is named
Question Answering NOK (QANOK) and it is described in detail in [12].
QANOK is a computer-based system whose primary function is to provide precise answers to questions set in natural language. For this purpose three algorithms were developed [12]:

- an algorithm which translates text expressed knowledge in natural language
  into FMTEK (Formalized Method for Text Expressed Knowledge) formalized
  notation,
- an algorithm which translates a question in natural language into QFMTEK
  (question FMTEK) formalized knowledge notation, and
- an algorithm which uses QFMTEK formalized question to find an answer
  within FMTEK formalized knowledge.

This grammar-based system is implemented in Python. It consists of two
phrase structure grammars (PSG). One is used for the syntactic analysis of a
non-formalized sentence (a sentence that is expressed using human language),
while the other is used in the formalized notation of the same sentence by the
application of parts that were identified during syntactic analysis. This formalized notation preserves the semantic relationships between words that were identified using wh-questions. The identification of possible semantic relationships
between words is preserved in the expanded MULTEXT-East lexicon.

The tests performed on the QANOK system showed that it can "understand"
simple sentences in English language and derive answers to the asked questions.
During the tests a lexicon composed of about a hundred chosen words was used.
The testing procedure consisted of several steps. Sentences in natural language
were described using regular grammar and equivalent regular expressions. Then
they were translated into their formal form (FNOK) using the syntax-controlled
translation into FNOK language. For each sentence one or more interrogative
sentences were formed. Each interrogative sentence was then transformed to the
QFNOK, using syntax-controlled translation. Unlike FNOK, QFNOK structure
uses variables in its expressions, but besides that, their structure is very similar. This similarity of FNOK and QFNOK structures simplifies the problem of
matching two expressions in the process of deriving the answer. Variables in the
QFNOK expression represent answers in the FNOK expressions.

The system was tested using 42 sentences in English (TENG – text in natural
English language). Based on these 42 TENG sentences, 88 questions were set
(QTENG – questions from a user in natural English language). For each sentence,
at least one question was set. Two question types were used - polar questions
and wh-questions. Polar questions require an answer in a form of Yes or No.
Wh-question is a question type that uses wh-words (who, whom, whose, what,
which, why, when, where, how etc.).

The analysis of answers has shown that the QA System and formalism for
storing knowledge will need some improvement. The problems recognized during
testing of the QANOK system are described in [12]. For example, for the question

set in singular form (*Who is on the beach?*), the answer can be in plural (*Girls are on the beach*). Unfortunately, the QANOK system will not provide an answer in this case.

To avoid these problems, authors of this paper developed a new system and new algorithms for its operation. The new system based on relational databases is briefly described in the following chapter.

2.2   The System Based on Relational Databases (RDBS)

Previous research [1] showed that concepts of the DNOK formalism can be translated into concepts of the entity relationship diagram. The whole NOK diagram can then be translated into relational database. For this purpose, authors developed an algorithm that transforms textual expressed knowledge into relational database.

On the basis of the MIRIS methodology [10,11] data models that describe a system for transformation are designed. On that ground, the system for storing knowledge (sentences expressed in natural human language) into relational databases, without any loss of semantics, is developed. The system was implemented using Oracle database.

The properties of relational databases ensure almost limitless usage of algorithms for question answering. Enriched with wh-questions that explain links between words, text stored into relational database (by following the prescribed procedures for text and question formalism) is not affected with any loss of semantics.

For this system authors developed improved algorithms for transformation of text expressed knowledge into formalized text shaped to be stored into relational database. This system integrated monolingual dictionary which enables to include different semantic relations, such as synonyms, homonyms etc.

3   Methodology

The grammar-based formalization method for text expressed knowledge was tested in the process of question answering. For a set of questions, the system tried to give a proper answer, according to the knowledge stored in the knowledge database.

After improving the formalization method and developing the system based on relational databases (RDBS), the same set of questions was used for the same text expressed knowledge. The goal was to test the results, compare them to answers received from the GBS system based on the first version of the formalization method and to compare and analyze the correctness of the answers received.

The knowledge in the system was based on 42 sentences expressed textually. Each system used its own method for storing the sentences formally. In the process of receiving the answers, a set of 88 questions was used. An excerpt of the

Table 1. An excerpt of the sentences

TENG	FNOK (GBS)	FNOK (RDBS)
The car drives.	drives("what?" the car)	drives ("what?" car ("art?" the))
A boy stares.	stares("who?" a boy)	stares ("who?" boy ("art?" a))
My car drives.	drives("what?" car("whose?" my))	drives ("what?" car ("whose?" my))
The red car drives.	drives("what?" the car("which?" red))	drives ("what?" car ("art?" the, "which?" red))
A tall boy stares.	stares("what?" a boy("which?" tall))	stares ("who?" boy ("art?" a, "which?" tall))
My incredibly tall boy stares.	stares("whose?" my "how?" incredibly "who?" boy("which?" tall))	stares ("who?" boy ("whose?" my, "which?" tall ("how?" incredibly)))

Table 2. An excerpt of the questions

QTENG	QFNOK (GBS)	QFNOK (RDBS)
Are girls on the beach?	are (_girls, _on _ the beach	are (_ girls, _ beach (_ on, _ the))
Who is on the beach?	is ("who?" X, & _ on _ the beach	is ("who?" X, _ beach (_ on, _ the))
Where are girls?	are (_ girls, & "where?" X	are (_ girls, "where?" X)
Who writes a letter to a friend?	writes ("who?" X, _ a letter _ to _ a friend	writes ("who?" X, _ letter (_ a), _ friend (_ to, _ a))
What does someone's electrical lamp do?	X (_ lamp (_ someone's, _ electrical)	X ((& does), (& do), _ lamp (_ someone's, _ electrical))
When does Tom drive a car?	drive (_ Tom, _ a car "when?" X	drive ((& does), _ Tom, _ car (& a), "when?" X)

sentences and questions for the Grammar-based system (GBS) and the System based on relational databases (RDBS) is given in Tables 1 and 2 respectively.

The new version of the formalization method offers improvement in several elements: a shift from linear to hierarchy method, articles, implementation of rules for transformation of different data types and introduction of dictionary. Examples are given in Tables 3 and 4.

A Shift from Linear to Hierarchy Method. The linear method used in the first formalization method starts with the verb and continues with other words in exact order they appear in the sentence. The new method also starts with the verb, but then continues with words directly connected to the verb creating hierarchy. Each word in the hierarchy is assigned to the object it refers to.

Articles. Instead of using articles attached to the word itself, the new formalization method uses the object role "art?" for each article in the sentence.

New Rules for Transformation of Different Data Types. New rules for transformation of different word types from TENG into FNOK and from QTENG into QFNOK are introduced. Rules for nouns, pronouns, adjectives, numbers, adverbs, prepositions, articles and verbs (including different tenses) are improved as well.

Dictionary. The new formalization method introduced dictionary in the process of storing data. For each word used in the sentence, synonyms and related forms are noted in the dictionary. For example, the verb *drive* is related to its forms *drives, drove* and *driven*. This enables the system to provide an answer, although a different form of word is used in the question.

Table 3. Changes introduced in FNOK

TENG	FNOK (GBS)	FNOK (RDBS)
Hierarchy		
My incredibly tall boy stares.	stares("whose?" my "how?" incredibly "who?" boy("which?" tall))	stares ("who?" boy ("whose?" my, "which?" tall ("how?" incredibly)))
Articles		
The red car drives.	drives("what?" the car("which?" red))	drives ("what?" car ("art?" the, "which?" red))
New rules for transformation of different word types		
The red car on the road drives.	drives("what?" the car("which?" red) "where?" on "what?" the road)	drives ("what?" car ("art?" the, "which?" red), "where?" road ("where?" on, "art?" the))
I talk about the solution with my colleague.	talk("who?" I, "where?" about "what?" the solution "where?" with "whose?" my colleague)	talk ("who?" I, "what?" solution ("what?" about, "art?" the), "who?" colleague ("who?" with, "whose?" my))
Tom has two cars.	has("who?" Tom, "how_many?" two "what?" cars)	has ("who?" Tom, "what?" cars ("how_many?" two))

Table 4. Changes introduced in FNOK - Dictionary

TENG	FNOK (GBS)	FNOK (RDBS)	DICT (RDBS)
Tom drives a car today.	drives("who?" Tom, "what?" a car "when?" today)	drives ("who?" Tom, "what?" car ("art?" a), "when?" today)	drive
This girl sings.	sings("who?" girl("which?" this))	sings ("who?" girl ("which?" this))	sing

Improvements are introduced in the QFNOK formalism, as well. Each word from QTENG now needs to become a node in QFNOK. Hierarchy in QFNOK is implemented and obligatory symbol) (closed parenthesis) is used as a mark for returning to the higher level of hierarchy. The biggest change is shown in

Table 5. Changes introduced in QFNOK

QTENG	QFNOK (GBS)	QFNOK (RDBS)
Are girls on the beach?	are (_ girls, _ on _ the beach	are (_ girls, _ beach (_ on, _ the))
Who is on the beach?	is ("who?" X, & _ on _ the beach	is ("who?" X, _ beach (_ on, _ the))
What does Julia do to a friend?	X (_ Julia & _ to _ a friend	X ((& does), (& do), _ Julia, _ friend (_ to, _ a))
What does Julia do on the pool?	X (_ Julia & _ on _ the pool	X ((& does), (& do), _ Julia, _ pool (_ on, _ the))
Where does Julia swim?	swim (_ Julia & "where?" X	swim ((& does), _ Julia, "where?" X)
How does Julia swim?	swim (_ Julia, "how?" X	swim ((& does), _ Julia, "how?" X)

Table 5, for example for the question *What does Julia do to a friend?*: forms of (& *does*) and (& *do*), and hierarchy _ *friend* (_ *to,* _ *a*).

For both systems tested, TENG sentences and QTENG questions were the same. Changes were introduced based on new rules for the new formalism in 29 FNOK records (12 minor changes and 17 major changes) and in every QFNOK record. Minor changes refer to articles (the new system considers each article a separate object) and changes of wh-questions used to determine roles (for example, *what* is sometimes changed into *which*; *what kind* into *what* or *which*, etc.). Major changes refer to hierarchy change or hierarchy introduction, and to the new rules concerning numbers and prepositions.

4   Research Results

After applying the same set of questions to the same text expressed knowledge, but by using two different methods for formalization of the text expressed knowledge, the quality of the provided answers is analyzed. Each answer is estimated for its completeness and correctness. Therefore, several types of answers are recognized (Table 6).

Correct Answer. The answer provided by the system is correct.

Incorrect Answer. The answer provided by the system is not based on the stored knowledge and the wrong set of words was used in the answer. For example, for the question *What does Tom do?* The system offers an answer: *has.* This answer cannot be accepted as correct. Another example: the system did not find an answer, although it had to and it offers an answer *I do not know.*

Missing Answer. The system did not provide an answer, but the answer was expected. For example, for the question *What does a pencil do?* the system will offer an answer *writes* because it uses knowledge expressed in the TENG *A pencil writes.* But, there is one more TENG that was supposed to be the origin for the answer: *Student's pencil writes.* Therefore, one more answer *writes* should have been offered.

Table 6. Answers received from the system

QTENG	ANSWER	TENG that contains the answer
Correct answer		
Are girls on the beach?	Yes	Girls are on the beach.
Who is on the beach?	girls	Girls are on the beach.
Where are girls?	on the beach	Girls are on the beach.
What does Julia do to a friend?	writes	Julia writes a letter to a friend.
What glows?	everything	Everything glows.
	someone's lamp	Someone's lamp glows.
	someone's electrical lamp	Someone's electrical lamp glows.
Incorrect answer		
Who is on the beach?	I do not know	Girls are on the beach.
Whose red car did he drive on Monday?	I do not know	He drove my red car on Monday.
What does Tom do?	has	Tom has two cars.
Missing answer		
What does a pencil do?	writes	A pencil writes.
		Student's pencil writes.
Semantically incomplete answer		
To whom does Julia write a letter?	friend	Julia writes a letter to a friend.
Semantically correct, but grammatically incorrect answer		
Whose car drives?	my	My car drives.

Semantically Incomplete Answer. The system answers the question partially (incompletely), i.e. the system finds an isolated, partial answer, elicited from the correct knowledge. Complete answer would be more native. For example, for the question *Who does Julia write a letter to?*, based on knowledge expressed in TENG *Julia writes a letter to a friend*, the system answers *friend*. This answer is considered semantically incomplete and it would be more accurate to get an answer *a friend* or even *to a friend*.

Semantically Correct, but Grammatically Incorrect Answer. The system provided an answer based on the proper knowledge, but the answer does not comply to Grammar rules. For example, for the question *Whose car drives?* the system responds *my*. The complete and correct answer would be *my car* or *mine*.

The analysis of answers received from both systems, according to previously described categories, is shown in Table 7. Based on the 42 TENG sentences and 88 QTENG questions, 144 answers were offered (some questions had more than one answer). The grammar-based QA system answered correctly to 118 out of 144 cases, while the QA system based on relational databases answered correctly to 134 out of 144 cases. The first system offered 7 incorrect answers, the second system offered 4. In the category: *Semantically incomplete answer*, the first system did not provide any answer, while the second system provided 2. In the category: *Semantically correct, but grammatically incorrect answer*, the first system offered one answer and the second offered 4. Although the answer was expected, the first system did not provide it in 18 cases, while the second system always provided an answer (but with different levels of correctness).

Table 7. Answer analysis

Answer category	GBS	RDBS
Correct answer	118	134
Incorrect answer	7	4
Semantically incomplete answer	0	2
Semantically correct, but grammatically incorrect answer	1	4
Missing answer	18	0
Total	144	144

Graphs in Fig. 2 show that the first system responded correctly in 82% of cases, while for the second system correctness increased to 93%.

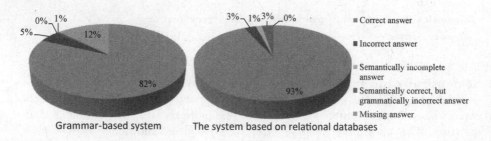

Fig. 2. Answer analysis for both QA systems

Comparison of individual answers received from both systems (Table 8) showed that for 4 cases RDBS showed worse results than the first system (it offered incorrect or semantically incomplete answer, while GBS offered correct answer). In 119 cases both systems offered equally qualitative answers (sometimes correct, sometimes incorrect). In 21 cases the second system was better (it offered correct or at least semantically correct, but grammatically incorrect answer, while the first system offered incorrect answer).

Table 8. The quality of new answers given by RDBS compared to GBS QA system

Answer quality	Number of answers	%
Worse	4	3%
The same	119	83%
Better	21	14%

5   Conclusion and Future Work

This paper presents a comparison of answers received from two QA systems developed on different versions of formalization methods. The first system uses grammars and it is written and implemented in Python. The system based on relational databases is an improved system and it is based on storing text in relational databases without the loss of semantics. It is implemented in Oracle.

The goal of this research was to evaluate both systems and to determine which of them should be further developed. In the evaluation process, the first QA system was tested using a set of 42 sentences and 88 questions. To achieve better testing results, the formalization method needed to be improved. After its improvement, the same testing procedure was performed by using the same set of sentences and questions on the second QA system. The sum of 144 answers were received from the system. They were classified in 5 categories: correct; incorrect; semantically incomplete; semantically correct, but grammatically incorrect and missing answer. The second system was compared to the first regarding the quality of each given answer (worse, the same, better). The second system offered better answers in 14%. Therefore, it will serve as a ground for further research and QA system development. QA system such as this one can be used for questioning of any text expressed knowledge, but also to receive answers based on data already stored in relational databases. QA systems such as this will enable fast access to any data stored within relational databases. For business companies this knowledge is very important for the process of making business decisions.

The need for improvement still exists for the questions that did not receive the answer, although the answer was expected. Also, further analysis is needed for questions that received better answer when using the first formalization method (for example, due to the changes in the structure of the formalization method, the first system offered a correct answer to a certain question (*a friend*), while the other system offered only one part of the answer (*friend*).

In our future research, special attention will be paid to sentences that express passive and active form (for example: *He drove his car on Monday. The car was driven on Monday.*)

It is necessary to improve the system for cases that require multiple passes through hierarchy, for different Tenses (for example, questions that uses Present Tense should deliver answers without Future or Past Tense). Answers that belong to the category *Semantically incomplete answers* and to the category *Semantically correct, but grammatically incorrect answers* should be improved.

Acknowledgments. The research has been conducted under the project "Extending the information system development methodology with artificial intelligence methods" (reference number 13.13.1.2.01) supported by University of Rijeka (Croatia).

References

1. Asenbrener Katic, M., Pavlic, M., Candrlic, S.: The representation of database content and structure using the NOK method. Procedia Eng. **100**, 1075–1081 (2015)
2. Hirschman, L., Gaizauskas, R.: Natural language question answering: the view from here. Nat. Lang. Eng. **7**, 275–300 (2001)
3. Jach, T., Xięski, T.: Inference in expert systems using natural language processing. In: Kozielski, S., Mrozek, D., Kasprowski, P., Małysiak-Mrozek, B., Kostrzewa, D. (eds.) BDAS 2015. CCIS, vol. 521, pp. 288–298. Springer, Cham (2015). doi:10.1007/978-3-319-18422-7_26
4. Jakupovic, A., Pavlic, M., Han, Z.D.: Formalisation method for the text expressed knowledge. Expert Syst. Appl. **41**, 5308–5322 (2014)
5. Jakupovic, A., Pavlic, M., Mestrovic, A., Jovanovic, V.: Comparison of the nodes of knowledge method with other graphical methods for knowledge representation. In: Proceedings of the 36th International Convention CIS MIPRO, pp. 1276–1280 (2013)
6. Kovács, L.: SQL generation for natural language interface. J. Comput. Sci. Control Syst. **2**(18), 19–22 (2009)
7. Kumova, Bİ.: Generating ontologies from relational data with fuzzy-syllogistic reasoning. In: Kozielski, S., Mrozek, D., Kasprowski, P., Małysiak-Mrozek, B., Kostrzewa, D. (eds.) BDAS 2015. CCIS, vol. 521, pp. 21–32. Springer, Cham (2015). doi:10.1007/978-3-319-18422-7_2
8. Mollá, D., Vicedo, J.L.: Question answering in restricted domains: an overview. Comput. Linguist. **33**, 41–61 (2007)
9. Oracle: SQL- the natural language for analysis, Oracle White Paper (2015)
10. Pavlic, M.: Informacijski sustavi (Information systems). Skolska knjiga, Zagreb (2011)
11. Pavlic, M.: Oblikovanje baza podataka (Database design). Odjel za informatiku, Sveuciliste u Rijeci (2011)
12. Pavlic, M., Dovedan Han, Z., Jakupovic, A.: Question answering with a conceptual framework for knowledge-based system development 'node of knowledge'. Expert Syst. Appl. **42**, 5264–5286 (2015)
13. Pavlic, M., Jakupovic, A., Mestrovic, A.: Nodes of knowledge method for knowledge representation. Informatologia **46**, 206–214 (2013)
14. Pavlic, M., Mestrovic, A., Jakupovic, A.: Graph-based formalisms for knowledge representation. In: Proceedings of the 17th World Multi-Conference on Systemic Cybernetics and Informatics WMSCI, pp. 200–204 (2013)

Influence of Similarity Measures for Rules and Clusters on the Efficiency of Knowledge Mining in Rule-Based Knowledge Bases

Agnieszka Nowak-Brzezińska(✉) and Tomasz Rybotycki

University of Silesia, ul. Bankowa 12, 40-007 Katowice, Poland
agnieszka.nowak@us.edu.pl
http://zsi.ii.us.edu.pl/~nowak/

Abstract. In this work the subject of the application of clustering as a knowledge extraction method from real-world data is discussed. The authors analyze the influence of different clustering parameters on the efficiency of the knowledge mining process for rules/rules clusters. In the course of the experiments, nine different objects similarity measures and four clusters similarity measures have been examined in order to verify their impact on the size of the created clusters and the size of their representatives. The experiments have revealed that there is a strong relationship between the parameters used in the clustering process and future efficiency levels of the knowledge mined from such structures: some parameters guarantee to produce shorter/longer representatives of the created rules clusters as well as smaller/greater clusters' sizes.

Keywords: Rule-based knowledge bases · Cluster analysis · Similarity measures · Clusters visualization · Validity index

1 Introduction

An enormous interest in integrating database and knowledge-based system technologies to create an infrastructure for modern advanced applications results in creating knowledge bases (KBs) - database systems extended with some kind of knowledge, usually expressed in the form of $rules$[1] - logical implications of the "if condition$_1$ & ... & condition$_n$ $then$ conclusion" type [12].

An example of the rule from one of the KB (nursery) is as follows:

```
(health = priority) & (finance = inconv) & (parents = great_pret)
& (children = more) => (class = spec_prior[180]) 180
```

---

[1] Rules have been extensively used in knowledge representation and reasoning. This is very space efficient as only a relatively small number of facts needs to be stored in the KB and the rest can be derived from the inference rules. Such a natural way of knowledge representation makes the rules easily understood by people not involved in the expert system building.

© Springer International Publishing AG 2017
S. Kozielski et al. (Eds.): BDAS 2017, CCIS 716, pp. 67–78, 2017.
DOI: 10.1007/978-3-319-58274-0_6

It means that if these 4 premises: (`health=priority`), (`finance=inconv`), (`parents=great_pret`), (`children=more`) are met at the same time in the context of a given case, then the conclusion (`class=spec_prior`) is considered true (180 cases in the database confirm this rule).

*KB*s are constantly increasing in volume, thus the knowledge stored as a set of rules is progressively getting more complex and when the rules are not organized into any structure, the system is inefficient. There is a growing research interest in searching for methods that manage large sets of rules using the clustering approach as well as joining and reducing the rules [8]. Numerous papers on clustering techniques and their applications in engineering, medical, and biological areas have appeared in the pattern recognition literature during the past decade [4]. The type of the data to cluster is very important. There are a lot of different approaches to clustering (usually used for the same data type), however clustering the rules in a *KBS* is not so popular. In [9] the authors explore a clustering algorithm based on the Hopfield neural net algorithm. The paper presents a tool that can aid the maintainer in maintaining a *KBS*. A different approach is presented in [15], where a type of neuro-symbolic rules (so called neurules), integrating neurocomputing and production rules, is proposed. They exhibit characteristics such as modularity, naturalness and ability to perform interactive and integrated inferences and provide explanations for the reached conclusions. The conversion process merges symbolic rules having the same conclusion into one or more neurules. In [6] the authors applied a clustering method to cluster association rules on a numeric attribute and proposed an algorithm to generate representative rules from the clusters. A large number of rules makes them difficult for us to interpret their meaning. The paper [3] focuses on clustering induction rules for large knowledge sets based on the k-means algorithm. In [2] the author provides a new methodology for organizing and grouping the association rules with the same consequent. It consists of finding metarules, rules that express the associations between the discovered rules themselves. In [7] an approach to rule clustering for the super-rule generation is presented. Authors use the k-means clustering to cluster a huge number of rules to generate many new super-rules, which then can be analyzed manually by a researcher. An other approach for clustering association rules is presented in [10]. Clustered association rules are helpful in reducing the large number of association rules that are typically computed by existing algorithms, thereby rendering the clustered rules much easier to interpret and visualize. A practical use of these clusters is to perform segmentation on large customer-oriented databases. Yet another approach to deal with large number of rules is presented in [16]. It describes the application of two clustering algorithms (k-medoids and Partitioning About Medoids (PAM)) for these rules, in order to identify sets of similar rules and to understand the data better.

In our approach rules are divided into a number of groups based on similar premises in order to improve the inference process efficiency - the process is called *rules partition*[2] [13]. A user wants to get an answer from the system as soon as possible, thus instead of searching within the whole set of rules (as it happens in case of traditional inference processes), only the representatives of the groups are compared with the set of facts and/or hypotheses to be proven. The most relevant group of rules is selected and an extensive searching is performed only within a given group. Given a set of rules, new knowledge may be derived using a standard forward chaining inference process, which can be described as follows: each cycle of deductions starts with matching the condition part of each rule with known facts. If at least one rule matches the facts to have been asserted into the rule base, it is activated.

The main goal of the article is to measure the influence of using different clustering parameters (inter/intra cluster similarity measures, methods of creating the cluster's representative etc.) on the efficiency of the structure of KB based on rules clusters using the CluVis system (described in detail in Sect. 3). To verify it, different similarity measures and methods of hierarchical clustering have been implemented and used in the experiments.

The rest of the paper is organized as follows: Firstly, we mention all related efforts in the field of rules clustering in Sect. 2, where the description of different inter (like Jaccard, Goodall, Gower etc. - Sect. 2.1) and intra cluster similarity measures (single linkage, complete linkage etc. - Sect. 2.2) is presented. In Sect. 3, we identify two algorithms for visualization of hierarchical structures. We describe our experimental setup, evaluation methodology and the results on public data sets in Sect. 4.

2   Rules Clustering

Clustering is one of the oldest and most common methods of organizing data sets. The goal of clustering is to maximize *intra-cluster* similarity and minimize *inter-cluster* similarity, thus we are looking for a partition that groups similar elements and separates different elements in the best possible way. During this unsupervised process, similar objects (in accordance to a given similarity measure) are joined into groups so it often reduces the amount of data required to be analyzed in order to extract knowledge from the examined data set. Several methods of clustering have been proposed, each of them includes a variety of different techniques. The authors have selected one of the most well-known algorithms from a popular group of techniques called hierarchical clustering (which can be considered as a very natural approach) and used it to organize several KBs. The pseudocode of this algorithm - namely Classic AHC (agglomerative hierarchical clustering) algorithm [4] - is presented as Pseudocode 1.

Pseudocode 1. Classic AHC Algorithm.
Input: stop condition *sc*, ungrouped set of objects *s*
Output: grouped tree-like structure of objects

---

[2] The idea is new but it is based on the authors' previous research, where the idea of *clustering rules* was introduced.

1. Place each object o from s into a separate cluster.
2. Build similarity matrix M that consists of every clusters pair similarity value.
3. Using M find the most similar pair of clusters and merge them into one.
4. Update M.
5. **IF** sc was met end the procedure.
6. **ELSE REPEAT** from step 3.
7. **RETURN** the resultant structure.

The main advantage of hierarchical clustering is that it does not impose any special methods of describing the clusters similarity. This feature makes it ideal for organizing sets of complex objects such as DNA, text files or rules. Moreover, stop condition also is not arbitrary hence there are many possible ways to define it. In this work clustering is stopped when given number of clusters is generated.

2.1   Intra-cluster Similarity Measures

Measuring a similarity or a distance between two data points is a core requirement for several data mining and knowledge discovery tasks that involve distance computation. The notion of similarity or distance for categorical data is not as straightforward as it is for continuous data. When both types of data are present, the problem is much more complicated. It is necessary to find a measure which deals with this issue. In some cases, similarity measures take into account the frequency distribution of different attribute values in a given data set to define a similarity between two categorical attribute values. In this paper, we study a variety of similarity measures [13], presented in Table 1.

Table 1. Similarity measures for rules/clusters

Measure's name	$s(r_{ji}, r_{ki}) = s_{jki}$
IOF	if $r_{ji} = r_{ki}$ then $s_{jki} = 1$ else $s_{jki} = \frac{1}{1+\log(f(r_{ji}))\cdot\log(f(r_{ki}))}$
OF	if $r_{ji} = r_{ki}$ then $s_{jki} = 1$ else $s_{jki} = \frac{1}{1+\log\frac{N}{f(r_{ji})}\cdot\log\frac{N}{f(r_{ki})}}$
G1	if $r_{ji} = r_{ki}$ then $s_{jki} = 1 - \sum_{q\in Q} p^2(q)$ else 0 $\{Q \subseteq A : \forall_{q\in Q} p_i(x) \leq p_i(r_{ji})\}$
G2	if $r_{ji} = r_{ki}$ then $s_{jki} = 1 - \sum_{q\in Q} p^2(q)$ else 0 $\{Q \subseteq A : \forall_{q\in Q} p_i(x) \geq p_i(r_{ji})\}$
G3	if $r_{ji} = r_{ki}$ then $s_{jki} = 1 - p^2(r_{j_i})$ else 0
G4	if $r_{ji} = r_{ki}$ then $s_{jki} = p^2(r_{j_i})$ else 0
Gower	if r_i is numeric $s_{jki} = 1 - \frac{|r_{ji}-r_{ki}|}{range(r_i)}$ if r_i is categorical and $r_{ji} = r_{ki}$ then $s_{jki} = 1$ else 0
Jaccard	if $r_{ji} = r_{ki}$ then $s_{jki} = \frac{1}{n}$ else 0, where n is number of attributes considered
SMC	if $r_{ji} = r_{ki}$ then $s_{jki} = 1$ else 0

Notations. Having a set of attributes A and their values V, rules premises and conclusions are built using pairs (a_i, v_i), where $a_i \in A, v_i \in V_a$. A pair (a_i, v_i) is called a descriptor. In a vector of such pairs, i-th position denotes the value of the i-th attribute of a rule. It is important to note that most of the rules do not consist of all attributes in A, thus constructed vectors (describing the rules) are of different lengths. The frequency (f) of a given descriptor d is equal to 0 if d is not included in any of the rules from a given knowledge base KB (if $x \notin KB$ then $f(x) = 0$). The distribution of d will be defined as follows: $p(x) = \frac{f(d)}{N}$ for N denoting the number of rules in a given set. It can be also defined as $p^2(d) = \frac{f(d)(f(d)-1)}{N(N-1)}$. A data set might contain attributes that take several values and attributes that take very few of them. A similarity measure might give more importance to attributes with smaller sets of values, while partially ignoring the others. It may also work the other way around. Almost all similarity measures assign a similarity value between two rules r_1 and r_2 belonging to the set of rules R as follows:

$$S(r_j, r_k) = \sum_{i=1}^{N} w_i s(r_{ji}, r_{ki})$$

where $s_i(r_{ji}, r_{ki})$ is the per-attribute (for i-th attribute) similarity between two values of descriptors of the rules r_j and r_k. The quantity w_i denotes the weight assigned to the attribute a_i and usually $w_i = \frac{1}{d}$, for $i = 1, \ldots, d$. The simplest measure is SMC (Simple Matching Coefficient)[3] - which calculates the number of attributes that match in the two rules. The range of the per-attribute SMC is $\{0; 1\}$, with a value of 0 occurring when there is no match, and a value of 1 occurring when the attribute values match. It treats all types of attributes in the same way. This may cause problems in some cases (e.g. Centroid Linkage as used in this work), as it tends to favour longer rules. *Jaccard* is similar to SMC - it is however, more advanced. Rather than only counting the number of attributes that have the same value for both compared rules, it also divides the result by the number of attributes of both objects so longer rules are not favoured any more.

Deriving from *retrieval information systems*, two measures could be also used to check the similarity between rules (or clusters of rules). The inverse occurrence frequency (IOF) measure assigns a lower similarity to mismatches on more frequent values while the occurrence frequency (OF) measure gives opposite weighting for mismatches when compared to the IOF measure, i.e., mismatches on less frequent values are assigned a lower similarity and mismatches on more frequent values are assigned a higher similarity. In the experiments four measures $(G1 - G4)$, which are variants of a measure that was proposed by Goodall [5], have also been used. $G1$ attempts to normalize the similarity between two

---

[3] If both compared objects have the same attribute and this attribute has the same value for both objects then add 1 to a given similarity measure. If otherwise, do nothing. To eliminate one of the problems of SMC, which favours the longest rules, the authors also used Jaccards Index.

rules (or clusters) by the probability that the similarity value could be observed in a random sample of two points. This measure assigns a higher similarity to a match if the value is infrequent than if the value is frequent. $G1$ and $G3$ assign a higher similarity to a match when the attribute value is rare (f is low), while $G2$ and $G4$ assign a higher similarity to a match when the attribute value is frequent (f is high). *Gower*'s similarity coefficient is the most complex of the all used inter-cluster similarity measures as it handles numeric attributes and symbolic attributes differently. For ordinal and continuous variables it defines the value of s_{jki} as $s_{jki} = 1 - \frac{|r_{ji} - r_{ki}|}{range(r_i)}$, where: $range(r_i)$ is the range of values for the i-th variable. For continuous variables s_{jki} ranges between 1, for identical values $r_{ji} = r_{ki}$ and 0 for the two extreme values $r_{max} - r_{min}$. It is vital to mention that some of the aforementioned inter-object similarity measures are intended for categorical attributes only. In these cases, a similarity between the numerical attributes of two rules is calculated in the same way as in Gower's measure.

2.2 Inter-cluster Similarity Measures

Different intercluster similarity measures will produce different results (hierarchies) in most cases. Therefore, it is important to use more than just one algorithm parameters setting in the process of extracting knowledge from data. In this paper the most popular inter-cluster similarity measures: Single Link (SL), Complete Link (CoL), Average Link (AL) and Centroid Link (CL) [14] have been used. They are well-known in the literature and will not be discussed in detail here. In the authors' previous research it was noticed that the method of creating the representatives is of comparable importance to the clustering algorithm and the similarity measures used in clustering. In this work, a centroid is considered to be a cluster's representative (described in Sect. 2.3) as it aims to be the most average/centered rule. This is because defining this the geometrical center of rule clusters is a non-trivial task.

2.3 Clusters Representation

It is very important for data to be presented in the most friendly way. Sole visualization of clustering (described further in Sect. 3) is not enough, as it reduces the whole pattern discovery process to examining an accumulation of shapes. There are many methods of creating representatives. Their creation algorithm has been presented as Pseudocode 2).

Pseudocode 2. Representative creation algorithm.
Input: cluster c, threshold t [%]
Output: cluster representative r

1. In the cluster attributes set A, find only those attributes that can be found in t percent of objects, and put them in set A'
2. **FOR EACH** attribute a in A'

3. **IF** a is symbolic, count its modal in c
4. **IF** a is numeric, count its average value in c
5. **RETURN** attribute-value pairs from A' **AS** r

In this work, each cluster is described by its representative. A representative can be simply described as an average rule. As can be seen, representatives created this way consist only of those attributes which are considered most important for the whole cluster (they are common enough). In this way, the examined group is well described by the minimal number of variables[4].

3   CluVis

An example of cluster's (group of rules) representative, for one of the analyzed KB (nursery), may looks like:

```
(housing=less_conv)&(social=nonprob)&(health=priority)&(form=foster)&
(parents=pretentious)&(has_nurs=improper)&(finance=convenient)=>
(class=spec_prior)
```

It contains 7 descriptors and covers 5 rules in a given KB (for 10 created clusters and the following parameters: Gower's similarity measure and the Average link method).

The CluVis [17] is an application designed to group sets of rules generated by the RSES [1] and visualize them using selected treemap methods [18,19]. It is the first application capable of working on raw KBs as generated from the *RSES*. It has been successfully used in previous research to group and visualize medical knowledge bases generated from artificial data sets available on [11] as well as one generated from real medical data [14]. Moreover, it aggregates functionalities of both clustering and visualization software, making it a universal tool for exploring KBs. Part of its UI, used to set clustering parameters, has been shown on Fig. 1. Along with its main functionalities, described in more detail in Sect. 2, the CluVis is capable of generating reports of grouping (to txt or special xml files which can be opened in such applications as e.g. Libre Calc) which contain detailed information about each obtained cluster and about clustering in general (number of nodes, min/max/avg representative length). It is also possible to save a generated visualization to png file or find the best clustering (according to the implemented cluster validation indexes -MDI and $MDBI$). The CluVis is an open source application written in C++11 using QT graphic libraries. It is available in English and Polish and its source code can be downloaded from https://github.com/Tomev/CluVis. It is constantly upgraded by the authors.

It is possible that automatically created KBs (e.g. creating rules from dataset using LEM2 algorithm) may contain some undesired rules (those which would

---

[4] However the authors see the necesssity to analyze the meaning of methods for creating clusters' representatives and their influence on the overall efficiency.

Fig. 1. Sample treemap visualization.

never be activated or are simply redundant). It is essential to maintain the simplicity of KBs, thus the $CluVis$ is used to transform rules into an organized one presented in the form of a responsive visualization. The whole process is performed as follows. After loading KB into the application and selecting the clustering parameters, grouping begins. Each rule from KB is transformed into a vector of attribute-value pairs[5]. Then, using similarity a matrix, two the most similar clusters are joined. During the merging of two smaller clusters, their representative is calculated. Then the matrix is updated and the process is repeated until the stop condition is met. The CluVis ends clustering after a given (as clustering parameter) number of clusters is reached.

---

[5] At this stage some data about sole attributes are also gathered (max. and min. of numerical attributes, the number of times when a categorical attribute had a given value), as they are used in some similarity measures.

4   Experiments

In this section, an experimental evaluation of 9 similarity measures and 4 cluster-ing methods on 7 different KBs is presented. All of them are originally available at the UCI Machine Learning Repository [11]. From the original data, using the LEM2 algorithm (in $RSES$ software) decision rules have been generated. The details about KBs are summarized in Table 2. Let us assume the follow-ing labels for examined paramteres: (a) $AttrN$ - is the attributes number, (b) $ObjN$ - objects number, (c) N - Nodes number (the number of nodes is the sum of the number of rules and clusters), (d) U - Ungrouped objects, (e) $BiggRepS$ - Biggest representative size, (f) $AvgRepSize$ - Average representative size, (g) $wAvgRepS$ - Weighted Average representative size (divided by $AttrN$), (h) $BiggCRepL$ - Biggest cluster representative length, CN - Clusters number, (i) $BiggCluS$ - Biggest cluster size, where representative's size denotes number of descriptors used to describe it.

At this point it is necessary to emphasize that the results in the tables are the averages for all the test data sets, thus in some cases the standard deviation is greater than the average value.

Table 2. Characteristics of examined KBs

	Balance	Diab	Diabetes	Breast	Autos	Audiology	Arythmia
(a)	5	9	9	10	26	70	280
(b)	278	483	490	125	60	42	154
(c)	$555 \pm 9{,}0$	$937{,}3 \pm 19{,}1$	$950{,}5 \pm 19{,}6$	239	$112{,}1 \pm 2{,}3$	$77{,}1 \pm 3{,}0$	295,5
(d)	$7{,}5 \pm 8{,}7$	$11{,}8 \pm 13{,}8$	$13{,}6 \pm 14{,}9$	$5{,}1 \pm 3{,}4$	$3{,}6 \pm 2{,}9$	$3{,}6 \pm 2{,}9$	$5{,}8 \pm 5{,}7$
(e)	4 ± 0	$5{,}4 \pm 0{,}5$	$5{,}6 \pm 0{,}7$	9 ± 0	$11{,}9 \pm 1{,}9$	$67{,}0 \pm 0{,}4$	$151{,}9 \pm 4{,}6$
(f)	$3{,}5 \pm 0{,}4$	$3{,}4 \pm 0{,}8$	$3{,}3 \pm 0{,}8$	$7{,}1 \pm 1{,}1$	$8{,}6 \pm 1{,}67$	$49{,}7 + 1{,}2$	$133{,}4 \pm 11{,}6$
(g)	$1{,}4 \pm 0{,}2$	$2{,}9 \pm 0{,}7$	$2{,}9 \pm 0{,}7$	$1{,}4 \pm 0{,}2$	$3{,}2 \pm 0{,}6$	$1{,}5 \pm 0{,}4$	$2{,}1 \pm 0{,}2$
(h)	4 ± 0	$4{,}8 \pm 0{,}6$	$4{,}9 \pm 0{,}3$	9 ± 0	$10{,}7 \pm 0{,}5$	$66{,}8 \pm 0{,}5$	$147{,}4 \pm 1{,}5$
(i)	$180{,}2 \pm 93{,}5$	$314{,}8 \pm 137{,}2$	$335{,}2 \pm 140{,}7$	$7{,}6 \pm 32{,}8$	$38{,}3 \pm 14{,}0$	$29{,}8 \pm 7{,}8$	$111{,}5 \pm 41{,}7$

The performance of the different similarity measures has been evaluated in the context of knowledge mining using information like: the number of rules clusters, the number of ungrouped rules, the sizes of the biggest cluster (as well as representative of it) and the representative most specific. More specific means more detailed, containing a higher number of descriptors. The optimal structure of KBs with rules clusters should contain well-separated groups of rules, and the number of such groups should not be too high. Moreover, the number of ungrouped rules should be minimal. Creating an optimal description of each cluster (representative) is very important because they are further used to select a proper group (and reject all the others) in the inference process, in order to mine knowledge hidden in rules (by accepting the conclusion of the given rule as a true fact). The results of the experiments verify the initial hypotheses about

Table 3. Influence of inter-cluster similarity measures (AVG±SD(Me))

	CN	BiggCluS	BiggRepL	U	BiggRepS	AvgRepS
G1	$16,6 \pm 14,2$	$154,4 \pm 148,2$	$35,6 \pm 51,0$	$7,7 \pm 10,1$	$36,3 \pm 52,1$	$29,5 \pm 45,3$
	(10,0)	(86,0)	(9,0)	(5,0)	(9,0)	(6,4)
G2	$16,6 \pm 14,2$	$155,4 \pm 147,6$	$35,4 \pm 50,8$	$7,6 \pm 10,1$	$36,2 \pm 51,5$	$29,0 \pm 45,1$
	(10,0)	(98,5)	(9,0)	(5,0)	(9,0)	(6,7)
G3	$16,6 \pm 14,2$	$156,5 \pm 147,5$	$35,3 \pm 50,6$	$7,9 \pm 10,1$	$36,6 \pm 51,7$	$30,1 \pm 45,2$
	(10,0)	(114,5)	(9,0)	(5,5)	(9,0)	(7,4)
G4	$16,6 \pm 14,2$	$159,3 \pm 145,2$	$35,4 \pm 50,7$	$7,8 \pm 10,0$	$36,6 \pm 52,4$	$29,5 \pm 45,1$
	(10,0)	(111,5)	(9,0)	(5,0)	(9,0)	(6,6)
Go	$16,6 \pm 14,1$	$157,2 \pm 147,4$	$35,7 \pm 51,2$	$8,1 \pm 9,9$	$36,4 \pm 52,3$	$30,7 \pm 47,1$
	(10,0)	(86,0)	(9,0)	(5,5)	(9,0)	(6,7)
IOF	$16,6 \pm 14,2$	$157,9 \pm 145,8$	$35,3 \pm 50,6$	$8,0 \pm 10,1$	$36,3 \pm 51,7$	$29,8 \pm 45,7$
	(10,0)	(114,0)	(9,0)	(5,0)	(9,0)	(5,8)
OF	$16,5 \pm 14,3$	$157,7 \pm 146,6$	$35,2 \pm 50,4$	$7,7 \pm 10,1$	$36,3 \pm 51,5$	$28,7 \pm 44,7$
	(10,0)	(114,0)	(9,0)	(4,5)	(9,0)	(6,9)
SMC	$16,4 \pm 14,3$	$155,6 \pm 143,6$	$35,4 \pm 50,8$	$5,6 \pm 6,7$	$36,3 \pm 52,4$	$30,2 \pm 45,6$
	(10,0)	(88,5)	(9,0)	(4,0)	(9,0)	(6,7)
J	$16,5 \pm 14,2$	$145,2 \pm 141,1$	$35,4 \pm 50,7$	$5,2 \pm 7,7$	$36,6 \pm 52,8$	$30,5 \pm 46,7$
	(10,0)	(91,0)	(9,0)	(3,5)	(9,0)	(6,9)

Table 4. Influence of intra-cluster similarity measures

	SL	CL	AL	CoL
CN	$16,5 \pm 14,2$	$16,5 \pm 14,1$	$16,6 \pm 14,1$	$16,6 \pm 14,1$
U	$12,1 \pm 11,8$	$2,1 \pm 3,1$	$4,5 \pm 5,2$	$10,4 \pm 10,9$
BigCluS	$213,5 \pm 166,2$	$84,0 \pm 89,8$	$157,2 \pm 139,8$	$167,8 \pm 142,6$
BigRepL	$35,4 \pm 50,4$	$35,1 \pm 50,6$	$35,4 \pm 50,4$	$35,4 \pm 50,5$
BigRepS	$36,0 \pm 51,8$	$36,4 \pm 51,5$	$36,5 \pm 51,5$	$36,5 \pm 52,2$
AvgRepS	$27,2 \pm 44,2$	$31,4 \pm 44,5$	$30,9 \pm 46,3$	$29,4 \pm 46,4$
WAvgRepS	$2,6 \pm 1,0$	$1,8 \pm 0,5$	$2,0 \pm 0,7$	$2,4 \pm 0,9$

similarity measures and clustering methods. As can be seen in Tables 3 and 4 no single measure is always superior or inferior. This is obvious since each KB to have been analyzed has different characteristics (a different number of attributes and/or rules) as well as a different type of attributes. The use of some measures however guarantees to produce more general or more specific representatives for the created rules clusters. There are some pairs of measures that exhibit complementary performance, i.e., one performs well where the other performs poorly and vice-versa.

AL clustering is a compromise between the sensitivity of CL clustering to outliers and the tendency of SL clustering to form long chains that do not correspond to the intuitive notion of clusters as compact, spherical objects.

5 Summary

The article presents how the exploration of complex KBs can be applied based on clustering and visualization of rules clusters. The article presents the application of clustering as a knowledge extraction method from real-world data. Clustering a large set of objects (rules in this case) is not enough when exploring such an enormous amount of data in order to find some hidden knowledge in it. The extraction of valuable knowledge from large data sets can be difficult or even impossible. Modularization of KBs (by clustering) helps to manage the domain knowledge stored in systems using the described method of knowledge representation because it divides rules into groups of similar forms, context, etc. The authors analyze an influence of different clustering parameters on the efficiency of the knowledge mining process for rules/rules clusters. In the course of the experiments, nine different similarity measures and four clustering measures have been examined in order to verify their impact on the size of the created clusters and the size of the representatives. The experiments have revealed that there is a strong relationship between the parameters used in the clustering process and future efficiency levels of the knowledge mined from such structures: some parameters guarantee to produce shorter/longer representatives of the created rules clusters as well as smaller/greater clusters' sizes.

The authors propose to use clusters of rules and visualize them using treemap algorithms and envisage that this two-phase way of rules representation allows the domain experts to explore the knowledge hidden in these rules faster and more efficiently than before. In the future, the authors plan to find alternative ways of describing a rule cluster's centroid and to extend the software's functionality, especially in the context of parameters used in clustering and visualizing procedures, as well as importing other types of data sources. It would be easier then to support human experts in their everyday work by using the created software (CluVis) in cooperation with various expert systems.

References

1. Bazan, J.G., Szczuka, M.S., Wróblewski, J.: A new version of rough set exploration system. In: Alpigini, J.J., Peters, J.F., Skowron, A., Zhong, N. (eds.) RSCTC 2002. LNCS (LNAI), vol. 2475, pp. 397–404. Springer, Heidelberg (2002). doi:10.1007/3-540-45813-1_52
2. Berrado, A., Runger, G.: Using metarules to organize and group discovered association rules. Data Min. Knowl. Discov. **14**(3), 409–431 (2007)
3. Chemchem, A., Drias, H., Djenouri, Y.: Multilevel clustering of induction rules for web meta-knowledge. Adv. Intell. Syst. Comput. **206**, 43–54 (2013)
4. Dubes, R., Jain, A.: Clustering techniques: the user's dilemma. Pattern Recogn. **8**(4), 247–260 (1976)

5. Goodall, D.: A new similarity index based on probability. Biometrics **22**, 882–907 (1966)

6. Hashizume, A., Yongguang, B., Du, X., Ishii, N.: Generating representative from clusters of association rules on numeric attributes. In: Liu, J., Cheung, Y., Yin, H. (eds.) IDEAL 2003. LNCS, vol. 2690, pp. 605–613. Springer, Heidelberg (2003). doi:10.1007/978-3-540-45080-1_82

7. He, J., Chen, B., Hu, H.J., Harrison, R., Tai, P., Dong, Y., Pan, Y.: Rule clustering and super-rule generation for transmembrane segments prediction. In: IEEE Computational Systems Bioinformatics Conference, Workshops and Poster Abstracts, pp. 224–227 (2005)

8. Latkowski, R., Mikołajczyk, M.: Data decomposition and decision rule joining for classification of data with missing values. In: Tsumoto, S., Słowiński, R., Komorowski, J., Grzymała-Busse, J.W. (eds.) RSCTC 2004. LNCS (LNAI), vol. 3066, pp. 254–263. Springer, Heidelberg (2004). doi:10.1007/978-3-540-25929-9_30

9. Lee, O., Gray, P.: Knowledge base clustering for KBS maintenance. J. Softw. Maint. Evol. **10**(6), 395–414 (1998)

10. Lenty, B., Swamix, A., Widomy, J.: Clustering association rules. Stanford University

11. Lichman, M.: UCI machine learning repository (2013). http://archive.ics.uci.edu/ml

12. Nowak-Brzezińska, A.: Mining rule-based knowledge bases. In: Kozielski, S., Mrozek, D., Kasprowski, P., Małysiak-Mrozek, B., Kostrzewa, D. (eds.) BDAS 2015-2016. CCIS, vol. 613, pp. 94–108. Springer, Cham (2016). doi:10.1007/978-3-319-34099-9_6

13. Nowak-Brzezińska, A.: Mining rule-based knowledge bases inspired by rough set theory. Fundam. Inform. **148**, 35–50 (2016)

14. Nowak-Brzezińska, A., Rybotycki, T.: Visualization of medical rule-based knowledge bases. J. Med. Inf. Technol. **24**, 91–98 (2015)

15. Prentzas, J., Hatzilygeroudis, I.: Improving efficiency of merging symbolic rules into integrated rules: splitting methods and mergability criteria. Expert Syst. **32**(2), 244–260 (2015)

16. Reynolds, A.P., Richards, G., Rayward-Smith, V.J.: The application of K-medoids and PAM to the clustering of rules. In: Yang, Z.R., Yin, H., Everson, R.M. (eds.) IDEAL 2004. LNCS, vol. 3177, pp. 173–178. Springer, Heidelberg (2004). doi:10.1007/978-3-540-28651-6_25

17. Rybotycki, T.: Wizualizacja struktur hierarchicznych dla regulowych baz wiedzy, Sosnowiec (2015)

18. Shneiderman, B.: Tree visualization with tree-maps: 2-d space-filling approach. Trans. Graphics (TOG) **11**, 92–99 (1992). Association for Computing Machinery, New York

19. Wetzel, K.: Pebbles - using circular treemaps to visualize disk usage (2004)

Attribute Reduction in a Dispersed Decision-Making System with Negotiations

Małgorzata Przybyła-Kasperek[✉]

Institute of Computer Science, University of Silesia,
Będzińska 39, 41-200 Sosnowiec, Poland
malgorzata.przybyla-kasperek@us.edu.pl
http://www.us.edu.pl

Abstract. The aim of the study was to apply rough set attribute reduction in a dispersed decision-making system. The system that was used was proposed by the author in a previous work. In this system, a global decision is taken based on the classifications that are by the base classifiers. In the process of decision-making, elements of conflict analysis and negotiations have been applied. Reduction of the set of conditional attributes in local decision tables was used in the paper. The aim of the study was to analyze and compare the results that were obtained after the reduction with the results that were obtained for the full set of attributes.

Keywords: Decision-making system · Dispersed knowledge · Conflict analysis · Attribute reduction

1 Introduction

The ability to generate decisions based on dispersed knowledge is an important issue. By dispersed knowledge, we understand the knowledge that is stored in the form of several decision tables. Many times, knowledge is stored in a dispersed form for various reasons. For example, when knowledge in the same field is accumulated by separate units (hospitals, medical centers, banks) or when knowledge is not available at the same time but rather at certain intervals, or when the data is too large to store them and to process them in the form of a single decision table. When knowledge is accumulated by separate units, it is difficult to require that the local decision tables have the same sets of attributes or the same sets of objects. It is also difficult to require that these sets are disjoint. Rather, a more general approach is needed. The problem that is considered in this paper concerns the use of dispersed knowledge in the process of decision making. The knowledge that is used is set in advance and does not meet the assumptions about the separability or equality of the sets of attributes or the sets of objects.

Many models and methods have been proposed for the issue of the multiple model approach [4,5,9]. In a multiple classifier system, an ensemble is constructed that is based on base classifiers. The system designer has control over

© Springer International Publishing AG 2017
S. Kozielski et al. (Eds.): BDAS 2017, CCIS 716, pp. 79–88, 2017.
DOI: 10.1007/978-3-319-58274-0_7

the form of the dispersed knowledge. In this approach, certain constraints regarding the sets of attributes or sets of objects are adopted. Moreover, areas such as distributed decision-making [16,19] and group decision making [1,2] deal with this issue. In these approaches, in order to disperse knowledge, the horizontal or vertical division of one decision table is used. This study presents a completely different approach to the solution of the problem, due to the general assumptions that are adopted and the methods that are used. In the system discussed in the paper, an extension of Pawlak's conflict model [7,8] was applied. The system was proposed in the paper [11]. In the system, we describe the views of classifiers by using probability vectors over the decision classes. Then, the process of combining classifiers in coalitions is implemented. Negotiation is used in the clustering process. For every cluster, we find a kind of combined information. Finally, we classify the given test object by voting among the clusters using the combined information from each of the cluster.

The aim of the study was to use the reduction of the set of attributes in such a system. Attribute reduction is very important in the processing of huge amounts of data. This allows the computation time to be reduced, and sometimes can improve the classification accuracy. In the paper, a reduction that is based on the rough sets [6] was applied. Comparison of the results that were obtained with the results that were obtained without the reduction was made and conclusions were drawn.

2   A Dispersed Decision-Making System with Negotiations-Basic Concepts

The issues related to decision-making that are based on dispersed knowledge have been considered by the author for several years. Since the beginning, the author has tried to solve these problems by analyzing conflicts and by the creation of coalitions. The first approach assumed that the system has a static structure, i.e. coalitions were created only once [10,21]. In the following approaches, a dynamic structure has been applied. Due to the modification of the definition of the relations given by Pawlak, different types of coalitions in the dynamic structure were obtained. Firstly, disjoint clusters were created [13,15], then inseparable clusters were used [12]. However, the most extensive process of conflict analysis was applied in the approach with negotiations [11,14]. In this paper, this system is used. The system is briefly described below.

We assume that the knowledge is available in a dispersed form, which means in the form of several decision tables. Based on each local knowledge base, a classifier classifies the objects. Such a classifier is called a resource agent. A resource agent ag in $Ag = \{ag_1, \ldots, ag_n\}$ has access to resources that are represented by a decision table $D_{ag} := (U_{ag}, A_{ag}, d_{ag})$, where U_{ag} is the universe; A_{ag} is a set of conditional attributes; V_{ag}^a is a set of attribute a values and d_{ag} is a decision attribute. We want to designate homogeneous groups of resource agents. When the agents agree on the classification of an object, then they should be combined into one group. A two-step process of creating groups is used. At first, initial coalitions are created. For this purpose, for each agent $ag_i \in Ag$, the classification is represented as a vector of the values $[\bar{\mu}_{i,1}(x), \ldots, \bar{\mu}_{i,c}(x)]$, where $c = card\{V^d\}$. This vector will be defined based on certain relevant objects. That is, m_1 objects from each decision class of the decision tables of the agents that carry the greatest similarity to the test object. The value $\bar{\mu}_{i,j}(x)$ is defined as follows:

$$\bar{\mu}_{i,j}(x) = \frac{\sum_{y \in U_{ag_i}^{rel} \cap X_{v_j}^{ag_i}} s(x,y)}{card\{U_{ag_i}^{rel} \cap X_{v_j}^{ag_i}\}}, i \in \{1, \ldots, n\}, j \in \{1, \ldots, c\},$$

where $U_{ag_i}^{rel}$ is the subset of relevant objects selected from the decision table D_{ag_i} of resource agent ag_i and $X_{v_j}^{ag_i}$ is the decision class of the decision table of resource agent ag_i and $s(x,y)$ is the measure of the similarity between objects x and y. Based on the vector of values defined above, a vector of the rank is specified. The vector of the rank is defined as follows: rank 1 is assigned to the values of the decision attribute that are taken with the maximum level of certainty. Rank 2 is assigned to the next most certain decisions, etc. Proceeding in this way for each resource agent $ag_i, i \in \{1, \ldots, n\}$, the vector of the rank $[\bar{r}_{i,1}(x), \ldots, \bar{r}_{i,c}(x)]$ will be defined. Relations between agents are defined on the basis of which groups of agents are created. For this purpose, we define the function $\phi_{v_j}^x$ for the test object x and each value of the decision attribute $v_j \in V^d$; $\phi_{v_j}^x : Ag \times Ag \rightarrow \{0,1\}$

$$\phi_{v_j}^x(ag_i, ag_k) = \begin{cases} 0 & \text{if } \bar{r}_{i,j}(x) = \bar{r}_{k,j}(x) \\ 1 & \text{if } \bar{r}_{i,j}(x) \neq \bar{r}_{k,j}(x) \end{cases}$$

where $ag_i, ag_k \in Ag$. We also define the intensity of the conflict between agents using a function of the distance between the agents. We define the distance between agents ρ^x for the test object x: $\rho^x : Ag \times Ag \rightarrow [0,1]$,

$$\rho^x(ag_i, ag_k) = \frac{\sum_{v_j \in V^d} \phi_{v_j}^x(ag_i, ag_k)}{card\{V^d\}}, \text{where } ag_i, ag_k \in Ag.$$

Definition 1. *Let p be a real number that belongs to the interval $[0, 0.5)$. We say that agents $ag_i, ag_k \in Ag$ are in a friendship relation due to the object x, which is written $R^+(ag_i, ag_k)$, if and only if $\rho^x(ag_i, ag_k) < 0.5 - p$. Agents $ag_i, ag_k \in Ag$ are in a conflict relation due to the object x, which is written $R^-(ag_i, ag_k)$, if and only if $\rho^x(ag_i, ag_k) > 0.5 + p$. Agents $ag_i, ag_k \in Ag$ are in a neutrality relation due to the object x, which is written $R^0(ag_i, ag_k)$, if and only if $0.5 - p \leq \rho^x(ag_i, ag_k) \leq 0.5 + p$.*

By using the relations defined above, we can create groups of resource agents that are not in a conflict relation. The initial cluster, due to the classification of object x, is the maximum, due to the inclusion relation, subset of resource agents $X \subseteq Ag$ such that $\forall_{ag_i, ag_k \in X} R^+(ag_i, ag_k)$.

Then the negotiation stage is implemented. At this stage, we reduce the requirements for determining the compliance between agents. We assume that during the negotiation, agents put the greatest emphasis on the compatibility of the ranks that were assigned to the decisions with the highest ranks. We define the function ϕ_G^x for the test object x; $\phi_G^x : Ag \times Ag \to [0, \infty)$

$$\phi_G^x(ag_i, ag_j) = \frac{\sum_{v_l \in Sign_{i,j}} |\bar{r}_{i,l}(x) - \bar{r}_{j,l}(x)|}{card\{Sign_{i,j}\}}$$

where $ag_i, ag_j \in Ag$ and $Sign_{i,j} \subseteq V^d$ is the set of significant decision values for the pair of agents ag_i, ag_j. In the set $Sign_{i,j}$, there are the values of the decision, which the agent ag_i or agent ag_j gave the highest rank. During the negotiation stage, the intensity of the conflict between the two groups of agents is determined by using the generalized distance. The generalized distance between the agents for the test object x is denoted by ρ_G^x; $\rho_G^x : 2^{Ag} \times 2^{Ag} \to [0, \infty)$,

$$\rho_G^x(X, Y) = \begin{cases} 0 & \text{if } card\{X \cup Y\} \leq 1 \\ \dfrac{\displaystyle\sum_{ag, ag' \in X \cup Y} \phi_G^x(ag, ag')}{card\{X \cup Y\} \cdot (card\{X \cup Y\} - 1)} & \text{else} \end{cases}$$

where $X, Y \subseteq Ag$. The value of the function can be seen as the average difference of the ranks that are assigned to significant decisions within the combined group of agents consisting of the sets X and Y. For each agent ag that has not been included in any initial clusters, the generalized distance value is determined for this agent and all of the initial clusters with which the agent ag is not in a conflict relation and for this agent and other agents without a coalition with which the agent ag is not in a conflict relation. Then, the agent ag is included in all of the initial clusters for which the generalized distance does not exceed a certain threshold, which is set by the system's user. Moreover, agents without a coalition for which the value of the generalized distance function does not exceed the threshold are combined into a new cluster. When this stage is finished, we get the final form of the clusters.

Based on the decision tables from one cluster, common knowledge is generated. Based on this knowledge, the classification of the object is made. This situation can be seen as the creation of a classifier that is based on the aggregated knowledge. Therefore, for each cluster, a synthesis agent, as_j, where j - the number of cluster, is defined. As_x is a finite set of synthesis agents that are defined for the clusters that are dynamically generated for test object x. By a dispersed decision-making system with dynamically generated clusters, we mean

$$WSD_{Ag}^{dyn} = \langle Ag, \{D_{ag} : ag \in Ag\}, \{As_x : x \text{ is a classified object}\},$$

$$\{\delta_x : x \text{ is a classified object}\}\rangle$$

where Ag is a finite set of resource agents; $\{D_{ag} : ag \in Ag\}$ is a set of the decision tables of resource agents; As_x is the set of synthesis agents and $\delta_x : As_x \to 2^{Ag}$ is an injective function that each synthesis agent assigns to a cluster.

A method for defining an aggregated knowledge is described in detail in the paper [11]. This method was called an approximated method of the aggregation of decision tables. The method consists in the creation of new objects based on the relevant objects from the tables of resource agents. In order to define the relevant objects, parameter m_2 is used. The new objects are created by combining the relevant objects that have the same values for the attributes. Based on the aggregated table of the synthesis agent, a vector of probabilities is again determined, which is the result of the classification at the measurement level. Then, some transformations are performed on these vectors (described in the paper [11]) and the set of decisions is generated. In this set, there are decisions which have the highest support among all of the agents. This set is generated using the DBSCAN algorithm; a description of the method can be found in the papers [11,12,15]. The DBSCAN algorithm was used to search for the decisions that were closest to the decision with the greatest support among the agents.

3 Reduction of a Set of Attributes

The reduction of a set of attributes that is used in this paper is based on the rough set theory, which was proposed by Pawlak [6]. This reduction method is very popular and widely used [3,18,20]. The two main definitions of this reduction method are given below.

Definition 2. *Let* $D = (U, A, d)$ *be a decision table,* $x, y \in U$ *be given objects and* $B \subseteq A$ *be a subset of conditional attributes. We say that objects* x *and* y *are discernible by* B *if and only if there exists* $u \in B$ *such that* $a(x) \neq a(y)$.

Definition 3. *Let* $D = (U, A, d)$ *be a decision table. A set of attributes* $B \subseteq A$ *is called a reduct of decision table* D *if and only if:*

1. *for any objects* $x, y \in U$ *if* $d(x) \neq d(y)$ *and* x, y *are discernible by* A, *then they are also discernible by* B,
2. B *is the minimal set that satisfies condition 1.*

The reduct of the decision table is a minimal subset of conditional attributes that provides exactly the same possibility for the classification of objects to the decision classes as the original set of conditional attributes. Sometimes in the decision table not all attributes are required to classify objects to the decision classes. In such cases, the reduction allows unnecessary attributes to be removed by generating a subtable in which a set of attributes is limited to the reduct.

The aim of this study was to analyze the results that were obtained using the reduction of a set of attributes in a dispersed decision-making system. In this paper, for each local decision table, a reduct of the set of attributes was generated and a subtable of the local table was defined. Then, these reduced tables were used in a dispersed decision-making system.

4   Description of the Experiments and a Comparison of the Results

In the experimental part, at first, the reduction of the sets of conditional attributes of the local knowledge bases was applied, and then the reduced knowledge bases were used in a dispersed decision-making system that had negotiations. The results obtained in this way were compared with the results obtained for a dispersed system and full sets of the conditional attributes.

The author did not have access to dispersed data that are stored in the form of a set of local knowledge bases and therefore some benchmark data that are stored in a single decision table were used. The division into a set of decision tables was made for the data that was used.

The data from the UCI repository were used in the experiments - Soybean data set and Landsat Satellite data set. Both data sets are available in the UCI repository in a form that is divided into a training set and a test set. Table 1 presents a numerical summary of the data sets.

Table 1. Data set summary

Data set	# The training set	# The test set	# Conditional attributes	# Decision classes
Soybean	307	376	35	19
Landsat Satellite	4435	1000	36	6

For both sets of data, the training set, which was originally written in the form of a single decision table, was dispersed, which means that it was divided into a set of decision tables. A dispersed decision-making system with five different versions (with 3, 5, 7, 9 and 11 decision tables) was considered. The following designations are used:

– WSD_{Ag1}^{dyn} - 3 decision tables;
– WSD_{Ag2}^{dyn} - 5 decision tables;
– WSD_{Ag3}^{dyn} - 7 decision tables;
– WSD_{Ag4}^{dyn} - 9 decision tables;
– WSD_{Ag5}^{dyn} - 11 decision tables.

The division into a set of decision tables was made in the following way. The author defined the number of conditional attributes in each of the local decision tables. Then, the attributes from the original table were randomly assigned to the local tables. As a result of this division, some local tables have common conditional attributes. The universes of the local tables are the same as the universe of the original table, but the identifiers of object are not stored in the local tables.

The measures for determining the quality of the classifications were:

- *the estimator of classification error* e in which an object is considered to be properly classified if the decision class that is used for the object belonged to the set of global decisions that were generated by the system;
- *the estimator of classification ambiguity error* e_{ONE} in which an object is considered to be properly classified if only one correct value of the decision was generated for this object;
- *the average size of the global decisions sets* $\overline{d}_{WSD_{Ag}^{dyn}}$ that was generated for a test set.

In the first stage of the experiments for each local decision tables, the reducts of the set of conditional attributes were generated. For this purpose, the Rough Set Exploration System (RSES [17]) program was used. The program was developed at the University of Warsaw, in Faculty of Mathematics, Informatics and Mechanics, under the direction of Professor Andrzej Skowron. For both of the analyzed data sets for each version of the dispersion (3, 5, 7, 9 and 11 decision tables) and for each local decision table a set of reducts was generated separately. For this purpose, the following settings of the RSES program were used: Discernibility matrix settings - Full discernibility, Modulo decision; Method - Exhaustive algorithm. Many reducts were generated for certain decision tables. For example, for the Landsat Satellite data set and a dispersed system with three local decision tables for one of the tables, 1,469 reducts were obtained and for another table 710 reducts were obtained. If more than one reduct was generated for a table, one reduct was randomly selected from the reducts that had the smallest number of attributes. Table 2 shows the number of conditional attributes that were deleted from the local decision tables by the reduction of knowledge. In the table, the following designations were applied: $\#Ag$ - is the number of the local decision tables (the number of agents) and A_{ag_i} - the set of conditional attributes of the ith local table (of the ith agent). As can be seen for the Landsat Satellite data set and the dispersed system with 7, 9 and 11 local

Table 2. The number of conditional attributes removed as a result of the reduction of knowledge

Data set, $\#Ag$	A_{ag_1}	A_{ag_2}	A_{ag_3}	A_{ag_4}	A_{ag_5}	A_{ag_6}	A_{ag_7}	A_{ag_8}	A_{ag_9}	$A_{ag_{10}}$	$A_{ag_{11}}$
Soybean, 3	4	5	2	-	-	-	-	-	-	-	-
Soybean, 5	4	0	0	1	0	-	-	-	-	-	-
Soybean, 7	0	0	0	1	1	0	0	-	-	-	-
Soybean, 9	0	0	0	0	0	1	1	0	0	-	-
Soybean, 11	0	0	0	0	0	0	0	0	1	0	0
Landsat Satellite, 3	1	9	8	-	-	-	-	-	-	-	-
Landsat Satellite, 5	0	0	1	0	0	-	-	-	-	-	-
Landsat Satellite, 7	0	0	0	0	0	0	0	-	-	-	-
Landsat Satellite, 9	0	0	0	0	0	0	0	0	0	-	-
Landsat Satellite, 11	0	0	0	0	0	0	0	0	0	0	0

decision tables, the reduction of the set of conditional attributes did not cause any changes. Therefore, these systems will no longer be considered in the rest of the paper.

In the second stage of experiments, for both sets of data and each version of a dispersed system, an optimization of the system's parameters was carried out in accordance with the following test scenario.

In the first step, the optimal values of the parameters:

- m_1 - parameter, which determines the number of relevant objects that are selected from each decision class of the decision table and are then used in the process of cluster generation;
- p - parameter, which occurs in the definition of friendship, conflict and neutrality relations;
- m_2 - parameter of the approximated method of the aggregation of decision tables;

were determined. This was done by performing tests for different parameter values: $p \in \{0.05, 0.1, 0.15, 0.2\}$ for both data sets; $m_1, m_2 \in \{1, \dots, 10\}$ for the Soybean data set and $m_1, m_2 \in \{1, \dots, 5\}$ for the Landsat Satellite data set. In order to determine the optimal values of these parameters, a dispersed decision-making system with negotiations and a voting method instead of the DBSCAN algorithm were used. Then, the minimum value of the parameters m_1, p and m_2 for which the lowest value of the estimator of the classification error was obtained were chosen. In the second step, parameters ε and $MinPts$ of the DBSCAN algorithm were optimized. For this purpose, the optimal parameter values m_1, p and m_2, which had previously been set were used. Parameter ε was optimized by performing a series of experiments with different values of this parameter. Then, the values that indicate the greatest improvement in the quality of classification were selected. The best results that were obtained for the optimal values of the parameter p, m_1, m_2, $MinPts$ and ε are presented below.

The results of the experiments with both data sets and for the reduction of the set of conditional attributes are presented in Table 3 (Results with reducts). For comparison, Table 3 also contains the results for the same data sets and dispersed systems but without the reduction of the set of conditional attributes (Results without reducts). In the table, the following information is given: the name of the dispersed decision-making system (System); the selected, optimal parameter values (Parameters); the estimator of the classification error e; the estimator of the classification ambiguity error e_{ONE} and the average size of the global decisions sets $\overline{d}_{WSD_{Ag}^{dyn}}$.

When the results given in Table 3 are compared, it can be seen that only in three cases were poorer results obtained after the reduction. Only for the Soybean data set and the systems with three and five agents did the application of reduction result in a deterioration in the accuracy of the classification. It can be concluded that for the Soybean data set a situation in which a large number of attributes is removed (as was the case for these systems) is unfavorable. For

Table 3. Results of experiments

Experiments results with the Soybean data set

System	Results with reducts				Results without reducts			
	Parameters $m_1/p/m_2$	e	e_{ONE}	$\overline{d}_{WSD_{Ag}^{dyn}}$	Parameters $m_1/p/m_2$	e	e_{ONE}	$\overline{d}_{WSD_{Ag}^{dyn}}$
WSD_{Ag1}^{dyn}	4/0.05/1/0.0033	0.059	0.274	1.811	5/0.1/1/0.0088	0.019	0.266	1.697
WSD_{Ag2}^{dyn}	5/0.1/2/0.0037	0.037	0.378	2.082	4/0.05/1/0.0144	0.008	0.290	2.082
	5/0.1/2/0.0024	0.053	0.309	1.566	4/0.05/1/0.0122	0.021	0.258	1.529
WSD_{Ag3}^{dyn}	6/0.05/1/0.00222	0.013	0.295	1.989	2/0.05/4/0.0156	0.013	0.301	1.899
	6/0.05/1/0.00147	0.024	0.255	1.566	2/0.05/4/0.0117	0.032	0.277	1.572
WSD_{Ag4}^{dyn}					4/0.1/2/0.0174	0.024	0.293	1.822
	8/0.2/1/0.00204	0.008	0.253	1.561	1/0.05/3/0.0103	0.043	0.242	1.521
WSD_{Ag5}^{dyn}	10/0.15/1/0.00252	0.035	0.340	1.904	2/0.05/2/0.0225	0.035	0.322	1.875
	10/0.15/1/0.00171	0.045	0.274	1.388	2/0.05/2/0.0123	0.080	0.263	1.303

Experiments results with the Landsat Satellite data set

System	Results with reducts				Results without reducts			
	Parameters $m_1/p/m_2$	e	e_{ONE}	$\overline{d}_{WSD_{Ag}^{dyn}}$	Parameters $m_1/p/m_2$	e	e_{ONE}	$\overline{d}_{WSD_{Ag}^{dyn}}$
WSD_{Ag1}^{dyn}	1/0.05/1/0.00201	0.009	0.445	1.788	1/0.05/4/0.0029	0.022	0.390	1.786
WSD_{Ag2}^{dyn}	2/0.05/2/0.00174	0.008	0.390	1.702	1/0.05/3/0.0046	0.011	0.367	1.618
	2/0.05/2/0.00084	0.034	0.230	1.260	1/0.05/3/0.0024	0.040	0.220	1.237

the other systems with 7, 9 and 11 agents, a much smaller number of attributes were removed, and this reduction improved the quality of the classification. The greatest improvement for the Soybean data set was observed for the system with nine agents. This system, for the value $\overline{d}_{WSD_{Ag}^{dyn}} \approx 1.5$ when applying the reduction produced a five times better result (e) than without the reduction. In the case of the Landsat Satellite data set for both dispersed systems in which the reduction of attributes brought some changes, a better classification accuracy was obtained.

5   Summary

In this article, the reduction of sets of the attributes in a dispersed system with negotiations was applied. In order to generate reducts, the RSES program was used and a dispersed system with negotiations that was proposed by the author in the paper [11] was applied. Two data sets: the Soybean data set and the Landsat Satellite data set and five different versions of a dispersed system were analyzed. Based on the presented results of the experiments, it can be concluded that in most cases, the replacement of the sets of attributes by reducts results in an improved of quality of classification.

References

1. Bregar, A.: Towards a framework for the measurement and reduction of user-perceivable complexity of group decision-making methods. IJDSST **6**(2), 21–45 (2014)
2. Cabrerizo, F.J., Herrera-Viedma, E., Pedrycz, W.: A method based on PSO and granular computing of linguistic information to solve group decision making problems defined in heterogeneous contexts. Eur. J. Oper. Res. **230**(3), 624–633 (2013)
3. Demri, S., Orlowska, E.: Incomplete Information: Structure, Inference. Complexity. Monographs in Theoretical Computer Science. An EATCS Series. Springer, Heidelberg (2002)
4. Gatnar, E.: Multiple-model approach to classification and regression. PWN, Warsaw (2008)
5. Kuncheva, L.I.: Combining Pattern Classifiers Methods and Algorithms. Wiley, Hoboken (2004)
6. Pawlak, Z.: Rough sets. Int. J. Comput. Inf. Sci. **11**, 341–356 (1982)
7. Pawlak, Z.: On conflicts. Int. J. Man-Mach. Stud. **21**(2), 127–134 (1984)
8. Pawlak, Z.: An inquiry into anatomy of conflicts. Inf. Sci. **109**(1–4), 65–78 (1998)
9. Polikar, R.: Ensemble based systems in decision making. IEEE Circuits Syst. Mag. **6**(3), 21–45 (2006)
10. Przybyła-Kasperek, M., Wakulicz-Deja, A.: Application of reduction of the set of conditional attributes in the process of global decision-making. Fundam. Inform. **122**(4), 327–355 (2013)
11. Przybyła-Kasperek, M., Wakulicz-Deja, A.: A dispersed decision-making system - the use of negotiations during the dynamic generation of a system's structure. Inf. Sci. **288**, 194–219 (2014)
12. Przybyła-Kasperek, M., Wakulicz-Deja, A.: Global decision-making system with dynamically generated clusters. Inf. Sci. **270**, 172–191 (2014)
13. Przybyła-Kasperek, M.: Decision making system with dynamically generated disjoint clusters. Stud. Informatica **34**(2A), 275–294 (2013)
14. Przybyła-Kasperek, M.: Global decisions taking process, including the stage of negotiation, on the basis of dispersed medical data. In: Kozielski, S., Mrozek, D., Kasprowski, P., Małysiak-Mrozek, B., Kostrzewa, D. (eds.) BDAS 2014. CCIS, vol. 424, pp. 290–299. Springer, Cham (2014). doi:10.1007/978-3-319-06932-6_28
15. Przybyła-Kasperek, M., Wakulicz-Deja, A.: Global decision-making in multi-agent decision-making system with dynamically generated disjoint clusters. Appl. Soft Comput. **40**, 603–615 (2016)
16. Schneeweiss, C.: Distributed decision making-a unified approach. Eur. J. Oper. Res. **150**(2), 237–252 (2003)
17. Skowron, A.: Rough Set Exploration System. http://logic.mimuw.edu.pl/~rses/. Accessed 03 Nov 2016
18. Skowron, A.: Rough sets and vague concepts. Fundam. Inform. **64**(1–4), 417–431 (2005)
19. Skowron, A., Wang, H., Wojna, A., Bazan, J.G.: Multimodal Classification: Case Studies, Transactions on Rough Sets V. LNCS, vol. 4100, pp. 224–239. Springer, Heidelberg (2006)
20. Susmaga, R., Slowinski, R.: Generation of rough sets reducts and constructs based on inter-class and intra-class information. Fuzzy Sets Syst. **274**, 124–142 (2015)
21. Wakulicz-Deja, A., Przybyła-Kasperek, M.: Multi-agent decision system & comparision of methods. Stud. Informatica **31**(2A), 173–188 (2010)

Adjusting Parameters of the Classifiers in Multiclass Classification

Daniel Kostrzewa$^{(\boxtimes)}$ (iD) and Robert Brzeski

Silesian University of Technology, Gliwice, Poland
{daniel.kostrzewa,robert.brzeski}@polsl.pl

Abstract. The article presents the results of the optimization process of classification for five selected data sets. These data sets contain the data for the realization of the multiclass classification. The article presents the results of initial classification, carried out by dozens of classifiers, as well as the results after the process of adjusting parameters, this time obtained for a set of selected classifiers. At the end of article, a summary and the possibility of further work are provided.

Keywords: Classification · Multiclass · Kappa · Weka · UCI · PEMS · GCM · SMARTPHONE · URBAN · DIGITS

1 Introduction

Process of classification is very important part of data mining. For many years, data mining is a huge area of data processing, where many research are carried out as well as new implementation and development. The process of classification is only a piece of that area and at the same time itself is also a large range of knowledge, and increasingly growing. The main task of supervised classification is to divide vectors of a data set, to independent class of data vectors. Generally, classification can be divided through assigning vectors into two or more classes. In this way there is a binary or multiclass classification, and hereby the corresponding classifier, designed for the type of classification. The current study was focused on the multiclass classification. Appropriate related work in this subject can be found in Sect. 2.

To carry out the classification process five data sets were selected. All of them are designed for the implementation of multiclass classification. Four of them are available on the UCI Machine Learning Repository [2], while one data set is available on the BioInformatics Research Group (BIGS) [1]. More on this subjects are in Sect. 3. Section 4 presents the initial study and the obtained results. The aim of this research was the selection of the best classifiers. Then, for those selected classifiers, appropriate study was conducted in the form of parameters optimization. The results and the description of the individual parameters of the classifiers is given in Sect. 5. Analysis of the results, conclusions, and the possibilities of further work are contained in Sect. 6.

© Springer International Publishing AG 2017
S. Kozielski et al. (Eds.): BDAS 2017, CCIS 716, pp. 89–101, 2017.
DOI: 10.1007/978-3-319-58274-0_8

Current research is a continuation of the research presented in [20,21], where the attention was focused on dimensionality reduction, made through feature selection, conducted by IWO algorithm [23,26,27]. There, the classification was executed only by 1NN classifier. The present aim of research is to execute the process of classification, by many classifiers and use different data sets. In the future, it will form the basis, for extension not only of the process of dimensionality reduction, but also the possibility of proper comparative analysis of the whole process of classification with different classifiers.

2   Related Work

For many years data mining [15,36] is intensively developed part of IT research. One of the classical aspects of data mining is classification [4], executed on data set. During the supervised classification process, every data vectors from data set, is assigned to the one of the sets of collections result. If during this process, classifier supposed to assign the vector, to one of two result sets, then it is binary classification [12,35], otherwise the multiclass classification [6,22,25], sometimes called multivariate or multinomial. To execute one of these classification type, it is necessary to use suitable classifier and data set. Presented research is focused on multiclass classification, and has been done on the data sets, where just such a grouping into many subsets is adequate. Data vector is classified into one of the result set, based on contained attributes.

There are a lots of various classifiers like SMO [28], LMT [34], Logit Boost [13], Simple Logistic [34], Hoeffding Tree [16], IBk [5], J48, Naive Bayes [17] and many more [3,36]. They are still developed to achieve a better accuracy [4,9,29, 30,33], higher speed and lower memory usage. The results of the classification process can be evaluated and compared by several criteria [4,8,9,29,30]. The suitable criteria is selected according to the nature of the classification purposes.

There are few studies on the performance of classification, such as statistical comparisons of different classifiers over multiple data sets [11], ways of evaluating and comparing classifiers [4,8,9,29,30], benchmarking state-of-the-art classification algorithms for credit scoring [24], but the authors were unable to find one, as presented in this article.

3   Data Sets

All data sets were selected intentionally for multiclass classification, as the result of the assumption, that current research is focused on that kind of classification only. Additionally, they are differ in the number of classes, the number of training and testing data vectors and different quantity of attributes in one vector (Table 1). Special attention was made for choosing different quantity of attributes in one vector, for different data set. In most cases have been chosen data sets, with rather enough large quantity of attributes, to be able in the future, to conduct process of dimensionality reduction and to compare the obtained results.

Table 1. The quantity of classes, data vectors and attributes in vector of the data sets.

Name of data set	Quantity of classes	Quantity of training vectors	Quantity of validation vectors	Attributes quantity
PEMS	7	267	173	138 672
GCM	14	144	46	16 063
SMARTPHONE	12	7 767	3 162	561
URBAN	9	168	507	148
DIGITS	10	3 823	1 797	64

Four of the data sets, used in presented research are available on the UCI Machine Learning Repository [2]. One of the dataset is from BioInformatics Research Group [1].

The first selected data set is **PEMS** [2,10]. 440 daily records of data from the California Department of Transportation - PEMS [http://pems.dot.ca.gov/]. The data describes the occupancy rate, between 0 and 1, of different car lanes of San Francisco bay area freeways across time. The task on this dataset is to classify each observed day as the correct day of the week, from Monday to Sunday [2,10].

The second data set is **GCM** [1,31]. Global Cancer Map is used for multiclass cancer diagnosis. This data set contains tumor gene expression signatures, with 14 common tumor types and expression levels of 16,063 genes and expressed sequence tags, used to evaluate the accuracy of a multiclass classifier [31].

The third data set is **SMARTPHONE** [2,32]. Smartphone-based recognition of human activities and postural transitions data set, built from the recordings of 30 subjects, performing basic activities and postural transitions, while carrying a waist-mounted smartphone, with embedded inertial sensors [www.smartlab.ws]. The performed protocol of activities is composed of six basic activities and also included six postural transitions, that occurred between the static postures [2,32]. The task on this dataset is to classify each activity into one of 12.

The fourth data set is **URBAN** [2,18,19]. Urban land cover data set contains training and testing data, for classifying a high resolution aerial image into 9 types of urban land cover (trees, grass, soil, concrete, asphalt, buildings, cars, pools, shadows). Intended to assist sustainable urban planning efforts. Source of this data: Brian Johnson; Institute for Global Environmental Strategies; 2108-11 Kamiyamaguchi, Hayama, Kanagawa, 240-0115 Japan [2,18,19].

The fifth data set is **DIGITS** [2]. Optical recognition of handwritten digits data set, created using preprocessing programs made available by NIST [14], to extract normalized bitmaps of handwritten digits from a preprinted form. From a total of 43 people, 30 contributed to the training set and different 13 to the test set. Source of this data: E. Alpaydin, C. Kaynak; Department of Computer Engineering; Bogazici University, 80815 Istanbul Turkey [2].

4   Overview of the Classifiers, Description of Research

The process of classification was executed using the Weka software [3] (Waikato Environment for Knowledge Analysis), as Java classes, each containing code that implements one of the classifiers. These classes were used to create specialized application, that allows in easy way to carry out the whole set of the research experiments. This original software allows to multiple executions of classification process in accordance with a predetermined set of classification parameters and finally saving the obtained results in an appropriate form.

At the beginning, with the set of more than twenty, very popular classifiers, the experiment was carried out, for all of the data sets, with default parameters. The quality of classification was measured and specified with used of kappa factor [7].

Table 2. Initial research/results - kappa factor.

Classifier / data set	PEMS	GCM	SMARTPHONE	URBAN	DIGITS
Ada boost M1	0.1358	0.0748	0.1967	0.2090	0.1028
Bayes net	–	0.3187	0.7558	0.7383	0.8912
Decision stump	0.1358	0.0748	0.1967	0.2090	0.1028
Filtered classifier	0.9122	0.4124	0.7307	0.6352	0.7279
Hoeffding Tree	0.7099	0.4810	0.7797	0.7431	0.8380
IBk	0.7096	0.4127	0.8388	0.6507	**0.9771**
Iterative classifier optimizer	**0.9797**	0.4591	0.8694	0.7435	0.8875
J48	0.9121	0.4800	0.7979	0.6232	0.8417
JRip	0.8649	0.2883	0.8079	0.5231	0.8547
K*	–	0	–	0.6049	0.9592
LMT	0.8380	0.4353	0.9355	0.7459	0.9474
Logistic	–	–	0.9109	0.5178	0.9128
Logistic Base	0.8380	0.4353	0.9355	0.7459	0.9474
Logit Boost	0.9662	0.4602	0.8694	**0.7550**	0.8875
Multilayer perceptron	–	–.	0.8687	0.7234	0.9616
Naive bayes	0.6966	0.4810	0.6997	0.7408	0.8825
Naive bayes multinomial	0.3340	–	–	–	0.8770
OneR	0.8244	0.1752	0.2720	0.4242	0.1448
PART	0.8445	0.3870	0.8304	0.6496	0.8837
Random tree	0.6621	0.2930	0.7682	0.5910	0.7762
Randomizable filtered classifier	0.5625	0.3159	0.6126	0.2457	0.7539
REP tree	0.7973	0.3183	0.8287	0.6511	0.8374
Simple logistic	0.8380	0.4353	0.9355	0.7459	0.9474
SMO	0.8918	**0.5054**	**0.9444**	0.7030	0.9610
Min	0.1358	0	0.1967	0.2090	0.1028
Max	0.9797	0.5054	0.9444	0.7550	0.9771
Average	0.7227	0.3449	0.7448	0.6052	0.7877

$$kappa = \frac{Po - Pe}{1 - Pe}$$

where: Po - the actual/measured accuracy [9,30] of the test classifier for a given data set; Pe - random classifier accuracy for a given set of data.

The kappa factor is kind of improvement over the random classifier.

Results can be found in Table 2. The best obtained result for a given set of data is bold marked and worth 3 points, the results very close to the best are italics and worth 2 points and finally results close to the best are underlined and worth 1 point. On this basis rating of classifiers was created and presented in Table 3. For further study, the classifiers giving pre-best results for these data sets were selected. These selected classifiers in Table 3 are marked in bold.

Lack of data in the tables means, that the process of classification was aborted. Such a situation occurred after exceeding the maximum size of the used memory, in this case 24 GB. It is about 100 times more than the largest size of used data set (250 MB for PEMS).

Table 3. Initial research - final rating.

Classifier/data set	PEMS	GCM	SMARTPHONE	URBAN	DIGITS	Sum
SMO	2	3	3	1	2	**11**
Iterative classifier optimizer	3	1	1	2	0	**7**
LMT	1	1	2	2	1	**7**
Logistic Base	1	1	2	2	1	**7**
Logit Boost	2	1	1	3	0	**7**
Simple Logistic	1	1	2	2	1	**7**
Hoeffding Tree	0	2	0	2	0	**4**
IBk	0	0	0	1	3	**4**
J48	2	2	0	0	0	**4**
Multilayer perceptron	–	–	1	1	2	**4**
Naive bayes	0	2	0	2	0	**4**
Logistic	–	–	2	0	1	3
Filtered classifier	2	0	0	0	0	2
K*	–	0	–	0	2	2
Bayes net	–	0	0	1	0	1
JRip	1	0	0	0	0	1
PART	1	0	0	0	0	1
REP tree	0	0	0	1	0	1
Ada boost M1	0	0	0	0	0	0
Decision stump	0	0	0	0	0	0
Naive bayes multinomial	0	–	–	–	0	0
OneR	0	0	0	0	0	0
Random tree	0	0	0	0	0	0
Randomizable filtered classifier	0	0	0	0	0	0

The summary of initial research (Table 3) shows the chosen classifiers, used further in the proper research. From this set, the Multilayer Perceptron classifier was removed. The reason is that for this classifier it wasn't possible to carry out the test for every data sets (for 2 of the data sets, the program exceeded the maximum size of the used memory). As the presented results shows, the best classification with default parameters gives SMO classifier.

5   Tests - Optimization of Classifiers

For a chosen classifiers the main research was executed. This time additionally with the whole set of appropriate parameters, used for the classification algorithm. True is that in total there are tens of thousands different tests and it is not possible to present them all. That is why in this paper is shown only final best obtained result. Finally, in Table 4 is presented comparison of kappa factor, before optimization process (column D) and after this process (column O), all together with parameters for which it was obtained. Column '%' contains percentage improvement of result, obtained during the optimization process - improvement comparing to the result for the default parameters.

In Table 4 value 'max' is the highest value from appropriate column, except classifier Logistic Base (for which there are no parameters optimization). In similar way is calculated 'average' of these values.

Lack of parameter in Table 4 means that the result did not depend on it.

During parameters optimization the following parameters were tested for specified classifiers: **SMO** [3] tested parameters:

- N – according to the value data are: 0 – normalize; 1 – standardize; 2 – neither (with default value $= 0$). In our study were tested all 3 values.
- V – the number of folds for the internal cross-validation (with default value $= -1$ – use training data). In our study were tested values $- 1$ and 10.
- C – the complexity constant C (with default value $= 1$). In our study were tested values from $C = 0.01$ to $C = 100.0$.
- L – the tolerance parameter (with default value $= 10^{-3}$). In our study were tested values $L = 10^{-6}$ to $L = 1.0$.

In our study were tested parameters C and L cross with parameters N and V. **Iterative classifier optimizer** [3] tested parameters:

- A – if set, average estimate is used rather than one estimate from pooled predictions (with default value $=$ off). In the study were tested values on and off.
- R – number of runs for cross-validation (with default value $= 1$). In the study were tested values: 1, 2, 5, 10.
- metric – evaluation metric to optimize (with default value $=$ RMSE). Available metrics: RMSE, Correct, Incorrect, Kappa, Total cost, Average cost, KB relative, KB information, Correlation, Complexity 0, Complexity scheme, Complexity improvement, MAE, RAE, RRSE, coverage, Region size, TP rate, FP rate, Precision, Recall, F-measure, MCC, ROC area, PRC area.

Table 4. Comparison of kappa factor before and after optimization process.

Classifier/ data set	PEMS			GCM			SMARTPHONE			URBAN			DIGITS		
	D	O	%	D	O	%	D	O	%	D	O	%	D	O	%
SMO	0.891	0.891	0	0.505	0.646	27.79	0.944	0.948	0.358	0.703	0.744	5.87	0.961	0.965	0.45
	N=0, other doesn't matter			N=2, L=0.1, M=off			N=2, M=on, V=10, C=20			N=0, V=-1, M=on, C=0.05			N=1, V=10, M=on, C=0.1		
Iterative classifier optimizer	0.98	**0.986**	0.691	0.459	0.623	35.69	0.869	0.934	7.427	0.743	0.808	8.708	0.887	0.952	7.247
	P=95, I=50			P=85, I=100, H=0.2, Z=1			P=90, I=50, H=1, Z=2			P=90, I=100, H=1, Z=3			P=85, I=100, H=1, Z=1		
LMT	0.838	0.899	7.225	0.435	0.527	21.18	0.935	0.939	0.437	0.746	0.741	-0.65	0.947	0.957	0.977
	I=20, W=0.01, R=off			I=50 or 100, W=0.0001			I=50, W=0, R=off			I=10, W=0.01			[R=off, I =50, W=0.01] or [I=100, W=0.001]		
Logistic Base	0.838	–	–	0.435	–	–	0.935	–	–	0.746	–	–	0.947	–	–
Logit Boost	0.966	**0.986**	2.10	0.460	**0.646**	40.32	0.869	0.934	7.43	0.755	**0.813**	7.67	0.887	0.952	7.25
	various parameter values			P=90, I=100, H=0.5, Z=10			P=95, I=100, H=1, Z=2			P=85, I=50, H=1, Z=3			P=85, I=100, H=0.1 to 1, Z=1 to 10		
Simple Logistic	0.838	0.899	7.23	0.435	0.507	16.38	0.935	0.933	-0.24	0.746	0.741	-0.65	0.947	0.951	0.392
	I=20, W=0.01			I=20 to 100, W=0.01			only worse			only worse			I=50, W=0.1		
Hoeffding Tree	0.71	0.71	0	0.481	0.481	0	0.78	0.785	0.717	0.743	0.743	0	0.838	0.838	0
	at most the same			at most the same			L=2, S=1, H=0.2, G=50			at most the same			at most the same		
IBk	0.71	0.838	18.1	0.413	0.458	10.98	0.839	0.875	4.26	0.651	0.736	13.15	0.977	**0.980**	0.32
	I=on, K=12			I=on or F=on, K=11			I=on or F=on, K=18			I=on, K=19			I=on, K=4		
J48	0.912	0.946	3.70	0.48	0.505	5.14	0.798	0.834	4.58	0.623	0.690	10.77	0.842	0.856	1.698
	[R=on, J=on, N=10, M=1] or [R=on, S=on, J=on, N=10, M=1] or [R=on, I=on, doN=on, N=10, M=1]			[J=on, M=1 or 2] or [S=on, J=on, M=1 or 2]			[R=on, M=1, N=2] or [R=on, S=on, M=1, N=2] or [R=on, doN=on, M=1, N=2]			R=on, J=on, doN=on, N=5, M=10			S=on, M=1, C=0.01 or 0.02		
Naive Bayes	0.697	0.94	34.8	0.481	0.481	0	0.7	0.783	11.96	0.741	0.764	3.18	0.883	0.893	1.12
	D=on			–			K=on			K=on			K=on		
Average	0.838	0.899	8.207	0.461	0.541	17.5	0.852	0.885	4.103	0.717	0.753	5.340	0.908	0.927	2.16
Max	0.98	0.986	34.83	0.505	0.646	40.33	0.944	0.948	11.96	0.755	0.813	13.15	0.977	0.980	7.247

In the study were tested values: RMSE, Correct, Incorrect, Kappa, Complexity scheme, Complexity improvement, MAE, RAE, RRSE, Region size, TP rate, FP rate, Precision, Recall, F-measure, MCC, ROC area, PRC area.

Options specific to used default classifier Logit Boost:

- P – percentage of weight mass to base training on (with default value = 100). In the study were tested values: 95, 90, 85.
- H – shrinkage parameter (with default value = 1). In the study were tested values: 0.1, 0.2, 0.5, 1.0, 2.0, 5.0, 10.0.
- Z – maximum threshold for responses (with default value = 3). In the study were tested values: 0.1, 0.2, 0.5, 1.0, 2.0, 5.0, 10.0.
- I – number of iterations (with default value = 10). In the study were tested values 50 and 100.

In our study was tested parameter A cross with parameter R cross with parameter metric. After that, for the parameter metric equal 'Region size', as giving the best result, series of tests were performed, with the parameters P, H, Z, I, in a way of each parameter with every.

LMT [3] tested parameters:

- I – set fixed number of iterations for Logit Boost (instead of using cross-validation as default setting). In our study were tested values I = 10, 20, 50 and 100.
- R – split on residuals instead of class values (with default value = off).
- M – set minimum number of instances at which a node can be split (default value = 15). In our study were tested values M = 1, 2, 5, 10, 15, 20, 50 and 100.
- W – set beta for weight trimming for Logit Boost (default value = 0 – no weight trimming). In our study were tested values W = 0, 0.00001, 0.0001, 0.001, 0.01, 0.1, 1.0, 10.0.

In our study was tested parameter I cross with parameter R cross with parameter M cross with parameter W.

Logistic Base [3] this classifier does not have any parameters. The obtained result can be considered as reference values.

Logit Boost [3] tested parameters:

- P – percentage of weight mass to base training on (with default value = 100). In the study were tested values: 100, 95, 90, 85, 80.
- H – shrinkage parameter (with default value = 1). In the study were tested values from 0.1 to 10.
- Z – maximum threshold for responses (with default value = 3). In the study were tested values 0.1 to 10.
- I – number of iterations (with default value = 10). In the study were tested values 10, 20, 50, 100.

In our study was tested parameter P cross H cross Z cross I.

Simple Logistic [3] tested parameters:

- I – set fixed number of iterations for Logit Boost (with default value = 0 (off)). In the study were tested values: 10, 20, 50, 100.
- S – use stopping criterion on training set (with default value = off – cross-validation). In the study were tested values on and off.
- P – use error on probabilities (RMSE) instead of misclassification error for stopping criterion (with default value = off). In the study were tested values on and off.
- M – set maximum number of boosting iterations (with default value = 500). In the study were tested values 1, 10, 100, 1000.
- W – set beta for weight trimming for Logit Boost (with default value = 0 – no weight trimming). In the study were tested values 0, 0.001, 0.01, 0.1.
- A – the AIC is used to choose the best iteration (instead of cross-validation or training error) (with default value = off). In the study were tested values on and off.

In our study were tested the parameters I, W, M, S, P, A in a way of each parameter with every.

Hoeffding Tree [3] tested parameters:

- L – the leaf prediction strategy to use. 0 = majority class, 1 = Naive Bayes, 2 = Naive Bayes adaptive (with default value = 0). In the study were tested values all 3 values.
- S – the splitting criterion to use 0 = Gini, 1 = Info gain (with default value = 0). In the study were tested values: 0 and 1.
- H – threshold below which a split will be forced to break ties (with default value = 0.05). In the study were tested values: 0.001, 0.002, 0.005, 0.01, 0.02, 0.05, 0.1, 0.2, 0.5, 1, 2, 5, 10.
- G – grace period – the number of instances a leaf should observe between split attempts (with default value = 200). In the study were tested values: 1, 2, 5, 10, 20, 50, 100, 200, 500, 1000, 2000, 5000, 10000.

In our study were tested parameter L cross with parameter S cross H cross G.

IBk [3] tested parameters:

- K – used in classification number of nearest neighbours (k) (with default value = 1). In the study were tested values from 1 up to 20.
- I – weight neighbours by the inverse of their distance (used when K > 1) (with default value = off). In the study was checked off and on.
- F – weight neighbours by 1 of their distance (used when K > 1) (with default value = off). In the study was checked off and on.

In our study was tested parameter K cross with I and F parameters.

J48 [3] tested parameters:

- U – use unpruned tree (with default value = off). In the study were tested values: on and off.
- O – do not collapse tree (with default value = off). In the study were tested values: on and off.
- C – set confidence threshold for pruning (with default value = 0.25). In the study were tested values: 0.01, 0.02, 0.05, 0.1, 0.2, 0.25, 0.5.
- M – set minimum number of instances per leaf (with default value = 2). In the study were tested values: 1, 2, 5, 10, 20.
- R – use reduced error pruning (with default value = off). In the study were tested values: on and off.
- N – set number of folds for reduced error pruning. One fold is used as pruning set (with default value = 3). In the study were tested values: 2, 3, 5, 10.
- S – do not perform subtree raising (with default value = off). In the study were tested values: on and off.
- J – do not use MDL correction for info gain on numeric attributes (with default value = off). In the study were tested values: on and off.
- doNotMakeSplitPointActualValue – do not make split point actual value (with default value = off). In the study were tested values: on and off.

In our study were tested all of these parameters, in a way of each parameter with every, wherever it was possible (sometimes parameters exclude each other).
Naive Bayes [3] tested parameters:

- D – use supervised discretization to process numeric attributes.
- K – use kernel density estimator instead of normal distribution for numeric attributes.

In our study were tested all 3 settings (off, D = on, K = on).

Of course, by having such many parameters, the optimization process could be carried out infinitely. Therefore, it was decided to use such values of parameters, as described above. Generally, for parameters where the possible values are 'on' and 'off', we were tested both of the values. For other parameters we tested values larger and smaller then the default.

Table 5. Pearson product – moment correlation coefficient.

	PEMS	GCM	SMARTPHONE	URBAN	DIGITS	Correlation coefficient
Training	*267*	*144*	*7767*	*168*	*3823*	–
Max of %	34.83	40.33	11.96	13.15	7.247	**-0.613**
Average %	8.207	17.5	4.103	5.34	2.16	**-0.555**
Average O	0.899	0.541	0.885	0.753	0.927	**0.516**

6   Conclusions

As can be seen in Table 4, the results obtained with default parameters, typically are not the best – what seems to be obvious. Only in 2 cases changing default parameters gives little worse result. This situation we have for URBAN data set and classifier LMT and Simple Logistic, where the result is worse by 0.65%. Also for Hoeffding Tree classifier, adjusting parameters usually did not give any improvement. In any other situation it was possible to get better result. The improvement was different and depend from used data set as well as classifiers. For data sets, the best effect of optimization was obtained for GCM – together with Logit Boost classifier, with the improvement of 40.32%.

Analysis of the results in Table 4 together with data in Table 1, allows to notice some convergence between quantity of training vectors (Table 5 – value: training) and improvement of classification (max of %, average %). The correlation was calculated and presented in Table 5. As we can see, the lower quantity of training vectors, the higher possibility of making the improvement. There is also some correlation, with the average value of kappa factor. For GCM and URBAN data set, the number of training vectors are quite small. So it is possible, that especially for data sets with small quantity of training vectors, the default parameters for classifiers are not able to give the best solution and particularly in that situation, it is worth to check other values of parameters. It also has to be noted, that we have a very small quantity of sample test for calculated correlation and to generalise these conclusion, it should be proved for appropriate quantity of different data sets.

In Table 4, best obtained kappa factor for given data set was marked bold. It can be noticed that after optimization process, two of the best classifiers are Logit Boost and SMO.

Generally, the parameters optimization gives better classification and it is obvious. In presented results average improvement is 7.46%, measured as kappa factor. But it is also worth to notice, that for different data sets, the best classification is given by different classifiers and also given classifier for different data sets is best optimized with used of different parameters. So there are no particular instructions of how to proceed the optimization, but the process is worth to do. Especially is worth to check for data sets with small quantity of training vectors.

The future work in this field will be study of dimensionality reduction of used data set (feature selection; dimensionality reduction – linear like PCA, and nonlinear, like: Isomap, LLE, GPLVM).

As a summary can be noticed, that the presented research is the continuation of the work presented in [20], but also the introduction to the planned one. This work will allow to perform comparisons to the result of the future research of reducing the dimensionality of the data sets. And to be able to compare, how the dimensionality reduction affects the quality of classification, in this research we wanted to obtain the very high result, through parameters optimization.

Finally, current paper presents both, the results of multiclass classification obtained by a number of different classifiers and also the optimization process for initially best classifiers.

Acknowledgements. This work was partly supported by BKM16/RAU2/507 and BK-219/RAU2/2016 grants from the Institute of Informatics, Silesian University of Technology, Poland.

References

1. GCM - Global Cancer Map dataset. http://eps.upo.es/bigs/datasets.html
2. UCI Machine Learning Repository. https://archive.ics.uci.edu/ml/datasets/
3. Weka 3. http://www.cs.waikato.ac.nz/ml/weka/
4. Agrawal, R., Imielinski, T., Swami, A.: Database mining: a performance perspective. IEEE Trans. Knowl. Data Eng. **5**(6), 914–925 (1993)
5. Aha, D., Kibler, D.: Instance-based learning algorithms. Mach. Learn. **6**, 37–66 (1991)
6. Aly, M.: Survey on multiclass classification methods, Technical report, Caltech (2005)
7. Arie, B.D.: Comparison of classification accuracy using cohen's weighted kappa. Expert Syst. Appl. **34**(2), 825–832 (2008)

8. Bach, M., Werner, A., Zywiec, J., Pluskiewicz, W.: The study of under- and over-sampling methods' utility in analysis of highly imbalanced data on osteoporosis. Inf. Sci. Life Sci. Data Anal. **381**, 174–190 (2016)
9. Costa, E., Lorena, A., Carvalho, A., Freitas, A.: A review of performance evaluation measures for hierarchical classifiers. In: Evaluation Methods for Machine Learning II, AAAI 2007 Workshop, pp. 182–196. AAAI Press (2007)
10. Cuturi, M.: Fast global alignment kernels. In: Proceedings of the International Conference on Machine Learning (2011)
11. Demsar, J.: Statistical comparisons of classifiers over multiple data sets. J. Mach. Learn. Res. **7**, 1–30 (2006)
12. Freeman, E., Moisen, G.: A comparison of the performance of threshold criteria for binary classification in terms of predicted prevalence and kappa. Ecol. Model. **217**(1–2), 48–58 (2008)
13. Friedman, J., Hastie, T., Tibshirani, R.: Additive logistic regression: a statistical view of boosting. Stanford University, Stanford (1998)
14. Garris, M., Blue, J., Candela, G., Dimmick, D., Geist, J., Grother, P., Janet, S., Wilson, C.: NIST form-based handprint recognition system. NISTIR 5469 (1994)
15. Haiyang, Z.: A short introduction to data mining and its applications (2011)
16. Hulten, G., Spencer, L., Domingos, P.: Mining time-changing data streams. In: ACM SIGKDD International Conference on Knowledge Discovery and Data Mining, pp. 97–106 (2001)
17. John, G., Langley, P.: Estimating continuous distributions in Bayesian classifiers. In: 11th Conference on Uncertainty in Artificial Intelligence, pp. 338–345, San Mateo (1995)
18. Johnson, B.: High resolution urban land cover classification using a competitive multi-scale object-based approach. Remote Sens. Lett. **4**(2), 131–140 (2013)
19. Johnson, B., Xie, Z.: Classifying a high resolution image of an urban area using super-object information. ISPRS J. Photogrammetry Remote Sens. **83**, 40–49 (2013)
20. Josinski, H., Kostrzewa, D., Michalczuk, A., Switonski, A.: The exIWO metaheuristic for solving continuous and discrete optimization problems. Sci. World J. 2014 (2014). 14 p. doi:10.1155/2014/831691. Article ID 831691
21. Świtoński, A., Polański, A., Wojciechowski, K.: Human identification based on gait paths. In: Blanc-Talon, J., Kleihorst, R., Philips, W., Popescu, D., Scheunders, P. (eds.) ACIVS 2011. LNCS, vol. 6915, pp. 531–542. Springer, Heidelberg (2011). doi:10.1007/978-3-642-23687-7_48
22. Kasprowski, P., Harezlak, K.: Using dissimilarity matrix for eye movement biometrics with a jumping point experiment. In: Czarnowski, I., Caballero, A.M., Howlett, R.J., Jain, L.C. (eds.) Intelligent Decision Technologies 2016. SIST, vol. 57, pp. 83–93. Springer, Cham (2016). doi:10.1007/978-3-319-39627-9_8
23. Kostrzewa, D., Josinski, H.: The exIWO metaheuristic - a recapitulation of the research on the join ordering problem. Commun. Comput. Inf. Sci. **424**, 10–19 (2014)
24. Lessmanna, S., Baesens, B., Seowd, H.V., Thomasc, L.: Benchmarking state-of-the-art classification algorithms for credit scoring: an update of research. Eur. J. Oper. Res. **247**(1), 124–136 (2015)
25. Mehra, N., Gupta, S.: Survey on multiclass classification methods. Int. J. Comput. Sci. Inf. Technol. **4**(4), 572–576 (2013)
26. Mehrabian, A., Lucas, C.: A novel numerical optimization algorithm inspired from weed colonization. Ecol. Inform. **1**(4), 355–366 (2006)

27. Pahlavani, P., Delavar, M., Frank, A.: Using a modified invasive weed optimization algorithm for a personalized urban multi-criteria path optimization problem. Int. J. Appl. Earth Obs. Geoinf. **18**, 313–328 (2012)
28. Platt, J.: Fast training of support vector machines using sequential minimal optimization. In: Schoelkopf, B., Burges, C., Smola, A. (eds.) Advances in Kernel Methods - Support Vector Learning (1998)
29. Powers, D.: Evaluation: from precision, recall and f-score to roc, informedness, markedness & correlation. J. Mach. Learn. Technol. **2**, 37–63 (2011)
30. Provost, F., Fawcett, T., Kohavi, R.: The case against accuracy estimation for comparing classifiers. In: Proceedings of the ICML 1998, pp. 445–453. Morgan Kaufmann, San Francisco (1998)
31. Ramaswamy, S., Tamayo, P., Rifkin, R.S.M., Yeang, C.H., Angelo, M., Ladd, C., Reich, M., Latulippe, E., Mesirov, J., Poggio, T., Gerald, W., Loda, M., Lander, E., Golub, T.: Multiclass cancer diagnosis using tumor gene expression signatures. PNAS **98**(26), 15149–15154 (2001)
32. Reyes-Ortiz, J.L., Oneto, L., Sama, A., Parra, X., Anguita, D.: Transition-aware human activity recognition using smartphones. Neurocomputing **171**, 754–767 (2016)
33. Smith, M., Martinez, T.: Improving classification accuracy by identifying and removing instances that should be misclassified. In: Proceedings of the IEEE International Joint Conference on Neural Networks, pp. 2690–2697 (2011)
34. Sumner, M., Frank, E., Hall, M.: Speeding up logistic model tree induction. In: 9th European Conference on Principles and Practice of Knowledge Discovery in Databases, pp. 675–683 (2005)
35. Unler, A., Murat, A.: A discrete particle swarm optimization method for feature selection in binary classification problems. Eur. J. Oper. Res. **206**(3), 528–539 (2010)
36. Wu, X., Kumar, V., Quinlan, J., Ghosh, J., Yang, Q., Motoda, H., McLachlan, G., Ng, A., Liu, B., Yu, P., Zhou, Z.H., Steinbach, M., Hand, D., Steinberg, D.: Top 10 algorithms in data mining. Knowl. Inf. Syst. **14**, 1–37 (2008)

Data Mining - A Tool for Migration Stock Prediction

Mirela Danubianu[⊠]

Faculty of Electrical Engineering and Computer Science,
"Stefan cel Mare" University, Universitatii str. 1, 720229 Suceava, Romania
mdanub@eed.usv.ro
http://www.eed.usv.ro

Abstract. The migration phenomenon is an important issue for most of the European Unions countries and it has a major socio-economic impact for all parts involved. After 1989, a massive migration process started to develop from Romania towards Western European countries. Beside qualified personnel in search of different and new opportunities, Roma people became more visible, as they were emigrating in countries with high living standards where they were generating significant integration problems along with costs. In order to identify the problems faced by the Roma community from Rennes, a group of sociologists developed a questionnaire, which contains, among other questions, one relating to the intention of returning home. This paper presents a research that aims to build various models, by data mining techniques, to predict that Roma people return to the home country after a five years interval. The second goal is to assess these models and to identify those aspects that have most influence in the decision-making process. The result is based on the data completed by more than 100 persons from Rennes.

Keywords: Data mining · Classification · CRISP-DM model · Migration phenomenon · Return prediction

1 Introduction

It is well known that after 1989, attracted by the prospect of a better life, a large number of Roma people have migrated from Eastern European countries to Western Europe. In big cities from France such as Lyon, Nanterre, Toulouse, Roubaix true centers of concentration of Roma have been constituted [7]. Lacking education and training, Roma people have often resorted to begging as an alternative to legal work. Treated with fear and hostility they were subject to forced evictions, violence, and discrimination coming from local population and political powers [7].

To identify problems faced by Roma community from Rennes, a group of Romanian sociologists developed a questionnaire, which contains, a set of questions regarding some important aspects of Roma migrants life in France, and regarding their intention to return to Romania.

© Springer International Publishing AG 2017
S. Kozielski et al. (Eds.): BDAS 2017, CCIS 716, pp. 102–114, 2017.
DOI: 10.1007/978-3-319-58274-0_9

This paper presents a research that aims to demonstrate if and how data mining techniques can be used as tools to predict the intention of Roma migrants to return home during a five years interval, and finally to estimate migration stock. We used data collected by the above mentioned questionaires completed by 101 persons who were living in Rennes. We conducted this study on Roma migrants because they form compact settlements and so, it was easier to collect the data requested.

In Sect. 2, the concept of Knowledge Discovery in Databases is presented, and it is shown that data mining (modelling) is one of the steps in the context of CRISP-DM industrial model. In Sect. 3, is described a case study regarding some models, built on CRISP-DM steps and tasks, that aim to allow the prediction of Roma migrants return home. Finally, some conclusions and future works are drawn.

2 Data Mining

As a consequence of information and communication technology development, very large volumes of various types of data were collected and stored. Their analysis is not possible anymore by using traditional database techniques. In this context, a new concept, aiming to identify *"valid, novel, potential useful and understandable patterns in data"* [4] emerged. It is called Knowledge Discovery in Databases (KDD), and is a complex, interactive and iterative process.

Over time, many models for this process have been proposed both by academia [1,4,5] and industry [8]. All of these models consist of a succession of steps. They start from understanding the domain and data used for its representation, continue with the preparation of data for data mining algorithms and their effective implementation in order to detect existing patterns, and end with visualisation and interpretation of these patterns.

In this research we used the most known industrial model - CRISP-DM model [8], presented in Fig. 1.

It starts with a business analysis in order to determine the KDD goals and continues with a data understanding stage which aims to collect, describe and verify data quality. In order to give to the data set the proper format for a certain data mining algorithm, a data preparation step is necessary. During this step data are filtered and the relevant features are selected. Data type transformation, discretization or sampling are also executed.

The central point of KDD process is the data mining stage. Data mining performs analysis of large volumes of data using specific algorithms. These are designed to offer good performances of calculation on large amounts of data, and produce a particular enumeration of patterns. Using patterns or rules with a specific meaning, data mining may facilitate the discovery, from apparently unrelated data, of relationships that are likely to anticipate future problems or might solve the problems under study. It involves the choice of the appropriate data mining task, and, taking into account specific conditions, the choice and the implementation of the proper data mining algorithm.

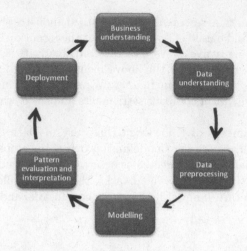

Fig. 1. CRISP-DM model for KDD process [8]

For the next stage, the mined models are evaluated against the goals defined in the first stage.

The last step in the process uses the knowledge discovered in order to simply generate a report or to deploy a repeatable data mining process that aims to improve the built models.

Finally, it should be noted that these steps may be iteratively and interactively performed, and human experts have an important role for continuous adjustment of the whole process.

3   Data Mining to Predict the Intention of Roma Migrants to Return to Their Home Country

Even if at the moment we have a small amount of data, in this paper we try to demonstrate that data mining techniques are suitable to build models that can predict the return from France to Romania of Roma migrants. We drive our research focusing on Rennes, and we will cover the steps required by the CRISP-DM model.

3.1   Business Understanding

The migration phenomenon is an important issue for most of the European Union's countries and it has a major socio-economic impact for all parts involved. Government decisions on issues rised by Roma integration both in Romania and in countries where they have migrated must consider, among other things, an estimated number of people. Besides demographic aspects, such as natality rate and mortality, Roma migration from one country to another and inverse phenomenon the return to the home country has a major influence on this issue.

For this resoan, we aim to develop models to predict if a certain Roma individual will return to Romania within a time frame of five years. Starting from this objective we propose a framework, as presented in Fig. 2.

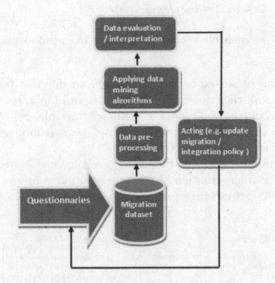

Fig. 2. Proposed workflow for prediction model

For the beginning we collected data from printed questionnaires aiming to build a migration data set. That data set is subject of data preprocessing tasks and at the end of this step data is in the proper format for a previously established data mining algorithm. It has to be noted, that for different data mining tasks data have to be prepared in various ways. At the end of applying data mining algorithms stage, prediction models are built, evaluated and interpreted by experts. If these models carry out valuable information they trigger authorities decisions which can adjusts, for example, integration policy.

3.2 Data Understanding

We used the data collected from a set of questionnaires completed by Roma persons living in Rennes. Each questionnaire contains questions about different aspects, such as: issues related to migration; aspects regarding to labor market insertion; issues related to family, health and children's education; data about French perception on Roma people; data corresponding to integration of Roma citizens in France, and personal data.

But, before transferring data from paper support to electronic form, it was mandatory to make a deep analysis of the questions and answers structures, because it might be necessary to split a single position from the questionnaire in more items in data set. An example is presented in Fig. 3.

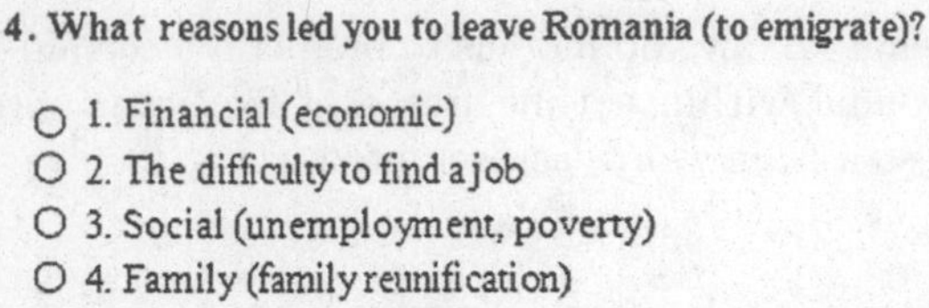

4. What reasons led you to leave Romania (to emigrate)?

◯ 1. Financial (economic)
◯ 2. The difficulty to find a job
◯ 3. Social (unemployment, poverty)
◯ 4. Family (family reunification)

Fig. 3. Example of question whose answers represent independent predictive variables

For the question presented in Fig. 3 we have four different possible answers. We could have put all these answers in a single item in migration data set, but as each of these could be considered a predictor variable, we transformed the initial data set in such a way that we have four items (characteristics) instead one, and their values are *YES* or *NO*.

There were completed 101 questionnaires, each having 72 items. Due to the existence of some similar situation as the one mentioned above, we obtained a data set having 102 features. Figure 4 shows a sample of data collected.

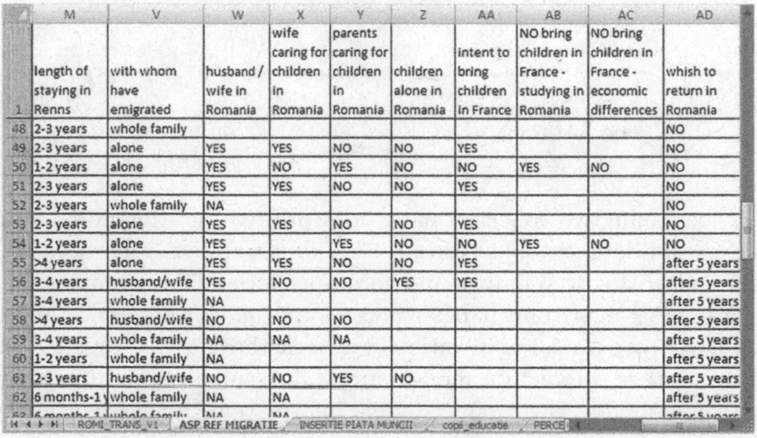

	M	V	W	X	Y	Z	AA	AB	AC	AD
	length of staying in Renns	with whom have emigrated	husband / wife in Romania	wife caring for children in Romania	parents caring for children in Romania	children alone in Romania	intent to bring children in France	NO bring children in France - studying in Romania	NO bring children in France - economic differences	whish to return in Romania
48	2-3 years	whole family								NO
49	2-3 years	alone	YES	YES	NO	NO	YES			NO
50	1-2 years	alone	YES	NO	YES	NO	NO	YES	NO	NO
51	2-3 years	alone	YES	YES	NO	NO	YES			NO
52	2-3 years	whole family	NA							NO
53	2-3 years	alone	YES	YES	NO	NO	YES			NO
54	1-2 years	alone	YES		YES	NO	NO	YES	NO	NO
55	>4 years	alone	YES	YES	NO	NO	YES			after 5 years
56	3-4 years	husband/wife	YES	NO	NO	YES	YES			after 5 years
57	3-4 years	whole family	NA							after 5 years
58	>4 years	husband/wife	NO	NO	NO					after 5 years
59	3-4 years	whole family	NA	NA	NA					after 5 years
60	1-2 years	whole family	NA							after 5 years
61	2-3 years	husband/wife	NO	NO	YES	NO				after 5 years
62	6 months-1 y	whole family	NA	NA						after 5 years

ROMI_TRANS_V1 ASP REF MIGRATIE INSERTIE PIATA MUNCII copii_educatie PERCE

Fig. 4. Sample of migration survey raw data set

The raw dataset has a poor quality. First, as it can be observed in Fig. 3 there are a lot of missing data. In order to perform a good analysis we have to fil the empty cells. Second, a part of features are expressed by numbers and others by descriptive data. Third, we found some empty columns, so those can't influence the prediction.

3.3 Data Preprocessing

Data preprocessing, is focused on two issues: first, the data must be organized into a proper form for data mining algorithms, and, secondly, the data sets used must lead to the best performance and quality for the models obtained by data mining operations [2]. As we noted in previous subsection, a first group of data preparation operations were made when we collected the data to build the raw migration data set. Further, the required tasks to improve data quality are presented in Fig. 5.

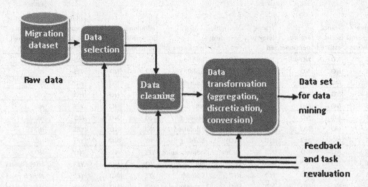

Fig. 5. Data preprocessing tasks

Data preparation stage begins with a useful data selection. As we noted above, there are empty columns and there is no way to find the proper values. These columns must be removed.

Then, after a careful analysis of the missing data in correlation with the existing one, we can design an algorithm that allows filling the gaps. In this way, together with the sociologist we completed the cells with a certain value, induced by earlier responses. For example a *NO* answer to *Do you have children?* question, enforces a *NA (Non-Aplicable)* answer to *Your children are in France?* or *Do you have children born in other country than Romania?* questions. After that, we inspected again the questionaires, and where we didn't found answers, we put *NR (Non-Response)* value.

Another issue we solved in this stage was related to data dimensionality and its impact on model building. As we noted in previous Section, the data set contains 101 cases and 102 attributes. In this circumstances, it is experimentally demonstrated that if we use the whole set of features we can't obtain good models.

This is the reason why we splitted the data set in subsets, each of them reffering one issue from those specified when we described data, i.e. issues related to migration; aspects of labor market insertion; issues related to family, health and children's education; data related to the French perception on Roma people; data corresponding to integration of Roma citizens in France, and personal data.

Finally we obtained six data sets, containing various predictor variable and a class label with two classes *NO* - for those who do not intend to return in their mother country and *after 5 years* - for those who intend to return in an interval of 5 years. These data sets allow us to build many models that intend to predict if a person will return or not in Romania.

Figure 6 presents a final data set that contains variables regarding the insertion of Roma people on the labor market in France.

H	I	J	K	L	M	N	O	P	AE
France occasional activities	France social assisted	France pensioner	France unemployed	currently unemployed	work area	additional income - social aid	additional income - pension	additional income - child allowance	whish to return in Romania
YES	NO	NO	NO	NO	building	NO	NO	NO	NO
YES	NO	NO	NO	NO	building	NO	NO	YES	NO
YES	NO	NO	NO	NO	housekeeping	YES	NO	NO	NO
YES	NO	NO	NO	NO	building	NO	NO	NO	NO
YES	NO	NO	NO	NO	NR	YES	NO	YES	NO
YES	NO	NO	NO	NO	building	NO	NO	NO	NO
YES	NO	NO	NO	YES	education	NO	NO	YES	NO
YES	NO	NO	NO	NO	building	NO	NO	YES	NO
YES	NO	NO	NO	NO	housekeeping	YES	NO	NO	NO
NO	NO	NO	NO	YES	NA	NO	NO	YES	after 5 years
YES	NO	NO	NO	NO	Comert	YES	NO	YES	after 5 years
YES	NO	NO	NO	NO	housekeeping	NO	NO	YES	after 5 years
NO	YES	NO	NO	NO	NA	YES	NO	NO	after 5 years
NO	YES	NO	NO	NO	NA	YES	NO	NO	after 5 years
NO	NO	NO	NO	YES	NA	NO	NO	NO	after 5 years
NO	YES	NO	NO	NO	NA	YES	NO	NO	after 5 years

S V1   ASP REF MIGRATIE   INSERTIE PIATA MUNCII   cooi educatie   PERCEPTI

Fig. 6. Sample data set for model building

Further we conduct our analysis on these data sets. More specifically, for each of the six data sets we have built models that allowed us to predict, depending on the predictor variables contained in those sets, the intention of returning to the home country.

3.4   Data Mining Stage

The most used data mining methods for prediction are classification and regression. We use RapidMiner [6] to implement processes that build classification models, because most of our data are descriptive.

Classification aims to extract models describing data class. These models, known as classifiers, are then used to assign cases to target classes. This is a way to predict the target class for each new, unlabelled case in the data. It is a two step process. In the first step, called learning step or training phase, a model which describes a predetermined set of data classes is built. For this stage a training data set, is used. The class label of each training case is provided, this is the reason why this method is known as supervised learning. Once built, the model is applied on other data set that is also labeled in order to determine its accuracy. If it is validated, the model will be further used to predict classes of unlabeled cases [3].

Related to migration area, classification places people, according to some predictor variables extracted from their answers, in two predefined classes: one with Roma people that do not intend to return in Romania and one with those that intend to return home.

Thus it is possible to track the size and structure of various groups and finally, with a certain accuracy it is possible to estimate a migration stock.

Below is presented the methodology we used on each data subset for classification:

Classification Methodology

```
Input: migration data subset
 Split migration data subset in training data set and test data
    set Establish training data set size
    Randomly extract examples for training data set construction
 Build classification model
 Validate classification model
    Apply model to test data set
    Estimate model performances
Output:Classification model and its performances
```

For each of the six data subsets we used for training between 40% and 60% of the existing cases. Such a classifying process is shown in Fig. 7.

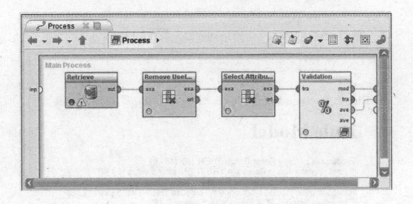

Fig. 7. Overview on RapidMiner modelling process

It is a process that achieves, simultaneously with the model construction, its validation on test data and an estimation of its performance. All these are embedded in Validation operator which actually is a subprocess, as shown in Fig. 8. In this example, we used *C4.5* algorithm for trees induction. We also built similar processes for *Random Forest* and *CHAID* algorithms. In all the cases we used *information gain* as splitting criterion for the decision tree nodes. Then, the built model was applied on test dataset, and at the end of this phase

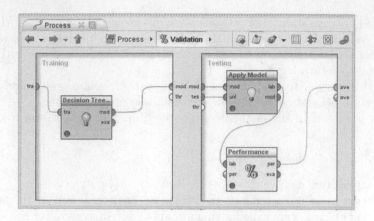

Fig. 8. RapidMiner validation subprocess

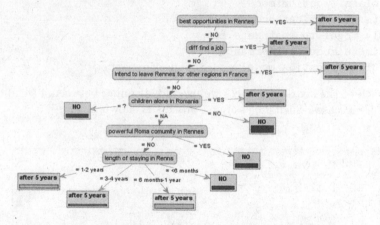

Fig. 9. Decision tree built on *"data migration"* data subset

Rule Model

IF countries before = Great Britain THEN NO (18 / 0)
IF best opportunities in Rennes = YES THEN after 5 years (0 / 12)
IF powerful Roma community in Rennes = YES THEN NO (17 / 0)
IF migrant other localities in France = Toulouse THEN after 5 years (0 / 8)
IF countries before = SUA THEN after 5 years (0 / 6)
IF length of staying in Rennes = <6 months THEN NO (10 / 0)
IF with whom have emigrated = whole family THEN after 5 years (0 / 16)
IF Arrival year > 2011.500 THEN NO (6 / 0)
IF Arrival year ≤ 2009.500 THEN NO (2 / 0)
ELSE after 5 years (0 / 1)

Fig. 10. Classification rules set built on *"data migration"* data subset

we obtained both the model and an evaluation of its performance. The decision tree, obtained as a model from the execution of this process, applied on migration issues dataset, is shown in Fig. 9.

We implemented this process considering both decision trees and classification rules as data mining methods. Such processes were build for all six data sets obtained as result of preprocessing step.

Figure 10 shows the classification rules based model built for the same data set.

4   Model Evaluation and Interpretation

Even though the central task of KDD process is data mining (known as modelling in CRISP-DM - Fig. 1), an important objective is the model evaluation.

A first observation is that, for each analyzed subset, a very small proportion of its features are used as splitting attributes for decision trees or as predictors for classification rules. Table 1 presents, for all migration data subsets, the total number of features and the number of features considered as predictors.

Table 1. Data sets characteristics.

	Dataset	Total number of features	Features used for model building
1	Migration issue	30	6
2	Labor market insertion	17	5
3	Family, health and children's education	9	2
4	French perception on Roma	8	3
5	Integration of Roma in France	31	5
6	Personal data	9	3

Quality prediction is very closely related to the built model quality. This is the reason why it is necessary to estimate some measures which allow to asses the built model quality and to validate it for future prediction.

Both for decision tree and classification rules a performance vector containing some measures such as: accuracy, absolute error, relative error, kappa, etc., must be used. In our work we used such a performance vector to establish the quality of models built on each of the six migration data subsets.

Table 2 presents achieved performances for decision trees, whereas Table 3 refers the same performances for classification rules sets. A quick look over the results presented in the two tables, indicates that, for all studied cases, classification rules based models have higher quality than those who are decision trees based.

Table 2. Decision trees performance vector

	Dataset	Accuracy	Absolute error	Relative error	Kappa
1	Migration issue	95,78	0,045	4,52	0,904
2	Labor market insertion	88,56	0,116	11,6	0,768
3	Family, health and children's education	57,11	0,491	49,07	0
4	French perception on Roma	54,78	0,496	49,6	0
5	Integration of Roma in France	55,22	0,496	49,63	0,026
6	Personal data	58,89	0,471	47,06	0,087

Table 3. Classification rules performance vector

	Dataset	Accuracy	Absolute error	Relative error	Kappa
1	Migration issue	97,94	0,038	3,81	0,959
2	Labor market insertion	91,75	0,146	14,6	0,834
3	Family, health and children's education	56,99	0,469	46,93	0
4	French perception on Roma	68,04	0,393	39,31	0,317
5	Integration of Roma in France	70,41	0,384	38,44	0,411
6	Personal data	72,16	0,353	35,33	0,442

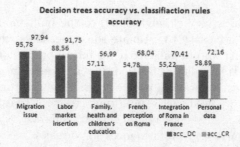

Fig. 11. Decision trees accuracy versus classification rules accuracy for migration data sets

Finally, Fig. 11 offers a comparative view about accuracy and absolute error obtained in tests both for decision trees and classification rules.

It is easy to see that models built on the data that refers to general migration issues have the highest accuracy and provide the best predictive tool.

A slightly lower quality, but high values for accuracy also show models that are based on the data set regarding labor market insertion. For all other data sets models are generated based on a very small number of predictors and their accuracy is significantly lower.

5   Conclusion and Future Works

This paper presents part of the research results regarding the possibility of building models that could allow to predict the intention of retuning to thei home country in a 5 year period of Roma migrats. We started from the data collected by sociologists, through questionnaires that aimed to detect problems faced by Roma migrants in France. We adapted this data to be suitable for the development of tools for predicting the intention of migrants to return in their home country. We considered data mining classification and we built some models both by decision tree and classification rules methods. Although we had a small volume of data, the results encourage us to say that this approach may lead to the development of tools that provide useful information to predict migration stock for Roma migrants.

Acknowledgments. The infrastructure used for this work was partially supported by the project Integrated Center for research, development and innovation in Advanced Materials, Nanotechnologies, and Distributed Systems for fabrication and control, Contract No. 671/09.04.2015, Sectorial Operational Program for Increase of the Economic Competitiveness co-funded from the European Regional Development Fund. We are grateful to Mrs. PhD. Ionela Galbau because she allowed us access to the raw data from questionnaires.

References

1. Brachman, R., Anand, T.: The process of knowledge discovery in databases: a human centered approach. In: Fayyad, U.M., Piatetsky-Shapiro, G., Smyth, P., Uthurusamy, R. (eds.) Advances in Knowledge Discovery and Data Mining, pp. 37–57. AAAI/MIT Press (1996)
2. Danubianu, M., Popa, V., Tobolcea, I.: Unsupervised information based feature selection for speech therapy optimization by data mining techniques. In: Proceedings of the Seventh International Multi-Conference on Computing in the Global Information Technology ICCGI 2012, pp. 29–39 (2012)
3. Danubianu, M., Tobolcea, I.: Using data mining approach for personalizing the therapy of dyslalia. In: Proceedings of the 3rd International Conference on E-Health and Bioengineering (EHB 2011), pp. 113–117 (2011)
4. Fayyad, U., Piatetsky-Shapiro, G., Smyth, P.: The KDD process for extracting useful knowledge from volumes of data. Commun. ACM **39**, 27–34 (1996)

5. Hipp, J., Untzer, U.G., Grimmer, U.: Integrating association rule mining algorithms with relational database systems. In: Proceedings of the 3rd International Conference on Enterprise Information Systems (ICEIS 2001), pp. 130–138 (2001)
6. RapidMiner: Rapidminer, the industry 1 open source data science platform. https://rapidminer.com/us/
7. Vasilcu, D., Sechet, R.: Vingt ans d'experience migratoire en roumanie postcommuniste. Espace Popul. Soc. **2**, 215–228 (2011)
8. Wirth, R., Hipp, J.: CRISP-DM: towards a standard process model for data mining. In: Proceedings of the 4th International Conference on the Practical Applications of Knowledge Discovery and Data Mining, pp. 29–39 (2000)

A Survey on Data Mining Methods for Clustering Complex Spatiotemporal Data

Piotr S. Maciąg[✉]

Institute of Computer Science, Warsaw University of Technology,
Nowowiejska 15/19, 00-665 Warsaw, Poland
pmaciag@ii.pw.edu.pl
http://www.ii.pw.edu.pl/ii_eng

Abstract. This publication presents a survey on the clustering algorithms proposed for spatiotemporal data. We begin our study with definitions of spatiotemporal datatypes. Next we provide a categorization of spatiotemporal datatypes with the special emphasis on the spatial representation and diversity in temporal aspect. We conduct our deliberation focusing mainly on the complex spatiotemporal objects. In particular, we review algorithms for two problems already proposed in literature: clustering complex spatiotemporal objects as polygons or geographical areas and measuring distances between complex spatial objects. In addition to description of the problems mentioned above, we also attempt to provide a comprehensive references review and provide a general look on the different problems related to the clustering spatiotemporal data.

Keywords: Data mining · Clustering spatiotemporal data · Clustering algorithms

1 Introduction

Exploration of spatiotemporal data is a key aspect in many areas of management, design and business. Rapid increase of collected spatiotemporal data is associated with an intensive development of wireless sensor networks, improving sensors design techniques and increasing transmission capacity in mobile networks. Spatiotemporal data may be related to the following areas of applications: collections of events generated by sensors deployed over certain geographical regions, information about trajectories of vehicles, animals or groups of people or evolutions of phenomenons in both spatial and temporal aspects. Analysis of changes in climate and weather is the field which may generate huge amounts of spatiotemporal data described not only by the sets of points, but also by the complex objects like polygons. The problem of discovering frequent patterns in spatiotemporal data is related to several applications tasks like: analysis of traffic in cities [20], movements prediction of celestial bodies in astronomy [6] or crime analysis [41].

In addition, standard methods used in canonical data mining problems like *apriori* based algorithms [1], efficient clustering [18], periodicity detection [39],

© Springer International Publishing AG 2017
S. Kozielski et al. (Eds.): BDAS 2017, CCIS 716, pp. 115–126, 2017.
DOI: 10.1007/978-3-319-58274-0_10

fast validation and data interpolation [38] need to be integrated into non-trivial, sophisticated approaches which can deal with data uncertainty, shifts in spatial or temporal dimensions and non-invariant scaling problems. In the publication, we attempt to provide a survey on the most recent methods developed for the spatiotemporal data clustering.

In particular, we propose a review of clustering methods for both spatiotemporal points and polygons (geographical areas). Previous reviews on that matter consider only the categorization of spatial objects [34, 48]. Additionally, we review distance measures proposed for complex spatial or geographical objects. To the best of our knowledge this is the first survey on the mentioned measures, gathering their properties and computing algorithms from multiple resources and recently proposed publications. In opposition to our paper, the survey proposed in [47] considers the frequent patterns discovery methods rather than clustering event-based spatiotemporal data. An attempt to provide a review of patterns discovery methods for trajectory-based spatiotemporal data has been proposed in [39], which do not consider at all clustering methods for complex geographical objects and focus mainly on moving objects.

The layout of the paper is as follows: in Sect. 2 we review a categorization of spatiotemporal datatypes in the view of their adaptations to clustering algorithms. Section 3 summarizes results in the area of clustering complex spatiotemporal objects as polygons and areas. Section 4 recalls the most important distance measures for complex spatial objects. Section 5 provides a survey on the recently proposed clustering algorithms for moving objects and trajectories. Conclusions to the survey are given in Sect. 6.

2 Spatiotemporal Datatypes

Spatiotemporal datatypes are dependent on the real-world applications. Based on literature [34, 39, 48], we can distinguish two types of spatiotemporal data: event-based (also known as location-based), collected from stationary deployed sensors and trajectory-based (also referred as ID-based [39]) used to describe movements of objects. For the event-based data case, each event may be associated with a property p which value is denoted by the function $f(x, y, t, p)$ where (x, y) is a location (usually expressed in terms of longitude and latitude), t is a time stamp during which the event has been collected. Considering the more complex spatiotemporal objects (as polygons or areas), the location of an object may be denoted by the set of its coordinates. In the case of trajectory-based spatiotemporal data, for a given set of n objects $o_1, o_2, \ldots, o_n$, a trajectory of an object o_i is represented by a sequence of geographical points $(x_1, y_1, t_1), (x_2, y_2, t_2), \ldots, (x_n, y_n, t_n)$, where (x_j, y_j) is a location at the time stamp t_j. The above distinction between event-based and trajectory-based spatiotemporal data has been mainly introduced in [39].

The categorization presented above can be in addition broaden with the specification of different spatial datatypes (points, lines and polygons) and their extensions to time domain: database may contain only the last snapshot of actual positions of observed objects or the whole history of an evolution of a spatiotemporal phenomenon. In some other cases only a stream of spatiotemporal data may be available.

The categorization of spatiotemporal datatypes is given in [34]. The authors of [34] distinguish the following categories: Fixed location denoting datasets containing occurrences of events of predefined types on geographical areas. Algorithms for clustering (and partially patterns discovery) for that datatype have been proposed in [9,23,41]. If the spatial dimension is extended to polygons or geographical areas, then one may refer to the clustering algorithms presented in [30,50]. On the other hand, category defined as Dynamic location refers to trajectory-based spatiotemporal data. Datatypes denoted by Updated snapshot and either by Dynamic location or Fixed location labels refer to spatiotemporal data streams. The former describes moving objects reporting only the current location or position. The latter denotes streams of events occurrences (with each event described as above). Algorithms for spatiotemporal data streams clustering and classification have been recently proposed in [7,31,32].

The forms denoted by Time Series (according to categorization presented in [34]) are particularly useful because provide means for adaptation of classical algorithms and similarity measures used in time series analysis. For example, [14] presents a new similarity measure (*Edit Distance on Real sequence, EDR*) developed for the comparison of trajectories of objects. On the other hand, the time series representations of event-based spatiotemporal data are still unknown and will be developed in the future years.

In Table 1 we provide references contributing to the spatiotemporal datatypes definitions and data mining techniques proposed for them.

First propositions of spatial and spatiotemporal clustering using statistical approach have been given in [24,36]. [36] gives a clustering method using spatial scan statistics (the approach has been improved in [27]), whereas [24] proposes an extension taking into account spatial shifts in the nature of evolving phenomenon. Due to this, proposed algorithm is able to detect clusters of spatiotemporal data which dynamically change their position and shape.

Clustering complex spatial and spatiotemporal objects is gaining attention of researchers nowadays [29,30,50]. The idea is to discover neighboring areas or geographical regions characterized by the same (or similar) value of non-spatiotemporal attribute (f.e. pollution). [26] adapts *Fuzzy C-Means* algorithm to spatiotemporal data. [25] raise the problem of anomaly detection in spatial time series using spatiotemporal clustering.

3   Clustering Spatiotemporal Events and Complex Geographical Objects

Proposed algorithms often operate on the more complex spatial objects f.e. polygons or lines. Classical density-based clustering algorithm - DBSCAN has been proposed in [18]. Many variations of the well-known density clustering algorithms like DBSCAN, OPTICS, NN were adapted to operate on spatiotemporal data.

Table 1. Summary of the publications on spatiotemporal data types and data mining techniques proposed for them.

No	First author	Spatiotemporal datatype	Data mining method
1	Erwig [17]	Datatypes definitions	
2	Wang [50]	Polygons (geographical areas)	Clustering
3	Birant [9]	Points (events)	Clustering
4	Joshi [30]	Polygons (geographical areas)	Clustering
5	Wang [49]	Points (events)	Clustering
6	Estvill-Castro [19]	Points (events)	Clustering
7	Wang [51]	Polygons	Clustering
8	Zhang [53]	Polygons (geographical areas)	Clustering
9	Damiani [15]	Points\trajectories	Clustering
10	Izakian [26,27]	Points (time series)	Clustering
11	Izakian [25]	Points (time series)	Clustering\anomaly detection
12	Kulldorff [36]	Discrete events	Clustering\statistical approach
13	Iyengar [24]	Discrete events	Clustering
14	Schubert [46]	Discrete events	Anomaly detection
15	Mohan [41]	Discrete events	Patterns analysis
16	Shekhar [47]	Discrete events	Patterns analysis\anomaly detection
17	Nanni [43]	Trajectories	Clustering
18	Li [39]	Trajectories\discrete events	Clustering\patterns analysis
19	Palma [45]	Trajectories	Clustering
20	Gudmundsson [21]	Moving objects	Clustering
21	Jeung [28]	Moving objects	Clustering
22	Li [40]	Moving objects	Clustering

In addition, some non-standard grouping algorithms have been proposed: f.e. Spatio-Temporal Polygonal Clustering (STPC). Clustering algorithms are categorized into five main domains: *Partitioning, Hierarchical, Grid-based, Model-based* and *Density-based* [22].

In *Partitioning* methods, clusters are computed according to the mean value in a cluster (K-means) or based on the selection of an object which is nearest to the cluster's center (K-Medoid). The name of the category: *Partitioning methods* is inspired by the fact, that each object in a dataset is assigned to one and only one cluster (there are no objects classified as a noise). Objects partitioning is performed according to the predefined optimization criterion - for a given number of clusters we would like to find assignments which minimizes the sum of distances between objects and centers of clusters (their mean values or the most central elements) to which they belong. *Hierarchical* clustering methods can be divided into two approaches: *ascending* and *descending*. In the ascending approach, each data object is initially assigned to its own cluster. Then, the algorithm gradually merges clusters until a predefined number of groups will be reached. On the other hand, descending approach divides one cluster (into which all data objects are initially assigned) until a predefined number of clusters will be reached.

Grid-based methods proceed with dividing data space into cells (a grid). Then, in the clustering phase, some cells are merged based on a predefined condition. For example: two dense, neighboring cells are merged into one (a cell is dense if it contains a number of objects greater than the predefined threshold). STING is an example of grid-based clustering algorithm [50]. *Model-based* methods try to fit clusters according to the predefined model of the data (like probability distribution). An example of a model-based method is the Expectation-Maximization algorithm [42]. *Denisity-based* methods try to find clusters according to distributions of density in a dataset. Due to this, appropriate density threshold should be specified by user: f.e. an estimated number of objects in a predefined neighborhood of an object [18]. Due to their simplicity, density-based clustering algorithms are widely used in data mining. Attempts to improve their efficiency and reduce needs of expert knowledge during parameter specification have been made in [4,19].

3.1   Algorithms for Clustering Complex Spatiotemporal Objects

In this section, we proceed with description of clustering algorithms for complex spatiotemporal objects.

ST-GRID (SpatioTemporal GRID) is a clustering method based on the partitioning spatiotemporal space into two separate grids: for spatial and temporal dimensions. In [49], the authors propose to compute the precision of a grid, based on the so-called *k-dist* graph which is constructed by random sampling of a dataset, calculating distance from each sample to its k-nearest neighbor and sorting calculated distances in decreasing order. The presence of clusters will be indicated by the easily noticeable threshold in the sorted distances. Calculated thresholds may be used as grid resolutions. As in the typical grid clustering algorithms, dense neighboring cells are merged to create spatiotemporal clusters. The above procedure has been originally developed only for spatiotemporal points.

ST-DBSCAN (SpatioTemporal-DBSCAN) is the algorithm developed on the conceptions of the well-known density clustering algorithm - DBSCAN (Density Based Clustering Algorithm with Noise) [18]. ST-DBSCAN has been introduced in [49] and then rearranged in [9]. ST-DBSCAN modifies DBSCAN to detect clusters according to their non-spatial, spatial and temporal dimensions. Before we describe ST-DBSCAN algorithm, we took a quick glance on the pure DBSCAN algorithm. One of the most important properties of DBSCAN is the ability to detect clusters with an arbitrary shape: circular, ellipsoidal, linear or even more complicated. However, need to specify density thresholds may result that the algorithm will do not detect proper but sparse clusters. That problem has been addressed in the another density based clustering algorithm - OPTICS [4]. ST-DBSCAN considers cluster densities according to both spatial and temporal thresholds (assuming that for many applications they are very different). Additionally, ST-DBSCAN is able to include or exclude an object from the cluster on the basis of its non-spatiotemporal attributes: f.e. if the represented temperature is very different from the cluster's average temperature.

ST-SNN and ST-SEP-SNN Algorithms are two variations of the well known Shared Nearest Neighbor (SNN) density-based clustering algorithm. It is a noteworthy fact, that both algorithms *ST-SNN and ST-SEP-SNN* have been originally presented (in [50]) for clustering sets of polygons rather than points. Similarity between two objects according to the SNN algorithm is defied as a number of nearest neighbors shared by these two objects.

- A list of spatiotemporal neighbors of any polygon p is denoted by $NN(p) = k\text{-}SPN\text{-}List(p) \cap k\text{-}TN\text{-}List(p)$ where $k\text{-}SPN\text{-}List(p)$ and $k\text{-}TN\text{-}List(p)$ are lists of k neighbors of a polygon p in respectively spatial and temporal dimensions.
- Similarity between a pair of polygons p and q is the number of nearest spatiotemporal neighbors that they share: $similarity(p,q) = |NN(p) \cap NN(q)|$.
- The density of a polygon p is defined as the number of polygons that share at least Eps neighbors with p - $density(p) = |\{q \in \mathcal{D} | similarity(p,q) \geq Eps\}|$.
- A core polygon is a polygon where $CoreP(\mathcal{D}) = \{p \in \mathcal{D} | denisty(p) \geq MinPs\}$ where $MinPs$ is a user specified threshold.

The above conceptions determine clustering spatiotemporal polygons according to the *ST-SEP-SNN* algorithm. After marking each polygon either as a core or non-core, the algorithm proceed with clusters creation by processing each polygon in the dataset. During processing step, if an unprocessed core polygon p has been encountered, a new cluster is created and all polygons in the $NN(p)$ list are assigned to the new cluster (the same is recursively applied to the unprocessed core polygons encountered in the $NN(p)$ list).

ST-SNN is an algorithm that proceeds similarly to the *ST-SEP-SNN* algorithm presented above, with exception that the list of nearest neighbors $NN(p)$ of a polygon p is created using slightly different method. Rather than separately compute and then intersect lists of k-nearest spatial and temporal neighbors, ST-SNN combines spatial and temporal dimensions into one measure and computes only one list of the k-nearest neighbors.

STPC [30] is another denisty-based clustering algorithm developed for spatiotemporal polygons or areas. Again, the algorithm has been developed on the basis of the conceptions of the DBSCAN algorithm. Referring to the above mentioned ST-SEP-SNN algorithm, STPC computes lists of spatial and temporal neighbors on the basis of predefined distances (rather than k-nearest neighbors). The union of both lists contain spatiotemporal neighborhood of a polygon. If the neighborhood is appropriately dense, then the polygon is marked as a core polygon and the algorithm proceeds similarly to the above.

4   Distance Measures for Complex Spatiotemporal Objects

It is noteworthy to recall spatial distance measures used for polygons or other complex geographical objects. Table 2 provides a comparison of developed distance measures for complex spatiotemporal objects: polygons and trajectories. In the case of polygons, m and n denote the numbers of their vertices and in the case of trajectories their constituting sequences of locations.

Table 2. A comparison of distance measures for complex spatiotemporal data types.

No.	Distance	Spatiotemporal data type	Computational complexity
1	Hausdorff distance	Polygons and trajectories	$\mathcal{O}(m + n)$ [5] - only for polygons $\mathcal{O}(m * n)$ [2] - both for polygons and trajectories
2	Fréchet distance	Polygons and trajectories	$\mathcal{O}((m + n)log(m + n))$ [10] - for convex polygons[a] $\mathcal{O}((m * n)log(m * n))$ [3] - for trajectories
3	Exact separation distance	Polygons	$\mathcal{O}(log(m) + log(n))$ [44]
4	Minimum vertices approximation	Polygons	$\mathcal{O}(m * n)$ [44]
5	Centroid distance	Polygons	$\mathcal{O}(m + n)$ [44]
6	Simplified hausdorff distance	Polygons	$\mathcal{O}(m + n)$ [29]
7	Discrete Fréchet distance	Trajectories	$\mathcal{O}(m * n)$ [16]
8	Edit distance on real sequence	Trajectories	$\mathcal{O}(m * n)$ [14]
9	Dynamic time warping	Trajectories	$\mathcal{O}(m * n)$ [52]
10	Edit distance with real penalty	Trajectories	$\mathcal{O}(m * n)$ [13]

[a]For non-convex but simple polygons one may refer also to [10] where the algorithm with a non-trivial complexity is given.

Figure 1 presents a comparison between `Minimum Vertices Approximation`, `Exact Separation Distance` and `Centroid Distance`. Also, in the figure we show the Hausdorff distance for two polygons. The simplified Hausdorff distance is computed using the same formula as shown in Fig. 1, but only between vertices of polygons. Formula 1 presents a method for computing the Hausdorff distance for two polygons.

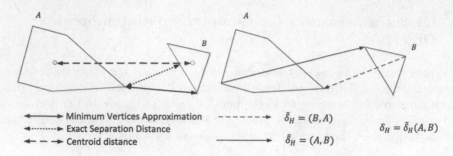

Fig. 1. Examples of distance measures for spatial polygons.

$$\delta_H = max\big(\tilde{\delta}_H(A,B), \tilde{\delta}_H(B,A)\big)$$
$$\tilde{\delta}_H(A,B) = \max_{a \in A} \min_{b \in B} \| (a,b) \| .$$
(1)

A few distance measures presented in Table 2 may be used both for polygons (geographical areas) and trajectories. If particular distance measure preserves triangle inequality property, then it is possible to reduce the time of computations performed during clustering: f.e. neighborhood search [35]. Also, the methods combining the above measures with spatial and metrical indexes have been proposed in [11, 33].

5 Other Clustering Problems in the Area of Spatiotemporal Data

In this section, we provide a look on the other clustering problems related to spatiotemporal data. In particular, we attempt to provide a general overview of the most important methods proposed for clustering trajectory-based data.

Finding groups of similar moving objects - let assume that for a set of objects $o_1, o_2, \ldots, o_n$ a database stores the trajectory of a movement of each object. Additionally, let assume that each trajectory is represented in the form of a sequence of points, each associated with a timestamp. The problem of discovering *flocks* in the dataset is described as the problem of finding those sets of objects which for a predefined time interval stay within a disk which radius length is a parameter specified by an expert. A time interval is expressed as the sequence of consecutive timestamps. The problem of finding *flocks* of objects have been introduced in [21] and also developed in [8]. The above problem has been extended to finding *convoys* [28] and *swarms* of moving objects [39]. A convoy is created from a flock by relaxing containment within a disk constraint i.e. rather than looking for the fixed disks of objects, the algorithm searches for dense regions using a clustering algorithm.

Clustering trajectories - the problem has been well studied in literature. Among most popular algorithms for clustering spatiotemporal trajectories and their similar segments are: Trajectory-OPTICS [43], TRACLUS [37] or DENTRAC [12]. The important property of these algorithms is the ability to cluster segments of trajectories rather than whole trajectories. The property is motivated by the fact that objects usually move together only in small segments of their trajectories. For example: TRACLUS proceeds with three phases: in the first, a trajectory represented in the form of a sequence of points is simplified. The number of points in the sequence is reduced and the resulting parts of each trajectory are replaced with line segments. The replacement should preserve trends and angles representing turns in movements. Then, in the second phase, clustering of similar line segments is performed. In the last step, for each discovered group of similar line segments a representative trajectory is computed. Due to the complex nature of trajectories, an appropriate similarity metric should be selected. The proposed distance measure between two trajectories contains the following components: perpendicular, parallel and angle. The components are computed as follows: the perpendicular components is computed as a Lehmer mean of the distances between ending points of one segment projected into another. The parallel component is computed as a minimum from the distances between endings of one line segment projected into another. The angle component of the distance measure is defined as a product of the length of one of line segments and sinus of an angle between line segments.

6   Conclusions

In the publication, we provide a descriptive review of recently proposed algorithms for clustering complex spatiotemporal objects. In particular, we conduct a survey on the algorithms for clustering complex spatial objects: polygons or dynamically changing areas. Among the reviewed algorithms are ST-GRID, ST-DBSCAN, ST-SNN and STPC. Additionally, we provide references and a brief summary of the distance measures proposed for complex spatial objects (i.e. the Hausdorff distance, simplified Hausdorff distance and the other recently proposed heuristics). We also attempt to provide a look on the other methods proposed for clustering spatiotemporal objects i.e. trajectory-based data. The categorization of spatiotemporal datatypes presented at the beginning of the paper provides a staring point for considering new research fields in the area of spatiotemporal data mining. In particular, the most promising directions are: developing algorithms for spatiotemporal data streams and adaptation of knowledge discovery methods proposed in time series analysis to spatiotemporal data.

References

1. Agrawal, R., Srikant, R.: Fast algorithms for mining association rules in large databases. In: Proceedings of the 20th International Conference on Very Large Data Bases, VLDB 1994, pp. 487–499. Morgan Kaufmann Publishers Inc., San Francisco (1994)
2. Alt, H., Behrends, B., Blömer, J.: Approximate matching of polygonal shapes. Ann. Math. Artif. Intell. 13(3), 251–265 (1995)
3. Alt, H., Godau, M.: Computing the frÉchet distance between two polygonal curves. Int. J. Comput. Geom. Appl. 05(01n02), 75–91 (1995)

4. Ankerst, M., Breunig, M.M., Kriegel, H.P., Sander, J.: OPTICS: ordering points to identify the clustering structure. In: Proceedings of the 1999 ACM SIGMOD International Conference on Management of Data, SIGMOD 1999, pp. 49–60. ACM, New York (1999)
5. Atallah, M.J.: A linear time algorithm for the hausdorff distance between convex polygons. Inf. Process. Lett. **17**(4), 207–209 (1983)
6. Aydin, B., Angryk, R.: Spatiotemporal frequent pattern mining on solar data: current algorithms and future directions. In: 2015 IEEE International Conference on Data Mining Workshop (ICDMW), pp. 575–581, November 2015
7. Bazan, J.G.: Hierarchical classifiers for complex spatio-temporal concepts. In: Peters, J.F., Skowron, A., Rybiński, H. (eds.) Transactions on Rough Sets IX. LNCS, vol. 5390, pp. 474–750. Springer, Heidelberg (2008). doi:10.1007/978-3-540-89876-4_26
8. Benkert, M., Gudmundsson, J., Hübner, F., Wolle, T.: Reporting flock patterns. Comput. Geom. **41**(3), 111–125 (2008)
9. Birant, D., Kut, A.: ST-DBSCAN: an algorithm for clustering spatial-temporal data. Data Knowl. Eng. **60**(1), 208–221 (2007). Intelligent Data Mining
10. Buchin, K., Buchin, M., Wenk, C.: Computing the fréchet distance between simple polygons. Comput. Geom. **41**(1–2), 2–20 (2008). special Issue on the 22nd European Workshop on Computational Geometry (EuroCG)22nd European Workshop on Computational Geometry
11. Chan, K.P., Fu, A.W.C.: Efficient time series matching by wavelets. In: Proceedings 15th International Conference on Data Engineering (Cat. No. 99CB36337), pp. 126–133, March 1999
12. Chen, C.-S., Eick, C.F., Rizk, N.J.: Mining spatial trajectories using nonparametric density functions. In: Perner, P. (ed.) MLDM 2011. LNCS (LNAI), vol. 6871, pp. 496–510. Springer, Heidelberg (2011). doi:10.1007/978-3-642-23199-5_37
13. Chen, L., Ng, R.: On the marriage of Lp-norms and edit distance. In: Proceedings of the Thirtieth International Conference on Very Large Data Bases, VLDB 2004, vol. 30, pp. 792–803. VLDB Endowment (2004)
14. Chen, L., Özsu, M.T., Oria, V.: Robust and fast similarity search for moving object trajectories. In: Proceedings of the 2005 ACM SIGMOD International Conference on Management of Data, SIGMOD 2005, pp. 491–502. ACM, New York (2005)
15. Damiani, M.L., Issa, H., Fotino, G., Heurich, M., Cagnacci, F.: Introducing 'presence' and 'stationarity index' to study partial migration patterns: an application of a spatio-temporal clustering technique. Int. J. Geogr. Inf. Sci. **30**(5), 907–928 (2016)
16. Eiter, T., Mannila, H.: Computing discrete fréchet distance. Technical report, Vienna University of Technology (1994)
17. Erwig, M., Güting, R.H., Schneider, M., Vazirgiannis, M.: Spatio-temporal data types: an approach to modeling and querying moving objects in databases. GeoInformatica **3**(3), 269–296 (1999)
18. Ester, M., Kriegel, H.P., Sander, J., Xu, X.: A density-based algorithm for discovering clusters in large spatial databases with noise. In: Second International Conference on Knowledge Discovery and Data Mining, pp. 226–231. AAAI Press (1996)
19. Estivill-Castro, V., Lee, I.: Autoclust: automatic clustering via boundary extraction for mining massive point-data sets. In: Proceedings of the 5th International Conference on Geocomputation, pp. 23–25 (2000)
20. Gora, P., Rüb, I.: Traffic models for self-driving connected cars. Transp. Res. Procedia **14**, 2207–2216 (2016). Transport Research Arena (TRA 2016)

21. Gudmundsson, J., van Kreveld, M.: Computing longest duration flocks in trajectory data. In: Proceedings of the 14th Annual ACM International Symposium on Advances in Geographic Information Systems, GIS 2006, pp. 35–42. ACM, New York (2006)

22. Han, J.: Data Mining: Concepts and Techniques. Morgan Kaufmann Publishers Inc., San Francisco (2005)

23. Huang, Y., Zhang, L., Zhang, P.: A framework for mining sequential patterns from spatio-temporal event data sets. IEEE Trans. Knowl. Data Eng. 20(4), 433–448 (2008)

24. Iyengar, V.S.: On detecting space-time clusters. In: Proceedings of the Tenth ACM SIGKDD International Conference on Knowledge Discovery and Data Mining, pp. 587–592. ACM (2004)

25. Izakian, H., Pedrycz, W.: Anomaly detection and characterization in spatial time series data: a cluster-centric approach. IEEE Trans. Fuzzy Syst. 22(6), 1612–1624 (2014)

26. Izakian, H., Pedrycz, W., Jamal, I.: Clustering spatiotemporal data: an augmented fuzzy c-means. IEEE Trans. Fuzzy Syst. 21(5), 855–868 (2013)

27. Izakian, H., Pedrycz, W.: A new PSO-optimized geometry of spatial and spatio-temporal scan statistics for disease outbreak detection. Swarm Evol. Comput. 4, 1–11 (2012)

28. Jeung, H., Yiu, M.L., Zhou, X., Jensen, C.S., Shen, H.T.: Discovery of convoys in trajectory databases. Proc. VLDB Endow. 1(1), 1068–1080 (2008)

29. Joshi, D., Samal, A., Soh, L.K.: A dissimilarity function for clustering geospatial polygons. In: Proceedings of the 17th ACM SIGSPATIAL International Conference on Advances in Geographic Information Systems, GIS 2009, pp. 384–387. ACM, New York (2009)

30. Joshi, D., Samal, A., Soh, L.K.: Spatio-temporal polygonal clustering with space and time as first-class citizens. Geoinformatica 17(2), 387–412 (2013)

31. Kasabov, N., Capecci, E.: Spiking neural network methodology for modelling, classification and understanding of EEG spatio-temporal data measuring cognitive processes. Inf. Sci. 294, 565–575 (2015). Innovative Applications of Artificial Neural Networks in Engineering

32. Kasabov, N., Scott, N.M., Tu, E., Marks, S., Sengupta, N., Capecci, E., Othman, M., Doborjeh, M.G., Murli, N., Hartono, R., Espinosa-Ramos, J.I., Zhou, L., Alvi, F.B., Wang, G., Taylor, D., Feigin, V., Gulyaev, S., Mahmoud, M., Hou, Z.G., Yang, J.: Evolving spatio-temporal data machines based on the neucube neuromorphic framework: design methodology and selected applications. Neural Netw. 78, 1–14 (2016). special Issue on "Neural Network Learning in Big Data"

33. Keogh, E., Ratanamahatana, C.A.: Exact indexing of dynamic time warping. Knowl. Inf. Syst. 7(3), 358–386 (2005)

34. Kisilevich, S., Mansmann, F., Nanni, M., Rinzivillo, S.: Spatio-temporal clustering. In: Maimon, O., Rokach, L. (eds.) Data Mining and Knowledge Discovery Handbook, pp. 855–874. Springer, Boston (2010)

35. Kryszkiewicz, M., Lasek, P.: TI-DBSCAN: clustering with DBSCAN by means of the triangle inequality. In: Szczuka, M., Kryszkiewicz, M., Ramanna, S., Jensen, R., Hu, Q. (eds.) RSCTC 2010. LNCS (LNAI), vol. 6086, pp. 60–69. Springer, Heidelberg (2010). doi:10.1007/978-3-642-13529-3_8

36. Kulldorff, M.: A spatial scan statistic. Commun. Stat. Theory Methods 26(6), 1481–1496 (1997)

37. Lee, J.G., Han, J., Whang, K.Y.: Trajectory clustering: a partition-and-group framework. In: Proceedings of the 2007 ACM SIGMOD International Conference on Management of Data, SIGMOD 2007, pp. 593–604. ACM, New York (2007)
38. Li, L., Revesz, P.: A comparison of spatio-temporal interpolation methods. In: Egenhofer, M.J., Mark, D.M. (eds.) GIScience 2002. LNCS, vol. 2478, pp. 145–160. Springer, Heidelberg (2002). doi:10.1007/3-540-45799-2_11
39. Li, Z.: Spatiotemporal pattern mining: algorithms and applications. In: Aggarwal, C.C., Han, J. (eds.) Frequent Pattern Mining, pp. 283–306. Springer, Cham (2014). doi:10.1007/978-3-319-07821-2_12
40. Li, Z., Ding, B., Han, J., Kays, R.: Swarm: mining relaxed temporal moving object clusters. Proc. VLDB Endow. 3(1–2), 723–734 (2010)
41. Mohan, P., Shekhar, S., Shine, J.A., Rogers, J.P.: Cascading spatio-temporal pattern discovery. IEEE Trans. Knowl. Data Eng. 24(11), 1977–1992 (2012)
42. Moon, T.K.: The expectation-maximization algorithm. IEEE Sig. Process. Mag. 13(6), 47–60 (1996)
43. Nanni, M., Pedreschi, D.: Time-focused clustering of trajectories of moving objects. J. Intell. Inf. Syst. 27(3), 267–289 (2006)
44. Ng, R.T., Han, J.: CLARANS: a method for clustering objects for spatial data mining. IEEE Trans. Knowl. Data Eng. 14(5), 1003–1016 (2002)
45. Palma, A.T., Bogorny, V., Kuijpers, B., Alvares, L.O.: A clustering-based approach for discovering interesting places in trajectories. In: Proceedings of the 2008 ACM Symposium on Applied Computing, SAC 2008, pp. 863–868. ACM, New York (2008)
46. Schubert, E., Zimek, A., Kriegel, H.P.: Local outlier detection reconsidered: a generalized view on locality with applications to spatial, video, and network outlier detection. Data Min. Knowl. Disc. 28(1), 190–237 (2014)
47. Shekhar, S., Evans, M.R., Kang, J.M., Mohan, P.: Identifying patterns in spatial information: a survey of methods. Wiley Interdisc. Rev.: Data Mining Knowl. Discov. 1(3), 193–214 (2011)
48. Tork, H.F.: Spatio-temporal clustering methods classification. In: Doctoral Symposium on Informatics Engineering, vol. 1, no. 1, pp. 199–209. FEUP (2012)
49. Wang, M., Wang, A., Li, A.: Mining spatial-temporal clusters from geo-databases. In: Li, X., Zaïane, O.R., Li, Z. (eds.) ADMA 2006. LNCS (LNAI), vol. 4093, pp. 263–270. Springer, Heidelberg (2006). doi:10.1007/11811305_29
50. Wang, S., Cai, T., Eick, C.F.: New spatiotemporal clustering algorithms and their applications to ozone pollution. In: Proceedings of the 2013 IEEE 13th International Conference on Data Mining Workshops, ICDMW 2013, pp. 1061–1068. IEEE Computer Society, Washington, DC (2013)
51. Wang, W., Du, S., Guo, Z., Luo, L.: Polygonal clustering analysis using multilevel graph-partition. Trans. GIS 19(5), 716–736 (2015)
52. Yi, B.K., Jagadish, H.V., Faloutsos, C.: Efficient retrieval of similar time sequences under time warping. In: Proceedings of the Fourteenth International Conference on Data Engineering, ICDE 1998, pp. 201–208. IEEE Computer Society, Washington, DC (1998)
53. Zhang, Y., Eick, C.F.: Novel clustering and analysis techniques for mining spatiotemporal data. In: Proceedings of the 1st ACM SIGSPATIAL PhD Workshop, SIGSPATIAL PhD 2014, pp. 2:1–2:5. ACM, New York (2014)

Architectures, Structures and
Algorithms for Efficient Data Processing

Multi-partition Distributed Transactions over Cassandra-Like Database with Tunable Contention Control

Marek Lewandowski[✉] and Jacek Lewandowski[✉]

Institute of Computer Science, Warsaw University of Technology,
Nowowiejska 15/19, 00-665 Warsaw, Poland
marek.m.lewandowski@gmail.com,lewandowski.jacek@gmail.com
http://www.ii.pw.edu.pl

Abstract. The amounts of data being processed today are enormous and they require specialized systems to store them, access them and do computations. Therefore, a number of NoSql databases and big data platforms were built to address this problem. They usually lack transaction support which features atomicity, consistency, isolation, durability and at the same time they are distributed, scalable, and fault tolerant. In this paper, we present a novel transaction processing framework based on Cassandra storage model. It uses Paxos protocol to provide atomicity and consistency of transactions and Cassandra specific read and write paths improvements to provide read committed isolation level and durability. Unlike built-in Light Weight Transactions (LWT) support in Cassandra, our algorithm can span multiple data partitions and provides tunable contention control. We verified correctness and efficiency both theoretically and by executing tests over different workloads. The results presented in this paper prove the usability and robustness of the designed system.

Keywords: Big Data · Transactions · Cassandra · Paxos · Consistency · NoSQL

1 Introduction

Big Data is the term used for describing sets of increasing volumes of data, which do not fit in a single machine. There are many sources of Big Data, such as system logs, user website clicks, financial transactions, weather measurements, data from Internet of Things, and many others.

NoSql databases were created to support storing Big Data and provide the means to analyze it. Databases differ in ways they represent the data, but the key principle remains the same: store the data for future analysis. NoSql databases span over hundreds, even thousands servers located in various physical locations and are designed to overcome individual node failures and network partitions. They provide guarantees for certain behaviour in face of such problems which usually means they eventually achieve the consistency of data at some point.

© Springer International Publishing AG 2017
S. Kozielski et al. (Eds.): BDAS 2017, CCIS 716, pp. 129–140, 2017.
DOI: 10.1007/978-3-319-58274-0_11

To this end, the ensemble of data is cut into groups of records which are called partitions. Partitions are data replication and sharding units, which means that they are distributed across the cluster in one or more copies so that the load can be balanced among different server nodes, and the availability of a distinct chunk of data is increased.

Properties of relational databases, such as ACID compliant transactions are usually sacrificed in NoSQL-like solutions in favour of high availability, fault tolerance and scalability [17]. Though, there exist some solutions which can help to overcome such deficiencies. Unfortunately, they come with either functional or performance limitations which make them difficult to adopt. The functional limitations include constraining a transaction to a single partition of data or the lack of certain isolation levels. The performance problems are usually related to the use of pessimistic locks to achieve a certain level of consistency.

We overcame those problems by modifying Cassandra database [10]. Cassandra is proven to be performant and provide some basic transaction support known as LWT. Although the usability of LWT is limited, the model of storing data makes it a perfect platform for building transaction support extension which features good performance and rich functionality.

2   Preliminaries

Database Ω consists of key-value pairs $\{(k_i, v_j), (k_k, v_m), ...\}, k_i \in K, v_i \in V$ stored by cluster of nodes $\{n_1, n_2, ...\}, n_i \in N$. Changes to data are represented by *mutations* $\delta(k, v)$. Each mutation is stored in nodes identified by topology τ (Definition 1). Transaction is a set of mutations $\Delta(\delta_1, \delta_2, ...)$, where mutation $\delta(k, v)$ is an operation on key k with value v.

Definition 1. Topology *is a partitioning function* $\tau : K \mapsto N^{\mathrm{RF}}$ *which for a given k returns a set of replica nodes of size* RF.

Client $c_i \in C$ is an actor, which performs transactions $\Delta_1, \Delta_2, ...$ and communicates to a cluster N using messages M. Client c begins a transaction Δ by sending a message $M(c, n, begin\_tx())$ to any node $n \in N$ followed by receiving a message $M(n, c, initial\_tx\_state(\Lambda_0))$, which includes initial *transaction state* (Definition 3), which will reflect all subsequent changes done in Δ.

During the transaction Δ, client c sends messages of two types: mutations and selects $M(c, n, select(k))$. All mutations within a transaction Δ are isolated from other transactions - they are private to the enclosing transaction until it is committed. For the duration of the transaction, mutations are stored in a *private transaction storage*, which is a local data structure at each n_i. Select operations query against current state of the database, thus transactions provide read-committed isolation level. This means that transactions always get most recent committed state of data, thus a subsequent query for the same key can yield different value if there was a successfully committed transaction in between select queries.

After each mutation client c receives a message $M(c, n, update\_tx\_state(\lambda))$, which includes a transaction item λ referencing that mutation. A client is responsible for tracking changes performed in a transaction, thus it has to append all the received transaction items to the initial transaction state Λ_0.

Transactions are decoupled from physical network connection and so is transaction state which means that if a client loses connection to a cluster due to some kind of network failure then transaction can be completed after the client connects back to the cluster.

Definition 2. Transaction item $\lambda(keyspace, table, token)$ *is a reference to a single mutation $\delta(k, v)$ of transaction Δ, which indicates that replica nodes of k store $\delta(k, v)$ in private transaction storage. It is a triple with (a) keyspace – name of a keyspace, (b) table – name of a table, (c) token – value of a token, which is a value computed by hash function from Cassandra key k_p, which is used by topology τ to determine replicas $N^{\mathrm{RF}}{}_k$. Cassandra key $k(k_p, k_c)$ consists of two keys: (a) k_p is a partitioning key, which is used by topology τ to determine N_k^{RF}, (b) k_c is a clustering key, which determines column in wide rows.*

Definition 3. Transaction state $\Lambda(id, \{\lambda_i, \lambda_j, ...\})$ *reflects changes done in transaction Δ. Transaction state allows to identify nodes $N' \subset N$, which are affected by Δ and store its $\{\delta_1, \delta_2, ...\}$. It is a pair of unique id of type timeuuid, which is UUID with encoded timestamp [3] and a set of transaction items (Definition 2).*

To perform rollback client sends $M(c, n, tx\_rollback(\Lambda))$, and then n sends out rollback messages to other nodes $M(n_i, n_j, rollback(\Lambda))$, where nodes $\{n_j, ..., n_k\}$ are identified by $\{\lambda_1, \lambda_2, ...\}$, which makes them remove private storage for that transaction and mark Δ as rolled back in an another local data structure called *transaction log*.

In order to commit transaction Δ a client c sends message $M(c, n, tx\_commit(\Lambda))$ and subsequently n_i becomes the proposer of the transaction and is responsible for orchestrating distributed consensus round among nodes affected by transaction Δ. When consensus is reached each node moves private data of transaction Δ from private transaction storage to the main storage and marks Δ, as committed in the transaction log. The client c receives message $M(n_i, c, tx\_commit\_resp(committed))$ about successfully committed transaction.

3   Multi Partition Transactions (MPT) Algorithm

There can be many concurrent transactions and some of them mutate values associated with the same keys, thus concurrency control guaranteeing that interfering transactions are not committed at the same time is required. Commit procedure solves this problem by grouping such transactions and performing distributed consensus round, which selects a single transaction that is committed and rollbacks the others. Commit procedure uses two local data structures *transaction index*, which is responsible for identifying interfering transactions and grouping them into the same consensus rounds and *transaction log*, which records committed or rolled back transactions.

3.1   Representation of a Transaction

Transaction state (Definition 3) is used to identify transaction Δ and to track mutations done in Δ, where each mutation is represented by *transaction item* λ (Definition 2) received in a response to a mutation message. A client c begins transaction by sending $M(c, n, begin\_tx())$ and receives empty transaction state Λ in a message $M(n, c, initial\_tx\_state(\Lambda_0))$. Values of $\{\delta_1(k_1, v_1), \delta_2(k_2, v_2), ...\}$ tracked by transaction items $\{\lambda_1, \lambda_2, ...\}$ are unknown to $\{\lambda_1, \lambda_2, ...\}$, but each value v can be found at replica nodes $N^{\mathbb{RF}}(n_i, nj, ...) \subset N$ by topology function τ applied to each λ.

Tracking Updates. Transaction state Λ changes as transaction Δ progresses and its set of transaction items have to be updated with new transaction items after each mutation message.

A client is responsible for keeping track of all performed operations within a transaction. Transaction state is assumed to be a source of knowledge about a transaction, which means that it has valid information about which replicas are affected by a transaction.

3.2   Isolation from Other Transactions

Transactions must run in isolation from each other and provide read-committed isolation level, thus transactions cannot see each other, nor anyone else can see effects of transactions before it completes with commit or rollback. Expected read-committed isolation is different than the one known from RDBMS, because in RDBMS such isolation holds write-locks for the duration of a transaction, therefore if a transaction writes then subsequent reads return written values. We want to avoid distributed locks, which decrease performance and cause deadlocks, therefore read-committed isolation in *MPT* algorithm does not hold any locks, thus each $M(c, n, select(k))$ receives current v, as it is in Ω.

Each mutation $\delta(k, v)$ done in transaction Δ should be stored on replicas responsible for given key keeping it private for that Δ. Each mutation should be isolated from transactions other than Δ and accessible only by Δ. In order to satisfy isolation requirement, *private transaction storage* has to be present on each node.

Each mutation message is executed by n_i, but instead of operating on current live data of Ω, mutations $\{\delta_1, \delta_2, ...\}$ are put into private storage presented in Definition 4.

Definition 4. *Private transaction storage ω is a local data structure on each node n_i which stores mutations $\{\delta_1, \delta_2, ...\}$ of transaction Δ in isolation from other transactions $\{\Delta_i, \Delta_j, ...\}$ and live data set of database Ω and provides operations listed below:*

- $\oplus(\Lambda, \delta(k, v)) \mapsto \Lambda k$ – *stores mutation $\delta(k, v)$ of transaction Δ_j represented by transaction state Λ and returns transaction item λ_k.*

– $clear(\Lambda)$ $\dashv$ *removes all mutations* $\{\delta_1, \delta_2, ...\}$ *associated with transaction* Δ_j
represented by transaction state Λ *on node* n_i
– $get(\Lambda) \mapsto \{\delta_1, \delta_2, ...\}$ – *returns set of mutations of transaction* Δ_j *represented*
by transaction state Λ *stored on replica* n_i

Definition 4 abstracts from memory and time, but an implementation of private transaction storage has to $clear(\Lambda)$ after certain timeout in order to free memory (assuming in-memory storage) and to be resilient to failures of clients, which can fail and abandon transaction Δ_j at any point in time.

Quorum Requirements. Each mutation $\delta(k, v)$ of transaction Δ_j should be written to at least quorum of replicas $N^{\mathrm{RF}} = \tau(k)$ (see Definition 1) in order to depend on majority during the commit process. If quorum is not available, then Δ_j needs to be rolled back.

3.3 Consensus by Paxos

We use *Paxos* to reach consensus among nodes N', which select a single transaction Δ to commit out of set of concurrent *conflicting* transactions $\{\Delta_i, \Delta_j, ...\}$ (Definition 5). *Paxos* provides consensus on a single value among many values proposed during the same *Paxos* round, therefore in order to commit a single transaction Δ, all conflicting $\{\Delta_i, \Delta_j, ...\}$ must participate in the same *Paxos* round, which is not trivial considering many mutations $\{\delta_1(k_1, v_1), \delta_2(k_2, v_2), ...\}$, as there is no longer single key k, which would identify *Paxos* round, as it is in *LWT*.

Definition 5. Conflicting transactions set - *denoted by* $\mathcal{C}\{\Lambda_i, \Lambda_j, ...\}$ *is a set of transactions where all* $\{\Delta_i, \Delta_j, ...\}$ *include at least single common mutation* $\delta(k)$, *which mutates value associated with the same key* k.

Proposed Transaction State. Our true value is transaction $\Delta(\delta_1, \delta_2, ...)$, however mutations $\{\delta_1(k_1, v_1), \delta_2(k_2, v_2), ...\}$ are distributed over nodes $(\tau(k_1) \cup \tau(k_2) \cup ...) = N' \in N$. As a result of that it is impossible to propose Δ itself, but it is possible to propose transaction state Λ, which references mutations $\{\delta_1, \delta_2, ...\}$, thus proposed *Paxos value* in *MPT* is Λ. Any node n_i can try to commit Δ given Λ, because n_i can check which nodes participate in Δ and where are private data stored.

Reaching the Same Paxos Round. Conflicting transactions $\mathcal{C}\{\Lambda_i, \Lambda_j, ...\}$ must participate in the same *Paxos* round identified by *Paxos* round id ι. If there are many transactions $\{\Delta_i, \Delta_j, ...\}$ being committed at the same time and those $\{\Delta_i, \Delta_j, ...\}$ are in conflict with each other, then only single transaction Δ should get committed and rest of them should be rolled back. Nodes N' must agree on *Paxos* value, which is transaction state Λ, thus agree on which Δ is committed. Rest of transactions $(\mathcal{C}\{\Lambda_i, \Lambda_j, ...\} - \Lambda)$ can be rolled back, as long as they participate in the same *Paxos* round ι. In case of *LWT* ι is determined by k, however *MPT* supports more than one key, therefore we need a function which maps $\Lambda \mapsto \iota$ with properties: (a) it maps conflicting transactions to the same ι, (b) it maps non-conflicting transactions to different ι'.

Conflict Function. Function $\Lambda \mapsto \iota$ can be a composition of two functions: $\Lambda \mapsto \mathcal{C}\{\Lambda_i, \Lambda_j, ...\}$ and $\mathcal{C}\{\Lambda_i, \Lambda_j, ...\} \mapsto \iota$, where the former groups transactions into conflicting sets and the latter assigns *Paxos* round id ι to each set. In order to group conflicting transactions we need a function, which compares transaction states Λ_1 and Λ_2 in pairs and detects whether Λ_1 and Λ_2 are in conflict. Definition 6 presents such function.

Definition 6. Conflict function *denoted, as* $\zeta(\Lambda_1, \Lambda_2) \mapsto (\mathcal{C}_1, \mathcal{C}_2)$, *where* $\mathcal{C}_1 = \mathcal{C}(\Lambda_1, \Lambda_2) \wedge \mathcal{C}_2 = \emptyset$ *or* $\mathcal{C}_1 = \mathcal{C}(\Lambda_1) \wedge \mathcal{C}_2 = \mathcal{C}(\Lambda_2)$, *the former case is when transactions are in conflict and contain at least single δ for the same k, the latter otherwise.*

3.4   Transaction Index

Transaction index denoted, as χ is a service, local to a node, which provides function $\Lambda \mapsto \iota$ required to reach the same *Paxos* round for conflicting transactions.

Registration of transaction state Λ in χ on each $n_i \in N'$ is precondition to the commit procedure of *MPT*. Since χ is local, each transaction Δ receives different *Paxos* round id ι on different nodes, but the value of ι is never used outside of a node, thus ι does not have to be globally the same. ι is used only to identify *Paxos* round for $\mathcal{C}\{\Lambda_i, \Lambda_j, ...\}$ at each n_i. If Δ participates in the same *Paxos* round identified by ι as other conflicting transactions, then only single Δ is committed after N' agree on the proposed Λ.

Transaction index stores sets of conflicting transactions $\mathcal{C}\{\Lambda_i, \Lambda_j, ...\}$ and assigns ι to each set. When a new transaction state Λ_{new} is registered, χ tries to find a set $\mathcal{C}\{\Lambda_i, \Lambda_j, ...\}$ to which Λ_{new} can be added by applying conflict function $\zeta(\Lambda_1, \Lambda_2) \mapsto (\mathcal{C}_1, \mathcal{C}_2)$ to Λ_{new} and each transaction Δ in each set. If there is a conflict, sets can be merged together, thus Λ_{new} joins $\mathcal{C}\{\Lambda_i, \Lambda_j, ...\}$ and receives the same ι.

During the commit procedure of *MPT* transaction $\Delta(\delta_1, \delta_2, ...)$ has to register in each quorum for each replica subset $(N^{\mathrm{RF}}_{k_1} \cup N^{\mathrm{RF}}_{k_2} \cup ...) = (\tau(k_1) \cup \tau(k_2) \cup ...) = N' \subset N$. Moreover transaction Δ cannot be registered in transaction index χ when it can obtain more than one *Paxos* round id ι. These two requirements guarantee that if transaction Δ_1 and transaction Δ_2 are conflicting for some key k, then there exists node $n_i \in N^{\mathrm{RF}}_k$ on which Δ_1 and Δ_2 receive the same ι, thus both transactions are part of the same *Paxos* round and eventually only one is committed.

Rolling Back Concurrent Transactions. Transaction index knows which *Paxos* round id ι to assign to which transaction state Λ and it knows conflicting transactions set $\mathcal{C}\{\Lambda_i, \Lambda_j, ...\}$ to which Λ is assigned. This knowledge is used to rollback conflicting transactions when *Paxos* round ι is finished at node n_i and Λ_{learnt} is learnt by n_i. The node calls *clear*(Λ) function of private transaction storage for each $\Lambda \in (\mathcal{C}\{\Lambda_i, \Lambda_j, ...\} - \Lambda_{learnt})$.

If a client c wants to rollback transaction Δ then that client sends a message $M(c, n, tx\_rollback(\Lambda))$ after which n_i identifies N' and broadcasts message $M(n_i, n_j, rollback(\Lambda))$ for each $n_j \in N'$. When n_j receives such message it uses *clear*(Λ) function of private transaction storage. Transaction Δ is rolled back after all $n \in N'$ clear private data.

3.5 Transaction Log

Transaction Log $\mathcal{L}$ is a local data structure present at each $n_i \in N$. Its responsibility is to record committed and rolled back transactions. It provides the following operations: (a) *record_as_committed*(Λ) – record a committed transaction (b) *record_as_rolled_back*(Λ) – record a rolled back transaction (c) *find*(Λ) $\mapsto$ *log_state* – finds information concerning a transaction. Returns *log_state* which can be one of three: *committed*, *rolled back* or *unknown* when transaction is not recorded in the log.

3.6 Pros and Cons

The most crucial advantage of *MPT* over LWT is its support for multi-partition transactions. As a consequence, *MPT* is less resilient to failures than LWT because it depends on quorums of distinct replica groups being available. It also requires more network messages. In terms of CAP theorem, both algorithms sacrifice consistency over availability and network partition tolerance. *MPT* is scalable by its design and it does not require distributed locks. Another benefit of *MPT* is its tunable contention level adjustable to specific workload needs. A downside is that *MPT* assumes that its clients follow a specific protocol and handle transaction state correctly, which makes it not resilient to Byzantine faults.

4   Related Work

There are a lot of NoSQL databases - a web page which collects the information about the known systems of this kind [2] reports over 225 existing NoSql databases. Although they are called NoSQL, not all of them fall into this category because of horizontal scalability and high availability support. We can find there the implementations which support transactions to some degree such as MarkLogic, IBM Informix, Datomic or Oracle. Wide column stores, which Cassandra belongs to, includes Hive, Accumulo, Microsoft Azure Tables Storage and other less popular engines (see [1] for categorized NoSQL database popularity ranking). We shortly summarize the algorithms used by NoSQL engines along with its strengths and weaknesses in comparison to MPT.

4.1 Multiversioning

Multi version concurrency control [5] is known in the area of distributed data-bases as well as in the area of software transactional memory [16]. It can out-perform 2-phase commit in terms of efficiency and it handles failures in a more efficient way [9]. The main idea behind this approach is to attach a version to each operation so that both reads and writes are applied on an exact version of data. This can significantly increase concurrency of transactions by reducing the number of conflicts.

As noted in [8], this approach makes performing serializable transactions difficult. Instead it optimizes for atomic updates and consistent view of data. MPT allows to perform serializable-like transaction by providing a way to specify commit condition, as well as, it provides similar guarantees as multiversioning. Though, atomicity and consistency is achieved at much higher cost because of the use of Paxos.

4.2 RAMP Transactions

Read Atomic Multi Partition [4] was designed to address the requirements of the following use cases: secondary indexing, foreign key enforcement, and materialized view maintenance. *RAMP* identifies new transaction isolation level *Read Atomic*, which assures consistent visibility of changes made by the transaction: either all or none of each transaction's updates are seen by other transactions.

The key principle behind *RAMP* is its ability to detect non-atomic partial read, and to repair it with additional rounds of communication with the nodes by the means of the metadata attached to each write and multi-versioning, which means that data are accessible in different versions and each modification to data creates a new version. Versions are either committed or not yet committed, in which case the data can be discarded by aborting the transaction [4, p. 6]. There are three variants of *RAMP* algorithm which differ in the type of the metadata. RAMP is linearly scalable, which was tested during trials with the number of nodes spanning up to 100 servers [4, p. 10].

Fault tolerance *RAMP* is resilient to failures of part of the nodes in the cluster – for the reason that *RAMP*'s transaction does not contact nodes that are not affected by the transaction. The number of rounds varies between 1 and 2 depending on the existence of the race and algorithm variant. *RAMP* does not block, read transactions race write transactions, but such races are detected and handled by *RAMP*.

RAMP optimizes for the similar uses as Mutliversioning approach and aims to provide similar guarantees. Therefore, write performance is higher than MPT can achieve. On the other hand, it would be difficult to implement serializable transactions within this model.

4.3   Consensus Algorithms

Paxos. *Paxos* is an algorithm to solve distributed consensus problem [11,12], used as part of the larger systems in [7,13], and more recently in the Google's Chubby service [6], which is a lock service intended to provide coarse-grained locking, as well as reliable storage. *Paxos* is resilient to $(\frac{n}{2} - 1)$ failures of nodes. It implements optimistic concurrency control thus it is nonblocking which means that a transaction can be interrupted when a conflict occurs. The message complexity is similar to 3PC.

Raft. The aim of *Raft* designers was to create a consensus algorithm which would be more understandable to its predecessor [15]. The main idea behind the algorithm is decomposition of the consensus problem into distinct phases (a) leader election, (b) log replication, (c) membership changes. *Raft* is fault tolerant, because the leader has to acquire only majority of acknowledges, thus is resilient to failure of $(\frac{n}{2} - 1)$ nodes, same like *Paxos*. However, failure of the leader causes delay because a new leader has to be elected before client's request is handled. In principle *Raft* does not block, because all client requests are handled by the same leader. However, election of the leader can be delayed in case of *split vote*, in which two candidates receive the same number of votes, thus neither of them obtained majority vote and an election has to start over in the next term after a certain timeout. Approximating message complexity of *Raft* is not that clear: assuming that the leader is already elected, a new value is learnt by followers in at least $3n$ messages. However, election process is an additional overhead. Moreover, heartbeat messages are exchanged, thus the total number of messages exchanged can be greater than in previous algorithms.

5   Complexity Analysis

Let us analyze the performance of the *MPT* algorithm considering independent increase of the following parameters: (a) p - modified partitions, (b) N - number of nodes in a cluster, (c) $\mathbb{RF}$ - replication factor, (d) concurrent conflicting transactions, (e) concurrent independent transactions.

5.1   Number of Requests

Assume transaction Δ, which modifies keys: k_1, k_2, k_3, which might be replicated on one up to three unique replica groups, thus size of N' spans from minimum of $\mathbb{RF}$ to maximum of $PC \cdot \mathbb{RF}$. Number of unique replica groups depends whether replica groups are independent, thus contain different nodes, of if they overlap in nodes.

Starting a transaction requires a single message to any node in the cluster, which responds with transaction state Λ. Each mutation of a key needs to be written to the private transaction storage ω and acknowledged by the quorum from a replica group which involves sending $[\mathbb{RF}, PC \cdot \mathbb{RF}]$ messages. Then, the transaction is committed, thus a single commit message is sent. Replicas need to transition to the setup phase, thus leader sends $[N, \mathbb{RF} \cdot N]$ messages and receive responses. Setup phase performs data consistency check, thus each node reads from a quorum of replicas, that is $[\mathbb{RF} \cdot (\frac{\mathbb{RF}}{2} + 1), PC \cdot \mathbb{RF} \cdot (\frac{\mathbb{RF}}{2} + 1)]$ messages. Next, the leader transitions replicas to prepare phase, with $[\mathbb{RF}, PC \cdot \mathbb{RF}]$ messages, followed by a propose phase with additional $[\mathbb{RF}, PC \cdot \mathbb{RF}]$ messages. Finally the nodes transition to a commit phase with $[\mathbb{RF}, PC \cdot \mathbb{RF}]$ messages.

The total number of messages, assuming no contention and no failures, is $2 + 5 \cdot PC \cdot \mathbb{RF} + PC \cdot \mathbb{RF} \cdot (\frac{\mathbb{RF}}{2} + 1)$ where $PC \in [1, \mathbb{RF}]$. An example evaluation results are shown in Fig. 1.

$\mathbb{RF}$	$PC = 1$	$PC = 2$	$PC = 3$	$PC = 4$	$PC = 5$
$\mathbb{RF} = 3$	23	44	65		
$\mathbb{RF} = 5$	37	72	107	142	177

Fig. 1. Message count given different $\mathbb{RF}$ and unique partitions count PC

Increasing Number of Partitions. When *MPT* spans more partitions, it is more likely that these partitions are stored on different replicas, thus the number of replica groups increases as well. Replica groups are bound in size by the total number of nodes in the cluster. If replica groups cover all nodes, then increase in modified partitions will not further affect number of requests performed. Therefore a number of requests increases linearly with the replica groups, but is has a limit, after which it remains constant.

Increasing Number of Nodes. Let us consider a cluster of N nodes and a transaction that modifies set of partitions S_p, which are replicated in the subset of nodes N'. When the nodes are added to the cluster token ranges are reassigned, thus nodes N' are responsible for smaller token ranges, which causes two possible outcomes: (a) N' increases to N'', since after reassignment S_p is replicated by more nodes, (b) N' remains the same, if the token ranges of S_p were not affected by the reassignment.

The former case increases the number of requests during *MPT*, as if number of partitions is increased Sect. 5.1.

Higher Replication Factor. In this case $\mathbb{RF}$ is increased to $\mathbb{RF}'$, thus replica groups contain more replicas. *MPT* relies on the quorum of each replica group, thus number of requests increases as well, but only if replica groups remain disjoint. In case they overlap, the number of requests stays the same, even if replication factor increases.

Increase in Concurrent Conflicting Transactions. Let us consider a partition k_p and an set of transactions S_Δ, in which each transaction modifies k_p, that increases to a larger set S'_Δ. Transactions participate in the same *Paxos* round, due to the fact that all of them modify the same partition k_p, and any conflict function detects it. Concurrent transactions could cause contention and leadership changes of the *Paxos* round, but only if transactions are committed at the same time. The contention is detected by rejecting proposed ballot, in which case the coordinator of a transaction waits until a timeout occurs and tries again with a higher ballot. The timeout provides the time window, in which the current leader proposes its transaction. Even if leadership changes, the proposed and accepted transaction is committed and other conflicting transactions are rolled back, thus if contention is resolved, then the algorithm finishes the one transaction from the set S'_Δ. Initial contention might have negative effect on the overall performance of the algorithm. Moreover, transactions S'_Δ will store their mutations of k_p on the same replica group $N^{RF}_{k_p}$, and register them in the transaction index χ, which increases memory footprint on the nodes in $N^{RF}_{k_p}$.

Increase in Concurrent Independent Transactions. Independent transactions, which do not mutate the same keys, do not cause contention on each other. Therefore the performance is not affected by that factor, likewise the memory footprint spreads out across the cluster.

6   Conclusions

The algorithm we designed and implemented solves the problem of performing transactions enclosing multiple updates across many rows belonging to different replica sets. Unlike the other algorithms, which base on multi-versioning, our solution does not create a snapshot and provides light-weight read-committed isolation level. We keep a copy of those items which are changed in the private transaction storage. Therefore, we have the current view of data items in the database until we change them. Also, the conditions which need to be satisfied for the transaction to be accepted are flexible in the way that we can specify a custom expression to be evaluated on a consistent view of a database during commit which is in principle an extended compare and swap functionality.

Moreover, our algorithm provides a way to define conflict resolution function so that the performance can be tuned for better handling large transactions or more precise conflict resolution which in turn leads to lower contention when a lot of concurrent transactions are committed.

The algorithm was implemented and tested for correctness, stability and performance and the results, along with more detailed description can be found in [14].

References

1. DbEngines. http://db-engines.com/en/ranking. Accessed 09 Nov 2016
2. NoSQL. http://nosql-database.org. Accessed 09 Nov 2016
3. UUID and timeuuid types. https://docs.datastax.com/en/cql/3.3/cql/cql_reference/uuid_type_r.html. Accessed 15 Apr 2016

4. Bailis, P., Fekete, A., Hellerstein, J.M., Ghodsi, A., Stoica, I.: Scalable atomic visibility with ramp transactions. In: Proceedings of the 2014 ACM SIGMOD International Conference on Management of Data, SIGMOD 2014, pp. 27–38. ACM, New York (2014). http://doi.acm.org/10.1145/2588555.2588562
5. Bernstein, P.A., Goodman, N.: Multiversion concurrency control—theory and algorithms. ACM Trans. Database Syst. **8**(4), 465–483 (1983). http://doi.acm.org/10.1145/319996.319998
6. Burrows, M.: The chubby lock service for loosely-coupled distributed systems. In: Proceedings of the 7th Symposium on Operating Systems Design and Implementation, pp. 335–350. USENIX Association (2006)
7. Chandra, T.D., Griesemer, R., Redstone, J.: Paxos made live: an engineering perspective. In: Proceedings of the Twenty-Sixth Annual ACM Symposium on Principles of Distributed Computing, PODC 2007, pp. 398–407. ACM, New York (2007). http://doi.acm.org/10.1145/1281100.1281103
8. Faleiro, J.M., Abadi, D.J.: Rethinking serializable multiversion concurrency control. Proc. VLDB Endow. **8**(11), 1190–1201 (2015). http://dx.doi.org/10.14778/2809974.2809981
9. Halici, U., Dogac, A.: Concurrency control for distributed multiversion databases through time intervals. In: Proceedings of the 19th Annual Conference on Computer Science, CSC 1991, pp. 365–374. ACM, New York (1991). http://doi.acm.org.doi.eczyt.bg.pw.edu.pl/10.1145/327164.327297
10. Lakshman, A., Malik, P.: Cassandra: a decentralized structured storage system. ACM SIGOPS Oper. Syst. Rev. **44**(2), 35–40 (2010)
11. Lamport, L.: The part-time parliament. ACM Trans. Comput. Syst. **16**(2), 133–169 (1998). http://doi.acm.org/10.1145/279227.279229
12. Lamport, L., et al.: Paxos made simple. ACM SIGACT News **32**(4), 18–25 (2001)
13. Lampson, B.W.: How to build a highly available system using consensus. In: Babaoğlu, Ö., Marzullo, K. (eds.) WDAG 1996. LNCS, vol. 1151, pp. 1–17. Springer, Heidelberg (1996). doi:10.1007/3-540-61769-8_1
14. Lewandowski, M.: Multi partition transactions in Cassandra. Technical report, Institute of Computer Science, Warsaw University of Technology. http://repo.bg.pw.edu.pl/index.php/pl/r#/info/master/WUTad49127820284899b17f93ada17d23d7/?r=affiliation
15. Ongaro, D., Ousterhout, J.: In search of an understandable consensus algorithm. In: 2014 USENIX Annual Technical Conference (USENIX ATC 2014), pp. 305–319 (2014)
16. Perelman, D., Fan, R., Keidar, I.: On maintaining multiple versions in STM. In: Proceedings of the 29th ACM SIGACT-SIGOPS Symposium on Principles of Distributed Computing, PODC 2010, pp. 16–25. ACM, New York (2010). http://doi.acm.org/10.1145/1835698.1835704
17. Stonebraker, M.: SQL databases v. NoSQL databases. Commun. ACM **53**(4), 10–11 (2010). http://doi.acm.org/10.1145/1721654.1721659

The Multi-model Databases – A Review

Ewa Płuciennik[(⊠)] and Kamil Zgorzałek

Institute of Computer Science, Silesian Technical University,
Akademicka 16, 44-100 Gliwice, Poland
Ewa.Pluciennik@polsl.pl,kamizgo506@student.polsl.pl
http://www.polsl.pl

Abstract. The following paper presents issues considering multi-model
databases. A multi-model database can be understood as a database
which is capable of storing data in different formats (relations, documents, graphs, objects, etc.) under one management system. This makes
it possible to store related data in a most appropriate (dedicated) format
as it comes to the structure of data itself and the processing performance.
The idea is not new but since its rising in late 1980s it was not successfully and widely put into practice. The realm of storing and retrieving
the data was dominated by the relational model. Nowadays this idea
becomes again up-to-date because of the growing popularity of NoSQL
movement and polyglot persistence. This article attempts to show the
state-of-the-art in multi-model databases area and possibilities of this
reconditioned idea.

Keywords: Database · Relational · NoSQL · Multi-model · Polyglot
persistence

1 Introduction

Every time a commercial application is built it has to persist (store) some data
so it can outlast even when application is not operating. So far choice of storage
layer was quite simple if it comes to data structure model – a relational database.
Of course most of the applications follow the object paradigm and to process
relational data they need some additional layer – object-relational mapping,
or force the programmer to use native methods to work with the underlying
database. For the few years this choice is becoming more complicated. NoSQL
movement has arisen. Now application designers should choose not only the
technique to work with persistent data but the data model as well.

Right now there are four main categories of NoSQL databases: key-value, column family, document and graph ones. In relational databases elementary unit
of data is a tuple (row). Tuples in relational model cannot be nested (it is impossible to create aggregates in sense of one object containing another object – for
example customer and his/her orders, or treating complex object as a unit) and
have very strict structure (list of fields). Relational data model does not define
concept of complex field – relationship between objects are modelled through

© Springer International Publishing AG 2017
S. Kozielski et al. (Eds.): BDAS 2017, CCIS 716, pp. 141–152, 2017.
DOI: 10.1007/978-3-319-58274-0_12

referential integrity. In contrast key-value, column-family and document databases are cut out for this kind of aggregates. In case of document databases these aggregates are also structured (documents/nested documents) and in column-family databases aggregates have two levels (column and row). Graph databases in turn are the best for entities with wide and complicated network of relationships [25]. The last model to mention is the object oriented one. This model follows the object oriented programming paradigm. Object databases emerged in 1990s. Because of strong integration with applications they did not gain much popularity but they are still on the market (e.g. Versant Object Database).

NoSQL Market Forecast 2015–2020 [16] states that key-value store is currently the best solution for scalable, high performance and robust large databases but the biggest obstacle to use NoSQL is transaction consistency and therefore relational databases will not fall into disuse. NoSQL databases should not be treated as replacement for relational databases but rather as its supplement. This report also forecasts graduated convergence of RDBMS and NoSQL into hybrid storage systems.

Having at their disposal such diversity of data models system architects face difficult decision of choosing which model would be the best and in what circumstances, if it comes to performance, security, consistency, etc. The multi-model databases seem good choice if flexibility in data models is our priority. This article attempts to review the state of the art in a multi-model data storage and more detailed comparison of chosen solutions, if it comes to functionality and performance.

2   Multi-model Databases

As was mentioned above the idea of combining different data models into one logical storage is not new. In article [29] from 1997 Garlic system was presented. Its architecture was based on wrappers that encapsulated the data sources (relational, object databases, images archive, etc.) and modelled the data as objects. Each object had an identifier composed of implementation ID (ID – defining an interface and a repository) and a key which identified an object within a given repository. Objects also had methods for searching and data manipulation. Query language was SQL, extended with such method invocation possibility and support for path expressions and nested collections. It is worth pointing out that nowadays such query language exists – JPQL, mentioned in the following part of this chapter. In 1998 patent no. US 5713014 A was published [27]. It defines "Multi-model database management system engine for database having complex data models". This system architecture is composed of physical storage layer, conceptual model layer (based on entity-relationship definition), logical data layer (consists of CODASYL, relational database and object oriented one) and external view layer (API/Language layer: SQL, C++, ODBC). Data unit is defined as a record. Records associations are stored in form of individual pointers or Dynamic Pointer Array (DPA). In article [26] from 2003 authors presented other approach to multi-model databases. The described system utilized different data models at different layers of database architecture: the highest, conceptual layer was based on object oriented and hierarchical, semi-structured models; underlying, logical layer was based on XNF2 (eXtended Non First Normal Form) model; physical layer consisted of RDBMS and MonetDB.

At present, if we look on the NoSQL market, there are quite a few databases which aspire to a multi-model type. ArangoDB stores documents in collections. Although documents can have different structure, documents which have the same attributes share common structure called "shape". This allows for reducing disk and memory space needed for storing data. ArangoDB uses JavaScript to write "actions" which can be compared to database stored procedures. To query the data ArrangoDB offers Query By Example (example document is in JSON[1] format) and its own query language AQL, which supports documents, graphs and joins [17].

Aerospike offers NoSQL key-value database (formerly Citrusleaf) which is optimized for flash memory to achieve real-time speed for managing terabytes of data. In late 2012 Aerospike has acquired AlchemyDB [1] which can be considered as a first (in modern NoSQL movement) database integrating different data models – relational, document and a graph on top of it. The basic unit of storage is a record uniquely identified by a so-called digest within a namespace (a top level container for the data). Aerospike query provides value-based lookup through the use of secondary indexes. Aerospike supports UDF (User-Defined Function) written in C or Lua scripting language [2].

OrientDB supports graph and document models. Documents consisting of key-value pairs (where a value can have simple or complex data type) are stored in classes and clusters (grouping documents together). A cluster or a class can be a vertex of a graph. Each record (e.g. document) has its own identifier (RID) which consists of a cluster ID and a position within a given cluster. RID describes physical position of a piece of data so it allows very fast data localisation. The database uses OrientDB SQL dialect to create, update or search for the data [19].

PostgreSQL from its beginning was considered as an object-relational database. Since version 9.2 it offers relational, object-relational, nested relational, array, key-value (hstore) and document (XML, JSON) store. What is more PostgreSQL offers possibility to nest other databases through Foreign Data Wrappers for relational and NoSQL (CouchDB, MongoDB, Redis, Neo4j) databases [9]. MarkLogic can store data in JSON, XML or RDF triples so it is considered multi-model. It serves as data integrator [15]. Virtuoso in turn offers RDF relational data management [18].

If it comes to relational databases, now they are all object-relational (Oracle, DB2, SQL Server, PostgreSQL) which means that user can define its own data type. What is more these databases also offer XML and JSON storage format. For example Oracle 12c offers JSON storage supported by SQL language extension allowing to use path expression and dedicated SQL functions (JSON_QUERY, JSON_VALUE and JSON_TABLE to name a few) [12].

---

[1] JSON - JavaScript Object Notation.

If we look at multi-modelness from programmer's point of view, we should consider languages/libraries that give us opportunity to work with different (if it comes to the model) databases from object application with minimum effort (the same interface and even query language). The solution with the widest list of cooperating databases is DataNucleus – implementation of JPA[2] and JDO[3] (which assumes arbitrary data model). At present it offers unified access by means of JPQL (Java Persistence Query Language – SQL-like query language for object data model defined in JPA) for the following databases: RDBMS, HBase, MongoDB, Cassandra, Neo4j, JSON, Amazon S3, GoogleStorage, LDAP, NeoDatis and among others JSON format (this list de facto covers all kinds of NoSQL databases) [8]. However, DataNucleus although quite universal suffers from a few practical disadvantages, for instance not all JPQL constructions work properly for all data stores [28].

Multi-modelness within the meaning of the polyglot persistence is now also appreciated by the big players in the commercial database and cloud processing markets. IBM have created a Next Generation Data Platform Architecture (based on Hadoop) which "combines the use of multiple persistence (storage) techniques to build, store, and manage data applications (apps) in the most efficient way possible" [31]. In this solution organizations that use cloud, dump their data into so-called landing zones without need to understand the schema or setup of the dedicated structures.

If we want to build a system that can work with multi-model data first we need to decide where to place a layer responsible for the data integration. We can put all the responsibility on the application, which then is obligated to dispatch the data into the fully separated so-called persistence lanes. Single persistence lane is responsible for cooperating with one database and contains an adequate data mapper. This architecture is quite simple to implement but (i) data should be divided into databases (and thereby data models) in advance, (ii) cross-lane queries are very hard to implement. Other solution is to complicate architecture but gain better functionality. The options are [24]:

- polyglot mapper placed between application and persistence lanes, which makes cross-lane queries possible and maintains a single object data model; as polyglot mapper DataNucleus, EclipseLink or Hibernate OGM can be used,
- nested database where there is one persistence lane but main database gives a possibility to map/connect other databases (e.g. Postgresql); cross lane queries are possible, but choice of databases is limited,
- omnipotent database which supports many models; main advantage of this approach is a single database with a single maintenance (backup, restore, etc.).

The idea of multi-model database seems to be a perfect solution for implementing polyglot persistence. But there are few challenges. Deciding which approach is the best is not straightforward. First one needs to consider (trade-off) two things:

---

[2] Java Persistence API.
[3] Java Data Objects.

– what kind of data model is needed – of course more models means more elastic (adaptive) data storage,
– how many different databases covers the requirements – each database needs to be maintained and it requires resources to operate.

Second decision to make is how to query this kind of structure. Of course one, universal language would be the best if it comes to cross-model queries but the question is:

– can already existing language/solution (e.g. JPQL) be used or a new one has to be created – which is not a simple task although is completely possible [17,26,29],
– are cross-model queries required; if not a better solution is to use native query languages for individual databases; if so existing multi-model databases have to be considered.

   If it comes to unified query language over different data structures (models) multi-model databases are the best.

3   Comparison of Chosen Multi-model Databases

For more detailed comparison three databases were chosen: ArangoDB and OrientDB (key-value, documents and graphs) and Couchbase (key-value, document). This choice is based on number of models maintained by databases and DB-Engines popularity ranking[4]. The ranking covers all database models. OrientDB is the highest classified open source multi-model database (three models, position no. 45[5]), ArangoDB is classified on position no. 76 and constantly moving up (see footnote 5). Couchbase was chosen as two-model database (since version 2.0 it covers document and key-value models [30]) with position no. 23 (see footnote 5) – highest out of two-model databases.

3.1   Data Model

First data models will be described. In ArangoDB the basic data structure is a document which can consists of any number of attributes with values (of simple or complex type). Each document has three special attributes [3]:

– _id – unique, unequivocal identifier of a document within the confines of the database,
– _key – identifies a document within the confines of a collection,
– _rev – document's version.

Documents are organized into collections which can be compared to tables in a relational database. Documents are schema-less so they can differ in terms of number and types of attributes. There are two kinds of collections: document (nodes) collection and edges collection. Edge is a special kind of document which has two peculiar attributes: _from and _to used to denote relation between documents. So documents are organized into a directed graph.

In OrientDB basic (the smallest) data unit is a record. There are four possible record types: byte record (BLOB), document, node (vertex) and edge. Document is the most flexible type of record because it can be schema-less. Alike ArangoDB, in OrientDB documents have special (obligatory) attributes [4]:

- *@class* – defines a class of a document (classes define types of records; they are schema-less, schema-full or mixed one),
- *@rid* – an automatically generated identifier for documents within the confines of the database,
- *@version* – version number.

Records are grouped into clusters. By default there is one cluster (physical or in-memory) per class. One class can be divided into more clusters (shards) to enable e.g. parallel query execution. Documents can be related in form of strong (embedded) or weak relationships. Nodes and edges have a form of documents but edges can be stored in a lightweight form. A lightweight edge is embedded into vertex record which can improve performance, but this type of edge can not have any additional properties [13].

Couchbase Server operates on two data models: document and key-value. Because these two models are very similar one can doubt if Couchbase can be perceived as multi-model. Nonetheless Couchbase uses key-value for searching to improve performance [22].

Table 1 contains summary of data model characteristics of covered databases. There are many similarities especially if it comes to JSON data format or indexing on key fields. Differences are more subtle and involve above all, data organization.

Table 1. Comparison of described databases

Characteristic	ArangoDB	OrientDB	Couchbase server
Data models	Key-value, document, graph		Key-value, document
Key-value model implementation	Automatically created index on key field		
Document model implementation	JSON data format		
Graph model implementation	Special edge document with attributes _from and _to		N/A
Data organization	Collections	Classes	Buckets
Special attributes	Identifier, version		Identifier

3.2   Query Language

This section presents the chosen databases query languages in comparison to SQL which is an unattainable ideal for many database designers.

ArangoDB creators are among them. AQL has a syntax very similar to SQL and it is declarative. The main difference is that it operates on collections and does not have data definition part (DDL).

OrientDB uses SQL operating on classes with some graph extensions added (SELECT phrase can be replaced with TRAVERSE phrase [21]). The main difference is that instead of joins, classes' associations are used.

Couchbase uses its own query language N1QL which is SQL for JSON [7]. There are two possible kinds of queries: indices management (DDL) and CRUD (DML).

Table 2 presents example basic CRUD queries for described databases. All languages are very similar to SQL if it comes to basic phrases (INSERT, SELECT, UPDATE, DELETE), all allow full CRUD functionality. There are of course some differences, especially in AQL where special loop and filter constructions are used for performing operation on elements of collection (*cars*).

Table 2. CRUD operation syntax comparison

AQL	OrientDB SQL	N1QL
Create data		
INSERT {make: "MakeA", model: "ModelB"} IN cars	INSERT INTO cars (make, model) VALUES ('MakeA', 'ModelB')	INSERT INTO cars (KEY, VALUE) VALUES (UUID(), {"make":"MakeA", "model":"ModelB"})
Retrieve data		
FOR t IN cars FILTER t.make == "MakeA" RETURN t	SELECT FROM cars WHERE make LIKE 'MakeA'	SELECT * FROM cars WHERE make = 'MakeA'
Update data		
UPDATE "1" WITH price: "3000" IN cars	UPDATE cars SET price = '3000' WHERE @rid = '1'	UPDATE cars USE KEYS "1" SET price = "3000"
Delete data		
FOR t IN cars FILTER t.make = 'MakeA' REMOVE t	DELETE FROM cars WHERE make = 'MakeA'	DELETE FROM cars t WHERE t.make = "MakeA"

3.3   Indices and Transaction

Indices and transactions are the elements which are very important in databases.

ArangoDB automatically creates primary hash index for each collection and for graph edges. This index is based on _key field and it is not sorted so it cannot be used for range selection. Except for primary index, user can create his/her own indices of types: hash, skiplist, geo, fulltext, sparse [11].

OrientDB uses indices in a similar way as in the relational databases. User has four types of indices to choose from: SB-Tree (based on B-tree with some optimization added), hash, auto-sharding (for distributed systems) and Lucene (fulltext and spatial) [10].

Couchbase offers two kinds of indices: global and local. Global indices minimize network processing during interactive query processing. There are four types of indices: Global Secondary Index, MapReduce views, spatial views and full text index [6].

Since version 1.3 ArangoDB offers transactions configurable by the user with full ACID at repeatable reads isolation level [14]. OrientDB also provides full ACID transaction with two isolation levels: read committed and repeatable reads [20]. Couchbase does not support transactions [23].

4   Performance Tests

Multi-model databases potentially have wider set of applicable use cases since they cover more than one data model under one management system. The question is if the diversity of maintained data models does not diminish the performance. One of the ArangoDB creators (Claus Weineberger) has worked out and conducted tests embracing ArangoDB, OrientDB, Neo4j, MongoDB and Postgres databases [5]. The main goal was to prove that native multi-model databases can successfully compete with one-model databases. Tests were performed on Pocked social network[6] with 1 632 803 nodes and 30 622 564 edges. Tests were divided in to two groups: read/write/ad-hoc aggregation queries and queries typical for a graph (shortest path and nearest neighbour). The tests results are very promising for multi-model databases – especially ArangoDB. For single read ArangoDB is slightly slower than MongoDB whereas OrientDB is 50% slower than ArangoDB. For single write both multi-model databases were faster than MongoDB. As for aggregation ArangoDB was 2.5 times faster than MongoDB and 18 times faster than OrientDB. Full tests results can be found at [5]. Of course these tests are vendor tests but all results and code (written in JavaScript) was published for public use and author explicitly states that ArangoDB currently works best when the data fits in the memory.

In our tests we have decided to concentrate on CRUD operation from Java developer perspective. We did not measure database performance in cluster or for large datasets. Because all three databases share ability to operate on document model, we wanted to answer the question if the multi-modelness diminishes performance of simple document operations (as a point of reference MongoDB was used). If so, to what degree? In tests we have measured number of operations per second. As for operations we have a single read, a single write, a single

---

[6] https://snap.stanford.edu/data/soc-pokec.html.

update, a single delete and an aggregation (number of documents satisfying some condition). All tests were performed in five series with 100 repeats (for example 500 document writes). No special database tuning was conducted. All databases have operated on default keys and indices. The dataset consisted of generated personal data (name, surname, profession, weight, etc.) with different age values.

Create operations were conducted by saving document using special methods offered by the databases. For retrieving filtering query was used (filtering about 1% of documents – 50 year old persons). Query was written in query language appropriate for a particular database:

- ArangoDB: *FOR t IN testCollection FILTER t.age == 50 RETURN t,*
- OrientDB: *SELECT FROM testDocument WHERE age = 50,*
- Couchbase: *SELECT * FROM default WHERE age == 50.*

Update operation was conducted by adding one field to each document using dedicated database methods. Similarly, for delete test each document was removed from a database. Aggregation test relied on counting documents representing persons 50 and more years old (about 50% of the dataset) using query:

- ArangoDB: *FOR t IN testCollection FILTER t.age >= @age RETURN COUNT(t),*
- OrientDB: *SELECT COUNT(*) FROM testDocument WHERE key >= 50,*
- Couchbase: *SELECT COUNT(*) FROM default WHERE age >= 50.*

Table 3 summaries the test results – number of operation per second is presented.

Table 3. Test results – number of operation per second

Operation	ArangoDB	OrientDB	Couchbase	MongoDB
Create data	377	372	423	573
Retrieve data	415	235	261	492
Update data	639	581	681	695
Delete data	865	818	1792	956
Aggregation	421	175	303	452

For greater readability the results are presented in a form of bar chart in Fig. 1. Results for MongoDB are presented as reference line. Bars that are fitting under that line denote less performance than the reference database.

First, most conspicuous observation is that all multi-model databases CRUD operation were slower than in MongoDB except one case – document deletion in Couchbase. Documents in Couchbase are organized into buckets. A bucket is not only a logical unit but also physical storage. Remove method in Couchbase is performed not directly on document but throughout a bucket and probably is natively optimized. Second conclusion is that the slowest database was OrientDB and ArangoDB has results that are most evenly distributed and most similar to

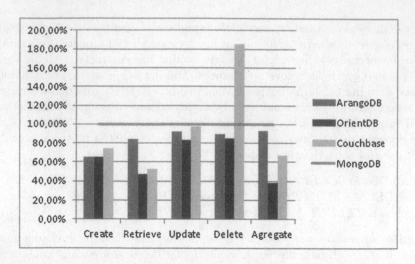

Fig. 1. CRUD test results - performance in relation to MongoDB

MongoDB. These tests are of course in their initial stage and need to be extended with additional scenarios because the subject is very interesting and very up-to-date.

5   Conclusion

There is no standard definition what multi-model means – support of more than one data model? It is obvious that this simple definition leads to one conclusion. Each database which supports UDT (User-Defined Type) or JSON is multi-model. Users can define types suited to their needs. In this case each modern database is multi-model, beginning from object-relational (Oracle, SQL Server, DB2) to ArangoDB for example. Perhaps we need stricter definition of multi-modelness of n-degree where n is a number of data models supported by a database out of the box. Right now it seems that maximum value of n is six.

If it comes to multi-model and adaptive database system, its designing and creating is complicated but possible and it already happens in commercial area of data storage market. Existing multi-model databases are promising technology to use when we need to create this kind of database system. It seems that it is worth to use them, provided they assure support for all or majority of the needed models. Using this kind of a database lighten resource and maintenance requirements and provide query language capable of operating on different data structures (for example documents and graphs). Multi-model databases are not as popular as other NoSQL solutions, but it seems that it will change although it is hard to anticipate in what degree. For now it can be stated that these databases are slightly slower that one-model solution – multi-model handling comes with a price. Is this price worth paying? To answer this question we certainly need more tests and to observe multi-model databases positions in popularity rankings.

Acknowledgement. This work was funded by statutory research of Faculty of Automatic Control, Electronics and Computer Science of Silesian University of Technology.

References

1. Aerospike Acquires AlchemyDB NewSQL Database. http://www.aerospike.com/pr ess-releases/aerospike-acquires-alchemydb-newsql-database-to-build-on-predictab le-speed-and-web-scale-data-management-of-aerospike-real-time-nosql-database-2/. Accessed 19 Nov 2016
2. Aerospike Documentation. http://www.aerospike.com/docs/. Accessed 19 Nov 2016
3. ArangoDB Data models and modelling. https://docs.arangodb.com/3.0/Manual/ DataModeling/index.html. Accessed 19 Nov 2016
4. Basic Concepts OrientDB. http://orientdb.com/docs/2.0/orientdb.wiki/Concepts. html. Accessed 19 Nov 2016
5. Benchmark: PostgreSQL, MongoDB, Neo4j, OrientDB and ArangoDB. https:// www.arangodb.com/2015/10/benchmark-postgresql-mongodb-arangodb/. Accessed 19 Nov 2016
6. Couchbase Server Indexing. http://developer.couchbase.com/documentation/ server/4.5/indexes/indexing-overview.html. Accessed 19 Nov 2016
7. Database Querying with N1QL. http://www.couchbase.com/n1ql. Accessed 19 Nov 2016
8. DataNucleus AccessPlatform 5.0 Documentation. http://www.datanucleus.org/ products/accessplatform_5_0/index.html. Accessed 19 Nov 2016
9. Foreign data wrappers. https://wiki.postgresql.org/wiki/Foreign_data_wrappers. Accessed 19 Nov 2016
10. Indexing - OrientDB Manual. http://orientdb.com/docs/last/Indexes.html. Accessed 19 Nov 2016
11. Indexing ArangoDB. https://docs.arangodb.com/3.0/Manual/Indexing/index. html. Accessed 19 Nov 2016
12. JSON in Oracle Database.https://docs.oracle.com/database/121/ADXDB/json. htm#ADXDB6246. Accessed 19 Nov 2016
13. Lightweight Edges - OrientDB. http://orientdb.com/docs/last/Lightweight-Edges. html. Accessed 19 Nov 2016
14. Locking and Isolation ArangoDB. https://docs.arangodb.com/3.1/Manual/ Transactions/LockingAndIsolation.html. Accessed 19 Nov 2016
15. MarkLogic Semantics. http://www.marklogic.com/wp-content/uploads/2016/09/ Semantics-Datasheet.pdf. Accessed 19 Nov 2016
16. NoSQL Market Forecast 2015–2020. http://www.marketresearchmedia.com/? p=568. Accessed 19 Nov 2016
17. On multi-model databases. Interview with Martin Schönert and Frank Celler. http://www.odbms.org/blog/2013/10/on-multi-model-databases-interview-with-martin-schonert-and-frank-celler/. Accessed 19 Nov 2016
18. Openlink Virtuoso Home. https://virtuoso.openlinksw.com/. Accessed 19 Nov 2016

19. OrientDB Manual - version 2.0, Document and Graph Models. http://www.ori
    entechnologies.com/docs/last/orientdb.wiki/Tutorial-Document-and-graph-model.
    html. Accessed 19 Nov 2016
20. Transactions - OrientDB Manual. http://orientdb.com/docs/last/Transactions.
    html. Accessed 19 Nov 2016
21. Traverse - OrientDB Manual. http://orientdb.com/docs/last/SQL-Traverse.html.
    Accessed 19 Nov 2016
22. Why Couchbase? http://developer.couchbase.com/documentation/server/current/
    introduction/intro.html. Accessed 19 Nov 2016
23. Couchbase: View and query examples. http://developer.couchbase.com/documenta
    tion/server/4.1/developer-guide/views-query-sample.html. Accessed 19 Nov 2016
24. Engelschall, R.S.: Polyglot Persistence Boon and Bane for Software Architects.
    https://docs.arangodb.com/3.0/Manual/DataModeling/index.html. Accessed 19
    Nov 2016
25. Fowler, M., Sadalage, P.: NoSQL Distilled: A Brief Guide to the Emerging World
    of Polyglot Persistence. Addison-Wesley, Upper Saddle River (2012)
26. van Keulen, M., Vonk, J., de Vries, A.P., Flokstra, J., Blok, H.E.: Moa and the
    multi-model architecture: a new perspective on NF^2. In: Mařík, V., Retschitzegger,
    W., Štěpánková, O. (eds.) DEXA 2003. LNCS, vol. 2736, pp. 67–76. Springer,
    Heidelberg (2003). doi:10.1007/978-3-540-45227-0_8
27. NoSQL Market Forecast 2015–2020: Multi-model database management system
    engine for database having complex data models US 5713014 A. http://www.
    google.com/patents/US5713014. Accessed 28 Nov 2016
28. Płuciennik-Psota, E.: Object (not only) relational interfaces survey. Stud. Inform.
    34, 301–310 (2012)
29. Roth, M., Schwarz, P.: Don't scrap it, wrap it! a wrapper architecture for legacy
    data sources. In: VLDB 1997 Proceedings of the 23rd International Conference on
    Very Large Data Bases, pp. 266–275 (1997)
30. Wiederhold, B.: Key-value or document database? Couchbase 2.0 bridges the gap.
    https://blog.couchbase.com/key-value-or-document-database-couchbase-2-dot-0-
    bridges-gap. Accessed 30 Jan 2017
31. Zikopoulos, P., deRoos, D., Bienko, C., Buglio, R., Andrews, M.: Big Data Beyond
    the Hype. A Guide to Conversation for Today's Data Center. McGraw Hill Edu-
    cation, New York (2014)

Comparative Analysis of Relational and Non-relational Databases in the Context of Performance in Web Applications

Konrad Fraczek and Malgorzata Plechawska-Wojcik[✉]

Institute of Computer Science, Lublin University of Technology,
Nadbystrzycka 36B, 20-618 Lublin, Poland
fraczek.konrad1@gmail.com, m.plechawska@pollub.pl

Abstract. This paper presents comparative analysis of relational and non-relational databases. For the purposes of this paper simple social-media web application was created. The application supports three types of databases: SQL (it was tested with PostgreSQL), MongoDB and Apache Cassandra. For each database the applied data model was described. The aim of the analysis was to compare the performance of these selected databases in the context of data reading and writing. Performance tests showed that MongoDB is the fastest when reading data and PostgreSQL is the fastest for writing. The test application is fully functional, however implementation occurred to be more challenging for Cassandra.

Keywords: Relational databases · NoSQL · MongoDB · Cassandra

1 Introduction

Since 1970, when Edgar Codd published his article [10], relational databases have dominated the database market. At present, in the most popular database systems rank, seven of the top ten positions are occupied by relational databases [20]. However, recent years have seen a dynamic growth of the Internet and mobile devices. This causes an enormous increase of the amount of generated data. Engineers started to look for alternatives to relational databases, which are not designed to effectively cope with such a large quantities of data. As a result, NoSQL databases have appeared. They offer better capabilities for performance scalability and a much more flexible data model than relational databases.

The aim of this study is to compare relational databases with selected non-relational databases: a document database (MongoDB), and a column-oriented database (Cassandra). For the purposes of this paper simple social-media web application was developed. The data models used in the application and performance of each database will be compared. The reason why we selected above-mentioned database management systems are as follows. Mongo and Cassanda are flagship NoSQL products, where MongoDB is the most popular document-oriented database whereas Cassanda - column-oriented database. As typical non

© Springer International Publishing AG 2017
S. Kozielski et al. (Eds.): BDAS 2017, CCIS 716, pp. 153–164, 2017.
DOI: 10.1007/978-3-319-58274-0_13

relational databases, Mongo and Cassanda are open-source. That is why among relational database we have chosen PostgreSQL, which is open-source, also available commercially.

The paper is a continuation of our previous work [21], where we performed analysis of data models and conducted some performance tests. As the IT market is developing rapidly, there is a need of verifying available database solutions and their adaptation to different conditions. Development of non-relational databases and a lack of extended research about the current state of art motivated us to continue the topic of NoSQL databases application.

This paper is organised as follows. Section 2 provides a review of related research. Section 3 contains a description of NoSQL databases. Section 4 introduces the implemented social-media web application. Section 5 presents the performance tests results. Section 6 is a summary of the paper.

2   Related Research

In the literature there is few research about the current state of art in the area of non-relational databases. In the paper [24] authors analysed the performance of non-relational databases based applications. A general comparison of relational and non-relational database was also discussed by Jatana and colleagues [15]. Characteristic of NoSQL databases background and data model was also discussed by Han and colleagues [12].

Lourenço et al. [19] have reviewed a number of NoSQL databases available on the market, including MongoDB, Cassandra and HBase. They compared them in terms of the consistency and durability of the data stored as well as with respect to thier performance and scalability. They concluded that the MongoDB database can be the successor of SQL databases, because it provides good stability and consistency of data. Cassandra is the best choice in cases when most of the operations are writes to the database.

Chandra in his publication [7] reviews the properties of BASE (Basically Available, Soft state, Eventual consistency) in NoSQL databases and compares them with the ACID (Atomicity, Consistency, Isolation, and Durability) properties. He also examines which databases are the most suitable for specific applications - in financial applications the relational databases are reported as the best choice. For the purposes of data analysis and data mining NoSQL technologies turn out to be better.

Choi et al. [9] compared the performance of Oracle and MongoDB. They found that the MongoDB database is several times faster than Oracle. The same database was compared by Boicea et al. [4]. The authors conclude their work with the claim that MongoDB is faster and easier to maintain. On the other hand, Oracle is the better choice when there is a need for mapping complex relationships between data.

Li and Manoharan [18] compared several NoSQL databases (including MongoDB, Cassandra, Hypertable, Couchbase) and SQL Server Express in the context of performance. They observed that NoSQL databases are not always faster than SQL. Lee and Zheng [16] compared the performance of HBase and MySQL. It turned out that when retrieving the same data, the NoSQL database is faster than relational ones.

Truica et al. [23] compared the performance of document databases (MongoDB, CouchDB and Couchbase), and relational databases (Microsoft SQL Server, MySQL and PostgreSQL). CouchDB proved to be the fastest during insertion, modification and deletion of data, and MongoDB while reading.

3   NoSQL Databases

The NoSQL term does not apply to a specific technology. It includes all non-relational databases. Almost all of them have the following common features:

- lack of support for SQL language, most of NoSQL databases define thier own query language, some of them have a syntax similar to SQL, for example CQL for Cassandra,
- lack of relations between data,
- designed for working in clusters,
- no ACID transactions,
- flexible data model.

One of the biggest problem related to storing data on many servers is ensuring data consistency. The CAP (Consistency, Availability, Partition tolerance) theorem, described by Brewer [5, 22] is related to this issue. It claims that a distributed database system can maintain only two of three conditions at the same time: consistency, availability and partition tolerance. Systems operating on a single machine are examples of CA systems - they are consistent (as there is no replication) and available. Systems operating on multiple machines are CP systems (MongoDB, HBase) or AP systems (Cassandra, CouchDB) [19].

Another term connected with NoSQL databases are BASE properties, which are equivalent to ACID properties known from relational databases [22]:

- Basically Available - if part of the servers fails, the rest of them should continue to respond to requests,
- Soft State - the state of the database can be changed, even if there are no writing operations performed at this moment,
- Eventual Consistency - after writing data on a single server, changes must be propagated to other machines; during this operation data are not consistent.

These days there are four types of NoSQL databases [22]:

- key-value stores - features offered by these databases are limited to the read, save and delete values for the specified key,
- document databases - they store data in documents with a dynamic structure such as JSON or XML,
- column-oriented databases - they store data in column families organised into rows; rows from the same column family can have different columns,
- graph databases - these are based on a mathematical model of the graph, they store data in graph vertices and relations between data in graph edges.

3.1 MongoDB

MongoDB [8] is an open source document-based database written in C++. It is the fourth most popular database. At the same time it is the most popular NoSQL database [20]. MongoDB stores data in BSON documents which are binary JSON documents. A single document is equivalent to a row in relational database. Documents are grouped into collections of documents. In contrast to the RDBMS, in MongoDB documents from the same collection may not have the same structure. MongoDB does not support ACID transactions. It offers atomic operations on single document only [17]. The maximum size of a single document is 16 MB. Mongo DB supports horizontal scalability through automatic sharding. Replication is implemented in master-slave mode - data are written to the master and then propagated to slaves [19]. MongoDB offers a very functional query language (which is based on JavaScript). It supports aggregate functions and MapReduce model [4]. MongoDB allows to define indexes to speed up queries [19].

3.2 Apache Cassandra

Cassandra [13] is an open source column-oriented database written in Java. It was developed by Facebook [18]. Cassandra stores data as relational databases, in the form of tables and rows. Each line consists of a primary key and columns. Rows in one table may have different columns. Each column consists of the name, value, and recording time values in milliseconds [19]. Just like MongoDB, Cassandra supports mechanisms of replication and partitioning. Unlike MongoDB, all servers are equal - there is no concept of master and slaves. Each server can handle write requests and propagate it to others. As data access interface Cassandra uses CQL (Cassandra Query Language) which is similar to SQL, however it offers much fewer functionalities.

4 Test Application

For the purposes of this work a social-media web application was made. The application at a particular moment can use one of the three databases - PostgreSQL, MongoDB or Cassandra, depending on the configuration. It provides such functionalities as sending posts, marking posts with hashtags, adding comments to posts, following other users, viewing the timeline which contains posts of followed users ordered by date in descending order and viewing all posts marked with a specific hashtag. One of the requirements was also the implementation of paging while retrieving messages. What is important, the pagination was carried out directly on the database and not in the application. We managed to achieve this goal for all selected databases.

The application was written in Java 8 and JavaScript. Following frameworks and libraries were used:

- Spring Boot [2] allows to create Java web application in a very simple way. The whole application is a single JAR file with embedded Tomcat, it can be run like standard Java console application.
- AngularJS [1] is a JavaScript framework providing such functionalities like automatic data-binding between view and model and dependency injection,
- Spring JDBC [3] - makes using JDBC driver easier by automatic opening and closing connections, result sets and statements, handling SQLException, handling transactions, iterating through result sets.

No ORM (Object-relational mapping) tool (like Hibernate) was used because it could affect the performance of the application.

4.1   SQL Implementation

Application was tested with PostgreSQL. Spring JDBC library was used for SQL data access. Figure 1 contains a data model for the relational database.

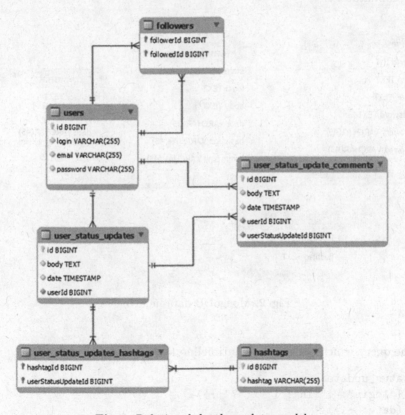

Fig. 1. Relational database data model

One of the most complex queries used in application was query which selects user's timeline. In case of SQL database it has the following structure:

```
SELECT user.login login, update.id id, update.date date,
 update.body body
FROM user_status_updates update
JOIN users user ON user.id = update.userId
JOIN followers f ON f.followedId = user.id
WHERE f.followerId = ?
ORDER BY update.date DESC
LIMIT 20 OFFSET (CURRENT_PAGE - 1) * 20
```

4.2   MongoDB Implementation

For MongoDB data access the official Java driver was used. As for relational database. Figure 2 contains data model for MongoDB database. It contains three document collections (comments documents are nested in status_updates documents). Nesting data results in a smaller number of data objects than in the relational database.

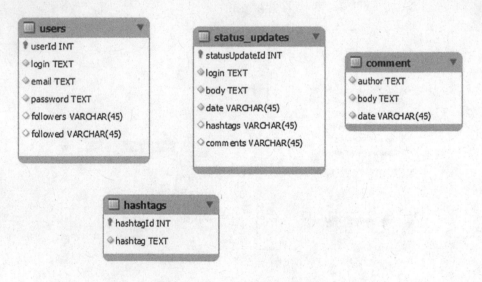

Fig. 2. MongoDB data model

The query which retrieves user's timeline looks as follows:

```
db.status_updates.
find({"login": {"$in": ["?","?"]}}).
sort({date: -1}).
skip((CURRENT_PAGE - 1) * 20).
limit(20);
```

Where in place of questions marks we put logins of followed users.

4.3 Cassandra Implementation

The DataStax driver was used for Cassandra data access. As for SQL and MongoDB databases. Figure 3 contains the data model schema. Yellow keys stand for partition keys and red keys for clustering keys. Arrows indicate the direction of sorting for the column defined during table creation. This data model is based on the model proposed by Brown [6].

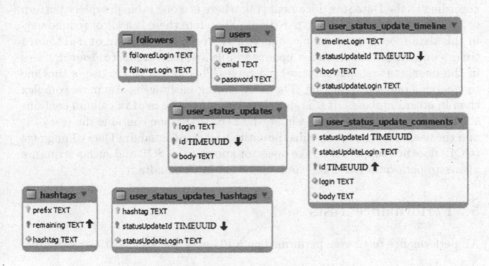

Fig. 3. Cassandra data model

In case of Cassandra, selecting user's timeline is more complex. For the first page of data query looks like this:

```
SELECT statusUpdateLogin, statusUpdateId,
toTimestamp(statusUpdateId) as date, body
FROM user_status_update_timeline WHERE timelineLogin = ?
```

For every subsequent page we had to add another condition in WHERE clause:

```
SELECT statusUpdateLogin, statusUpdateId,
toTimestamp(statusUpdateId) as date, body
FROM user_status_update_timeline
WHERE timelineLogin = ? and statusUpdateId < ?
```

Where in place of questions marks we put id of last status update from previous page, for example for second page it would be id of last status update from the first page (20th status update).

4.4   Comparison of Models

Data models for compared databases are entirely different. The SQL data model was designed to avoid redundancy and use relations between data. Therefore in queries there are many joins which can be very inefficient for large data sets.

The data model for MongoDB is the simplest one. By using features like nested documents and arrays, it consists of three collections of documents.

The data model for Cassandra is the most complex one. It was designed according to the DataStax document [14], where is a one table per query pattern to avoid reading from multiple partitions. Therefore there is a lot of redundancy in this data model. For example, one post is stored 1 + number_of_followers times - once in the user_status_updates table and number_of_followers times in the user_status_update_timeline table. This allows to get user's timeline by querying only one partition. The table storing hashtags is also more complex than in other databases. It contains three columns - the prefix column contains the first two characters of a hashtag, the remaining one contains the rest of it and the hashtag column contains the entire hashtag. Cassandra Query Language (CQL) does not support the like operator known from SQL and such a structure allows to perform a full-text search operation in Cassandra.

5   Performance Tests

All performance tests were performed on a PC with the specifications involving:

- Intel Core i5-4460 3,2 GHz processor,
- 16 GB RAM DDR3,
- Western Digital Blue 1TB SATA 3 7200rpm,
- Windows 10 Home Edition.

The test uses the following databases:

- PostgreSQL 9.5 for Windows x64,
- MongoDB 3.2 for Windows x64,
- Apache Cassandra 3.7.

For maximum reliability before every test defragmentation was performed. JMeter was chosen as a tool supporting the tests. For MongoDB and Cassandra the writing options were set in such a way that a write was successful only after saving data on the physical disk. To maximise the speed of reading the data in the databases, indexes were defined on the columns used in the query conditions.

5.1   Simulating Users Traffic

The first type of tests were those simulating the use of the application by 100 users simultaneously. Each test lasted for 5 min. The test plan was as follows:

- login to the application,
- view the first four pages of posts sent by current user,
- view the first four pages of posts from current user's timeline,
- send new post marked with two hashtags.

Each database was tested on four different data sets. Each data set contained a different number of users and posts: 1000 users and 1 million posts, 5000 users and 5 million posts, 10 000 users and 10 million posts, 15 000 and 15 million posts. For each data set, every user followed 100 other users.

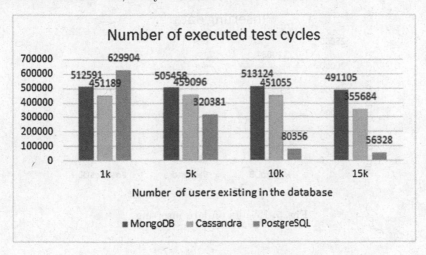

Fig. 4. Number of cycles performed during the 5-min test

Table 1. Average execution time of individual operations

Average execution time [ms]	PostgreSQL				MongoDB				Cassandra			
Number of users [*10³]	1	5	10	15	1	5	10	15	1	5	10	15
Log in	30	49	36	24	4	3	4	2	29	27	26	31
Read one page of posts	36	66	47	24	3	2	3	2	24	23	23	30
Read one page of timeline	43	99	828	1236	6	5	6	5	24	24	25	37
Send new post	122	214	196	264	535	552	537	570	421	414	424	520

Figure 4 contains information about the number of test cycles executed during a 5-min test. For a small data set PostgreSQL is the fastest database, but its efficiency drops dramatically with an increasing amount of data. For largest data sets the number of executed test cycles is several times smaller than for the other databases. Table 1 shows that the slowest operation of PostgreSQL is reading posts from timeline. MongoDB is the most efficient for large data sets. Its performance slightly decreased only for the largest dataset. Cassandra recorded a significant drop in performance only for 15 000 users.

5.2 Data Inserting

Another operation examined was inserting data. A single test consisted of inserting 1000 records to a table/collection that stores posts. Figure 5 contains the results. It shows that MongoDB is the slowest when adding data. This is the effect of using the journalled write concern which causes database return success status only after saving data on the physical disk.

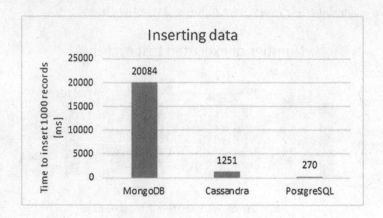

Fig. 5. Results of data inserting

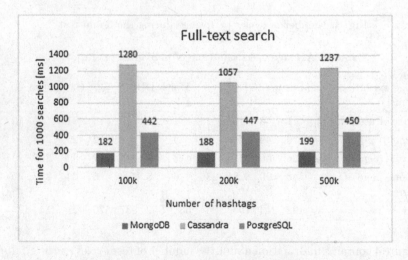

Fig. 6. Results of full-text searching

5.3 Full-Text Search

The last test was searching for hashtags that start with a specified pattern. For each database three tests were performed, each for different number of hashtags.

Figure 6 shows how long it takes to perform 1000 full-text searches. It turns out that the slowest is Cassandra which needs several times more time to perform this task than the other databases. MongoDB is about twice as fast as PostgreSQL.

6 Summary

The aim of this paper was to compare relational and non-relational databases. For the purpose of this work a social-media web application was created. The application was used to examine the performance of the selected databases.

All the databases provide a convenient interface for Java. Implementation of certain functions, such as pagination and full-text search is more complicated for Cassandra due to the fact that the query language is not as rich the SQL or MongoDB data access interface.

For selected data models the results show the performance advantages of non-relational databases over relational ones. For sufficiently large sets, the number of operations performed by a relational database is several times smaller. MongoDB was the fastest database in the context of reading. Only in the case of writing data was SQL the fastest.

The status of relational databases on the market is not at risk and it is hard to imagine that this will soon change. NoSQL databases are currently still a new and little-known solution. However, further development of the Internet and mobile devices will force software developers into increasing use of NoSQL databases.

Our future plans cover performance analysis of application NoSQL databases in BigData. This area grows rapidly and recent research [11] show that this trend is promising.

References

1. AngularJS. https://angularjs.org
2. Spring Boot. https://projects.spring.io/spring-boot/
3. Spring JDBC. https://docs.spring.io/spring/docs/current/spring-framework-refer ence/html/jdbc.html
4. Boicea, A., Radulescu, F., Agapin, L.I.: MongoDB vs Oracle - database comparison. In: Proceedings of Third International Conference on Emerging Intelligent Data and Web Technologies (EIDWT), pp. 330–335 (2012)
5. Brewer, E.: Cap twelve years later: how the rules have changed. Computer 45(2), 23–29 (2012)
6. Brown, M.: Learning Apache Cassandra. Packt Publishing, Birmingham (2015)
7. Chandra, D.G.: BASE analysis of NoSQL database. Future Gener. Comput. Syst. 52, 13–21 (2015)
8. Chodorow, K., Dirolf, M.: MongoDB: The Definitive Guide, 1st edn. O'Reilly Media, Sebastopol (2010)
9. Choi, Y.L., Jeon, W.S., Yoon, S.H.: Improving database system performance by applying NoSQL. JIPS 10, 355–364 (2014)

10. Codd, E.F.: A relational model of data for large shared data banks. Commun. ACM **13**(6), 377–387 (1970)
11. Gupta, S., Narsimha, G.: Efficient query analysis and performance evaluation of the NoSQL data store for bigdata. In: Satapathy, S.C., Prasad, V.K., Rani, B.P., Udgata, S.K., Raju, K.S. (eds.) Proceedings of the First International Conference on Computational Intelligence and Informatics. AISC, vol. 507, pp. 549–558. Springer, Singapore (2017). doi:10.1007/978-981-10-2471-9_53
12. Han, J., Haihong, E., Le, G., Du, J.: Survey on NoSQL database. In: 6th International Conference on Pervasive Computing and Applications (ICPCA), pp. 363–366 (2011)
13. Hewitt, E.: Cassandra: The Definitive Guide, 1st edn. O'Reilly Media, Sebastopol (2010)
14. Hobbs, T.: Basic Rules of Cassandra Data Modeling. http://www.datastax.com/dev/blog/basic-rules-of-cassandra-data-modeling
15. Jatana, N., Puri, S., Ahuja, M., Kathuria, I., Gosain, D.: A survey and comparison of relational and non-relational database. Int. J. Eng. Res. Technol. **1**(6) (2012)
16. Lee, C.H., Zheng, Y.L.: SQL-to-NoSQL schema denormalization and migration: a study on content management systems. In: Proceedings of IEEE International Conference on Systems, Man, and Cybernetics (SMC), pp. 2022–2026 (2015)
17. Li, X., Ma, Z., Chen, H.: QODM: a query-oriented data modeling approach for NoSQL databases. In: Advanced Research and Technology in Industry Applications (WARTIA), pp. 338–345 (2014)
18. Li, Y., Manoharan, S.: A performance comparison of SQL and NoSQL databases. In: Proceedings of IEEE Pacific Rim Conference on Communications, Computers and Signal Processing (PACRIM), pp. 15–19 (2013)
19. Lourenço, J.R., Cabral, B., Carreiro, P., Vieira, M., Bernardino, J.: Choosing the right NoSQL database for the job: a quality attribute evaluation. J. Big Data **2**, 1–26 (2015)
20. NVidia Corporation: Db engines ranking. http://db-engines.com/en/ranking
21. Plechawska-Wójcik, M., Rykowski, D.: Comparison of relational, document and graph databases in the context of the web application development. In: Grzech, A., Borzemski, L., Świątek, J., Wilimowska, Z. (eds.) Information Systems Architecture and Technology: Proceedings of 36th International Conference on Information Systems Architecture and Technology – ISAT 2015 – Part II. AISC, vol. 430, pp. 3–13. Springer, Cham (2016). doi:10.1007/978-3-319-28561-0_1
22. Sullivan, D.: NoSQL for Mere Mortals. Addison-Wesley, Boston (2015)
23. Truica, C.O., Radulescu, F., Boicea, A., Bucur, I.: Performance evaluation for crud operations in asynchronously replicated document oriented database. In: Proceedings of 20th International Conference on Control Systems and Computer Science, pp. 191–196 (2015)
24. Vokorokos, L., Uchnar, M., Lescisin, L.: Performance optimization of applications based on non-relational databases. In: International Conference on Emerging eLearning Technologies and Applications (ICETA), pp. 371–376 (2016)

Using Genetic Algorithms to Optimize Redundant Data

Iwona Szulc, Krzysztof Stencel$^{(\boxtimes)}$, and Piotr Wiśniewski

Faculty of Mathematics and Computer Science,
Nicolaus Copernicus University, Toruń, Poland
{iwa,stencel,pikonrad}@mat.umk.pl

Abstract. Analytic queries can exhaust resources of the DBMS at hand. Since the nature of such queries can be foreseen, a database administrator can prepare the DBMS so that it serves such queries efficiently. Materialization of partial results (aggregates) is perhaps the most important method to reduce the resource consumption of such queries. The number of possible aggregates of a fact table is exponential in the number of its dimensions. The administrator has to choose a reasonable subset of all possible materialized aggregates. If an aggregate is materialized, it may produce benefits during a query execution but also instigate a cost during data maintenance (not to mention the space needed). Thus, the administrator faces an optimisation problem: knowing the workload (i.e. the queries and updates to be performed), what is the subset of all aggregates that gives the maximal net benefit? In this paper we present a cost model that defines the framework of this optimisation problem. Then, we compare two methods to compute the optimal subset of aggregates: a complete search and a genetic algorithm. We tested these metaheuristics on a fact table with 30 dimensions. The results are promising. The genetic algorithm runs significantly faster while yielding solutions within 10% margin of the optimal solution found by the complete search.

Keywords: Materialized views · Aggregations · Complete search · Genetic algorithm

1 Introduction

A series of articles [6,7,20,21] discusses materialization of partial aggregations. Their purpose is to accelerate the execution of analytic queries. In our research, we assume a small-to-medium enterprise that uses simple open-source tools and unsophisticated inexpensive engines. Such an enterprise usually uses an object-relational mapper (ORM) to interface with a single open-source DBMS instance. Therefore, we have aimed at extending ORM tools with facilities for analytical querying. We developed algorithms that analyse workloads of HQL queries and identify all aggregations (called *metagranulas*) on fact tables that appear in the workload. A subset of these metagranulas is picked to be materialized in the database. We also presented a cost model that quantifies various subsets of

S. Kozielski et al. (Eds.): BDAS 2017, CCIS 716, pp. 165–176, 2017.
DOI: 10.1007/978-3-319-58274-0_14

metagranulas and allows picking the cheapest one with respect to the given work-load. We tested the complete search among all subsets of metagranulas. Finally, we prepared rewriting methods of HQL queries so that they use materialized metagranulas instead of base data.

In our solution that has been prototypically implemented as proof-of-concept, all peculiarities of the solution are hidden from application programmers. The ORM tool at hand (1) analyses the workload, (2) picks the optimal subset of metagranulas to materialize, (3) creates necessary database objects, and (4) rewrites arriving queries to use these materializations. However, there is one more problem to be solved. It is the cost of the complete search in the space of all subsets of metagranulas. The size of this space is exponential in the number of dimensions of the fact table (i.e. possible aggregations). In this paper we address the problem by applying a meta-heuristic called genetic programming. We use it to speed the search in the space of all subsets of metagranulas. The results of experiments show that genetic algorithms yield results close to optimal at significantly reduced computational complexity.

The contributions of this article are the following:

– It proposes a genetic algorithm to find the optimal subset of metagranulas.
– It show experimental results proving that such an algorithm is efficient, yet accurate.
– It also analyses the stability of this algorithm.

There are numerous granula-based architectures. They usually are columnar storages (see e.g. [1]) with profound examples of such database engines being Brighthouse [17–19] and Netezza [8]. In these systems granulas are used e.g. (1) to dynamically (i.e. at run-time) eliminate granulas that are not relevant and/or (2) to estimate the final or partial result. Our proposal is purely compile-time—it is applied only at the query rewrite and workload analysis.

Materialized aggregates can also be treated as partial stored results. A example of an implementation of such views are *FlexViews* [5] within MySQL based on the results described in [15,16]. FlexViews rely on applying changes that have been written to the change log. In some systems partial results are dynamically reusable [10,13]. In our solution the aggregates are statically chosen upfront and maintained automatically. Granula metadata may also be related to multi-dimensional histograms used for a better optimisation of a query plan [2,9]. Such metadata can be used in adaptive query processing [4] and/or self-tuning [3]. So far, our solution is an off-line tuning advisor. However, it can be also used as an on-line self-tuning mechanism or even an adaptive driver. It can have the form of a corrective query planner [11] or mid-query reoptimizer [12,14].

This article is organized as follows. Section 2 rolls out the problem and intro-duces a motivating example. Section 3 presents the cost model. Section 4 presents the use genetic programming and shows the results of its application. Section 5 concludes and discusses possible future work.

2   Graph of Views and Materialized Views with Fast Refresh

In this article a *materialized view* is assumed to be a materialized view with *fast refresh on commit*. If the discussed techniques are used with a DBMS that does not support such an option, such views are emulated by redundant tables with appropriate triggers that keep the materialized aggregates up to date. Prototype tools that automate the process to create such redundant tables and their triggers are presented [6,7,20,21].

As a running example in this article we use a simple database schema presented on Fig. 1. A database of such schema stores transactional data on sales. The amount of data is big enough to prohibit the naïve (direct) execution of analytical queries. On the other hand their size (and also the size and budget of the owning company) does not justify usage of heavy expensive solutions such as data warehouses.

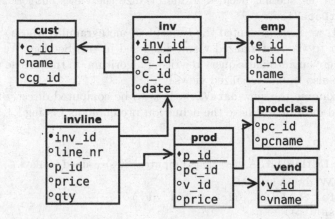

Fig. 1. The schema of a sample sales database

In such circumstances we can think of an intermediate solution. It can consist in using a number of materialized views with fast refresh on commit or redundant tables with triggers that keep their contents up to date (in case of a DBMS that does not support them).

Assume that a business user need to query the following views:

- **invoiceView (i)** that shows total sales for each invoice.
- **productDateView (pd)** that aggregates sales for each day and product.
- **customerDateView (cd)** that provides sums of sales grouped by a day and a customer.
- **dateView (d)** that computes total sales for each day.
- **customerMonthView (cgm)** that sums sales in each month and for each customer.
- **productMonthView (pm)** contains sales grouped by month and product.

These views form a graph of a natural partial order as shown on Fig. 2. A view A is smaller that another view B, if and only if we can compute the content of B by querying A. For example, the content of `customerMonthView` can be computed from the rows of `customerDateView`. It can also be computed from the content of `invoiceView` and the base table `inv_line`. Therefore, in the abovementioned natural order the view `customerMonthView` is bigger that views `customerDateView`, `invoiceView` and the base table `inv_line`.

If in our example application the number of database queries is significantly bigger that the number of updates, the simplest approach will be to create a materialized view with fast refresh on commit for all six listed views. However, one must take into account also the cost of the maintenance of these views that can annihilate the benefits caused by faster query execution. In practice, materializing all needed aggregates is hardly possible.

Assume that the database administrator has estimated the costs and benefits for his/her choice of views to materialize (see Sect. 3 that defines the cost model). Then, he/she has decided to materialize the views `customerDateView` and `productDateView`.

In [20, 21] we have presented the concept of metagranulas whose functionality is similar to the materialized views discussed in this Section. We have also described there an automatic query rewriting algorithm. In this article we extend it to rewrite also the materialized views.

Let us examine the view `dateView`. It is to be computed directly from base transactional data, it will have the definition presented as Listing 1.1.

Listing 1.1. A view that computes sales grouped by days

```
CREATE VIEW dateView AS
SELECT date, SUM( inl . price * inl . qty )
    AS sum_pricexqty
  FROM inv JOIN invline inl USING ( inv_id )
  GROUP BY inv . date
```

However, we have assumed to materialize the view `productDateView` (see Listing 1.2). If such a materialized view is set up, it will be reasonable to consider putting some indices on it. Our algorithm will analyse the partial order of views and suggest to create the view `dateView` according to Listing 1.3. This same mechanism will be used to rewrite queries that address various levels of aggregation. Thus, the query for total sales on best five days in 2016 formulated as shown on Listing 1.4 will be rewritten to the form presented as Listing 1.5.

3   Cost Model

Example queries from Listings 1.2 and 1.5 are based on the same materialized view, because both queries have the same groupings. In our discussion on the

Listing 1.2. A view that computes sales grouped by days and product

```
CREATE MATERIALIZED VIEW productDateView
WITH FAST REFRESH ON COMMIT
 AS
SELECT pr_id , date, SUM( inl . price * inl . qty )
      AS sum_pricexqty
 FROM inv JOIN invline inl USING ( inv_id )
 GROUP BY inv . date ;
```

Listing 1.3. A view that computes sales grouped by days based on `productDateView`

```
CREATE VIEW dateView AS
SELECT date , SUM( pd . sum_pricexqty )
      AS sum_pricexqty
 FROM productDateView pd
 GROUP BY pd . date
```

cost of query execution and data maintenance, we will assume that queries based on the same aggregation level are identical. The class of queries based on the same grouping level is called a *metagranula* [21]. The natural order discussed in Sect. 2 is analogously applied to metagranulas. Thus, the graph shown on Fig. 2 becomes a diagram of ordered metagranulas.

We identify two kind of metagranulas. A metagranula is called *proper* if the corresponding view is materialized with fast refresh on commit. Thus, queries of such a metagranula will be answered using this materialized view, i.e. very efficiently. If a metagranula is not proper, it will be called *virtual*.

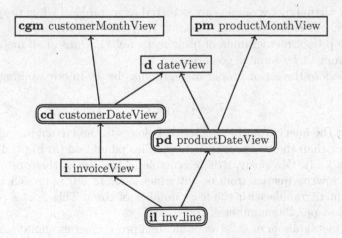

Fig. 2. The graph of the partial order of metagranulas (classes of views).

Listing 1.4. A query for 5 best sale days in 2016.

```
select * from (
  select date, sum(inl.price * inl.qty)
  from inv join invline inl USING (inv_id)
  where EXTRACT(YEAR FROM date) = 2016
  group by date1
  order by sum(inl.price * inl.qty) DESC
)
where rownum <= 5
```

Listing 1.5. A query for 5 best sale days in 2016.

```
select * from (
  select date, sum(pd.sum_pricexqty)
  from productDateView pd
  where EXTRACT(YEAR FROM date) = 2016
  group by date
  order by sum(pd.sum_pricexqty) DESC
)
where rownum <= 5
```

Let $\mathcal{M}$ denote the set of all metagranulas corresponding to the views from Sect. 2:

$$\mathcal{M} = \{\mathtt{il}, \mathtt{i}, \mathtt{pd}, \mathtt{cd}, \mathtt{d}, \mathtt{cgm}, \mathtt{pm}\}$$

For a given metagranula $m \in \mathcal{M}$ the symbol $|m|$ denotes the number of rows in the materialized view associated with the metagranula m. Obviously, if $m, n \in \mathcal{M}$ and $m < n$, then $|n| \leq |m|$.

In Fig. 2 virtual metagranulas are depicted as rectangles, while proper metagranulas are portrayed as ovals. Observe that each metagranula is bigger or equal to the proper metagranula of basic facts, i.e. $\mathtt{il}$. Data in all metagranulas is derived from $\mathtt{il}$ by some aggregation.

Let us denote the set of proper metagranulas by $\mathcal{P}$. In our running example it will be:

$$\mathcal{P} = \{\mathtt{il}, \mathtt{pd}, \mathtt{cd}\}$$

Consider the query from Listing 1.4. It belongs to the virtual metagranula $\mathtt{d}$, that is bigger than the two proper metagranulas $\mathtt{pd}$ and $\mathtt{cd}$ (in Fig. 2 $\mathtt{d}$ is placed above $\mathtt{pd}$ and $\mathtt{cd}$). Obviously, it is reasonable to use one of these proper metagranulas to rewrite queries from $\mathtt{d}$, but which one? In our approach we always choose the metagranula with the least number of rows. This causes processing of the smallest possible number of rows.

For any metagranula $m \in \mathcal{M}$ we indicate a proper metagranula $\widehat{m}$ such that:

1. $\widehat{m} \leq m$
2. if $n \in \mathcal{M}$ and $n \leq m$ then $|\widehat{m}| \leq |n|$

In particular, if $m \in \mathcal{P}$, it obviously holds that $\widehat{m} = m$.

Using the notions introduced above we propose a cost function that takes into account the cost of view maintenance for a given subset of metagranulas. Let $m \in \mathcal{M}$. Then for a given workload the cost of this metagranula is proportional to:

- the number of executions of queries of this metagranula denoted by $fr(m)$, and
- the cardinality of the associated proper metagranula, i.e. $|\widehat{m}|$.

On the other hand, the cost of the maintenance of materializations is proportional to the number of proper metagranulas. This thoughts lead to the following cost function.

$$\Phi(\mathcal{P}) = |\mathcal{P}| \sum_{m \in M} |(\widehat{m})| fr(m))$$

The proposed cost model has been tested on a sample database of schema shown on Fig. 1. The size of this database is about 500 GiB. The sizes of all tables are listed in Table 1.

Table 1. Row counts of tables from the example schema

Table name	Row count
Cust	1 800 000
Inv	739 678 300
Invline	7 490 614 234
Prod	48 000

In experiments we use the relational database management system PostgreSQL. Since it does not support materialized views with fast refresh on commit, the data materialization was emulated using redundant tables and triggers that keep derived data up to date. We tested three following database instances:

- **plain** that contains no proper mategranulas (apart from the base table that always must be stored);
- **medium** that corresponds to Fig. 2, i.e. with proper metagranulas cd and pd.
- **full** that has all metagranulas from Fig. 2 made proper.

Table 2 shows the sizes of the three databases after their proper metagranulas has been established.

As we have noted earlier, the choice of proper metagranulas directly impacts both the execution time of analytical queries and the time needed for redundant data maintenance.

Table 3 presents execution times of queries from particular metagranulas. Let us take a closer look at the executions times for metagranula d. The same query

Table 2. Proper metagranulas and volumes of tested database instances

Database name	Proper meta-granulas	Volume
Plain	il	424 GiB
Med	cd, il, pd	460 GiB
Full	cd, cgm, d, i, il, pd, pm	510 GiB

rewritten to use the proper metagranula cd executes significantly slower than rewritten to the metagranula pd. When we examine the sizes of metagranulas, we will find out that the cardinality of the metagranula cd is about 500 000 000, while the metagranula pd has about 60 000 000 rows. This observations attests the reasonability of our method to choose the proper metagranula for a given virtual metagranula. Table 4 shows the increase of resource consumption during data loading caused by the increase of the number of proper metagranulas. The results indicate that the proposed cost model is not flawless. However, it amounts to be a useful tool to find the optimal set of metagranulas to be materialized. If the number of metagranulas is reasonably small, the cost model can be used in a complete search for this task.

Table 3. The summary of query execution times for tested database instances.

Database	Query in cm	Query in pm	Query in i	Query in d	
Plain	>2 h	>2 h	>2 h	>2 h	
Med	152 s	159 s	>2 h	56.2 s (cd)	13 s (pd)
Full	<1 s	<1 s	215 s	<1 s	

Table 4. The time spent on inserting new invoices and synchronizing proper metagranulas

Invoices	Plain	Med	Full
5 000	1.62 s	1 m 33 s	6 m 35 s
10 000	3.15 s	2 m 51 s	13 m 05 s
20 000	5.28 s	5 m 29 s	26 m 08 s
50 000	13.39 s	14 m 50 s	65 m 23 s

4   Genetic Solution

Assume a slightly bigger set of metagranulas. It could be for example the set of 31 metagranulas ordered as Fig. 3 shows. In such circumstances the complete search could be too slow.

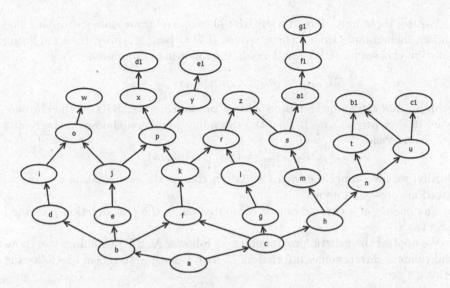

Fig. 3. A big sample graph of metagranulas.

The choice of proper metagranulas to be materialized out of 31 metagranulas from Fig. 3 is a selection of a subset of a set with 31 items. Obviously, this set must contain the base table, so we have "only" 30 metagranulas to choose from. Thus, we have to search in the space of 2^{30} subsets. We implemented the complete *brute-force* search that checks all possible subsets. The execution of it took ten hours on an average desktop computer. As the result we obtained one hundred best subsets of metagranulas.

Of course, if the set of metagranulas were even bigger (say 40, 50, 60 and so on), the time to execute the complete search would be unacceptable. Therefore, we switched to approximate methods. We chose the genetic programming approach as described below.

The goal of the algorithm is again the selection of the optimal (with respect to the cost function presented in Sect. 3) set of metagranulas, i.e. a subset $\mathcal{P} \subset M$. Below we define all notions necessary to create a genetic algorithm, namely the chromosome, the crossover, the mutations and the quality measure.

A *chromosome* is any subset $\mathcal{P} = \{m_1, \ldots, m_k\} \subset M$. For two chromosomes $\mathcal{P}' = \{m'_1, \ldots, m'_{k'}\} \subset M, \mathcal{P}'' = \{m''_1, \ldots, m''_{k''}\} \subset M$ we define the *crossover* operation to be:

$$\mathcal{C} = (\mathcal{P}' \cap \mathcal{P}'') \cup \{m_1, \ldots, m_k\},$$

where $\{m_1, \ldots, m_k\}$ is a randomly chosen subset such that both following conditions are satisfied simultaneously:

$$\{m_1, \ldots, m_k\} \subseteq \mathcal{P}'' \cup \mathcal{P}'' - \mathcal{P}'' \cap \mathcal{P}''$$

$$\min(|\mathcal{P}'|, |\mathcal{P}''|) \le |\mathcal{C}| \le \max(|\mathcal{P}'|, |\mathcal{P}''|)$$

We use three mutations. Firstly, the *loosing mutation* causes dropping one random metagranula from a chromosome. If $\mathcal{P} = \{m_1, \ldots, m_k\}$, then the losing mutation chooses an index i and yields the following chromosome:

$$M_{losing} = \{m_1, \ldots, m_i, m_{i+1}, \ldots, m_k\}$$

Secondly, the *extending mutation* adds a random metagranula to the chromosome. If $\mathcal{P} = \{m_1, \ldots, m_k\}$, then the extending mutation chooses metagranula $m \notin \mathcal{P}$ and yields:

$$M_{extending} = \{m_1, \ldots, m_k, m\}$$

Thirdly, we also apply the *mixed mutation* that is the combination of the two mutations presented above.

The *quality* of a chromosome is computed using the cost function presented in Sect. 3.

We applied the genetic programming as follows. At the beginning, one thousand random chromosomes are chosen. In each iteration we repeat the following actions:

1. Sort the chromosomes with respect to their quality.
2. Leave 9 best chromosomes intact.
3. Out of remaining chromosomes, we choose a group to be subjects of mutations (each chromosome undergoes at most one mutation).
4. Choose a random number of pairs of chromosomes and crossover them. For each selected pair the worst chromosome of the three (two original ones and the crossover) is dropped.

The algorithm is stopped after one hundred iterations.

We implemented this procedure in Java and tested on an average desktop computer (the same as in case of the brute-force method). We used the graph of metagranulas as presented on Fig. 3. In each test run we randomly modified the sizes and usage frequencies of metagranulas.

The average time to complete the brute-force algorithm was 10 h. As a result, we collected 100 best metagranulas and their costs. In case of the genetic program described above its run-time was always below one minute.

We completed 20 such test runs. In 17 runs the genetic algorithm found the optimal solution. In 2 runs the genetic algorithm yielded a solution from top twenty as indicated by the brute-force. They were the second and sixteenth best metagranulas. The costs slightly diverged from the optimal cost. Once the genetic algorithm produced a significantly worse chromosome that placed out of top hundred solutions. In vast majority of cases our genetic method produced an optimal or at worst an acceptable solution. Therefore, in our opinion the presented genetic solution is robust.

5   Conclusions and Future Work

In this article we presented a new method to select the set of aggregates to be materialized in a relational database. Instead of a complete search in the space of all possible subsets of dimensions to aggregate, we employ a genetic algorithm. The experiments have shown that the genetic algorithm produces a solution that is at most slightly worse that the optimal solution depending on the nature of the ordering of metagranulas.

The presented results inspire further research. Firstly, the experiments should be repeated on bigger sets of metagranulas. It could be interesting, how TCP-H looks like if we analyse it using the metagranula approach.

Secondly, the metagranula diagrams considered in this article have only one minimal element (one peak), since all metagranulas are induced by one fact table. It could be interesting to examine production databases whose workloads have interesting aggregate queries based on more complex database schemata.

Thirdly, the proposed graph analysis methods are worth integrating with schema management tools (e.g. *Liquibase*) and code analysis tools that can identify most interesting analytical queries.

References

1. Boncz, P.A., Manegold, S., Kersten, M.L.: Database architecture evolution: mammals flourished long before dinosaurs became extinct. PVLDB **2**(2), 1648–1653 (2009). http://www.vldb.org/pvldb/2/vldb09-10years.pdf
2. Bruno, N., Chaudhuri, S., Gravano, L.: STHoles: a multidimensional workload-aware histogram. In: Proceedings of the 2001 ACM SIGMOD International Conference on Management of data, Santa Barbara, CA, USA, 21–24 May 2001, pp. 211–222 (2001). http://doi.acm.org/10.1145/375663.375686
3. Chaudhuri, S., Narasayya, V.R.: Self-tuning database systems: a decade of progress. In: Proceedings of the 33rd International Conference on Very Large Data Bases, University of Vienna, Austria, 23–27 September 2007, pp. 3–14 (2007). http://www.vldb.org/conf/2007/papers/special/p3-chaudhuri.pdf
4. Deshpande, A., Ives, Z.G., Raman, V.: Adaptive query processing. Found. Trends Databases **1**(1), 1–140 (2007). http://dx.doi.org/10.1561/1900000001
5. Flexviews: Incrementally refreshable materialized views for MySQL, January 2012. http://code.google.com/p/flexviews/
6. Gawarkiewicz, M., Wiśniewski, P.: Partial aggregation using hibernate. In: Kim, T., Adeli, H., Slezak, D., Sandnes, F.E., Song, X., Chung, K., Arnett, K.P. (eds.) FGIT 2011. LNCS, vol. 7105, pp. 90–99. Springer, Heidelberg (2011). doi:10.1007/978-3-642-27142-7_11
7. Gawarkiewicz, M., Wiśniewski, P., Stencel, K.: Granular indices for HQL analytic queries. In: Kozielski, S., Mrozek, D., Kasprowski, P., Małysiak-Mrozek, B., Kostrzewa, D. (eds.) BDAS 2014. CCIS, vol. 424, pp. 30–39. Springer, Cham (2014). doi:10.1007/978-3-319-06932-6_4
8. Hindshaw, F., Metzger, J., Zane, B.: Optimized Database Appliance, Patent No. U.S. 7,010,521 B2, Assignee: Netezza Corporation, Framingham, MA, issued 7 March 2006
9. Ioannidis, Y.E.: The history of histograms (abridged). In: VLDB, pp. 19–30 (2003). http://www.vldb.org/conf/2003/papers/S02P01.pdf
10. Ivanova, M., Kersten, M.L., Nes, N.J., Goncalves, R.: An architecture for recycling intermediates in a column-store. ACM Trans. Database Syst. **35**(4), 24 (2010). http://dx.doi.org/10.1145/1862919.1862921

11. Ives, Z.G., Halevy, A.Y., Weld, D.S.: Adapting to source properties in process-ing data integration queries. In: Proceedings of the ACM SIGMOD International Conference on Management of Data, Paris, France, 13–18 June 2004, pp. 395–406 (2004). http://doi.acm.org/10.1145/1007568.1007613
12. Kabra, N., DeWitt, D.J.: Efficient mid-query re-optimization of sub-optimal query execution plans. In: SIGMOD 1998, Proceedings ACM SIGMOD International Conference on Management of Data, 2–4 June 1998, Seattle, Washington, USA, pp. 106–117 (1998). http://doi.acm.org/10.1145/276304.276315
13. Kalyvianaki, E., Wiesemann, W., Vu, Q.H., Kuhn, D., Pietzuch, P.: SQPR: stream query planning with reuse. In: Proceedings of the 27th International Conference on Data Engineering, ICDE 2011, 11–16 April 2011, Hannover, Germany, pp. 840–851 (2011). http://dx.doi.org/10.1109/ICDE.2011.5767851
14. Markl, V., Raman, V., Simmen, D.E., Lohman, G.M., Pirahesh, H.: Robust query processing through progressive optimization. In: Proceedings of the ACM SIGMOD International Conference on Management of Data, Paris, France, 13–18 June 2004, pp. 659–670 (2004). http://doi.acm.org/10.1145/1007568.1007642
15. Mumick, I.S., Quass, D., Mumick, B.S.: Maintenance of data cubes and summary tables in a warehouse. In: SIGMOD Conference, pp. 100–111 (1997)
16. Salem, K., Beyer, K., Lindsay, B., Cochrane, R.: How to roll a join: asyn-chronous incremental view maintenance. SIGMOD Rec. 29(2), 129–140 (2000). http://doi.acm.org/10.1145/335191.335393
17. Slezak, D., Synak, P., Borkowski, J., Wroblewski, J., Toppin, G.: A rough-columnar RDBMS engine - a case study of correlated subqueries. IEEE Data Eng. Bull. 35(1), 34–39 (2012). http://sites.computer.org/debull/A12mar/infobright1.pdf
18. Slezak, D., Synak, P., Wojna, A., Wroblewski, J.: Two database related interpre-tations of rough approximations: data organization and query execution. Fundam. Inform. 127(1–4), 445–459 (2013). http://dx.doi.org/10.3233/FI-2013-920
19. Slezak, D., Wroblewski, J., Eastwood, V., Synak, P.: Brighthouse: an ana-lytic data warehouse for ad-hoc queries. PVLDB 1(2), 1337–1345 (2008). http://www.vldb.org/pvldb/1/1454174.pdf
20. Wisniewski, P., Stencel, K.: Query rewriting based on meta-granular aggregation. In: CS&P, pp. 457–468 (2013)
21. Wisniewski, P., Stencel, K.: Query rewriting based on meta-granular aggregation. Fundam. Inform. 135(4), 537–551 (2014). http://dx.doi.org/10.3233/FI-2014-1139

Interoperable SQLite for a Bare PC

William Thompson, Ramesh Karne, Alexander Wijesinha^(✉), and Hojin Chang

Towson University, Towson, MD 21252, USA
awijesinha@towson.edu

Abstract. SQLite, a widely used database engine, has been previously transformed to run on a bare PC without the support of any OS or kernel. However, the transformed SQLite database was stored in main memory i.e., it had no file system. This paper extends the transformation process to enable bare PC SQLite to work with standard file system interfaces based on the FAT32 file specification. It further presents mechanisms and programming interfaces for a bare machine file system integrated with SQLite that uses a removable USB flash drive. The bare SQLite database and file system can interoperate with conventional OS-based database systems. It can be adapted in the future to work with bare Web browsers, large bare databases, other bare applications, and bare mobile devices.

Keywords: SQLite · Transformation · Interoperability · Performance

1 Introduction

SQLite is a self-contained, zero-configuration, stand-alone (not client/server) lean database management system. It is commonly used in Web browsers, mobile devices and embedded systems. A SQLite amalgamation [1] consisting of about 130K lines of code has been transformed to run on a bare PC with no operating system (OS) or kernel [15,16]. The bare PC SQLite version uses the :memory option, where the database is stored in real memory with no standard file interfaces. Potential advantages of running SQLite or applications such as Web servers or VoIP clients on a bare PC include the elimination of OS overhead and OS-related vulnerabilities.

This paper discusses the addition of a standard FAT32 file system [14] to the bare PC SQLite version so that it can interoperate with a conventional OS-based SQLite database engine. The lean bare PC FAT32 file system [13], which is intertwined with the associated application, uses a removable USB flash drive.

2 Bare Machine Computing

Approaches to reduce OS overhead and improve application performance include Exokernel [5], OS-Kit [21] and Palacio/Kitten [12]. In a bare machine computing or bare PC application, all intermediary software in the form of an OS, lean

© Springer International Publishing AG 2017
S. Kozielski et al. (Eds.): BDAS 2017, CCIS 716, pp. 177–188, 2017.
DOI: 10.1007/978-3-319-58274-0_15

kernel, or external libraries is eliminated. Bare applications are written in single programming language such as C/C++ and directly communicate to hardware without middleware or a centralized kernel [9,10]. Bare PC Web servers [7], Webmail servers [4], email servers [6] and split-servers [18,19] have been developed previously.

3   SQLite Transformation

The SQLite amalgamation package used in the transformation [16] has two source files (shell.c and sqlite3.c) and two header files. It runs on Windows/Visual Studio (VS). Detailed documentation on SQLite amalgamation is given in [2]. SQLite provides a command line interface and a file interface for user input. SQL queries can be run in single command mode or as a transaction. The database is stored using a standard SQLite file format in Windows or Linux. The transformation process is described in detail in [15,16]. It eliminated 85 system calls and replaced them with direct bare PC hardware interfaces. These system calls can be classified as file, timer, data types, process, memory, and standard I/O. The file system calls were not replaced as the database was intended to run in main memory. The remaining calls were replaced with equivalent bare PC calls, which are much simpler and do not require any centralized OS or kernel. These calls run in single user mode along with its application. Conventional database management systems use standard file systems that are provided by the host OS, and it is possible to port a database from one OS platform to another by using middleware tools [20]. However, these tools are themselves platform dependent and cannot be used to adapt an OS-based database/file system to run on a bare PC.

4   Design and Interfaces

Figure 1 shows respective architectural views of conventional and bare PC environments for running SQLite. We assume USB mass storage is used to store the SQLite database file based on the FAT32 file system. In a conventional environment such as Visual Studio (VS) on Windows, SQLite runs on top of the OS, which provides the necessary interfaces for virtual memory, file management and device drivers. In a bare PC, four objects MemObj, FileObj, UsbFileObj and UsbObj provide the complete functionality needed for the operation of SQLite. The MemObj provides real memory allocation and deallocation needed for SQLite (called by malloc() and free()).

The FileObj provides the above file API. A USB file object (UsbFileObj) provides initialization, reset and plug-and-play features for all USB ports. Finally, a USB device driver object (UsbObj) provides driver functionality [11] specific to bare PC applications, which uses USB 2.0 standard specification [17], enhanced host controller specification [8], and USB mass storage specification [3]. These four objects are an integral part of SQLite and a database application running on a bare PC.

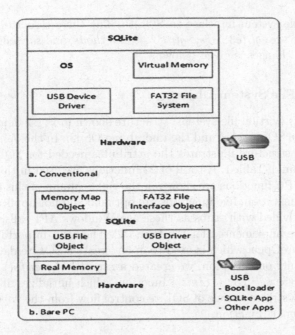

Fig. 1. Software architecture

We enhanced the bare PC file system in [13] to be fully compatible with the FAT32 specification so that it can interoperate with any OS platform. The enhanced file API has five functions, namely: createFile(), deleteFile(), resize-File(), flushFile(), and flushAll(). These are used to interface with the SQLite database. The createFile() function has a file name, memory address pointer, file size and file attributes. It returns a file handle. The file handle is the index value of the file in a file table structure, which has all the control information of a file. This approach enables a direct index into the file table to be used without the need for searching. The deleteFile() function uses the file handle to delete a file. The resizeFile() function is used to increase or decrease a previously allocated file size. The flushFile() function updates the USB mass storage device from its related data structures and memory. The flushAll() interface is used to flush all files and related structures onto the USB drive. A single executable includes all the bare PC code. SQLite runs as a separate task within the application. As SQLite is written in C, the code is wrapped in C++ to communicate with the object-oriented code in the bare PC. In a Windows environment, the USB contains the SQLite database; in a bare PC, it contains the SQLite database along with boot, load, SQLite application, and other applications such as a Web server if needed.

SQLite provides a separate wrapper for a given OS interface or virtual file system (VFS) [2]. This wrapper approach in SQLite motivated us to develop a bare PC API that substitutes for a given OS and enables the development of a bare PC file system interface to SQLite. The three structures changed in the SQLite code are *sqlite3_vfs*, *sqlite3_io_methods*, and *sqlite3_file*. A brief overview of the changes is given below.

4.1 Virtual File System Object

The *sqlite3_vfs* (virtual file system) structure shown in Fig. 2 defines the interface between the SQLite core and the underlying OS [2]. In the Windows OS, an instance of this structure illustrates the attributes needed for SQLite as shown in the left column of Table 1. A total of 22 functions are used in this object. The equivalent bare PC functions are shown in the right column of this table. For the bare PC implementation, five functions are optional and not applicable, and five methods are provided with stubs as these are Windows API calls. The remaining functions are implemented for the bare PC. The most important method in this object is bareOpen, which is used to open/create a SQLite database file. In order to substitute our function, we created a *register_barevfs()* function and inserted it into the *sqlite3_os_init()* function, which initializes all OS parameters. Figure 3 illustrates a trace of SQLite control flow from the "*open_db()*" call to the "bareOpen()" function.

```
struct sqlite3_vfs {
    int iVersion;              /* Structure version
number (currently 3) */
    int szOsFile;             /* Size of subclassed
sqlite3_file */
    int mxPathname;           /* Maximum file
pathname length */
    sqlite3_vfs *pNext;       /* Next registered
VFS */
    const char *zName;        /* Name of this
virtual file system */
    void *pAppData;           /* Pointer to
application-specific data */
    int (*xOpen)(sqlite3_vfs*, const char *zName,
sqlite3_file*, int flags, int *pOutFlags);
    int (*xDelete)(sqlite3_vfs*, const char *zName,
int syncDir);
    int (*xAccess)(sqlite3_vfs*, const char *zName,
int flags, int *);
    int (*xFullPathname)(sqlite3_vfs*, const char
*zName, int nOut, char *zOut);
    void *(*xDlOpen)(sqlite3_vfs*, const char
*zFilename);
    ...
    int (*xCurrentTime)(sqlite3_vfs*, double*);
    int (*xGetLastError)(sqlite3_vfs*, int, char *);
    int (*xCurrentTimeInt64)(sqlite3_vfs*,
sqlite3_int64*);
    int (*xSetSystemCall)(sqlite3_vfs*, const char
*zName, sqlite3_syscall_ptr);
    sqlite3_syscall_ptr
(*xGetSystemCall)(sqlite3_vfs*, const char
```

Fig. 2. SQLite virtual file system object

Table 1. Attributes for $sqlite3\_vfs$

Windows	Bare PC
iVersion	iVersion -1
sizeof(winFile)	sizeof(bareFile)
MAX_PATH	$MAXPATHNAME$ -512
pNext	pNext -0
win32	"bare"
pAppData	pAppData - 0
winOpen	bareOpen
winDelete	bareDelete
winAccess	bareAccess
winFullPathname stub	only
winDlOpen	stub only
winDlError	stub only
winDlSym	stub only
winDlClose	stub only
winRandomness	bareRandomness
winSleep	bareSleep
winCurrentTime	bareCurrentTime
winGetLastError	N/A (optional)
winCurrentTimeInt64	N/A
xSetSystemCall	N/A
xGetSystemCall	N/A
xNextSystemCall	N/A

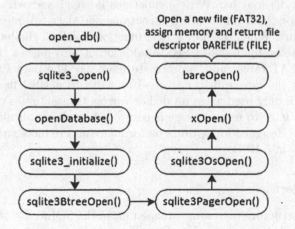

Fig. 3. Trace of SQLite control flow

4.2 I/O Method Object

SQLite manipulates the contents of the file system using a combination of four types of file operations: create, delete, truncate and write. The SQLite object named *sqlite3_io_method* consists of I/O functions as shown in the left column of Table 2. There are 17 functions in the Windows API related to the SQLite VFS. Eight functions are not applicable to a bare PC environment and the other nine are implemented for bare PC applications as shown in the table.

Table 2. I/O methods for *sqlite3_file*

Windows	Bare PC
iVersion	iVersion -1
winSync	bareSync
winFileSize	bareFileSize
winLock bareLock	N/A
winUnlock	bareUnlock -N/A
winCheckReservedLock	bareCheckReservedLock -N/A
winFileControl	bareFileControl -N/A
winSectorSize	bareSectorSize
winDeviceCharacteristics	bareDeviceCharacteristics
winShmMap	N/A (optional)
winShmLock	N/A
winShmBarrier	N/A
winShmUnmap	N/A

The bareRead() and bareWrite() functions do read and write respectively using real memory, and all the USB structures and data are memory mapped into main memory. When the bareSync() function is called, the bare PC system flushes the database file to the USB. In addition, it also flushes the root directory, modified FAT tables and appropriate file data. The bare PC application directly invokes the device driver to read or write to a USB flash drive. The flush operation is only used when needed or when a transaction is complete. The frequency of updates to mass storage is controlled by the application. The bare PC file system is designed for optimal performance and reduces write operations as they are slow in a USB.

4.3 File Object

The file object structure represents an open file in the SQLite OS interface layer. We have extended this object in bareFile. The extended structure consists of the *_iobuf* structure which contains parameters that are needed to implement the bare PC file system. For example, cacheStartAddr is the real memory address provided by the memory object. The index value points to the entry in the file table. The openFile method also has an instance of a bareFile which is a *sqlite3_file* type. This bare PC file instance is linked with bareio which points to all the functions needed in a bare PC. These are the *sqlite3_io_methods* described in the previous section. The implementation of the above functions is done in C. The size of the new code is approximately 1,300 lines including comments. The new SQLite runs on a bare PC FAT32 file system.

5   Interoperability and Performance

The SQLite database file that runs on a bare PC is interoperable with one running on VS/Windows. Thus, a database can be created in Windows or on a bare PC and used in either environment. The same can be done with data updates. In conventional systems, interoperability is achieved by porting the database management system to run on a different platform. For example, an Oracle DBMS running on Linux can be ported to run on Windows.

While such database systems have OS-specific dependencies (e.g. Oracle Linux, Oracle Solaris, Oracle Windows), the bare PC SQLite database system is independent of any OS platform; it can run on any x86 based machine with a USB 2.0 interface and create a database file (FAT32 format) that can be used by SQLite running on any platform. Also, one can easily design and implement interfaces that work with other file system formats for different computer hardware architectures.

5.1   Interoperability

This section describes experiments to demonstrate the interoperability of SQLite. The measurements were conducted on Dell Optiplex 960 models with a 3.16 GHz dual-core system. However, we only ran this on a single core processor to compare with a single core bare PC application. The SQLite database was run on Windows 7 with Visual Studio (VS) 2010 and also on a bare PC with no OS or hard disk. The USB flash drive contains the bare application suite including boot and load programs. We also ran SQLite and a Web server application in a multi-threaded manner on the bare PC.

Initially, a USB is created with the bare application suite and no database file on it. This is a bootable and executable USB for the bare PC (it contains bare boot sector, loader, and application). The same USB is then used to save a SQLite database file created by VS in Windows. Figure 4 shows a screenshot on the Windows machine, where the database was created and tested. Here, one table (name: s5k) containing 5,000 inserts was edited in a file (5k.sql) formed as a single transaction. This file was read by SQLite using the .read command and executed. The .tables command shows the name of the table (s5k) in VS. One new record was inserted into the above table (123 VS) and the count(*) SQL statement shows 5,001 records in the database. At this point, we use .quit from VS and obtained the created database rkktest.sdb, which was saved in the USB.

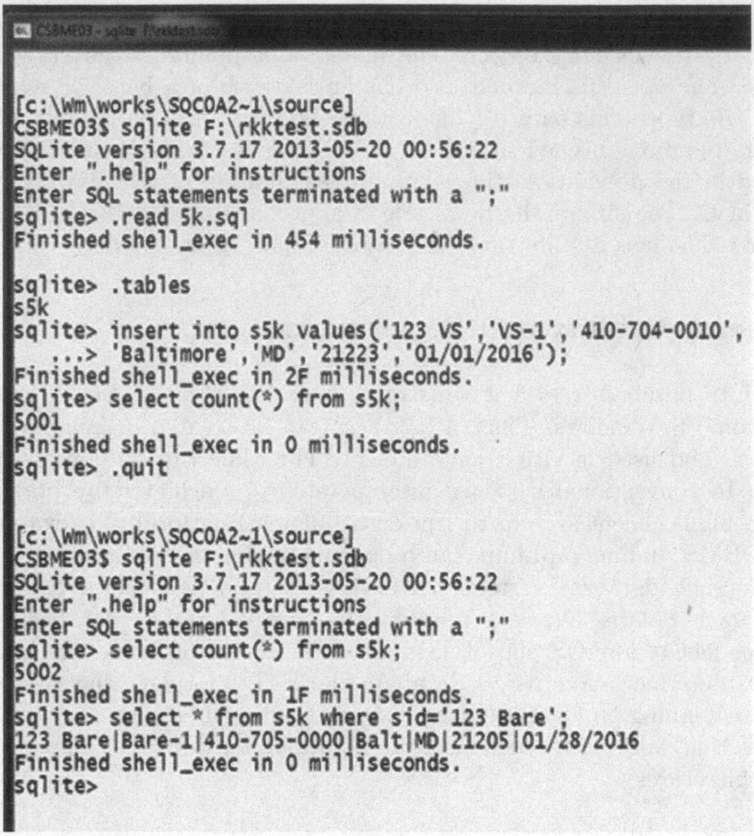

```
[c:\Wm\works\SQCOA2~1\source]
CSBME03$ sqlite F:\rkktest.sdb
SQLite version 3.7.17 2013-05-20 00:56:22
Enter ".help" for instructions
Enter SQL statements terminated with a ";"
sqlite> .read 5k.sql
Finished shell_exec in 454 milliseconds.

sqlite> .tables
s5k
sqlite> insert into s5k values('123 VS','VS-1','410-704-0010',
   ...> 'Baltimore','MD','21223','01/01/2016');
Finished shell_exec in 2F milliseconds.
sqlite> select count(*) from s5k;
5001
Finished shell_exec in 0 milliseconds.
sqlite> .quit

[c:\Wm\works\SQCOA2~1\source]
CSBME03$ sqlite F:\rkktest.sdb
SQLite version 3.7.17 2013-05-20 00:56:22
Enter ".help" for instructions
Enter SQL statements terminated with a ";"
sqlite> select count(*) from s5k;
5002
Finished shell_exec in 1F milliseconds.
sqlite> select * from s5k where sid='123 Bare';
123 Bare|Bare-1|410-705-0000|Balt|MD|21205|01/28/2016
Finished shell_exec in 0 milliseconds.
sqlite>
```

Fig. 4. SQLite on Windows

This USB with the database file is used to boot the bare PC. The bare PC, after initialization and loading, reads the database file into memory as a memory mapped file. It recognizes the existing database which came from VS/Windows environment. Figure 5 shows a bare PC screenshot showing the newly inserted record in VS (123 VS). Now, we have inserted a new record 123 Bare in the bare PC database. The new count(*) shows 5,002 records indicating that it loaded the database successfully and added a new record. Figure 6 shows the database activities performed in the bare PC; at the end it flushes the database.

This bare PC database is used in VS/Windows. As shown in Fig. 4, the count(*) shows 5,002 records and the 123 Bare record. The above tests show that bare SQLite is interoperable with conventional OS-based SQLite in addition to having the same basic management and storage capabilities for data and files. While SQLite interoperability is easily provided by conventional systems with OS support, we have shown that an interoperable database can run on a bare PC with no OS or kernel overhead.

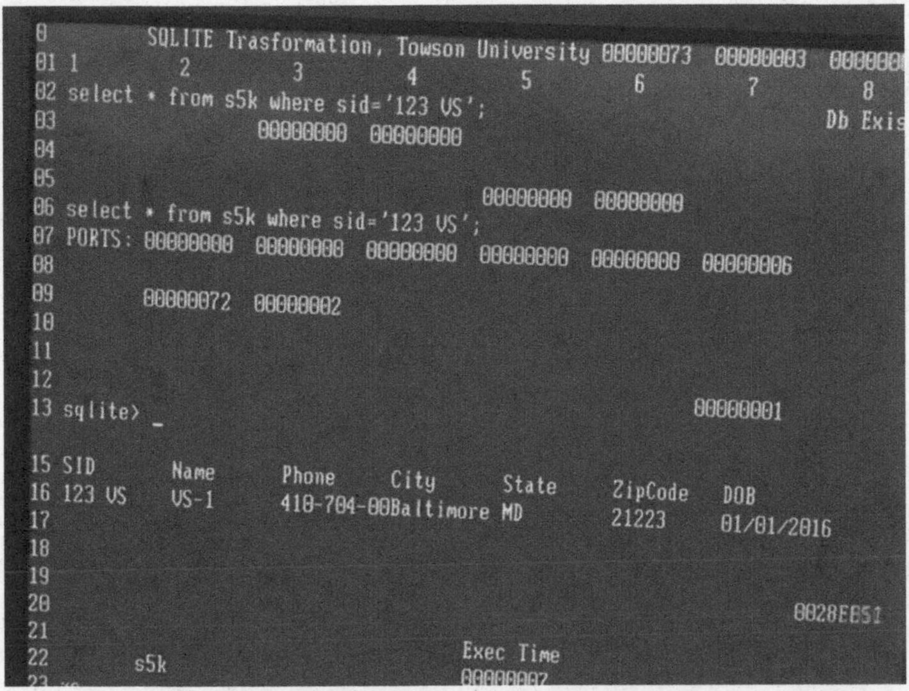

Fig. 5. Bare PC: read database created in VS

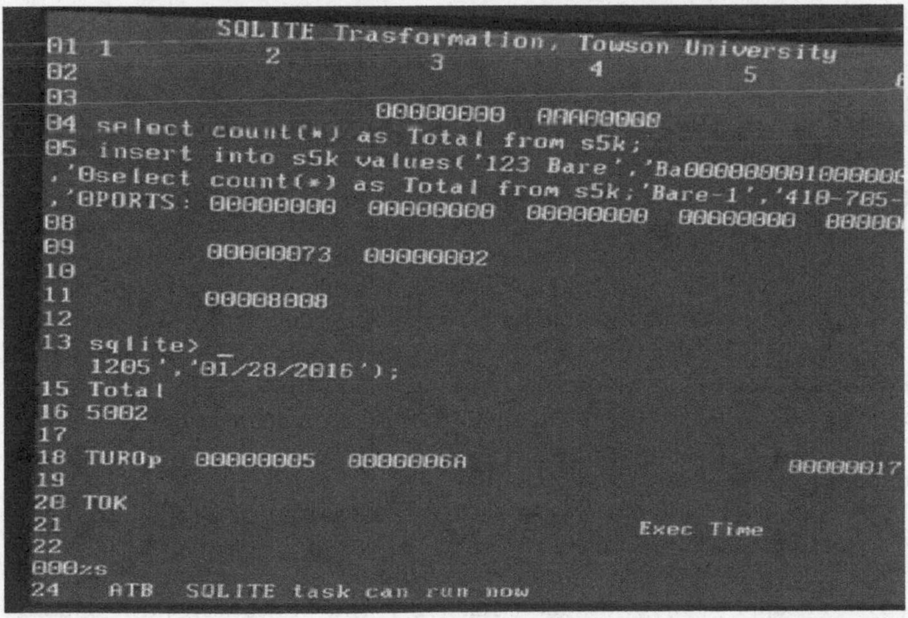

Fig. 6. Bare PC: inserting one record

5.2   Performance

This section shows some basic performance data collected to illustrate the potential gains due to eliminating OS overhead in bare SQLite. The database queries in this study were done by using a single SQL statement (one at a time), or by collecting a set of SQL statements in a single transaction using BEGIN and COMMIT.

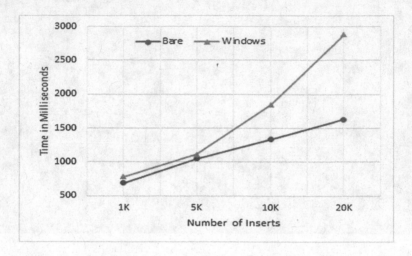

Fig. 7. Inserts with transactions

Figure 7 shows the performance when the number of inserts into a single table is varied from 1,000 to 20,000 records. These inserts were placed in a file and run by using the .read meta-command in a single transaction. SQLite on a bare PC SQLite performs much better than on VS/Windows as expected. The bare PC performance improvements are attributed to less overhead in the hardware interfaces compared to system calls in OS. Figure 8 shows run times for inserting records without transactions when the number of inserts is varied from 10 to 100. Notice that the run times for Windows are very large as SQLite creates and updates a journal file for each SQL insert statement. In transaction mode (Fig. 7), the times for Windows were less because SQLite only flushes the file at the end of a transaction. Without transactions (Fig. 8), the bare PC system performs much faster than Windows for individual SQL statements as it updates the main database and journal in memory and does a final flush after running all the queries from a given file.

To make a fair performance comparison between Windows and bare SQLite without transactions, a future study will need to run the above experiment when both systems use the journal file approach and when both do not. The journal file approach provides more reliability as it provides frequent updates, but requires more time to run. Note that the journal files are accessed multiple times depending on the number of SQL statements or transactions.

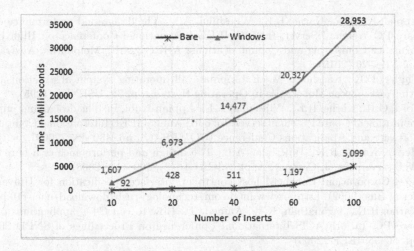

Fig. 8. Inserts without transactions

6   Conclusion

We described a SQLite database with a file system that runs on a bare PC and is interoperable with OS-based systems. The design and implementation details were provided to show how the bare PC file system interfaces to the SQLite virtual file system. We also tested interoperability of the bare SQLite system with a conventional Windows SQLite system and showed how a VS/Windows database can be used in a bare PC environment and vice versa. The performance results suggest the feasibility of building scalable bare database systems.

A similar approach can be used to transform other OS-based databases such as MySQL and PostgreSQL server to run on bare machines. Also, the database management functions can be split into a user interface and a database execution engine. This will enable standard and familiar user interfaces to be used with a secure bare database engine that can be hidden behind a conventional server. For example, one could use bare PC SQLite to create and manage databases, and use the Windows OS to provide user interfaces. Database (and other) servers are naturally suited for bare PC or bare machine applications as they focus on a single monolithic executable and are easily tailored for the backend. Future studies could investigate the pros and cons of splitting database management functions in this manner.

References

1. SQLite. http://www.sqlite.org/download.html
2. The SQLite OS interface or VFS. http://www.sqlite.org/vfs.html
3. Universal Serial Bus mass storage class, bulk only transport, revision 1.0 (1999). http://www.usb.org

4. Appiah-Kubi, P., Karne, R.K., Wijesinha, A.L.: The design and performance of a bare PC Webmail Server. In: 12th IEEE International Conference on High Performance Computing and Communications, AHPCC 2010, Melbourne, Australia, pp. 521–526 (2010)

5. Engler, D.R., Kaashoek, M.: Exterminate all operating system abstractions. In: 5th Workshop on Hot Topics in Operating Systems, p. 78. USENIX (1995)

6. Ford, G.H., Karne, R.K., Wijesinha, A.L., Appiah-Kubi, P.: The design and implementation of a bare PC email server. In: 33rd Annual IEEE International Computer Software and Applications Conference (COMPSAC), pp. 480–485 (2009)

7. He, L., Karne, R.K., Wijesinha, A.L.: The design and performance of a bare PC Web server. Int. J. Comput. Appl. (IJCA) **15**(2), 100–112 (2008)

8. Intel Corporation: Enhanced host controller interface specification for Universal Serial Bus (2002). http://www.intel.com/technology/usb/download/ehci-r10.pdf

9. Karne, R.K., Jaganathan, K.V., Ahmed, T.: How to run C++ applications on a bare PC. In: 6th ACIS International Conference on Proceedings of SNPD 2005, pp. 50–55, IEEE (2005)

10. Karne, R.K., Jaganathan, K.V., Rosa, N., Ahmed, T.: DOSC: dispersed operating system computing. In: 20th Annual ACM Conference on Object Oriented Programming, Systems, Languages, and Applications (OOPSLA), pp. 55–61 (2005)

11. Karne, R.K., Liang, S., Wijesinha, A.L., Appiah-Kubi, P.: A bare PC mass storage USB device driver. Int. J. Comput. Appl. **20**(1), 32–45 (2013)

12. Lange, J., et al.: Palacios and Kitten: new high performance operating systems for scalable virtualized and native supercomputing. In: 24th IEEE International Parallel and Distributed Processing Symposium (2010)

13. Liang, S., Karne, R.K., Wijesinha, A.L.: A lean USB file system for bare machine applications. In: Proceedings of 21st International Conference on Software Engineering and Data Engineering, ISCA, pp. 191–196 (2012)

14. Microsoft Corp.: FAT32 file system specification (2000). http://microsoft.com/whdc/system/platform/firmware/fatgn.rnspx

15. Okafor, U., Karne, R., Wijesinha, A., Appiah-Kubi, P.: A methodology to transform an OS-based application to a bare machine application. In: 12th IEEE International Conference on Ubiquitous Computing and Communications (IUCC-2013), Melbourne, Australia (2013)

16. Okafor, U., Karne, R.K., Wijesinha, A.L., Rawal, B.: Transforming SQLITE to run on a bare PC. In: Proceedings of 7th International Conference on Software Paradigm Trends, Rome, Italy, pp. 311–314 (2012)

17. Perisoft Corp.: Universal serial bus specification 2.0. http://www.perisoft.net/engineer/usb_-20.pdf

18. Rawal, B., Karne, R., Wijesinha, A.L.: Splitting HTTP requests on two servers. In: 3rd Conference on Communication Systems and Networks (COMSNETS) (2011)

19. Rawal, B., Karne, R.K., Wijesinha, A.L.: Mini web server clusters for HTTP request splitting. In: IEEE International Conference on High Performance, Computing and Communications (HPCC), pp. 94–100 (2011)

20. Rezende, F.F., Hergula, K.: The heterogeneity problem and middleware technology: experiences with and performance of database gateway. In: International Conference on Very Large Databases (VLDB 1998), pp. 146–157 (1998)

21. The OS Kit Project: School of computing. University of Utah, Salt Lake, UT (2002). http://www.cs.utah.edu/flux/oskit

FM-index for Dummies

Szymon Grabowski[1], Marcin Raniszewski[1], and Sebastian Deorowicz[2($\boxtimes$)]

[1] Institute of Applied Computer Science, Lodz University of Technology,
Al. Politechniki 11, 90–924 łódź, Poland
{sgrabow,mranisz}@kis.p.lodz.pl
[2] Institute of Informatics, Silesian University of Technology,
Akademicka 16, 44-100 Gliwice, Poland
sebastian.deorowicz@polsl.pl

Abstract. Full-text search refers to techniques for searching a document, or a document collection, in a full-text database. To speed up such searches, the given text should be indexed. The FM-index is a celebrated compressed data structure for full-text pattern searching. After the first wave of interest in its theoretical developments, we can observe a surge of interest in practical FM-index variants in the last few years. These enhancements are often related to a bit-vector representation, augmented with an efficient rank-handling data structure. In this work, we propose a new, cache-friendly, implementation of the rank primitive and advocate for a very simple architecture of the FM-index, which trades compression ratio for speed. Experimental results show that our variants are 2–3 times faster than the fastest known ones, for the price of using typically 1.5–5 times more space.

Keywords: Data structures · Text indexing · FM-index

1 Introduction

A full-text index is a data structure allowing to quickly find any substring of the index text. The functionality of full-text searching is currently supported by most major database vendors, both in the SQL and no-SQL world: MS SQL Server, PostgresSQL, Oracle, MySQL, MongoDB, MariaDB and many more. From the algorithmic point of view, it is interesting to observe possible interplays between the performance and space usage of various full-text index solutions.

The rapid development of compressed data structures in the first decade of our century changed the landscape of modern algorithmics. Prominent examples of those achievements are compressed indexes for unstructured and semi-structured texts, compressed trees, graphs, binary relations, RDF triples and color range counting. Real applications of these sophisticated data structures, however, lag behind, with a notable exception of bioinformatics [4,26]. In this work we revisit one of the most celebrated concepts in stringology in recent years, the FM-index by Ferragina and Manzini [6]. The key component of virtually any of multiple variants of this index is the operation *rank*, usually performed on a

© Springer International Publishing AG 2017
S. Kozielski et al. (Eds.): BDAS 2017, CCIS 716, pp. 189–201, 2017.
DOI: 10.1007/978-3-319-58274-0_16

bit-vector B, which, for a given integer j, returns the number of set bits in B's prefix of length j. We propose a new, cache-friendly, implementation of the rank primitive and advocate for a very simple architecture of the FM-index, which trades compression ratio for speed.

We use the following notation. An index will be built for a text $T[1\ldots n]$ over an integer alphabet $\Sigma = \{1,\ldots,\sigma\}$. The index will be queried with patterns $P[1\ldots m]$ over the same alphabet. The rank operation will be calculated for the bit vector $B[1\ldots n]$. We assume the CPU cache line is of size $L = 512\,(\text{bits})$. The colloquial term "popcount" (population count) will often be used for the operation of counting the number of bits 1 in a given bit sequence.

2   The FM-index Architecture

The FM-index is basically the result of the Burrows–Wheeler transform (BWT) of text T, denoted as T^{bwt}, with two helper structures: a count array $C[1\ldots\sigma]$ such that $C[i] = |\{T[j] : T[j] \leq i \text{ and } 1 \leq j \leq n\}|$, and a data structure answering $Occ(T^{\text{bwt}}, c, pos) = |\{T^{\text{bwt}}[j] : T^{\text{bwt}}[j] = c \text{ and } 1 \leq j \leq pos\}|$ queries. This allows to count the occurrences of a pattern P, finding the ranges of the (implicit) suffix array (SA) for T starting with successive suffixes of P in the successive loop iterations, see Fig. 1. If the function $Occ(\cdot)$ is realized with a wavelet tree (WT) [21,22], the count queries run in $O(m \log \sigma)$ worst-case time. Plenty of FM-index variants exist, with different space-time complexity tradeoffs and different practical performances. For a survey, see [22]; important recent results were presented in [2,15,20]. Experimental comparisons can be found, e.g., in [8].

Count-Occs(T^{bwt}, n, P, m)

1	$i \leftarrow m$; $sp \leftarrow 1$; $ep \leftarrow n$
2	**while** $(sp \leq ep)$ **and** $(i \geq 1)$ **do**
3	$\quad c \leftarrow P[i]$
4	$\quad sp \leftarrow C[c] + Occ(T^{\text{bwt}}, c, sp - 1) + 1$
5	$\quad ep \leftarrow C[c] + Occ(T^{\text{bwt}}, c, ep)$
6	$\quad i \leftarrow i - 1$
7	**if** $(ep < sp)$ **then return** "not found" **else return** "found $(ep - sp + 1)$ occs"

Fig. 1. Counting the number of occurrences of pattern P in T with the FM-index.

As the wavelet tree, a sequence representation based on alphabet decomposition, is used in most proposed and tested FM variants, we briefly describe this data structure. Given sequence S of length n and integer alphabet $\Sigma = \{1, 2, \ldots, \sigma\}$, the (standard) wavelet tree $WT(S)$ is a binary balanced tree with σ leaves. Its root stores a bit-string of length n telling if the successive symbols of S belong to the subalphabet $\{1, \ldots, \sigma/2\}$ or to the subalphabet

$\{\sigma/2 + 1, \ldots, \sigma\}$. Similarly, both children of the root store binary sequences of total length n, where the left (right) child decomposes the first (second) half of the alphabet into its halves. The decomposition continues in the same manner down to the leaves. The wavelet tree supports efficient rank and access queries (among others) for the input sequence S.

3   Rank with One Cache Miss

Jacobson [13] showed that the rank operation for a bit vector of length n can be implemented in constant time using $O(n \log \log n / \log^2 n)$ extra bits. This, however, requires three memory accesses (to one superblock counter and one block counter, plus a lookup into a table with precomputed popcount answers), therefore more practical ideas were later presented [8,25]. Following the idea of Gog and Petri [8] (who in turn extended the approach of Vigna [25]), we use one level of counters, interleaving the bit vector data with the counters, to improve the locality of memory accesses. If one counter and an interval of bits from B takes exactly one (aligned) cache line, we can calculate the rank with one cache miss in the worst case. In contrast, Gog and Petri [8] interleave 64-bit counters with bit vector data of 256 bits in their RANK-1L variant. Note that their structure is logically divided into chunks of $64 + 256 = 320$ bits, which are usually not aligned to the cache line.

We come back to our variant. Assume for clarity that n is a multiple of $512 - 64 = 448$. More precisely, we maintain a bit table $B'[1 \ldots n']$, where $n' = \lceil 512n/448 \rceil$, and $B'[512i + 1 \ldots 512i + 64]$ is a 64-bit counter $R[i]$ storing the value of $rank_1(B, 448i)$, while $B'[512i + 65 \ldots 512i + 512] = B[448i + 1 \ldots 448i + 448]$, for any valid $i \geq 0$. We also require that B' is aligned to a multiple of the cache line size. Now, $rank_1(B, j) = R[\lfloor j/448 \rfloor] + popcnt(B'[512\lfloor j/448 \rfloor + 65 \ldots 512\lfloor j/448 \rfloor + 64 + (j \bmod 448)])$. The popcount operation is performed using the hardware 64-bit opcode POPCNT (known as the __builtin_popcountll function in gcc), which seems fastest on the Intel Nehalem and later CPUs.

Note that for $n < 2^{32}$, 32-bit counters are enough, yet using 64-bit counters provides proper alignment for calling the __builtin_popcountll instruction (7 times). Alternatively, we could use a 32-bit counter, then call the 32-bit __builtin_popcount once and finally __builtin_popcountll 7 times (see Fig. 2, the top scheme (a)). Yet another pair of variants (with 64-bit and 32-bit counters, respectively), reducing the number of popcount instructions, but using more space, maintains 256-bit rather than 512-bit blocks in B'. The space overhead of the four possible variants is $64/448 = 14.3\%$, $32/480 = 6.7\%$, $64/192 = 33.3\%$ and $32/224 = 14.3\%$.

A drawback of our approach is that it involves a division by a number not being a power of 2 (e.g., 448). On the other hand, modern compilers (including gcc) convert integer division by a constant into a multiplication and a few additions and shifts, which is several times faster than general division.

Additionally, we employ software prefetching to reduce the access time to a memory cell. In the variants which involve a wavelet tree, we interleave the WT

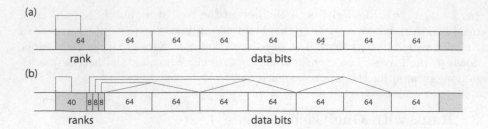

Fig. 2. Rank with one cache miss. Blocks have 512 bits: 64 bits for rank data, 448 bits for data bits. Variant (a): up to 7 popcount instruction calls are needed to calculate the rank. Variant (b): only 40 bits are spent for the rank up to the block beginning, and thanks to three following 8-bit subfields up to 2 popcount calls are enough.

accesses for both boundaries, in order to increase the delay between the prefetch instruction and actual memory access, which in turn reduces the overall time.

4   FM-dummy

Dealing with the dependence on the alphabet size is one of the key issues in FM-index design. We propose several variants of the FM-index. Although these ideas are hardly novel, we are not aware of their implementations (with one exception). In the experimental section we will, however, show that these schemes, together with our rank implementations, offer attractive time-space tradeoffs.

In the first variant, for a small alphabet, we maintain σ bit vectors of length n, one per alphabet symbol, together with the corresponding rank data. We propose to use it if $\sigma \leq 16$. Let us denote this algorithm as FM-dummy1. We admit that this scheme, with compressed rank, was proposed by Mäkinen and Navarro in 2004 in a technical report [18], in Sect. 3.2 appropriately entitled "Replacing Occ Structure by Individual Bit Arrays". Their idea was to obtain $O(m)$ count time (i.e., with no dependence on the alphabet size) with $O(H_0 n)$ bits of space, yet in an erratum note dated 9th Dec. 2004 they noticed an error in analysis. In theoretical terms, the desired properties of this algorithm are obtained only for $\sigma = O(\text{polylog}(n))$. The same solution is also used in ABySS [24], a well-known de novo genome assembler.

The second variant is suggested for the case of $\sigma > 16$. Before applying the BWT, we encode the text using a dense code. We work on triples of bits of the input symbols, and the code space of 8 values is divided into disjoint subsets of sizes: (i), b—the number of distinct prefixes of codewords ("beginners") of length ≥ 1, (ii), c—the number of distinct following symbols of codewords ("continuers") of length ≥ 2. In this way, we obtain $\sum_{i=0}^{j-1} bc^i$ codewords of length up to $j \geq 2$ (cf. [9, Sect. 4]). There are two issues with the used encoding though: (i) any match in the encoded text (except at the very end of the text) must be followed with a symbol from the beginners, hence the (backward) search over the pattern must start with a dummy "any-beginner symbol" in the FM-index; fortunately it is easy to simulate it with setting appropriately the initial suffix range (the *sp* and *ep* variables in Fig. 1), (*ii*) *bidirectional* searches over the FM-index, which have applications in approximate index string matching [17] and some DNA sequence analysis problems (e.g., maximal unique matches) [1], become problematic. In the experimental section this variant is dubbed FM-dummy2. We also tried the SCBDC [5] encoding, which is both prefix- and suffix-free and thus requires no verifications, yet its results were not competitive.

In FM-dummy1 and FM-dummy2 we also try out eliminating part of the linear scan (several popcounts) over a block. More precisely, with blocks of 256 bits we use 48 bits for the counter and two bytes of its 64-bit word are spent for storing the ranks for the 64- and the 128-bit prefix of the block data (which reduces the maximal number of POPCNT operations in a block from 3 to 1). Similarly, with blocks of 512 bits we use 40 bits for the counter and three bytes storing the number of ones in three successive subblocks of 128 bits each (Fig. 2, the bottom scheme (b)). Implementations involving this idea have letter 'c' in the name, e.g., FM-dummy1_256c.

DNA is an important target of text indexes and the FM-dummy1 variant presented above may not always be preferred since it is not quite succinct. To address this issue, we propose FM-dummy3, which assumes the alphabet of size 5 (ACGTN), where the symbol N stands for any symbol not from ACGT in the text. It is also assumed that the patterns are from the ACGT (sub)alphabet, otherwise they would make little sense from the biological point of view. In a block of 512 (1024) bits, there are four 32-bit counters for the four valid pattern symbols, followed by 384 (896) bits of data. The block data consist of symbols packed into bytes in triples (which can be easily done, since $5^3 \leq 256$). To obtain a rank for a given symbol and a given position in block, we scan the data bytes with a reference to a lookup table having 125×4 entries.

Finally, we implemented an FM-index with a Huffman-shaped multiary wavelet tree, namely with arity 4 and 8 (FM-HWT4 and FM-HWT8). (For completeness, we added also a variant with a Huffman-shaped binary wavelet tree.) In the 4-ary (8-ary) case, each block contains 4 (8) 32-bit counters, followed by packed data as a sequence of pairs (triples) of bits. The block size is a parameter, set to 512 or 1024 bits. For example, if the 8-ary variant is chosen and 512-bit blocks, we have $512 - 8 \times 32 = 256$ bits for the data, which are grouped in four 64-bit words, each containing 21 triples of bits (1 bit per 64-bit word is then "wasted"). Counting the rank for a symbol from the 8-ary alphabet for the data sequence is performed using simple bitwise operations followed by the hardware popcount.

5   Boosting Short Pattern Search with a Hash Table

In [10] we showed how to augment the standard suffix array with a hash table (HT), to start the binary search from a much narrower interval. The start and end position in the suffix array for each range of suffixes with a common k-long prefix was inserted into the HT, with the hash function calculated for the prefix string. The same function was applied to the pattern's prefix and after an HT lookup the binary search was continued with reduced number of steps.

Now we propose to use this idea with an FM index. First, the pattern's *suffix* of length k is sought in the HT and then the search continues in a standard manner. The number of symbols submitted to a standard FM-index backward search is reduced from m to $m - k$ (note the requirement of $m \geq k$).

The extra data structure comprises three components. One is a lookup table over all σ^2 2-symbol strings (LUT2), whose entries are the suffix array intervals (represented as pairs of integers) corresponding to the input 2-grams. Another is the actual HT, where each entry stores the left and the right boundary of the suffix array interval for the corresponding string of length k. Yet another component stores the hashed strings without their first two symbols (i.e., of length $k - 2$), at positions corresponding to the entries of HT. Collisions in HT are resolved with linear probing. Note that, contrary to the SA-hash solution [10], we cannot avoid storing the $k - 2$ symbols since we do not have an explicit suffix array and fast access to arbitrary text position, to resolve collisions. Using a perfect hashing scheme does not fix this issue either, since looking for a k-gram *not occurring in the text* gives a "random" position in the hash table, which could imply spurious matches.

The search for the pattern's suffix $P[m - k + 1 \ldots m]$ more precisely proceeds as follows. First the symbols $P[m - k + 1 \ldots m - k + 2]$ are checked in LUT2; an empty SA interval implies no occurrences of P in T. The hash value for $P[m - k + 1 \ldots m]$ is calculated and possible collisions are first checked against LUT2 and only if this test is passed, the HT component is inspected.

In the experimental section we set $k = 5$ and run experiments only on short patterns (of length from 6 to 10). This allows to speed up searches significantly for a price of only moderate increase in the space use for real texts.

6   Estimating an Upper Bound for the Number of Pattern Occurrences

In some mining applications, e.g., on the usage of foreign words from a specified set in an English text collection, we may be satisfied with an approximate result of the count query. In this section we present a simple solution returning an optimistic estimation (i.e., an upper bound) for the number of pattern occurrences in the text. A compensation for the inexact results are (i) faster execution of the count query and (ii) lower space requirements, compared to a standard FM-index.

To this end, we quantize the alphabet to reduce it to a specified size σ', where $3 \leq \sigma' < \sigma$. Basically, we use the simple $T'[i] \leftarrow T[i] \bmod \sigma'$ formula. The same modification is applied at runtime for the pattern. This allows, for example, to use the FM-dummy1 scheme for the new alphabet, which is faster than, e.g., wavelet tree based variants for the original (presumably, quite large) alphabet. Naturally, due to possible collisions the returned count value is likely to be larger than the true one, yet we expect that with growing m the ratio between the returned and the true count tends to 1. We admit that the general alphabet reduction idea for speeding up pattern search is not new; Külekci et al. [16] applied it with success for online search by decomposing the text symbols into a k-bit filter part and the remaining bits; only if the pattern, transformed in the same way, matches the filter part, it is compared on the remaining bits. They also mentioned that it was possible to create an index over the k-bit filtered data as well, yet we are not aware of any progress on this idea.

We also note here an ingenious FM-index based algorithm by Orlandi and Venturini [23] to return an approximate number of occurrences of a pattern in the text, using little space. This algorithm provides an *additive* error, that is, given (beforehand) any error parameter ℓ, it guarantees to report the number of occurrences of pattern P in the range $[count(P), count(P) + \ell - 1]$, where $count(P)$ is the correct number of occurrences of P in the text.

7   Experimental Results

All experiments were run on a machine equipped with a 6-core 3.4 GHz Intel i7 CPU (4930K) with 64 GB of RAM, running Ubuntu 15.10 LTS 64-bit. The RAM modules were 8×8 GB DDR3-1600 with the timings 11-11-11. One CPU core was used for the computations. All codes were written in C++ and compiled with 64-bit gcc 5.2.1 (with `-O3 -mpopcnt`). The source codes for our implementations are available at https://github.com/mranisz/fmdummy/releases/tag/v3.0.0.

The test datasets were taken from the Pizza & Chili site[1]. We used the 200-megabyte versions of the files dna, english, proteins, sources, and xml.

In order to evaluate the search algorithms, we generated 1 million patterns of length $m = 20$; the patterns were extracted randomly from the corresponding datasets (i.e., each pattern returns at least one match), with a special procedure for the DNA dataset, where only patterns over the subalphabet ACGT were allowed. The performance of count queries only was measured. Actually, the current implementations of our indexes have no support for the locate query, but for a possibly honest comparison we reduced the sampling in the other indexes (where it is available) to very large values (at least 1 million), in order to make the space overhead totally negligible.

We compared the following FM-index variants (Fig. 3):

- FM-HF-V5 based on a Huffman-shaped binary wavelet tree with uncompressed bit-vectors; it is called V5 in [8],
- FM-FB-V5, presented in [7], as a refinement of the fixed-block boosting technique from [15],
- FM-hybrid-HF_8, related to FM-HF-V5, but with the wavelet tree bit-vectors divided into blocks for which one of three simple encoding methods is separately chosen [14]; the superblock size of 8 was always chosen, as the sizes of

---

[1] http://pizzachili.dcc.uchile.cl/.

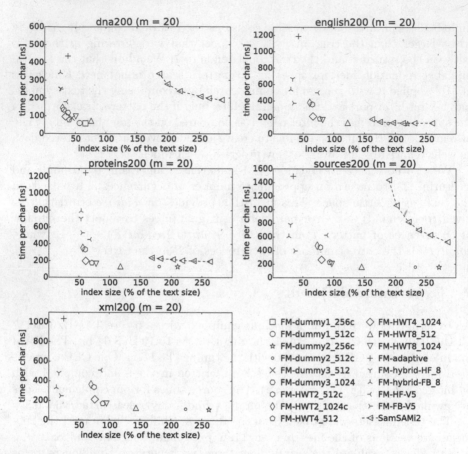

Fig. 3. Count query. 1M patterns of length 20 were used. Times are averages in ns per character.

16, 32, and 64 gave similar results (the index with parameter 8 was the fastest yet using slightly more space than the other choices),

- FM-hybrid-FB_8, combining the hybrid technique from [14] with fixed-block boosting [15],
- FM-adaptive[2], a recently proposed algorithm [12] related to [14], yet with the main modification of using a variable-length coding (Gamma coding) in blocks rather than fixed length coding,
- FM-dummy1, our variant for a small alphabet (dna200 only from the test collection), with block sizes of 256 or 512 bits,
- FM-dummy3, our variant for DNA specifically, with block sizes of 512 or 1024 bits,
- FM-dummy2, our variant using the (c, b) dense coding on triples of bits, with block sizes of 256 or 512 bits,

---

[2] https://github.com/chenlonggang/Adaptive-FM-index.

– FM-HWT2, FM-HWT4, and FM-HWT8, our variants based, respectively, on the 2-, 4-, and the 8-ary Huffman-shaped wavelet tree, with block sizes of 512 or 1024 bits; note that for FM-HWT2 part of the linear scan over a block is eliminated just like in the FM-dummy1 and FM-dummy2 variants denoted with letter 'c' in their name.

The implementations of the variants FM-HF-V5 and FM-hybrid-HF_8 were taken from the SDSL library (https://github.com/simongog/sdsl-lite).

Additionally, we included in the comparison one index of another type, Sam-SAMi [11], which is an efficient modification of a sampled suffix array. The tested variant is denoted as SamSAMi2 and each series corresponds to six parameter settings: (q, p) equal to $(4, 1)$, $(5, 1)$, $(6, 1)$, $(8, 2)$, $(10, 2)$ and $(12, 2)$, respectively.

As can be seen, our variants are significantly faster than existing FM-index implementations, yet they also take up much more space. Among our variants the ones based on the multiary wavelet tree are most compact and may often be preferred. If more search speed is required, we can use FM-dummy2. Among FM-dummy1, FM-dummy2, and FM-HWT2 variants, the ones labeled with 'c' were always faster (while using the same amount of memory) than their simpler counterparts, and we omitted showing the results from the slower implementations.

The sampled suffix array variant, SamSAMi2, is not competitive in count queries (at least if the patterns are not very long), and its additional drawback is a requirement on the minimal pattern length: $m \geq q - p + 1$. Note that SamSAMi2 results are not shown for xml200 since the times are by about two orders of magnitude longer than of the other indexes.

We also checked if our 1-cache-miss rank is indeed faster than RANK-1L [8], used in FM-HF-V5. To this end, FM-HWT2_512c (with blocks of 448 bits and 64 bits for counter data) was changed to work with RANK-1L, containing chunks of 512 bits. The original FM-HWT2_512c variant was faster by 4–10%.

In the next experiment, we compared a few variants involving a hash table (see Sect. 5). The load factor for the hash tables was set to 0.9, with the exception of dna200, where 0.5 was used, and the parameter k was set to 5 in all cases. The results are presented in Fig. 4. From each variant group (dummy1, dummy2, dummy3 and HWT-based) we chose one representative (with best speed), since generally the results in groups were close and presenting them all would make the graphs very crowded. Although the extra space is not negligible in most cases, the search speedup for short patterns is very significant, often by a factor of around 2–3. Note that the space overhead for dna200 even for the load factor of 0.5 is so small that it encouraged us to prefer more speed in this case.

For the final experiment, we quantized the alphabet for each given dataset and counted the occurrences of the transformed pattern in the transformed text (see Sect. 6). Obviously, the obtained counts are never smaller than the true counts, and the results in Table 1 (restricted to three datasets, to save space) present the ratio between the reported counts and the true counts, in percent (where, e.g., 200% means that the reported counts over all the test patterns are twice greater in total that the true counts over these patterns). In this experiment, as the patterns we took all substrings of length k from the text.

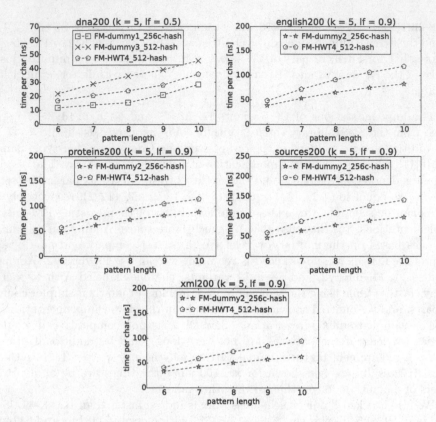

Fig. 4. Count query for the variants with a hash table. 1M patterns of length $\{6, 7, \ldots, 10\}$ were used. Times are averages in ns per character. The extra space added by the hash table component was $0.12784n$, $0.21367n$, $0.30863n$, and $0.16082n$ for english, proteins, sources, and xml, respectively, in all the variants. For dna the corresponding extra space was $0.00261n$ in the FM-dummy1 and FM-dummy3 variants, and $0.00312n$ in FM-HWT4.

This means that if a given substring of length k occurs several times in the text, it will also have a stronger "impact" on the computed statistics. Such solution goes in accord with the methodological guidances given in [19]. We can see that using $\sigma' \geq 6$ gives a moderate average increase in the number of reported matches, in most cases, yet a strange phenomenon can be observed that using a larger σ' may yield a worse result (e.g., when σ' increased from 6 to 7 for english200).

Although the dominating trend in FM-indexes is to reduce their space at acceptable search speed, we are aware of one scheme in which space is sacrificed for higher speed [3]. This solution works on q-grams (DNA only is supported) and stores at regular intervals rank counts for all σ^q supersymbols. Alas, we weren't able to obtain the software from the authors for own testing.

Table 1. The ratios of the reported pattern counts to the true pattern counts for patterns of varying length m and with the alphabet quantized to the new size of σ'.

dna200						
m	$\sigma' = 3$	$\sigma' = 4$	$\sigma' = 5$	$\sigma' = 6$	$\sigma' = 7$	$\sigma' = 8$
8	2109.13	604.45	100.01	2109.05	100.00	100.00
12	1846.54	263.88	100.00	1846.47	100.00	100.00
16	920.62	134.75	100.00	920.60	100.00	100.00
24	931.71	136.41	100.00	931.69	100.00	100.00
english200						
8	3459.66	802.80	313.87	235.69	236.50	163.76
12	847.71	208.59	147.39	121.68	231.66	112.74
16	263.98	121.74	124.10	104.63	253.65	103.79
24	213.91	109.86	121.26	101.97	237.36	101.61
xml200						
8	150.00	117.28	113.35	106.64	107.99	106.38
12	120.08	114.14	110.00	106.11	108.04	103.97
16	120.99	112.55	108.25	106.42	105.95	103.32
24	118.81	110.22	106.10	103.83	103.77	102.15

8   Conclusions

We presented several simple FM-index variants, with preference to search speed rather than succinctness. While most of the applied ideas are hardly novel, we believe some of them have not been experimentally verified before. We show that the fastest known FM-index variants on most datasets can be sped up by a factor of 2 or more for the price of 1.5–5 times more space. While the space overhead is significant, our solutions are still compact, e.g. need around $1n$ bytes of space, which is comparable to the text size (which can be discarded in an FM-index). Perhaps the most important building brick that we introduce is the (uncompressed) rank with one cache miss in the worst case. Also, we note that Navarro in his survey on wavelet trees [21, p. 7] claims about the multiary variants of this data structure that "although theoretically attractive, it is not easy to translate their advantages to practice". Our results suggest, however, that Huffman-shaped 4- and 8-ary wavelet trees offer interesting space-time tradeoffs.

Acknowledgments. We thank Simon Gog for providing the FM-FB-V5 and FM-hybrid-FB_8 sources and helping us in running sdsl-lite, and Shaun D. Jackman for a remark concerning the ABySS de novo genome assembler.

The work was supported by the Polish National Science Centre upon decision DEC-2013/09/B/ST6/03117.

References

1. Belazzougui, D., Cunial, F., Kärkkäinen, J., Mäkinen, V.: Versatile succinct representations of the bidirectional burrows-wheeler transform. In: Bodlaender, H.L., Italiano, G.F. (eds.) ESA 2013. LNCS, vol. 8125, pp. 133–144. Springer, Heidelberg (2013). doi:10.1007/978-3-642-40450-4_12
2. Belazzougui, D., Navarro, G.: Alphabet-independent compressed text indexing. ACM Trans. Algorithms 10(4), 23 (2014). Article 23
3. Chacón, A., Moure, J.C., Espinosa, A., Hernández, P.: n-step FM-index for faster pattern matching. Proc. Comput. Sci. 18, 70–79 (2013)
4. Deorowicz, S., Grabowski, S.: Data compression for sequencing data. Algorithms Mol. Biol. 8(1), 25 (2013)
5. Fariña, A., Navarro, G., Paramá, J.: Boosting text compression with word-based statistical encoding. Comput. J. 55(1), 111–131 (2012)
6. Ferragina, P., Manzini, G.: Opportunistic data structures with applications. In: Proceedings of FOCS, pp. 390–398. IEEE (2000)
7. Gog, S., Kärkkäinen, J., Kempa, D., Petri, M., Puglisi, S.J.: Faster, minuter. In: Proceedings of DCC, pp. 53–62. IEEE (2016)
8. Gog, S., Petri, M.: Optimized succinct data structures for massive data. Softw.: Pract. Exp. 44(11), 1287–1314 (2014)
9. Grabowski, S.: Making dense codes even denser. AGH Automatyka 12(3), 769–779 (2008)
10. Grabowski, S., Raniszewski, M.: Two simple full-text indexes based on the suffix array. In: Proceedings of PSC, pp. 179–191 (2014)
11. Grabowski, S., Raniszewski, M.: Sampling the suffix array with minimizers. In: Iliopoulos, C., Puglisi, S., Yilmaz, E. (eds.) SPIRE 2015. LNCS, vol. 9309, pp. 287–298. Springer, Cham (2015). doi:10.1007/978-3-319-23826-5_28
12. Huo, H., Chen, L., Zhao, H., Vitter, J.S., Nekrich, Y., Yu, Q.: A data-aware FM-index. In: Proceedings of ALENEX, pp. 10–23. SIAM (2015)
13. Jacobson, G.: Succinct static data structures. Ph.D. thesis, Carnegie Mellon University (1989)
14. Kärkkäinen, J., Kempa, D., Puglisi, S.J.: Hybrid compression of bitvectors for the FM-index. In: Proceedings of DCC, pp. 302–311. IEEE (2014)
15. Kärkkäinen, J., Puglisi, S.J.: Fixed block compression boosting in FM-indexes. In: Grossi, R., Sebastiani, F., Silvestri, F. (eds.) SPIRE 2011. LNCS, vol. 7024, pp. 174–184. Springer, Heidelberg (2011). doi:10.1007/978-3-642-24583-1_18
16. Külekci, M.O., Vitter, J.S., Xu, B.: Fast pattern-matching via k-bit filtering based text decomposition. Comput. J. 55(1), 62–68 (2010)
17. Lam, T.W., Li, R., Tam, A., Wong, S., Wu, E., Yiu, S.M.: High throughput short read alignment via bi-directional BWT. In: Proceedings of BIBM, pp. 31–36. IEEE (2009)
18. Mäkinen, V., Navarro, G.: New search algorithms and time/space tradeoffs for succinct suffix arrays. Technical report C-2004-20, University of Helsinki, Finland (2004)
19. Moffat, A., Gog, S.: String search experimentation using massive data. Philos. Trans. Roy. Soc. Lond. A: Math. Phys. Eng. Sci. 372(2016), 20130135 (2014)
20. Munro, J.I., Navarro, G., Nekrich, Y.: Space-efficient construction of compressed indexes in deterministic linear time. In: Proceeding of SODA (2017, to appear)
21. Navarro, G.: Wavelet trees for all. J. Discret. Algorithms 25, 2–20 (2014)

22. Navarro, G., Mäkinen, V.: Compressed full-text indexes. ACM Comput. Surv. **39**(1), 2 (2007)
23. Orlandi, A., Venturini, R.: Space-efficient substring occurrence estimation. Algorithmica **74**(1), 65–90 (2016)
24. Simpson, J.T., Wong, K., Jackman, S.D., Schein, J.E., Jones, S.J., Birol, I.: ABySS: a parallel assembler for short read sequence data. Genome Res. **19**(6), 1117–1123 (2009)
25. Vigna, S.: Broadword implementation of rank/select queries. In: McGeoch, C.C. (ed.) WEA 2008. LNCS, vol. 5038, pp. 154–168. Springer, Heidelberg (2008). doi:10.1007/978-3-540-68552-4_12
26. Vyverman, M., De Baets, B., Fack, V., Dawyndt, P.: Prospects and limitations of full-text index structures in genome analysis. Nucleic Acids Res. **40**(15), 6993–7015 (2012)

Lattice Based Consistent Slicer and Topological Cut for Distributed Computation in Monotone Spaces

Susmit Bagchi[✉]

Department of Aerospace and Software Engineering (Informatics),
Gyeongsang National University, Jinju, South Korea
profsbagchi@gmail.com

Abstract. The distributed database systems are increasingly employing distributed systems platforms for data deployment and query based computation. The models of distributed systems play a role in determining data partitioning and placement in distributed database systems. The applications of concepts of topological spaces are gaining research attention for modeling structures of distributed systems. In a distributed system, the slicer of distributed computation partitions a set of processes into subsets maintaining consistency property. In this paper, a lattice based slicer model of distributed computation is presented considering monotone topological spaces. The model considers state-space of asynchronous distributed computation. The proposed monotone slicer model of computation preserves the lattice cover of Birkhoff's representation. A set of analytical properties of the monotone slicer model is formulated. Furthermore, the topological cut of an event-based asynchronous distributed computation is formulated as a set of axioms.

Keywords: Distributed database · Distributed computing · Monotone spaces · Topology · Slicer · Lattice

1 Introduction

The big data analytics and dense graph based computations require combinations of distributed databases and distributed computations. In general, the deployment of distributed database systems require distributed systems as the underlying computational platforms. Storing and retrieving large data sets in distributed databases improve the overall system performance. However, the traditional distributed computing platforms and models are not fully adequate to support specific requirements such as, supporting iterative algorithms on data sets and, require the combinations of concepts of databases and distributed systems [17]. The distributed computational models play a role in determining the placement and retrieval of data sets. In majority cases, the distributed computing on large data sets assumes that data sets are shared between multiple processes and, computation is iterative in nature [4]. The shared data based computation

© Springer International Publishing AG 2017
S. Kozielski et al. (Eds.): BDAS 2017, CCIS 716, pp. 202–211, 2017.
DOI: 10.1007/978-3-319-58274-0_17

requires consistency maintenance of shared data under asynchrony. The complexity is increased if the shared data sets are partitioned and placed into distributed database, where partitioned data elements may be located at multiple locations. Interestingly, the application of concepts of topological spaces helps in modeling and analyzing asynchronous distributed computation involving shared data [5,7,19]. Researchers have proposed that, the generalized form of topological spaces is the monotone spaces having ending property [9]. In this paper, a model of lattice based consistent distributed computation and, associated monotone slicer of computation are presented considering state-based monotone topological spaces. The model considers asynchronous distributed computation, where the distributed states of processes are partially ordered. The consistency property of computation is maintained while constructing the topological model of slicer in monotone spaces. It is illustrated that Birkhoff's lattice cover and isomorphism are maintained in the proposed state-based monotone slicer model of distributed computation. A set of analytical properties of the model is formulated. Moreover, the definition of topological cut and the associated model of consistent cuts of asynchronous event-based distributed computation are formulated as axioms.

Rest of the paper is organized as follows. Section 2 presents interplay of distributed computation and distributed databases involving big data and data analytics applications. Section 3 represents related work. Section 4 presents formulation of models of asynchronous distributed computation in monotone spaces, monotone slicer of computation and associated consistency properties. Section 5 presents a set of analytical properties of the proposed models. Finally, Sect. 6 concludes the paper.

2   Distributed Database and Computation

Nowadays, the big data and data analytics based applications are widely employed for data storage, rendering, filtering and performing numerical analysis. The storage of big data requires distributed databases, where data is partitioned into subsets and is distributed to multiple nodes. The big data based applications require three computational steps as illustrated in Fig. 1.

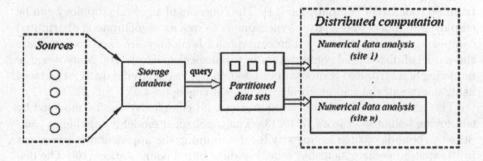

Fig. 1. Distributed database and computation stages.

In first step, the data is gathered into a database. Next, the data is queried and associatively partitioned into distributed databases based on data locality. Lastly, the distributed computational nodes (sites) collect specific data within large data sets and provide numerical analysis results. Evidently, the distributed databases storing partitioned data sets require the support from the underlying distributed file systems (DFS). For example, the Hadoop DFS is widely employed to provide file system support for distributed data storage and, query [17]. However, the integration of Hadoop DFS and high performance data analytics computations (i.e. distributed computation) remains a challenge involving distributed databases [17]. The system design employing combination of concepts of distributed databases and distributed computing is desired and, the models of distributed systems architectures affect the overall system performance [3,12,18]. The slicer model of distributed computation is required to partition computation involving disjoint data sets in asynchronous distributed systems.

3   Related Work

The hybridization of concepts of distributed databases and distributed systems are required to support big data and associated data analytics applications. In general, the distributed systems provide the supporting infrastructure to distributed database systems to store large data sets and, to execute computation on data. The model of distributed computing structures and associated properties play an important role in determining the performance of the distributed database applications. For example, the pMatlab and D4M are message passing library and data structure, which are designed to support high-performance data analytics applications involving distributed databases and distributed computation [3,18]. The peer-to-peer (P2P) distributed systems architecture is widely employed for supporting distributed applications such as, VoIP and data backup. The improvements of system performance are dependent on storage and computing capabilities of nodes in P2P systems.

The distributed computation slicer is designed in distributed P2P systems in order to achieve node partitioning [12]. The traditional models of slicer consider partial ordering or causal ordering of events. However, the topological structures executing higher dimensional automata are formulated to model and analyze distributed concurrent computation [11]. The concepts of algebraic topology can be employed to model and analyze synchronous as well as asynchronous distributed computing systems [1,14,15]. The concepts of homology are used to formulate the model of distributed computation in topological spaces [2,20]. Moreover, the modeling of distributed computation based on iterated shared data is formulated using combinatorial and algebraic topological concepts [4,5].

The stability of distributed computation and complexity can be modeled by employing homotopy theory [7,10,13]. The topological model is capable to compute the bounds of time complexity for determining the approximate agreement in iterated immediate snapshot model of distributed computation [16]. The distributed computation in topology is modeled as simplexes, where simplexes are a set of complex in topological spaces [8,14]. The properties of object-based distributed systems are formulated by employing topological structures [6]. The topological spaces can be generalized by introducing monotone function. The monotone spaces are the generalized forms of underlying topological spaces [9]. The slicer model of distributed computation can be formulated in monotone topological spaces considering the consistency property of computation.

4   Monotone Slicing in Distributed Computation

Let a distributed computation be denoted as D comprised of a set of distributed processes given by $P = \{p_j : 1 \leq j \leq N, j \in Z^+\}$. The distributed processes are considered to be deterministic finite state machines. A process $p_j \in P$ has a set of deterministic execution states denoted by S_j, where $\phi \notin S_j$ and, E_j be a set of local events of $p_j \in P$. The global set of events in the distributed system is given by, $\overline{E} = \bigcup_{j=1}^{N} E_j$. The entire state-space of D can be identified by,

$$S(D) = \bigcup_{j=1}^{N} S_j \tag{1}$$

The distributed processes invoke inter-process communications and state transitions occur in a process due to transmission/reception of messages. In the proposed model, the processing of events is considered to be internal to processes and, there exists a global function determining the computational dynamics of the overall system. Thus, the global state transition related to a process is governed by a function given by,

$$f : S(D) \to S(D),$$
$$f(s_j \subset S_j) \subset S_j \setminus \{s_j\} \tag{2}$$

The distributed computation is considered to be deterministic if it converges to a state-space and such converging state-space of D is defined as,

$$\overline{S(D)} = \bigcap_{j=1}^{N} S_j \tag{3}$$

Let $\Omega(X)$ be a power set of set X. The set of all possible cuts on distributed computation D is denoted by $\mathcal{C}(D) \subset \Omega(S(D))$. The set of boundary elements B_j of a process $p_j \in P$ is defined as,

$$\forall p_j \in P : f(B_j \subset S_j) \in \overline{S(D)} \tag{4}$$

Considering $B_j \subset S_j$ in a distributed system D, a computation in D is defined as,

$$\vec{D} = S(D) \setminus f(\bigcup_{j=1}^{N} B_j) \tag{5}$$

The above equation illustrates that, the distributed execution considers a set of global states excluding the convergent region in the state-space. Moreover, the computation includes the elements of boundary state-space. A monotone space over $\vec{D}$ is $(\vec{D}, g)$ equipped with a monotone function $g(.)$ having following properties:

$$g : \Omega(\vec{D}) \to \Omega(\vec{D}),$$
$$g(\phi) = \phi,$$
$$\forall A \in \Omega(\vec{D}) : A \subset g(A), \tag{6}$$
$$\forall A, B \in \Omega(\vec{D}) : A \subset B \Rightarrow g(A) \subset g(B).$$

This definition represents the monotone property of a distributed computation and the mapping of structural forms within the state-space of distributed computation under consideration.

4.1 Concept of Topological Cut

In this section, the concept of topological cut is introduced. Let, X_P be a point-set and a topological space over X_P is (X_P, τ_X) where, $\tau_X \subseteq \Omega(X_P)$. If $d_X : X_P \times X_P \to \Re$, where $d_X \in [0, +\infty)$ is a distance metric in X_P then (X_P, d_X, τ_X) is a metric space having topological structure. The space is segmented by a topological cut $S_X \subset X_P$ such that, the following axioms are satisfied,

$$S_X \neq \phi,$$
$$\exists X_1 \subset X_P, \exists X_2 \subset X_P : X_1 \cap X_2 = \phi,$$
$$X_P \setminus S_X = X_1 \cup X_2, \tag{7}$$
$$\forall x_1 \in X_1, \forall x_2 \in X_2, \forall s \in S_X : d_X(x_1, x_2) \leq d_X(x_1, s) + d_X(x_2, s)$$

The set S_X is a topological cut of (X_P, d_X, τ_X). The cut S_X is called topologically symmetric cut of (X_P, d_X, τ_X) if the following axioms are satisfied,

$$A \subset X_1, B \subset X_2 : [a \in A \wedge b \in B] \Rightarrow [\exists s \in S_X : d_X(a, s) = d_X(b, s)],$$
$$\mid A \mid = \mid B \mid = \mid S_X \mid \tag{8}$$

The computed set of distances of all elements of set X_1 and X_2 with respect to $\forall s \in S_X$ is given by,

$$\triangle^1_{\forall s \in S_X} = \{d_X(s, x) : \forall x \in X_1\},$$
$$\triangle^2_{\forall s \in S_X} = \{d_X(s, x) : \forall x \in X_2\} \tag{9}$$

The corresponding extremals of a topological cut S_X are,

$$\omega(X_1) = \{x : x \in X_1 \wedge d_X(s, x) = \max(\triangle^1_{\forall s \in S_X})\},$$
$$\omega(X_2) = \{x : x \in X_2 \wedge d_X(s, x) = \max(\triangle^2_{\forall s \in S_X})\} \tag{10}$$

The extremals of a topological cut are unique if, $\mid \omega(X_1) \mid = \mid \omega(X_2) \mid = 1$.

4.1.1   Definition of ||

As stated earlier, it is assumed that the distributed processes are executing as finite state machines in a distributed system and, the entire state-space is $X_P = S(D)$. A partial ordering relation ($\xrightarrow{m}$) is defined in global state-space of the respective distributed system by following axioms,

$$\forall p_i, p_j \in P : \xrightarrow{m} \subset (S_i \times S_j) \cup (S_j \times S_i),$$

$$\forall s_i \in S_i, \forall s_j \in S_j : [(s_i, s_j) \in \xrightarrow{m}] \Rightarrow [(s_j, s_i) \notin \xrightarrow{m}], \qquad (11)$$

$$[(s_i, s_j) \in \xrightarrow{m} \vee (s_{j+1}, s_{i+1}) \in \xrightarrow{m}] \Rightarrow [s_{i+1} = f(s_i)] \wedge [s_{j+1} = f(s_j)]$$

A relation $|| \subset S(D)^2$ between any two states within state-space of the distributed computation under consideration is defined by following axioms,

$$\forall p_i \in P : s_{i-1} \in A \subset X_1, s_{i+1} \in B \subset X_2 : [s_i = f(s_{i-1}) \wedge s_{i+1} = f(s_i)],$$

$$\forall (s_i || s_j) \Rightarrow [(s_i, s_j) \notin \xrightarrow{m}] \wedge [(s_j, s_i) \notin \xrightarrow{m}],$$

$$\forall (s_i || s_j) \Rightarrow \neg [\exists s_{j+1} \in B : (s_{j+1}, s_i) \in \xrightarrow{m}], \qquad (12)$$

$$(s_i || s_j) \Leftrightarrow (s_j || s_i)$$

Thus, the relation $||$ maintains the consistency of cuts in execution topological state-space of distributed processes.

4.2   Monotone Slicer Model

In order to formulate an observable distributed computation, it is necessary to restrict the dynamics of distributed computation in monotone topological spaces. Let $A_j \in \Omega(\overrightarrow{D})$ and $A_j \subset g(A_j)$. A set of cuts $C \subset \mathcal{C}(D)$ in $\overrightarrow{D}$ for $K < |\Omega(\overrightarrow{D})|$ is given by, $C - \bigcup_{j=1}^{K} y(A_j)$ such that following implication holds, where $\forall c \in C, |c| = N$ and, $u \neq v$,

$$\forall p_u, p_v \in P : [\forall x, \forall y : x, y \in c] \Rightarrow [x \in S_u \wedge y \in S_v] \qquad (13)$$

The set of consistent cuts forms a lattice $(L, \leq)$ in $\overrightarrow{D}$ where, $L \subset C$ and, $\forall c \in L, [\forall x, \forall y : x, y \in c] \Rightarrow (x || y)$. A set of lattice join-irreducible elements of L is given by, $J(L) \subset L$. A distributed computation in monotone topological spaces is consistent if the run R in $\overrightarrow{D}$ is defined as a sequence $R = < A_k : A_k \in \Omega(\overrightarrow{D}), k = 0, 1, 2, \ldots >$, where $A_k, A_m \in \Omega(\overrightarrow{D})$ such that, $(g(A_k), g(A_m)) \in \leq$ and, $m = k + 1$.

Let X be a set of Boolean variables given by, $X = \{x_n : n = 1, 2, 3, \ldots, I\}$ where, $x_n \in \{0, 1\}$. The predicate $\Gamma(X|c) \in \{0, 1\}$ is assumed to be computable at cut-state $c \in C$ in the corresponding distributed system. Thus, the slice of a distributed computation is defined as,

$$J(L|\Gamma) = \{c : (\Gamma(X|c) = 1) \wedge (c \in J(L))\} \qquad (14)$$

The lean slice of a distributed computation is $l = \{c : c \in J(L|\Gamma)\}$ where, $|l| = 1$. A state e is called critical state if $\exists \{e\} \in \Omega(\vec{D})$ and, $l = g(\{e\})$. The elements of $g(.)$ satisfying stable Boolean predicate $\Gamma(.)$ in a valid distributed computation are verified by,

$$(g(x), \Gamma) \Rightarrow [(x \in \Omega(\vec{D})) \wedge (g(x) \in L) \wedge (\Gamma(X|g(x)) = 1)] \qquad (15)$$

Hence, the set of consistent and stable executions in a distributed computation can be found if the corresponding sequences of executions satisfy Eq. (15).

The Birkhoff's lattice isomorphism model considers the application of multi-valued mapping function given by, $\beta : L \to C$. Thus, the Birkhoff's lattice isomorphism model can be incorporated or preserved in the monotone slicer model proposed in this paper by following the definition given as, $\beta(c \in L) = \{x : ((x, c) \in \leq) \wedge (x \in J(L))\}$. It immediately follows that, considering e be a critical state of a distributed computation in monotone spaces, $(g(\{e\}), \Gamma) \Rightarrow [\exists y \in L : g(\{e\}) \in \beta(y)]$. Moreover, if $A_k \in \Omega(\vec{D})$ such that, $g(A_k) \in L$ then the non-commutative composition is formulated as, $\beta \circ g \in J(L)$. This indicates that, the Birkhoff's lattice isomorphism as well as cover can be preserved in the proposed monotone slicer model of distributed computation.

4.3 Topological Cut in Event-Based Distributed Systems

The earlier section considered state-based model of distributed systems. However, a distributed system can be formulated as an event-based system. Following the event-based model of distributed computation, it is considered in this section that X_P is an entire event-space of the distributed computation under consideration having $X_P = \overline{E}$ and, the corresponding logical clock function is $r : \overline{E} \to Z^+$. Thus, the distance metric of (X_P, d_X, τ_X) in distributed computational model can be defined as,

$$\forall a \in \overline{E}, \forall b \in \overline{E} : d_X(a, b) = |r(a) - r(b)|,$$
$$d_X(a, b) \in [0, +\infty) \qquad (16)$$

The partial ordering relation $(\overset{h}{\to})$ is defined in event-space of distributed system under consideration as,

$$\forall a \in \overline{E}, \forall b \in \overline{E} : [(a, b) \in \overset{h}{\to}] \Rightarrow [r(a) < r(b)] \wedge [d_X(a, b) > 0] \qquad (17)$$

The topological cut $S_X \subset X_P$ is consistent in the point-set topological model of distributed computation if the following axioms are satisfied over partial ordering relation of events,

$$\forall a \in S_X, \forall b \in S_X : [\neg(a \overset{h}{\to} b)] \wedge [\neg(b \overset{h}{\to} a)],$$
$$|S_X| = N \qquad (18)$$

5   Properties of Monotone Slicer

The determination of consistency and observability of a distributed computation are two important factors in the runs of concurrent interacting processes. Identifications of critical state of computation and its property are required to form a lean slice. This section constructs a set of analytical properties of monotone slicer and proposes the existence of generating function of a slice in the execution lattice.

5.1   Theorem 1:

$g(.)$ is strictly consistent if $\forall s_k \in \overrightarrow{D}, \exists A_k \in \Omega(\overrightarrow{D}), g(A_k) \in L$.

Proof: Let in a distributed computation be, $\forall s_k \in \bigcup_{j=1}^{N} S_j, \exists A_k \in \Omega(\overrightarrow{D})$ such that, $g(A_k) \in L$. Thus, $\exists s_m \in \bigcup_{j=1}^{N} S_j, m = k + 1$ where, $(g(A_k), g(A_m)) \in \leq$. Hence, the run R is always consistent in the corresponding distributed computation. Thus, $g(.)$ is strictly consistent in distributed computation in monotone spaces.

5.2   Theorem 2:

If $V \subset \Omega(\overrightarrow{D})$ such that, $g(V) = C$ then, $g(V) \setminus L$ may not be consistent.

Proof: Let $V \subset \Omega(\overrightarrow{D})$ and $\exists \{x\} \in V$ such that, $g(\{x\}) \in C$. Now, $g(V) \setminus L \subset C$ and, if $g(\{x\}) \in L$ then, $g(\{x\}) \notin g(V) \setminus L$. Thus, it can be said that,

$$\forall \{x\} \in V : [g(\{x\}) \in L] \Rightarrow [g(\{x\}) \notin g(V) \setminus L] \tag{19}$$

Hence, $g(V) \setminus L$ is not consistent.

5.3   Theorem 3:

$g(.)$ is a generating function of $(L, \leq)$.

Proof: Let $E \subset \Omega(\overrightarrow{D})$ such that, $\forall x \in E, |x| < N$. This indicates, $\exists g(a \in E), \exists g(b \in E)$ such that, $(g(a), g(b)) \in \leq$. If $F \subset E$ such that, $\forall x \in F, g(x) \in L$ then, $g(F)$ is a generating function of lattice $(L, \leq)$.

5.4   Theorem 4:

If $E \subset \Omega(\overrightarrow{D})$ such that, $\forall x \in E, |x| < N$ and, $g(x) \in L$ then, $(g(F), \Gamma)$ induces a sub-lattice where, $F \subset E$.

Proof: Let $E \subset \Omega(\overrightarrow{D})$ such that, $\forall x \in E, |x| < N$ and, $g(x) \in L$. Let $F \subset E$ such that, the following implication holds,

$$\forall x \in F : [(g(x \in F), \Gamma)] \Rightarrow [\neg(g(x \in E \setminus F), \Gamma)] \tag{20}$$

However, $\forall x \in E$, following condition is satisfied in the distributed computation under consideration,

$$[(g(x) \in L) \wedge (L = g(E))] \Rightarrow [(g(E), \leq)] \tag{21}$$

Moreover, $g(x \in F) \subset L$. Hence, $(g(F), \Gamma)$ induces a sub-lattice of $(L, \leq)$.

6 Conclusions

Distributed database and distributed computing systems support executions of big data based data analytics applications. The model of underlying distributed computing structure plays a role in determining overall performance and stability of the respective system. The structure of distributed computation can be modeled in monotone topological spaces. The modeling of consistent monotone slicer of distributed computation in monotone topological spaces helps in determining consistent cuts in execution lattice preserving Birkhoff's lattice isomorphism. The monotone slicer model considers evaluation of stable predicate in consistent cuts in the lattice of distributed computation. In case of event-based distributed systems, the topological model of consistent cuts can be formulated considering the global event-space of the distributed systems.

References

1. Alpern, B., Schneider, F.B.: Defining liveness. Inf. Process. Lett. **21**(4), 181–185 (1985)
2. Bauer, U., Kerber, M., Reininghaus, J.: Distributed computation of persistent homology. In: Proceedings of Meeting on Algorithm Engineering and Experiments, SIAM, pp. 31–38 (2014)
3. Bliss, T.N., Kepner, J.: pMatlab parallel Matlab library. Int. J. High Perform. Comput. Appl. **21**(3), 336–359 (2007)
4. Borowsky, E., Gafni, E.: A simple algorithmically reasoned characterization of wait-free computation. In: Proceedings of the Sixteenth Annual ACM Symposium on Principles of Distributed Computing, pp. 189–198 (1997)
5. Conde, R., Rajsbaum, S.: An introduction to topological theory of distributed computing with safe-consensus. Electron. Notes Theor. Comput. Sci. **283**, 29–51 (2012)
6. Duarte, C.H.C.: Mathematical models of object-based distributed systems. In: Agha, G., Danvy, O., Meseguer, J. (eds.) Talcott Festschrift. LNCS, vol. 7000, pp. 57–73. Springer, Heidelberg (2011). doi:10.1007/978-3-642-24933-4_4
7. Fajstrup, L., Rauben, M., Goubault, E.: Algebraic topology and concurrency. Theoret. Comput. Sci. **357**(1–3), 241–278 (2006)
8. Fraigniaud, P., Rajsbaum, S., Travers, C.: Locality and checkability in wait-free computing. In: Peleg, D. (ed.) DISC 2011. LNCS, vol. 6950, pp. 333–347. Springer, Heidelberg (2011). doi:10.1007/978-3-642-24100-0_34
9. Ghosh, S.R., Dasgupta, H.: Connectedness in monotone spaces. Bull. Malays. Math. Sci. Soc. **27**(2), 129–148 (2004)
10. Goubault, E.: Some geometric perspectives in concurrency theory. Homol. Homotopy Appl. **5**(2), 95–136 (2003)
11. Goubault, E., Jensen, T.P.: Homology of higher dimensional automata. In: Cleaveland, W.R. (ed.) CONCUR 1992. LNCS, vol. 630, pp. 254–268. Springer, Heidelberg (1992). doi:10.1007/BFb0084796

12. Gramoli, V., et al.: Slicing distributed systems. IEEE Trans. Comput. **58**(11), 1444–1455 (2009)
13. Gunawardena, J.: Homotopy and concurrency. Bull. EATCS **54**, 184–193 (1994)
14. Herlihy, M., Kozlov, D., Rajsbaum, S.: Distributed Computing Through Combinatorial Topology. Elsevier, Amsterdam (2014)
15. Herlihy, M., Rajsbaum, S.: New perspectives in distributed computing. In: Kutyłowski, M., Pacholski, L., Wierzbicki, T. (eds.) MFCS 1999. LNCS, vol. 1672, pp. 170–186. Springer, Heidelberg (1999). doi:10.1007/3-540-48340-3_16
16. Hoest, G., Shavit, N.: Toward a topological characterization of asynchronous complexity. SIAM J. Comput. **36**(2), 457–497 (2006)
17. Huang, Y., Yesha, Y., Zhou, S.: A database based distributed computation architecture with Accumulo and D4M: an application of eigensolver for large sparse matrix. In: IEEE International Conference on Big Data (2015)
18. Kepner, J.: Massive database analysis on the cloud with D4M. In: Proceedings of HPEC Workshop (2011)
19. Saks, M., Zaharoglou, F.: Wait-free k-set agreement is impossible: the topology of public knowledge. SIAM J. Comput. **29**(5), 1449–1483 (2000)
20. Zomorodian, A., Carlsson, G.: Computing persistent homology. Discret. Comput. Geom. **33**(2), 249–274 (2005)

Storage Efficiency of LOB Structures for Free RDBMSs on Example of PostgreSQL and Oracle Platforms

Lukasz Wycislik[✉]

Institute of Informatics, Silesian University of Technology,
16 Akademicka St., 44-100 Gliwice, Poland
lukasz.wycislik@polsl.pl

Abstract. The article is a study upon storage efficiency of LOB structures for systems consisting of PostgreSQL or Oracle relational database management systems. Despite the fact that recently several NoSQL DBMS concepts were born (in particular document-oriented or key-value), relational databases still do not lose their popularity. The main content is focused on sparse data such as XML or base64 encoded data that stored in raw form consume significant volume of data storage. Studies cover both performance (data volume stored per unit of time) and savings (ability to save data storage) aspects.

Keywords: Storage efficiency · LOB · Oracle · PostgreSQL · Cloud computing

1 Introduction

The idea of the relational database management systems was created with the needs of flexible and efficient handling of collected structured data. At the beginning only the basic data types were being collected and processed. But along with the development of IT application there was a need to process raw/binary data that could encode audio, video, serialized documents of any format, etc. Initially these specific types of data have been stored outside of files managed by RDBMSs (Relational Database Management Systems) but directly in file systems. This led to several problems with the fulfillment of the ACID (Atomicity, Consistency, Isolation, Durability) properties and with implementation of backup mechanisms. Currently, most RDBMSs allow to store and to manipulate binary data the same way as structured data in terms of the ACID properties, authorization, access control and backup mechanisms.

The volume of binary data, that often comes from digitizers and represents audio, video and others is a separate issue. The volume of this type of data is usually managed at application level by defining the quality requirements of stored data and the use of domain-specific compression algorithms. For any other type of data it is also possible to use general compression algorithms (e.g. lzo, zlib, etc.) [14].

© Springer International Publishing AG 2017
S. Kozielski et al. (Eds.): BDAS 2017, CCIS 716, pp. 212–223, 2017.
DOI: 10.1007/978-3-319-58274-0_18

Some of the RDBMs enable transparent compression of stored content that is based on trial data compression at the time of storing, and in the event of a satisfactory compression rate to save so compressed content. It has to be mentioned here that the use of compression can have both positive and negative impact on the performance of the entire solution and it will depend on what is the bottleneck of processing in a particular system (CPU time vs. IO bandwidth). In the case of enterprise-class systems, there are many possible scenarios that can be implemented by different levels of the technology stack for compressing stored content from transparent compression made by disk arrays through compression enabled file systems and database systems to compression made at the application level. But most of these scenarios entail additional costs, which do not always justify their choice.

This article presents research on the possibility of saving disk storage (while controlling performance) for systems with relational databases build on top of JBOD (i.e. just a bunch of disks) or 'virtual JBOD' storage for data of variable 'information density'. Such an architecture is characteristic for low-cost on-premise systems but also, what is more important, for cloud services served in the IaaS model [19].

2   Related Work

This interest for data compression is unrelenting since the rapid development of computer technology. Initial struggle for space-saving use of floppy disks or slow modem connections continues to this day and the strategies and approaches are changing with the development of technology.

Contemporary trends moving computation tasks to the cloud seem to result in leaving as infrastructural aspects as compression to providers of cloud services. In fact, providers of such services, aiming to reduce the overall cost of infrastructure operating, are also interested in the subject of compression stored [3,6,7] or transmitted data [4]. But the decrease in the volume of compressed data at the side of the service provider usually does not mean a reduction in the cost of operating the system by the tenant – motto 'pay as you go' refers to the volume of data seen on the client side. Perhaps this will change in the future, but for now the tenant should be interested in data compression by himself. Of course, possible approaches depend on the model in which services are provided and this article focuses on the RDMSs and relates to IaaS model.

General considerations on the topics of data compression can be found in [15].

3   Implementations of Storing Binary Data in RDBMSs

A **Binary Large OBject** (BLOB) is a collection of binary data stored as a single entity in a database management systems. This data usually represents images, audio or other multimedia, serialized state of application objects or even executable code. Database support for BLOBs is not universal and each database platform offers its own solution.

3.1   Oracle

Oracle database is known and leading database platform that is being used both in small and very big applications. It is respected both for its reliability and performance. Its architecture assumes to take maximum control over the data files implementing its own buffers and making it even possible to store on raw partitions or Oracle self–made database file system named *ASM* (Automatic Storage Management) [9,11].

Oracle database started to support BLOB functionality in version 8 and besides BLOB stored in database files there was also BFILE datatype that allowed to store binary data outside of the database, as operating system files. System architect had to decide which of the above two types to use, because BLOB met ACID features but access time to data was much longer than using BFILE. With version 11g Oracle provided *SecureFiles* mechanism that completely redefined method of storing binary data inside of database files making it up to 10x faster than before and even faster than using BFILE [8].

The use of the enterprise version features allows not only for transparent compression, but even for data deduplication. These features however are extra paid so they cannot be taken into account during this study as they are not included in Oracle Express Edition 11g. Also partitioning, which greatly simplifies the management of large volumes of data, is available only on the enterprise platform, but there are solutions for standard platform, which for certain applications prove themselves as well [20]. Although compression and partitioning are orthogonal approaches, the best results in management of the data retention are obtained by the complementary usage - rarely requested data is stored on slower, cheaper, compressed media, while sensitive data is stored on faster, more reliable and more expensive media.

3.2   PostgreSQL

PostgreSQL (PgSQL) is the world's most advanced open source database. Developed over 25 years by a vibrant and independent open source community, PostgreSQL was born from the same research as Oracle and DB2 and contains comparable enterprise-class features such as full ACID compliance for outstanding transaction reliability and Multi-Version Concurrency Control (MVCC) for supporting high concurrent loads [2].

Developers of PostgreSQL, despite being inspired by the architecture of Oracle database, decided on a completely opposite approach to managing of data files. They rely on the host operating system much more, leaving under its control managing of data files and buffers.

PostgreSQL strategy for storing large objects is quite different from that developed by Oracle. There is no single datatype dedicated for storing large objects but any datatype that has a variable-length representation, in which the first 32-bit word of any stored value contains the total lenght of the value in bytes, is capable to store lots of data. This technique, introduced in version 7, was called *TOAST* (The Oversized-Attribute Storage Technique). PostgreSQL uses a fixed page size (commonly 8 KB), and does not allow tuples to span multiple pages. Therefore, it is not possible to store very large field values directly. To overcome this limitation, large field values are compressed and/or broken up into multiple physical rows. This happens transparently to the user, with only small impact on most of the backend code [16]. To store BLOB data one simple has to use BYTEA datatype that among others is also *TOAST* enabled.

PostgreSQL has also a large object facility, which provides stream-style access to user data that is stored in a special large-object structure what enables random seeking over entries. Unfortunately, a serious drawback of this datatype is the lack of support for ACID and a completely different interface to access data then SQL. Wider discussion on the advantages and disadvantages of the use of one of these datatypes can be found in [12].

4   Survey

4.1   Research Environment and Methods of Measurement

The study was conducted on an environment virtualized by VMware ESXi (supervising Dell Poweredge T420 server) and is shown in the Fig. 1.

Listing 1.1. Database load method

```
public long insertblob(InputStream str, boolean getvolume)
throws SQLException
{
    psInsert.setBinaryStream(1, str);
    psInsert.execute();
    if( getvolume ) {
        ResultSet rset = psSize.executeQuery();
        rset.next();
        return rset.getLong(1);
    }
    else return 0;
}
```

Guest operating system (Ubuntu distribution) hosts LXC containers [1] using Docker [17]. Both database platforms run with default configuration parameters as LXC containers with its own file systems for binaries. Database files however are placed on linked volumes that are mapped to dedicated disk drive mounted on operating system. Such architecture allows on the one hand to easily manage containers (that can be regarded as non persistent resources) and on the other hand to easily manage database volumes and measure changes in the volume of files.

Testing load of databases was conducted by dedicated java application being executed in java runtime environment. Access to the databases was implemented using JDBC (Java Database Connectivity) mechanisms. The method of loading data into both databases was presented in Listing 1.1.

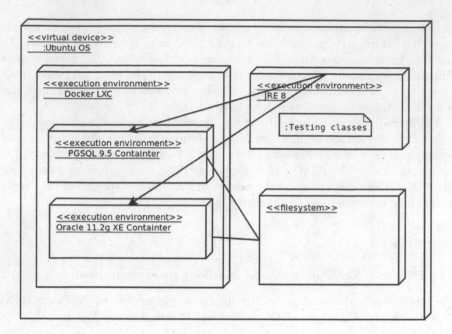

Fig. 1. Diagram of the test environment

As you can see the *insertblob* method operates on private objects *psInsert* and *psSize* of *PreparedStatement* class. Because the SQL statement is prepared only once during class initialization, database servers do not have to parse statement and develop execution plan each time the method is called letting it focus only on binding and executing of insert statement. Calling parameters are *str* that contains binary data to be stored and *getvolume* that is a flag indicating the need for database segment size calculation (since such calculations are quite time consuming they are not performed each time).

4.2 Measurement Methodology

These studies were inspired by the need for validation of database platforms for an electronic document archiving system. Therefore, measurements concentrate mostly on write performance and actual size (taking into account the overhead of concrete database platform) of stored files of different volume and different 'information density'.

The size of files was set at 7 KB, 100 KB and 10 MB and to examine the files of different 'information density' XML format (as sparse), base64 encoded (as medium) and random-byte (as dense) files have been chosen.

During the implementation of the research methods there was a need to measure the space allocated for storage of test files. It is important to distinguish between the logical volume size of files (resulting directly from the file size and the number of stored files) from the physical size of the disk space allocated by the files of the concrete DBMS. As the logical volume size can be calculated a priori, it is interesting to measure only the physical disk space allocation. This was done on three levels - as SQL query (of tablespace size allocation) on data dictionary of given DBMS, as the OS command that calculates the size of file or directory and as OS command that calculates free space of a disk. The last one lets to indirectly calculate the increase of space allocated as the difference between the final and the initial measurement. Although the use of this method is the most problematic (because along with subsequent files storing disk space consumption results not only from the increase in the given tablespace, but also from the increase in other supporting structures, such as database logs), it was necessary for research on BTRFS, as it so far in the OS commands displaying size of files the option of displaying the physical size of transparently compressed files is not implemented. So for tests on a standard file system (ext4) the second level method was used (after making sure that the results are the same as for the first level method) and for tests on a BTRFS file system the third level method was used.

All presented results are for volume up to 11 GB as it is the upper limit for user data in Oracle Express Edition 11g. For PostgreSQL database also further studies (results are not presented in charts) were conducted for up to 1 TB volume of data and the database platform acted stable and predictable.

4.3   Results

In the Fig. 2 increases of physical data volume for 7 KB files of different type for Oracle platform are shown. On the horizontal axis is the number of stored documents while on vertical axis is the volume of data in bytes. On the chart legend are symbols for each file type (b64 - base64, rnd - random bytes, xml - XML format) and one additional called *unitary* to show the logical data size (that is calculated as multiplication of file size and the number of stored files) as a reference line.

As we can see all measurements for all types of files are arranged almost in a collinear way on a line above *unitary* what is caused by the additional overhead for the organization of data by database server. In this particular case (for 7 KB files in Oracle database) the overhead is about 15.4% (calculated as 100% * (10.4 GB − 9.01 GB)/9.01 GB).

Similar measurements for PgSQL platform are shown in Fig. 3.

In this case, for base-64 and random files the database platform overhead is the same − 16.2% (calculated as 100% * (11.64 GB − 10.02 GB)/10.02 GB)) but for XML files we can see values that are much smaller than *unitary* ones. This is due to the use of transparent compression that saves storage space by 60.5%. Surely this is not the maximum savings that could be achieved [18], but we get it absolutely for free and without any additional effort.

The following figures (Figs. 4 and 5) show the final aggregated results of measurements for all types, sizes and database platforms. Trying to create a 'level playing field' the third deployment variant was introduced where Oracle database platform (that in the free version lacks compression) has been coupled

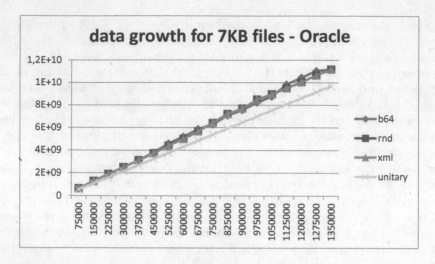

Fig. 2. The size of data for 7 KB files of different type for Oracle database

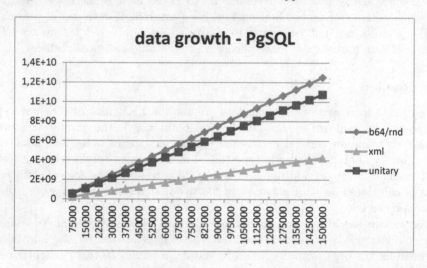

Fig. 3. The size of data for 7 KB files of different type for PgSQL database

with modern, open source Btrfs file system [13]. Btrfs, besides many other valuable features, has the capability of transparent compression and so it has been configured.

As shown in many studies (e.g. [5]), rich Btrfs functionality affects sometimes the performance degradation. Figure 4 (research for smallest files) seems to confirm this - the performance overhead is above 153% compared to Oracle coupled with 'plain' ext4 file system. PgSQL showed 30.5% better performance compared to Oracle on ext4.

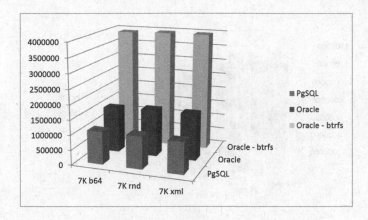

Fig. 4. The storage performance for 7 KB files of different type

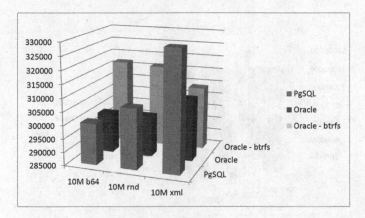

Fig. 5. The storage performance for 10 MB files of different type

The situation changes for the testing of largest (10 MB) files (Fig. 5). Previous huge disparities disappear and the differences in measures are emphasized only thanks to scaled axes. We can see that both database platforms coupled with ext4 file system are processing a little bit longer XML files. The opposite situation can be observed for the Oracle database coupled with Btrfs file system.

Further figures (Figs. 6, 7 and 8) show storage efficiency in term of data footprint. Each of these three figures (showing the results of measurement for 7 KB, 200 KB and 10 MB files) allow to draw similar conclusions.

Results for the Oracle database that uses ext4 and Btrfs are quite similar but environment with Btrfs always needs a little more storage (for 7 KB files it is an average of 2.8% and growing to an average of 14.3% for 10 MB files). It is quite surprising because it turns out that even for XML files the Btrfs transparent compression mechanism is not compatible with Oracle database and only makes processing time longer.

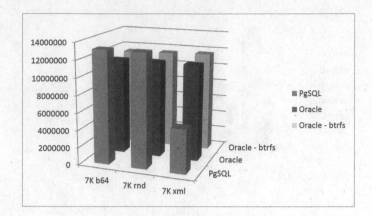

Fig. 6. The size of data for 7 KB files of different type

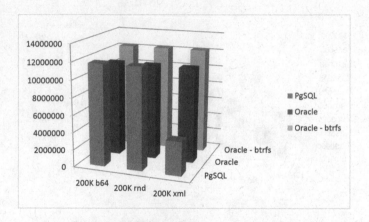

Fig. 7. The size of data for 200 KB files of different type

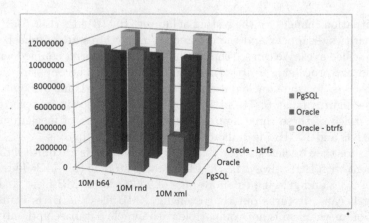

Fig. 8. The size of data for 10 MB files of different type

PgSQL has higher storing overhead (from 6.9% to 17.4%) than Oracle for files of base64 and random type. But for XML files PgSQL significantly outperforms Oracle database (from 54.6% to 64.5% what means that data volume is smaller from 2.2 to 2.81 times for PgSQL).

5 Summary

This paper presents a survey about storing efficiency of LOB structures for free RDBMSs. Despite the fact that recently several NoSQL DBMS concepts were born (in particular document-oriented or key-value), relational databases still do not lose their popularity. If there is a need for processing binary objects one has to ask oneself a few questions such as:

- Are binary objects the part of broader data model?
- Is there a need to manipulate binary objects using the metadata that describe it?
- Should binary objects be processed uniformly with the other data in terms of transaction management?

A positive answer to any of these questions should lead to consideration of the use of RDBMS.

Conducted research indicates that the footprint of stored binary data is not tremendous, but it should be noted that there are some differences between database platforms. We examined the well-known, industry leading Oracle database and open source PostgreSQL platform that is often considered (due to many similarities in operating and architecture) as a 'free replacement' for Oracle.

Oracle showed the lowest overhead for the stored data in raw representation, but in the case of 'sparse data' such as XML, PgSQL thanks to transparent compression option, can greatly save volume of stored data. It should be noted that the Oracle platform in the paid version (as Oracle Advanced Compression feature of Oracle Database Enterprise Edition [10]) also has the ability to compress data.

When assessing the performance of platforms (for the number of documents stored per unit of time) it should be noted that for small files PgSQL platform was more efficient but with the increase in file size differences are blurred and even pointed to the superiority of the Oracle.

Unfortunately attempts to apply the compression to Oracle coupled with Btrfs failed, despite the fact that for Btrfs transparent compression of the same files processed at the level of operating system worked correctly. Since such a configuration could be very cost-effective, future consideration should be given to examine Oracle coupled with other modern file systems that are transparent compression enabled.

References

1. Virtualization and Containerization of Application Infrastructure: A Comparison, vol. 21, University of Twente (2014)
2. EnterpriseDB: Postgresql overview. http://www.enterprisedb.com/products-services-training/products/postgresql-overview
3. Hovestadt, M., Kao, O., Kliem, A., Warneke, D.: Adaptive online compression in clouds—making informed decisions in virtual machine environments. J. Grid Comput. 11(2), 167–186 (2013). http://dx.doi.org/10.1007/s10723-013-9249-4
4. Jägemar, M., Eldh, S., Ermedahl, A., Lisper, B.: Adaptive online feedback controlled message compression. In: 2014 IEEE 38th Annual Computer Software and Applications Conference (COMPSAC), pp. 558–567, July 2014
5. Kljajic, J., Bogdanovic, N., Nankovski, M., Toncev, M., Djordjevic, B.: Performance analysis of 64-bit EXT4, XFS and BTRFS filesystems on the solid-state disk technology. INFOTEH-JAHORINA 15, 563–566 (2013). http://infoteh.etf.unssa.rs.ba/zbornik/2016/radovi/RSS-2/RSS-2-8.pdf
6. Nicolae, B.: High throughput data-compression for cloud storage. In: Hameurlain, A., Morvan, F., Tjoa, A.M. (eds.) Globe 2010. LNCS, vol. 6265, pp. 1–12. Springer, Heidelberg (2010). doi:10.1007/978-3-642-15108-8_1
7. Nicolae, B.: On the benefits of transparent compression for cost-effective cloud data storage. In: Hameurlain, A., Küng, J., Wagner, R. (eds.) TLDKS III. LNCS, vol. 6790, pp. 167–184. Springer, Heidelberg (2011). doi:10.1007/978-3-642-23074-5_7
8. Oracle: Securefiles performance, November 2007. http://www.oracle.com/technetwork/database/database-technologies/performance/securefiles performancepaper-130360.pdf
9. Oracle: A technical overview of new features for automatic storage management in oracle database 12c, January 2013. http://www.oracle.com/technetwork/products/cloud-storage/oracle-12c-asm-overview-1965430.pdf
10. Oracle: Oracle advanced compression with oracle database 12c release 2, November 2016. http://www.oracle.com/technetwork/database/options/compression/advanced-compression-wp-12c-1896128.pdf
11. Oracle: Oracle asm considerations for exadata upgrades, April 2016. http://www.oracle.com/technetwork/database/availability/maa-exadata-upgrade-asm-2339100.pdf
12. PostgreSQL: Binaryfilesindb https://wiki.postgresql.org/wiki/BinaryFilesInDB
13. Rodeh, O., Bacik, J., Mason, C.: BTRFS: the linux B-tree filesystem. TOS 9(3), 9 (2013). http://doi.acm.org/10.1145/2501620.2501623
14. Salomon, D.: Data Compression: The Complete Reference. Springer, New York Inc. (2006)
15. Salomon, D., Motta, G.: Handbook of Data Compression, 5th edn. Springer Publishing Company, Incorporated, London (2009)
16. Schönig, H.: PostgreSQL Administration Essentials. Community Experience Distilled. Packt Publishing, Birmingham (2014). https://books.google.pl/books?id=BMLUBAAAQBAJ
17. Vohra, D.: Hello Docker, pp. 1–18. Apress, Berkeley (2016). http://dx.doi.org/10.1007/978-1-4842-1830-3_1
18. Wycislik, L.: Performance issues in data extraction methods of ETL process for XML. In: Kozielski, S., Mrozek, D., Kasprowski, P., Małysiak-Mrozek, B., Kostrzewa, D. (eds.) Beyond Databases, Architectures, and Structures. CCIS, vol. 424, pp. 581–589. Springer International Publishing, Switzerland (2014)

19. Wycislik, L.: Features of SQL databases for multi-tenant applications based on Oracle DBMS. In: Kozielski, S., Mrozek, D., Kasprowski, P., Małysiak-Mrozek, B., Kostrzewa, D. (eds.) BDAS 2015-2016. CCIS, vol. 613, pp. 507–517. Springer, Cham (2016). doi:10.1007/978-3-319-34099-9_39
20. Wycislik, L.: Independent data partitioning in oracle databases for LOB structures. In: Gruca, A., Brachman, A., Kozielski, S., Czachórski, T. (eds.) Man–Machine Interactions 4. AISC, vol. 391, pp. 687–696. Springer, Cham (2016). doi:10.1007/978-3-319-23437-3_59

Optimization of Memory Operations
in Generalized Search Trees of PostgreSQL

Andrey Borodin[1]([⊠]), Sergey Mirvoda[1], Ilia Kulikov[2], and Sergey Porshnev[1]

[1] Ural Federal University, Yekaterinburg, Russia
amborodin@acm.org
[2] RWTH Aachen, Aachen, Germany

Abstract. Our team is working on new algorithms for intra-page indexing in PostgreSQL generalized search trees. During this work, we encountered that slight modification of the algorithm for modification of a tuple on a page can significantly affect the performance. This effect is caused by optimization of page compaction operations and speeds up inserts and updates of a data. Most important performance improvement is gained using sorted data insertion, time to insert data into an index can be reduced by a factor of 3. For a randomized data performance increase is around 15%. Size of the index also significantly reduced. This paper describes implementation and evaluation of the technique in PostgreSQL codebase. Proposed patch is committed to upstream and expected to be released with the PostgreSQL 10.

Keywords: PosgreSQL · GiST · Memory operations · Multidimensional index

1 Introduction

Nowadays databases solve many different technical challenges [11,15]. They store data compactly, resolve data write conflicts and allow to query data efficiently. The latter part of database tasks is undertaken by so called access methods (AM), i.e. algorithms and data structures enabling quick retrieval of data by certain conditions. The paramount of AMs for modern relational databases is B-tree [1] allowing fast retrieval of data by the primary key and many other useful operations [9]. Generalized index search tree (GiST) is an AM technique, which allows to abstract significant part of data access methods structured as a balanced tree. Use of the GiST allows AM developer to concentrate on his own case-specific details of AM and skip common work on the tree structure implementation within database engine, a query language integration, a query planner support etc.

GiST was first proposed by Hellerstein et al. in [12], further researches were undertaken by Kornacker et al. [13]. Later GiST was implemented in PostgreSQL with large contribution by Chilingarian et al. [8]. Current PostgreSQL GiST implementation accepts different trees as a datatype (and so called *opclass*). Opclass developer must specify 4 core operations to make a type GiST-indexable:

© Springer International Publishing AG 2017
S. Kozielski et al. (Eds.): BDAS 2017, CCIS 716, pp. 224–232, 2017.
DOI: 10.1007/978-3-319-58274-0_19

1. Split: a function to split set of datatype instances into two parts.
2. Penalty calculation: a function to measure "penalty" for unification of two keys.
3. Collision check: a function which determines whether two keys may have "overlap" or are not intersecting.
4. Unification: a function to combine two keys into one so that combined key collides with both input keys.

Operations 1, 2 and 4 are responsible for construction of index tree, while operation 3 is used to query constructed tree. In general case, tree construction can be seen as serial insertion of dataset into an index tree. Insertion is a recursive algorithm, executed for every node starting from root of the tree. This algorithm searches within a node for an entry (also called downlink) with a minimal penalty of insertion of an item being inserted. For a chosen downlink key is updated with operation 4 and algorithm is invoked recursively. If algorithm is invoked on a leaf page it just places item being inserted, if the node is overflown then upward sequence of splits with operation 1 is started.

For example, if for operations 1–4 we pick rectilinear rectangles, we get regular R-tree [10], though many different indexing schemes are possible. The PostreSQL GiST implementation assumes one node of the generalized tree is one page. It allows to manipulate with data larger than RAM: while some pages reside in RAM buffer others reside in persistent memory. Also, PostgreSQL implementation do not use stack recursion in algorithms and allows to use in one index multiple different datatypes with different opclasses. Our team was working on different improvements for spatial indexing [3,4] and specifically on intra-page indexing [7]. Research of the GiST implementation showed us that there is a room for insert\update performance improvements without radical codebase modification and introduction of new algorithms.

2 Optimization of Inserts and Updates in GiST

2.1 Currently Employed Update Method

In the PostgreSQL implementation items are placed on page by the function *gistplacetopage*. Official PostgreSQL documentation mentions it as a "workhorse function that performs one step of the insertion". It receives an array of items to be placed on the page and an index of the position where items should be placed. Index of the position either can be so called *InvalidOffsetNumber* to indicate that items should be placed at the end of page.

This function is doing 4 steps:

1. It checks whether items can be accommodated on page without overflow. If not, the function performs split according to GiST algorithm.
2. If index of allocation is valid, then existing item is deleted from page with system routine *PageIndexTupleDelete*. Only one tuple is deleted at once, this behavior was formed due to many callers of the function with different requirements and expectations.

3. Items are placed at the end of a page.
4. Performed actions are registered in write-ahead log (WAL). The WAL is responsible for recovery operations. For example, if power is cycled unexpectedly database will apply WAL records on most recent existing database snapshot, which was saved on disk.

Obviously, recently updated items are always at the end of a page. This fact has several negative consequences which can be immediately observed:

- It affects cache locality negatively;
- Delete of item almost always causes move of big regions on a page since tuples are stored compactly.

2.2 Proposed Method

Core PostgreSQL functions for buffer pages allows to add an item at the end, add an item at an arbitrary index, delete items in multiple ways and some other operations. Some AMs like B-tree have specific page functions according to their need. But all these functions don't include routine for item overwrite. In our patch for PostgreSQL core we implemented function *PageIndexTupleOverwrite*, which replaces an old tuple with provided one. Data is moved across page only size of the old tuple is not equal to the size of the new tuple. If there is not enough place to accommodate new tuple instead of old query execution is terminated with error, thus caller must check this precondition. This function has the following properties:

- Keeps item index unchanged;
- Performs no memory moves on page if memory-aligned sizes of old and new items are equal.

Also, the patch forks steps 2 and 3 of *gistplacetopage* into one *PageIndex TupleOverwrite* call in case when item array has one item and passed index is valid (delete required).

2.3 Effect on Write-Ahead Logging

Change in allocation process caused the need for change in WAL replay functions. Same conditional logic has to be applied to them. Fortunately, WAL registration process has not changed, we still store in WAL records same information: array of items and index of an item on the page. But now interpretation of this information is more complicated.

So, in order to disable WAL compatibility with previous log, patch application have to be accompanied with increase of constant $XLOG\_PAGE\_MAGIC$ which is responsible for version of WAL records format. If this step is omitted, index may be corrupted by mischief usage of WAL records.

2.4   Other Indices

The Block Range Index (BRIN) in PostgreSQL also has item deleted followed by item placement at the same place causing two memory moves. PostgreSQL Global Development Group members also stated that usage of *PageIndexTupleOverwrite* would make code more readable. But logic of WAL records in BRIN is far more complicated and performance benefits are not that viable. Thus, final version of our patch didn't include interventions into BRIN, but all code assertions were done taking into account usage in BRIN. Despite we could not find any statistically significant performance improvements, patch committer, Tom Lane, modified patch to update BRIN functions. Following is the code of data accommodation in BRIN

```
PageIndexDeleteNoCompact(oldpage, &oldoff, 1);
if (PageAddItemExtended
    (oldpage, (Item) newtup, newsz, oldoff,
    PAI_OVERWRITE | PAI_ALLOW_FAR_OFFSET)
    == InvalidOffsetNumber)
    elog(ERROR, ''failed to add BRIN tuple'');
```

This code was replaced with

```
if (!PageIndexTupleOverwrite(oldpage, oldoff,
    (Item) newtup, newsz))
    elog(ERROR, ''failed to replace BRIN tuple'');
```

Committer made this replacement just for the sake of code clearance, without proofs of performance increase. Worth noting that *PageIndexDeleteNoCompact* does not compact page item numbers (so called "line pointers" on page), but almost always causes tuple data defragmentation on page. This data defragmentation does not provide any actual optimization, but is bad for cache. Currently we couldn't construct condition and design test to show viable BRIN performance improvement, but certainly some operations are improved.

3   Evaluation

GiST key update essentially is insert of a new record. Hence performance evaluation was centered around inserts into table that has GiST index. For testing purposes, we used opclass from the *cube* extension. This extension implements regular R-tree for rectilinear multidimensional boxes with Guttman polynomial time split algorithm [10]. Tests were conducted on Ubuntu 14 LTS virtual machine under Microsoft Hyper-V hypervisor with Intel Core i5 processors, SSD disks and unlimited dynamic RAM. RAM consumption during tests never exceeded 9 Gb. All tested operations are sequential and do not depend from number of CPUs significantly.

3.1   Ordered Data Insertion

Following script creates a table, an index and inserts one million rows into the table. Each row contains only one attribute – 3-dimensional dot (box with zero margin). Rows are inserted in order of computation of the Cartesian product of three incremental sequences.

```
create table dataTable(c cube);
create index idx on dataTable using gist(c);
insert into dataTable(c)
    select cube(array[x/100,y/100,z/100])
    from
    generate_series(1,100,1) x,
    generate_series(1,100,1) y,
    generate_series(1,100,1) z;
```

Application of our patch reduces time of execution of this test script by a factor of 2. It is worth noting that this time includes not only RAM insert operations but also accommodation of the table in PostgreSQL heap and committing WAL records to a persistent storage. This dramatic performance improvement triggered further investigation of a patch effect. GiST performance for ordered data insertion, isolated from heap time, is improved more than 3 times on average (with the relative standard deviation around 10%). But it was also noted that size of the resulting index was reduced by a factor of 3.25. The *cube* extension, as some others, use Guttman's polynomial time split algorithm. It is known to generate "needle-like" minimum bounding boxes (MBBs) for sorted data due to imbalanced splits (splitting 100 tuples as 98 to 2). Due to previous throw-to-the-end behavior of GiST this imbalance was further amplified (most of inserts were going to bigger part after split). That is why GiST inserts for sorted data were slower before this patch.

3.2   Randomized Data Insertion

For the randomized data insertion, previous described test showed no statistically confident improvement. To measure performance more precisely we used following script:

```
create table dataTable(c cube);
create index idx0 on dataTable using gist(c);
create index idx1 on dataTable using gist(c);
create index idx2 on dataTable using gist(c);
create index idx3 on dataTable using gist(c);
insert into dataTable(c)
    select cube(array[random(), random(), random()])
    from
    generate_series(1, 100, 1) x,
    generate_series(1, 100, 1) y,
    generate_series(1, 30, 1) z;
```

This script creates one table with 4 indices to reduce effect of volatility of time spent in heap data insertion. The script inserts 300 000 rows into the table. For this script, average performance improvement of our patch is about 15%. Additional test result, hardware and software configuration and resources for reproduction can be found on pgsql-hackers mailing list [6].

Average index build time before patch is 78 s. Average index build time after patch is 68 s.

T-test shows statistically significant difference between the current (M = 77866.75 ms, SD = 8754.71) and patched (M = 67885.50 ms, SD = 7157.02) versions, p = 0.0256 with a 95% CI of 1406.54 to 18555.96.

4 Analysis

4.1 Randomized Data Insertion

Performance effect of the patch with randomized data is easily explainable and was expected before implementation. Every insert causes $h-1$ tuples to be changed and 1 tuple is placed to leaf page, where h is tree height. Tuples on internal pages always have nonzero perimeter: probability to have full page of equal $3D$ vectors chosen at random is negligible, thus internal page tuples unite many different points of leaf page or pages. MBB bytes size change may happen only due to transition from point to bounding box, which never happens for internal page tuples. Thus, $h-1$ updates are not changing tuples size and will not trigger memory move across page. This work is optimized away and less time is consumed for index creation.

4.2 Ordered Data Insertion

First, let's describe what we call throw-to-the-end behavior. Before the patch every update of a tuple changes order of tuple on a page, so updated tuple will have maximum ItemId number. This affects enumeration order: recently updated tuples are last in enumeration. Besides computational optimization this is the only thing that is changes by the patch in GiST behavior. But this change reduces three times size of an index. Research of this topic showed that *cube* extension does not fully implement Guttman's algorithm for page split [10]. It has three steps:

QS1 [Pick first entry for each group]
Apply algorithm PICK SEEDS, to choose two entries to be the first elements of the groups. Assign each to a group.

QS2 [Check if done]
If all entries have been assigned , stop. If one group has so few entries that all the rest must be assigned to it in order for it to have the minimum number m, assign them and stop.

QS3 [Select entry to assign]
Invoke Algorithm PICK NEXT to choose the next entry to
assign. Add it to the group whose covering rectangle
will have to be enlarged least to accommodate it.
Resolve ties by adding the entry to the group with the
smaller volume, then to the one with fewer entries, then
to either.
Repeat QS2.

Mentioned algorithm PICK SEEDS is omitted here, since it's purpose is clear
and it's implementation is correct from our point of view. *cube* extension implements step QS2 wrong: it does not check minimum group size (parameter m of
algorithm). This error makes unbalanced splits unfused possibility. Algorithm
PICK NEXT also differs from originally proposed. Here is Guttman version:

PN1 [Determine cost of putting each entry in group]
For each entry e not yet in a group, calculate d1 = area
increase required in the covering rectangle of group 1 to
include E. Calculate d2 similarly for group 2.

PN2 [Find entry with most preference for one group]
Choose any entry with the maximum difference
between d1 and d2.

   cube extension does not pick next tuples in order from best fits to worst fits.
It processes tuples in order of their location on page, for every tuple picking
most fitting group (1 or 2). This small difference makes possible situation, when
group 2 (right group, group with highest *ItemId's*) attracts many unchanged
tuples leaving just one opposite tuple in a group 1. And finally, there are two
important features of behavior of *cube* split algorithm:

1. PICK NEXT algorithm's choice of left or right part is based on volume extension. If all inserted points are located on same grid-aligned plane, volume of
   bounding box is zero.
2. *cube* cut ties of equal volume extensions by adding tuple to right (highest
   ItemId) part of split.

   This features are not errors per se, technically they are interpretation of undefined behavior of pseudocode from original paper. But together with two errors
and throw-at-the-end behavior of GiST they produce bad splits while inserted
data is located on same plane. This situation is inevitable during sequential
insertion of ordered data.
   Technically, before patch and after patch computational complexity is the
same. For insertion sequence with n elements it is $O(n\ log\ n)$ for worst case and
for average amortized case. But the difference is in the hidden constant.

Developed patch, among other benefits, prevents this behavior of *cube* extension inside GiST, but does not fix errors in *cube*. We could not construct a test to trigger this behavior after patch application, but neither we can prove that it is not possible.

4.3 Possible Fix of *cube* extension

Discovered flaws in *cube* source code create relatively long chain of coincidence leading to degradation of performance. Though each individual flaw is neither critical, nor performance significant. From our point of view first thing to fix in *cube* extension is to implement more modern split algorithm, such as Korotkov split [14] or RR* split [2].

But full implementation of modern indices like RR*-tree will require GiST API extension. From our point of view implementation of ChooseSubtree algorithm via penalty calculation is important limitation for complex data structures.

5 Conclusion

We proposed change in a technique employed by current PostgreSQL GiST implementation, summarized this technique as a core PostgreSQL patch and measured its performance improvement. However, most significant performance improvement for sorted data insertion and updates is not related to improved memory operations but is attributed to mitigation of weak split algorithm behavior. Developed patch was reviewed on a commitfest [5] by members of the community and now it has the status "Committed". We expect that results of this work will be available to users with the release of the PostgreSQL 10.

References

1. Bayer, R.: Binary b-trees for virtual memory. In: Proceedings of the 1971 ACM SIGFIDET (Now SIGMOD) Workshop on Data Description, Access and Control, pp. 219–235. ACM (1971)
2. Beckmann, N., Seeger, B.: A revised r*-tree in comparison with related index structures. In: SIGMOD 2009 Proceedings of the 2009 ACM SIGMOD International Conference on Management of Data (2009)
3. Borodin, A., Kiselev, Y., Mirvoda, S., Porshnev, S.: On design of domain-specific query language for the metallurgical industry. In: Kozielski, S., Mrozek, D., Kasprowski, P., Małysiak-Mrozek, B., Kostrzewa, D. (eds.) BDAS 2015. CCIS, vol. 521, pp. 505–515. Springer, Cham (2015). doi:10.1007/978-3-319-18422-7_45
4. Borodin, A., Mirvoda, S., Porshnev, S.: Database index debug techniques: a case study. In: Kozielski, S., Mrozek, D., Kasprowski, P., Małysiak-Mrozek, B., Kostrzewa, D. (eds.) BDAS 2015-2016. CCIS, vol. 613, pp. 648–658. Springer, Cham (2016). doi:10.1007/978-3-319-34099-9_50
5. Borodin, A.: Gist inserts optimization with pageindextupleoverwrite. https://commitfest.postgresql.org/10/661

6. Borodin, A.: [PoC] GiST optimizing memmoves in gistplacetopage for fixed-size updates. https://www.postgresql.org/message-id/CAJEAwVGQjGGOj6mMSgM wGvtFd5Kwe6VFAxY%3DuEPZWMDjzbn4VQ%40mail.gmail.com

7. Borodin, A.: [proposal] improvement of GiST page layout. https://www.postg resql.org/message-id/CAJEAwVE0rrr+OBT-P0gDCtXbVDkBBG%5FWcXwCB K=GHo4fewu3Yg@mail.gmail.com

8. Chilingarian, I., Bartunov, O., Richter, J., Sigaev, T.: Postgresql: the suitable DBMS solution for astronomy and astrophysics. In: Astronomical Data Analysis Software and Systems (ADASS) XIII, vol. 314, p. 225 (2004)

9. Garcia-Molina, H., Ullman, J.D., Widom, J.: Database System Implementation, vol. 654. Prentice Hall, Upper Saddle River (2000)

10. Guttman, A.: R-trees: A Dynamic Index Structure for Spatial Searching, vol. 14. ACM, New York (1984)

11. Hameurlain, A., Morvan, F.: Big data management in the cloud: evolution or crossroad? In: Kozielski, S., Mrozek, D., Kasprowski, P., Małysiak-Mrozek, B., Kostrzewa, D. (eds.) BDAS 2015-2016. CCIS, vol. 613, pp. 23–38. Springer, Cham (2016). doi:10.1007/978-3-319-34099-9_2

12. Hellerstein, J.M., Naughton, J.F., Pfeffer, A.: Generalized search trees for database systems, September 1995

13. Kornacker, M., Mohan, C., Hellerstein, J.M.: Concurrency and recovery in generalized search trees. In: ACM SIGMOD Record, vol. 26, pp. 62–72. ACM (1997)

14. Korotkov, A.: A new double sorting-based node splitting algorithm for r-tree. In: Proceedings of the Spring/Summer Young Researchers Colloquium on Software Engineering. vol. 5 (2011)

15. Krechowicz, A., Deniziak, S., Łukawski, G., Bedla, M.: Preserving data consistency in scalable distributed two layer data structures. In: Kozielski, S., Mrozek, D., Kasprowski, P., Małysiak-Mrozek, B., Kostrzewa, D. (eds.) BDAS 2015. CCIS, vol. 521, pp. 126–135. Springer, Cham (2015). doi:10.1007/978-3-319-18422-7_11

Text Mining, Natural Language Processing, Ontologies and Semantic Web

Sorting Data on Ultra-Large Scale with RADULS

New Incarnation of Radix Sort

Marek Kokot, Sebastian Deorowicz[✉], and Agnieszka Debudaj-Grabysz

Institute of Informatics, Silesian University of Technology,
Akademicka 16, 44-100 Gliwice, Poland
{marek.kokot,sebastian.deorowicz,agnieszka.grabysz}@polsl.pl

Abstract. The paper introduces RADULS, a new parallel sorter based on radix sort algorithm, intended to organize ultra-large data sets efficiently. For example 4 G 16-byte records can be sorted with 16 threads in less than 15 s on Intel Xeon-based workstation. The implementation of RADULS is not only highly optimized to gain such an excellent performance, but also parallelized in a cache friendly manner to make the most of modern multicore architectures. Besides, our parallel scheduler launches a few different procedures at runtime, according to the current parameters of the execution, for proper workload management. All experiments show RADULS to be superior to competing algorithms.

Keywords: Radix sort · Thread-level parallelization

1 Introduction

Although the area of sorting algorithms has been investigated from time immemorial, there is still the need for developing faster implementations as well as the place for improvement. The demand for even faster sorters is the result of accumulation of increasing amounts of data in all areas of life. Starting from user applications, through industry, to strictly scientific applications, organizing the data is required everywhere.

It seems that the possibility of constructing *new* sorting algorithms in the strict sense has been exhausted, that is why faster sorters are constructed by using the following techniques:

- widening the range of applications of internal sorting,
- parallelization,
- hardware friendliness,
- hybrids of the above mentioned.

The range of applications of internal sorting algorithms become wider naturally, due to the availability of increasingly larger memory sizes.

At the same time, the prevalence of multicore architectures gives the possibility of using thread-level parallelism to increase computing power. The programming environments that supports developing of multithreaded programs become

© Springer International Publishing AG 2017
S. Kozielski et al. (Eds.): BDAS 2017, CCIS 716, pp. 235–245, 2017.
DOI: 10.1007/978-3-319-58274-0_20

more and more popular. Unfortunately, the effort in the parallelization of a code could be of no effect if the algorithm were not architecture friendly, especially cache friendly. The subject is crucial for single-threaded algorithms, but becomes even more critical for parallel ones.

Concerning modern architectures it is known, that the performance bottleneck is accessing the main memory. The problem is being solved by equipping CPUs with cache systems to reduce the time to access data. Most often cache memories have a hierarchical structure. When accessing data a processor tries to find it in the first-level (L1) cache, then in case of a fail—at higher levels (L2, L3). If the data was not stored in the cache, it is necessary to tap into the main memory with much longer latency. This is called *a cache miss*.

For multicore architectures individual cores have their own, private first-level caches, while last-level cache is typically shared. Memory access policy have to coordinate several problems, e.g., keeping coherent data of private caches, controlling shared data accesses, especially in case of cache misses. As caches are organized in lines, there is a conflict when separate threads residing on separate cores request access to separate data for modification, but the data falls in the same cache line. This is called *false sharing* [7]. In such a situation a synchronization protocol forces unnecessary memory update to keep coherency.

The sorting algorithm which is very well suited for parallelization is radix sort. It represents non-comparison based sorts [12] with the low computational complexity of $O(N)$, where N is the number of elements to sort. Keys to be sorted are viewed as sequences of fixed-sized pieces, e.g., decimal numbers are treated as sequences of digits while binary numbers are treated as sequences of bits. Generally these pieces correspond to *digits* that creates numbers represented in a base-R number system, where R is its *radix*.

Still, there are two basic approaches to radix sort. According to the first variant, digits of the keys are examined from the most to the least significant ones (LSD). In the second variant, digits are processed in the opposite direction, i.e. from the least to the most significant ones (MSD). The general idea of the algorithm is that at each radix pass the array of keys is sorted according to every consecutive digit. The number of passes corresponds to the length of a key. The most often a counting sort is selected as the inner sorter. In case of LSD sorting, it is not intuitive, that the final distribution is properly ordered. In fact, the usage of a stable method like the counting sort (preserving the relative order of items with duplicated keys in the array), guaranties accuracy of results. Two traversals through the array are required at each pass. During the first one, a histogram of the number of occurrences of each possible digit value is obtained. Next, for each key, the number of keys with smaller (or the same) values of digits on investigated position is calculated. Finally the keys are distributed to their appropriate positions on the basis of the histogram, during the second traversal through the array. The idea of MSD method is the same, although it is worth to mention, that after the first pass of sorting, the array of keys is partitioned into a maximum of R different *bins*. Every single bin contains keys for one possible value of the most significant byte. During every consecutive pass the keys are sorted within bins from the previous pass and they are not distributed among separate bins.

In this paper we propose RADULS, which is a parallel radix sorter capable to sort 4 G 16-byte records in less than 15 s using 16-cores. A few parallelization strategies are cooperating to assure load balancing. Additionally, RADULS is cache friendly to reduce long latencies of accesses to the main memory.

The paper is organized as follows. In Sect. 2 a brief description of the state-of-the-art radix sorters is given. Section 3 describes our algorithm, which is experimentally evaluated against the top parallel sorters (radix-based and comparison-based) in Sect. 4. In Sect. 5 we discuss some of the applications of radix sort. The last section concludes the paper.

2   Related Works

The area of sorters using thread-level parallelism was broadly investigated in recent decades. Our inspiration was the paper by Satish *et al.* [11], where, among others, an architecture-friendly LSD radix sorter for CPU was proposed. Firstly, the authors identified bottlenecks in common, parallel implementations, such as irregular memory accesses or conflicts in cache or shared memory accesses. Secondly, they proposed the ways to avoid them. The idea is to maintain buffers in local storages for collecting elements, that belong to the same radix. The buffers for $B \times R$ bytes in local memory for each thread is reserved, where R is the radix, and B bytes are buffered for each radix. The B value was selected as a multiple of 64 bytes (cache line size). Buffers' contents are written to global memory when enough elements were accumulated and then is reused for other entries. Such an approach avoids cache misses, as buffers occupy contiguous space and fit the cache memory.

In [2] an in-place parallel MSD radix sorter was proposed. There is no auxiliary array available in the distribution phase, that is why swap operations are performed in order to place keys to their proper bins. This phase is especially challenging when parallelizing, because of existing dependencies between reading and writing within swaps, while the array of keys is partitioned among threads. Hence, the authors solve the problem in two stages called *speculative permutation* and *repair*. The stages are iterated until all the keys are rearranged and placed in their target bins. Next, the bins can be sorted independently. However, these sub-tasks can be heavily imbalanced, because of differences in bin sizes. That is why the authors use adaptive thread reallocation scheme to gain proper load balancing.

3   Our Algorithm

3.1   General Idea

The algorithm follows the MSD approach with radix $R = 256$. Our parallel scheduler launches four different procedures for single radix pass depending on the current digit position and the size of the bin. It also uses some special treatment of large bins for better balancing of work among threads. Finally, sufficiently small bins are handled by comparison-based sorting routines. In the following subsections we describe each of the procedures in details.

3.2   First-Digit Pass

Figure 1 presents a general scheme of sorting according to the first digit. At the beginning the keys are distributed into bins by T threads using buffered radix split algorithm inspired by a single pass of Satish et $al.$ [11] LSD radix sort. For better load balancing we initially divide the input array into $8T$ chunks of sizes linearly growing from $N/64T$, where N is the array size (constants chosen experimentally).

```
1: function FIRSTPASS(data_start, data_end, current_byte, T)
2:     bins ← BUFFEREDRADIXSPLIT(data_start, data_end, current_byte)
3:     if current_byte > 0 then
4:         [small_bins, big_bins] ← SPLITBINS(bins)
5:         for all bin in small_bins do
6:             TASK_QUEUE.PUT(bin, current_byte − 1)
7:         T_big ← min(T, 1.25 × N_big/N)            ▷ N_big—no. of rec. in all big bins
8:         T_small ← T − T_big
9:         for i = 1 to T_small do
10:            Run PROCESSBINS in a new thread
11:        for all bin in big_bins do
12:            BIGBINPASS(bin.start, bin.end, current_byte − 1, T_big)
13:        for i = 1 to T_big do
14:            Run PROCESSBINS() in a new thread
```

Fig. 1. Pseudocode of the first pass of RADIUS algorithm

When the distribution is over, the bins are marked as small or big. The threshold is set to $2N/3T$. The idea behind this is to assign sufficiently large number of threads for handling big bins, i.e. to avoid the situation in which threads assigned to processing of small bins completed their work, when the big bin are still processed. Small bins are intended for processing by a different procedure than big ones. That is why the priority queue is created for keeping tasks describing small bins. Tasks are ordered from the largest size of a small bin to the lowest one. T_{small} of newly created threads are to handle the mentioned queue. Furthermore, when big-bin processing is over, T_{big} threads for small-bin processing are created (see Fig. 1) to replace the released ones.

3.3   Next Passes for Big Bins

The pass processing big bins (Fig. 2) is quite similar to the first-digit pass. The only difference is that after marking bins as big and small the first of them are

```
1: function BIGBINPASS(data_start, data_end, current_byte, T)
2:     bins ← BUFFEREDRADIXSPLIT(data_start, data_end, current_byte)
3:     if current_byte > 0 then
4:         for all bin in bins do
5:             if ISBIG(bin) then
6:                 BIGBINPASS(bin.start, bin.end, current_byte − 1, T)
7:             else
8:                 TASK_QUEUE.PUT(bin, current_byte − 1)
9:         for i = 1 to T do
10:            Run PROCESSBINS in a new thread
```

Fig. 2. Pseudocode of the algorithm handling big bins produced in the first stage

processed immediately in a recursive manner, while the later are added to the priority queue. The queue of small bins from discussed passes is the separate one. This implies it is handled only with the threads previously assigned to big bins (after the first pass).

3.4 Next Phases for Small Bins

Descriptions of small bins generated in previous passes are kept in the form of tasks in a priority queue. Each available thread processes these tasks one by one. Figure 3 shows the pseudocode of a single-threaded algorithm handling the tasks from the queue.

To sort a bin according to some digit one of following methods is chosen relating on the size of the bin. Simple counting sort without buffering the keys is used when the bin fits a half of L2 cache. In the opposite case, the buffering algorithm inspired by single pass of Satish *et al.* algorithm is used. The case of tiny bins is discussed in the following subsection.

The new bins obtained in this place can be handled in three ways. If the bin size is smaller than $N/4096$, it is processed recursively to avoid too many (potentially costly) operations on the queue (which is shared by many threads). Larger bins are inserted into the priority queue.

3.5 Handling Tiny Bins

The bins containing smaller than 384 keys (value chosen experimentally) are processed by comparison sorters to avoid relatively costly passes of radix sort. We experimented with several comparison algorithms, but finally picked three of them: introspective [10] (implemented as part of the standard C++ library), Shell sort [13] (with sequence of increments reduced only to $1, 8$), and insertion sort [9]. For the smallest arrays ($N \leq 32$) we use insertion sort. The threshold between introspective sort (a hybrid of quick sort [6] and heap sort [15]) and Shell sort depends on the key size (expressed in bytes), but usually is in the range 100–180.

```
1: function MSDRADIXBINS(data_start, data_end, current_byte)
2:    if data fit in cache then
3:        bins ← RADIXSPLIT(data_start, data_end, current_byte)
4:    else
5:        bins ← BUFFEREDRADIXSPLIT(data_start, data_end, current_byte)
6:    if current_byte > 0 then
7:        for all bin in bins do
8:            if ISTINYBIN(bin) then
9:                COMPARISONSORT(bin)
10:           else
11:               if TOOSMALLFORQUEUE(bin) then
12:                   MSDRADIXBIN(bin.start, bin.end, current_byte − 1)
13:               else
14:                   TASK_QUEUE.PUT(bin, current_byte − 1)
15: function PROCESSBINS
16:    while [bin, byte] ← TASKS_QUEUE.POP() do
17:        MSDRADIXBINS(bin.start, bin.end, byte)
```

Fig. 3. Pseudocode of the algorithm handling small bins produced in previous stages

While deciding whether the current bin is tiny or not we also monitor the "narrowing factor" defined as the number of keys in the "parent" bin divided by the number of keys in the current bin. If this factor is larger than $\sqrt{R} = 16$ we speculate that the next pass of the radix sort should be more profitable than using introspective or Shell sort. Thus in such a situation the tiny bin threshold is set to 32.

4   Experimental Results

RADULS was implemented in the C++14 programming language and uses native C++ threads. A few SSE2 instructions were used for fast transfers of buffered memory to the main memory without cache pollution. For compilation we used GCC 6.2.0. All experiments were performed at workstation equipped with two Intel Xeon E5-2670v3 CPUs (12 cores each, 2.3 GHz) and 128 GB RAM.

We compared RADULUS with the following parallel sorting algorithms:

- TBB—the parallel comparison sort of $O(N \log N)$ average time complexity implemented in the Intel Threading Building Blocks [8] (2017 Update 3 release),
- MCSTL—the parallel hybrid sort [1,14], now included in GNU's libstdc++library,
- Satish-1—our implementation of the buffered LSD radix sort introduced by Satish et al. [11] with the buffer size for a specific digit equal to the cache line size ($B = 64$),
- Satish-4—the same as Saitsh-1, but with $B = 256$,
- PARADIS—the state-of-the-art in-place radix sort algorithm by Cho et al. [2].

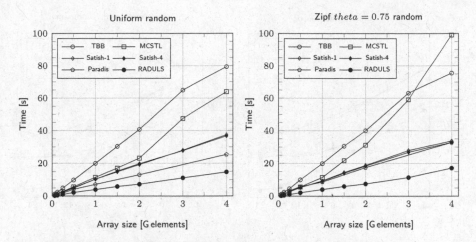

Fig. 4. Experimental results for random input: uniform distribution (left) and Zipf $\theta = 0.75$ distribution (right). The sorted records are of size 16 bytes (8 bytes for key and 8 bytes for data). The number of threads was set to 16.

Unfortunately, we were not able to obtain either the PARADIS source codes or library delivering it. Since the algorithm is far from being trivial to implement we decided to include in our comparison the running times just from the PARADIS paper without any time scaling (although PARADIS was evaluated at Intel Xeon E7-8837 CPU clocked at higher rate, i.e. 2.67 GHz).

In the first set of experiments we compared the running times of sorting algorithms for array sizes in the range from 62.5 M to 4 G records. The records of length 16 bytes consisted of 8-bytes-key and 8-bytes-data fields (to allow indirect comparison with PARADIS). The keys were produced randomly with uniform distribution and Zipf distribution [5] with $\theta = 0.75$ (once again to allow comparison with PARADIS results).

Figure 4 shows that RADULS clearly outperforms the competitive algorithms when run for 16 threads. PARADIS was the second best for uniform distribution. The second place for Zipf data are, however, shared by PARADIS and Satish algorithms. The difference between Satish-1 and Satish-4 is marginal in both cases.

In the second experiment, we evaluated the influence of number of running threads. Figure 5 shows both the absolute running times and the relative speedup of the algorithms. As it can be observed, the relative speedups of Satish-1 is better than of Satish-4, but the later is faster for smaller number of threads, especially for a single thread. Both Satish algorithms and TBB scales well only for less than 8 threads. Then their speedups saturates below 9. RADULS scales better and for 16 threads the relative speedup is about 11.5. MCSTL performs even better (in term of scalability) for uniformly distributed keys. Nevertheless, its absolute running times are much longer than RADULS times. The inspection of PARADIS paper (Fig. 7) shows that for 16 threads its speedups is almost 10.

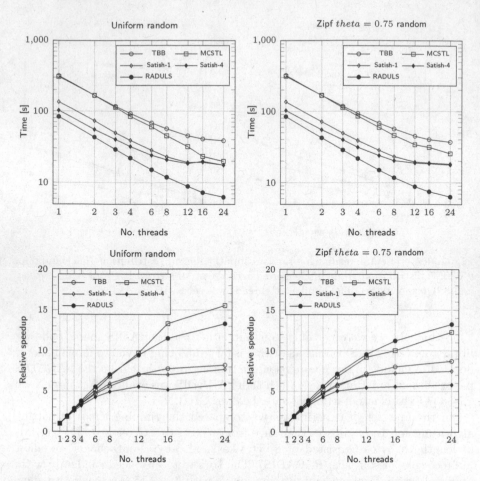

Fig. 5. Experimental results for random input: uniform distribution (left) and Zipf $\theta = 0.75$ distribution (right). The array contained 2 G elements of size 16 bytes (8 bytes for key and 8 bytes for data). The number of threads was: 1, 2, 3, 4, 6, 8, 12, 24.

Finally, we experimented with various record sizes and types of data. The upper chart in Fig. 6 shows the running times for records from 8 to 32 bytes with 8-byte (or 16-byte in one case) keys. It can be noticed that RADULS is always the fastest. The running times grows from 4.82 s to 12.97 s when key size is 8-bytes long and the record size grows from 8 to 32 bytes. The sublinear time increase was possible due to use of comparison sorting routines for tiny bins. The lower chart in Fig. 6 presents the results for three sets of k-mers, being the data from large sequencing project (see Sect. 5 for details). Once again RADULS appeared to be the winner.

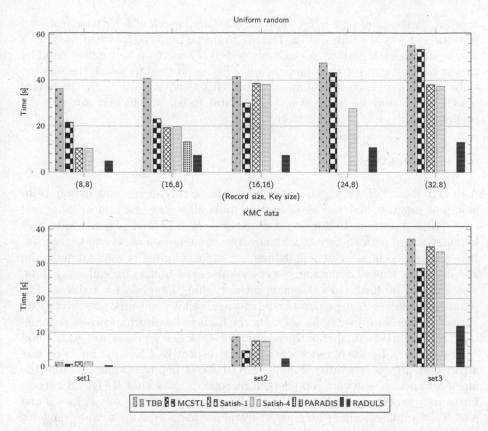

Fig. 6. Experimental results for records of various size (up) and type (down). The sets of KMC data contained 16-byte records with 16-byte keys of sizes 63 M, 413 M, 1887 M.

5 Possible Applications

RADULS is a general purpose sorter, but the two of its possible applications are especially worth mentioning. We have already tested the first one, as it consisted in joining our sorter with existing k-mer counter software, i.e. KMC [3, 4]. By k-mers we mean unique substrings of length k in a set of reads from sequencing projects. The procedure of determining k-mers is often used in initial stages of sequencing data processing. The input data can be larger than 1 TB. Therefore, modern k-mer counters usually process in two stages. In the first stage the extracted k-mers are distributed into several hundred disk files, which are then processed separately (the second stage). One of the possibilities of handling a single file is to sort the strings and then remove duplicates. In fact, this solution was used in KMC.

Another application of the proposed algorithm could be to sort short strings. Let's consider searching multiple patterns in a text with the aid of a suffix array, a classical full-text index applying a binary search of a pattern against a collection of sorted suffixes of the text. It could be performed in two ways. In a naive solution each pattern is sought separately. However, a better way could be to sort the patterns (or their prefixes only) first, and then to search them in the suffix array in an incremental manner, i.e. with a reduced range for the binary search, which may be faster overall. It is vital to use an efficient sorter in the preliminary phase of this procedure.

6 Conclusions

Although the art of sorting algorithms seems to be thoroughly understood, technological progress and new development tools allow the creation of more and more efficient sorters. In out paper we propose radix-based sorter, which owes its outstanding performance to an innovative combination of several techniques. It is parallelized in a cache friendly manner—thus adapted to modern multicore architectures. Due to maintaining software-managed buffers for collecting data, which are to be flushed to the main memory, long latencies are reduced. The parallel scheduler—being a part of out software—allows sub-task-size-driven execution to avoid workload imbalance. On the basis of monitoring current parameters an appropriate number of threads per sub-task and a proper sorting method can be selected. Beside of radix algorithm, introspective, Shell and insertion sort algorithms are incorporated. Finally, RADULS is highly optimized using the latest advances in software compilers. Experiments show that RADULS outperforms its competitors for both uniformly distributed data as well as for skewed one. That implies it may become an irreplaceable sorter for a wide range of applications.

Acknowledgments. The work was supported by the Polish National Science Centre under the project DEC-2013/09/B/ST6/03117 (SD, ADG) and by Silesian University of Technology grant no. BKM507/RAU2/2016 (MK).

References

1. MCSTL: The multi-core standard template library (2008). http://algo2.iti.kit.edu/singler/mcstl/
2. Cho, M., Brand, D., Bordawekar, R., Finkler, U., Kulandaisamy, V., Puri, R.: PARADIS: an efficient parallel algorithm for in-place radix sort. In: Proceedings of the VLDB Endowment—Proceedings of the 41st International Conference on Very, pp. 1518–1529 (2015)
3. Deorowicz, S., Kokot, M., Grabowski, S., Debudaj-Grabysz, A.: KMC 2: fast and resource-frugal k-mer counting. Bioinformatics **31**(10), 1569–1576 (2015). http://dx.doi.org/10.1093/bioinformatics/btv022
4. Deorowicz, S., Debudaj-Grabysz, A., Grabowski, S.: Disk-based k-mer counting on a PC. BMC Bioinform. **14**(1), 160 (2013). http://dx.doi.org/10.1186/1471-2105-14-160

5. Gray, J., Sundaresan, P., Englert, S., Baclawski, K., Weinberger, P.: Quickly generating billion-record synthetic databases. In: Proceedings of the SIGMOD, pp. 243–252 (1994)
6. Hoare, C.: Quicksort. Comput. J. **5**(1), 10–15 (1962)
7. Intel: Intel Guide for Developing Multithreaded Application, Intel (2011). http://www.intel.com/software/threading-guide
8. Intel: Threading Building Blocks (2016). https://www.threadingbuildingblocks.org/
9. Knuth, D.: The Art of Computer Programming. Addison-Wesley, Boston (1968)
10. Musser, D.: Introspective sorting and selection algorithms. Softw.: Pract. Exp. **27**(8), 983–993 (1997)
11. Satish, N., Kim, C., Chhugani, J., Nguyen, AD., Lee, V., Kim, D., Dubey, P.: Fast sort on CPUs and GPUs: a case for bandwidth oblivious simd sort. In: Proceedings of the 2010 International Conference on Management of Data, pp. 351–362 (2010)
12. Sedgewick, R.: Algorithms in C++, Parts 1–4: Fundamentals, Data Structure, Sorting, Searching. Addison-Wesley-Longman, Harlow (1998)
13. Shell, D.: A high-speed sorting procedure. Commun. ACM **2**(7), 30–32 (1959)
14. Singler, J., Sanders, P., Putze, F.: MCSTL: the multi-core standard template library. In: Kermarrec, A.-M., Bougé, L., Priol, T. (eds.) Euro-Par 2007. LNCS, vol. 4641, pp. 682–694. Springer, Heidelberg (2007). doi:10.1007/978-3-540-74466-5_72
15. Williams, J.: Algorithm 232: Heapsort. Commun. ACM **7**(6), 347–348 (1964)

Serendipitous Recommendations Through Ontology-Based Contextual Pre-filtering

Aleksandra Karpus[1(✉)], Iacopo Vagliano[2], and Krzysztof Goczyła[1]

[1] Faculty of Electronics Telecommunication and Informatics,
Gdańsk University of Technology,
ul. Gabriela Narutowicza 11/12, 80-233 Gdańsk-Wrzeszcz, Poland
{aleksandra.karpus,krzysztof.goczyla}@eti.pg.gda.pl
[2] Department of Control and Computer Engineering,
Politecnico di Torino, C.so Duca degli Abruzzi, 24, 10129 Turin, Italy
iacopo.vagliano@polito.it

Abstract. Context-aware Recommender Systems aim to provide users with better recommendations for their current situation. Although evaluations of recommender systems often focus on accuracy, it is not the only important aspect. Often recommendations are overspecialized, i.e. all of the same kind. To deal with this problem, other properties can be considered, such as serendipity. In this paper, we study how an ontology-based and context-aware pre-filtering technique which can be combined with existing recommendation algorithm performs in ranking tasks. We also investigate the impact of our method on the serendipity of the recommendations. We evaluated our approach through an offline study which showed that when used with well-known recommendation algorithms it can improve the accuracy and serendipity.

Keywords: Recommender systems · Ontologies · Context-awareness · Serendipity

1 Introduction

Recommender systems aim at providing suggestions for items to be of use to a user. An item could be a movie, a book or even a friend in some social recommender. Context-Aware Recommender Systems (CARS) are a particular category of recommender systems which exploits contextual information to provide more adequate recommendations. For example, in a temporal context, vacation recommendations in winter should be very different from those provided in summer. Or a restaurant recommendation for a Saturday evening with your friends should be different from that suggested for a workday lunch with co-workers [29].

Typically, recommendation algorithms are evaluated according to some accuracy measure, such as *mean absolute error*, *precision* or *recall*. However, accuracy is not the only important aspect of a recommender system. Overspecialized

---

The second author was supported by a fellowship from TIM.

recommendations may be unsatisfactory for a user [16]. Recommendations are overspecialized if they are all of the same kind, for example, all movies of the same genre. To deal with this problem, many other properties can be considered, e.g. *novelty*, *serendipity* and *diversity*. *Novelty* describes how many items unseen by the user appeared in the recommendation list. *Serendipity* measures the number of unexpected and interesting items recommended, while *diversity* assesses how much items in the list differ from each other.

In particular, serendipity is useful because users do not want to receive recommendations about items they already knew or consumed. It also does not make a lot of sense to recommend to a user very popular items, e.g. a bestseller book, which he could discover by seeing a commercial or going to the nearest bookstore. For this reason, it is important to propose items that are interesting and unexpected [16].

In a previous study, we proposed an ontology-based and context-aware technique which can be combined with existing recommendation algorithms [20]. In this paper, we examine how it performs in ranking tasks. We also investigate the impact of our method on the quality of recommendations according to *precision*, *recall* and *serendipity* measures, trying to answer following research question:

– *Can incorporating contextual information in the recommendation process improve not only accuracy but also serendipity of recommendations?*

To answer this question, we performed an off-line study on the ConcertTweets data set [1], which describes users interests in musical events. The evaluation of the obtained recommendations confirmed that the use of contextual information can improve the serendipity of recommendations.

The rest of the paper is organized as follows. Section 2 describes related work. Section 3 presents our ontology-based and context-aware approach. Section 4 discusses the evaluation approach and the results obtained. Conclusions and directions for future work close the paper.

2   Related Work

In this section, we focus on the state-of-the-art of the two main topics related to this paper. Section 2.1 describes work concerning the usage of ontologies in the recommendation processes. Section 2.2 focuses on the meaning of the word *serendipity*, its etymology and different definitions in the field of recommender systems.

2.1   Ontology Based Recommender System

A number of ontology-based and context-aware recommender system have been proposed. AMAYA allows management of contextual preferences and contextual recommendations [27]. AMAYA also uses an ontology-based content categorization scheme to map user preferences to entities to recommend. News@hand [5] is a hybrid personalized and context-aware recommender system, which retrieves news via RSS feed and annotates by using system domain ontologies. User context is represented by a weighted set of classes from the domain ontology. Rodriguez et al. [30] proposed a CARS which recommends Web services. They use a multi-dimensional ontology model to describe Web services, a user context, and an application domain. The multi-dimensional ontology model consists of a three independent ontologies: a user context ontology, a Web service ontology, and an application domain ontology, which are combined into one ontology by some properties between classes from different ontologies. The recommendation process consists in assigning a weight to the items based on a list of interests in the user ontology. Our work is somehow similar to this approach, because we also use more than one ontology and one of them represents the context dimensions. However, all those works focus on a specific domain and an ad-hoc algorithm, while our approach for representing user preferences is cross-domain and can be applied to different recommendation algorithm.

Hawalah and Fasli [15] suggest that each context dimension should be described by its own taxonomy. Time, date, location, and device are considered as default context parameters in the movie domain. It is possible to add other domain specific context variables as long as they have clear hierarchical representations. Besides context taxonomies, this approach uses a reference ontology to build contextual personalized ontological profiles. The key feature of this profile is the possibility of assigning user interests in groups, if these interests are directly associated with each other by a direct relation, sharing the same superclass, or sharing the same property. Similarly, we model context-dependent user preferences using ontology. They are kept in the form of modules, which represent specific context situations, so we actually also group user interests. However, we have one ontology for all context dimensions, in contradiction to one taxonomy per each context dimension, which is a crucial conceptual simplification.

Other works use ontologies and taxonomy to improve the quality of recommendations. Su et al. proved that ontological user profile improves recommendation accuracy and diversity [31]. Middleton et al. [24] use an ontological user profile to recommend research papers. Both research papers and user profiles are represented through a taxonomy of topics, and the recommendations are generated considering topics of interest for the user and papers classified in those topics. Mobasher et al. [25] proposed a measure which combines semantic knowledge about items and user-item rating, while Anand et al. [4] inferred user preferences from rating data using an item ontology. Their approach recommends items using the ontology and inferred preferences while computing similarities. A more detailed description of ontology-based techniques is available in [8,22].

2.2 Serendipity

According to the Oxford dictionary[1], *serendipity* is "the occurrence and development of events by chance in a happy or beneficial way". It was coined in 1754 by the English author Horace Walpole in one of his letters, in which he describes his

---

[1] https://en.oxforddictionaries.com/definition/serendipity.

unexpected discovery by referring to "a silly fairy tale, called *The Three Princes of Serendip*: as their highnesses travelled, they were always making discoveries, by accidents and sagacity, of things which they were not in quest of" [32].

The common definition of serendipity in recommender systems does not exist yet, since it is challenging to say which items are serendipitous and why [18].

Ziegler et al. described serendipitous items as those with a low popularity [34]. Results obtained by Maksai et al. confirmed this intuition. They have proved that the most popular items have serendipity equal to zero. However, the definition of the serendipity by Maksai et al. differs from the previous one. "Serendipity is the quality of being both unexpected and useful" [23].

Iaquinta et al. [18] extended previous definitions of the serendipity. Serendipitous items are novel, unexpected and interesting to a user. Adamopoulos and Tuzhilin also require that items have to be novel and unexpected to the user, but they add a third feature: a positive emotional response. "Serendipity, the most closely related concept to unexpectedness, involves a positive emotional response of the user about a previously unknown (novel) item and measures how surprising these recommendations are" [2].

Simpler definition was proposed by Zhang et al. "Serendipity represents the *unusualness* or *surprise* of recommendations" [33].

Kotkov et al. emphasized the problem of technical understanding of concepts used in the prior definitions, i.e. novelty and unexpectedness. "Publications dedicated to serendipity in recommender systems do not often elaborate the components of serendipity [...]. It is not entirely clear in what sense items should be novel and unexpected to a user" [21].

3 Recommendation Approach

Our recommendation approach is based on two ontologies: Recommender System Context (RSCtx)[2] which represents the context, and Contextual Ontological User Profile (COUP), which represents user preferences. In the following, we firstly introduce each ontology, and then we describe the recommendation process.

3.1 The Recommender System Context Ontology

Recommender System Context (RSCtx) extends PRISSMA[3], a vocabulary based on Dey's definition of context [7]. PRISSMA relies on the W3C Model-Based User Interface Incubator Group proposal[4], which describes mobile context as an encompassing term, defined as the sum of three different dimensions: user model and preferences, device features, and the environment in which the action is performed. A graph-based representation of PRISSMA is provided in Fig. 1.

---

[2] http://softeng.polito.it/rsctx/.

[3] http://ns.inria.fr/prissma.

[4] http://www.w3.org/2005/Incubator/model-based-ui/XGR-mbui/.

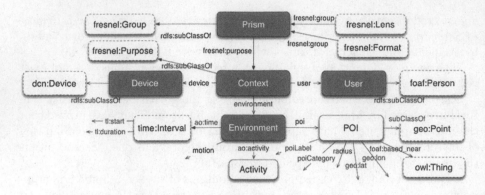

Fig. 1. The PRISSMA vocabulary [6].

We extended the time and location representations. We needed a more expressive model of these two dimensions, since asking for recommendations which have the same time stamp and the coordinates of the actual context is too restrictive and the recommender system may not have enough data. On the contrary, by generalizing the context (for example distinguishing among weekend and working day, or considering the city or neighborhood instead of the actual user position) may enable the recommender system to provide recommendations. The concept prissma:POI has been extended with various properties to represent the location in the context of a specific site by integrating the Buildings and Rooms vocabulary[5]. Furthermore, other properties related to the hierarchical organization of the location (such as the neighborhood, the city and the province of the current user position) have been added, and some concepts from the Juso ontology[6] have been reused. Figure 2(a) depicts relations and attributes which characterize a location. Gray rectangles indicate concepts from Juso and rooms vocabulary. The representation of time augments time:Instant defined in the Time ontology[7]. Some time intervals have been defined: the hours and the parts of the day (morning, afternoon, etc.). In addition, days of the week are classified in weekdays or weekend and seasons are represented. Figure 2(b) illustrates how time is represented and the relations with PRISSMA and the Time ontology.

3.2   The Contextual User Profile Ontology

To model user profiles, we used the Structured-Interpretation Model (SIM) [11,12], which consists of two types of ontological modules, i.e. *context types* and *context instances*. Context types describe the terminological part of an ontology (TBox) and are arranged in a hierarchy of inheritance. Context instances describe assertional part of an ontology (ABox) and are connected with corresponding context types through a relation of instantiation. There is another

---

[5] http://vocab.deri.ie/rooms.

[6] rdfs.co/juso/latest/html.

[7] https://www.w3.org/2006/time.

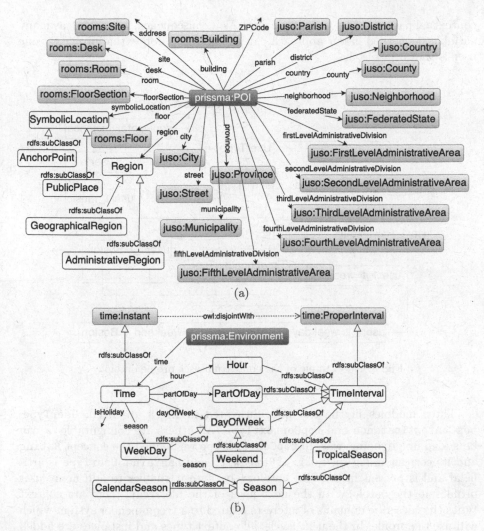

Fig. 2. Concepts and relations of RSCtx representing the location dimension (a) and the time dimension (b).

kind of relation, i.e. aggregation, which links context instances of more specific context types to a context instance of a more general context type. In the class hierarchy in a classical ontology there always exists a top concept, i.e. Thing. In a SIM ontology there is a top context type and a top context instance connected by instantiation. It is possible to add multiple context instances to one context type and aggregate multiple context instances into one context instance. Details can be found in [9,10,13].

We allow storage of many user profiles in one SIM ontology. We also support a storage of preferences from multiple domains by adding context types related to different domains. We add context types and context instances related to contextual parameters in a dynamic way. As a consequence, we can use as many variables as needed in our approach. An example of a contextual profile for one user is shown in Fig. 3.

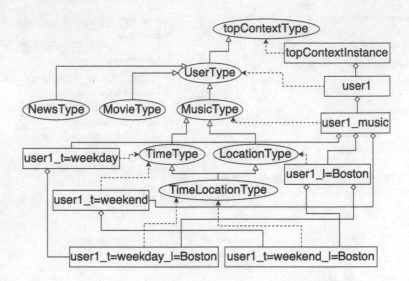

Fig. 3. An example of the contextual user profile ontology

Three modules in the example illustrated in Fig. 3 are fixed: `UserType`, `topContextInstance` and `topContextType`. All others are configurable or can be added in a dynamic way. In `topContextType` we defined the concept `Rating` and its corresponding roles, e.g. `isRatedWith` and `hasValue`. `UserType` is artificial and is present in the SIM ontology because it enables to add many user profiles to the ontology. In the next level of the hierarchy, there are context types that describe domains of interests related to a recommender system which will use the profile. In the next levels, all context types and instances are added to the contextual user profile during the learning phase or later, when a new context situation occurs.

3.3   The Recommendation Process

We use the ontologies previously presented for pre-filtering in the recommendation process. *Pre-filtering* approaches use the current context to select a relevant subset of historical data on which a recommendation algorithm is applied [3]. The aim is to provide a general ontology-based pre-filtering approach, which can be used with existing algorithms.

The system consists of three main functional parts: context detection and generalization, user profile and pre-filtering, and recommendation. In the first part, we used the RSCtx ontology to identify the user context from raw data and generalize it in the desired granularity level. The second part is responsible for building the user profile, finding a context instances that fits the actual user context, and returning only relevant preferences. Because of the lack of similarity measure for SIM ontologies, we get the data from all users who have rated an item in currently considered context (i.e. all context instances for the same values of contextual parameters from different users). However, taking only k nearest neighbors would be more efficient. The last part uses well-known algorithms, e.g. Item k-Nearest Neighbors (kNN), for providing recommendations. For this task we exploit implementations from the LibRec[8] library.

The general recommendation process proceeds as follows. Given a user and his current situation, a proper generalization of his context is generated by using the RSCtx ontology. Then, an appropriate context instances from COUP is identified by using the generalized context. If a context instance is not found in the user profile, the generalization step is repeated to search for a module that corresponds to the new context. If it is found, preferences from considered user and all other users who have the context instance with the same value for the same contextual parameters are prepared to be used with a recommendation algorithm.

4   Evaluation

We conducted offline experiments in order to evaluate the impact of contextual information on the serendipity of recommendations. We selected a number of algorithms, and we compared the accuracy and serendipity of each algorithm when used as is and when combined with the proposed pre-filtering technique. We aimed to answer the following question: *Does the context improve serendipity of items in the recommendation list?*

We relied on ConcertTweets[9] data set, which combines implicit and explicit user ratings with rich content as well as spatiotemporal contextual dimensions. It contains ratings that refer to musical shows and concerts of various artists and bands. Since the data set was generated automatically, there are two rating scales: one numerical scale with ratings in the range [0.0, 5.0] and one descriptive scale with possible values equal to *yes*, *maybe* and *no*, although *no* never occurred. We decided to split the dataset into two separate sets according to the scale type and we mapped the descriptive values *yes*, *maybe* and *no* with the numerical values 2, 1 and 0. Table 1 presents some statistics about the data by considering the whole data set and each of the sets generated when splitting by scale type. We prepared two pairs (one for each scale) of training and test sets for hold-out validation. In each test set, we put 20% of the newest ratings of each user. All other ratings were placed in each training set. The split was performed based on rating time value.

---

[8] http://www.librec.net/.
[9] https://github.com/padamop/ConcertTweets.

Table 1. Statistics on the data contained in ConcertTweets data set at the time of the experiment

	Ratings		
	All	Descriptive	Numeric
Number of users	61803	56519	16479
Number of musical events	116320	110207	21366
Number of pairs artist and musical events	137382	129989	23383
Number of ratings	250000	219967	30033
Maximum number of ratings per user	1423	1419	92
Minimum number of ratings per user	1	1	1
Average number of ratings per user	4.045	3.892	1.823
Maximum number of ratings per item	218	216	38
Minimum number of ratings per item	1	1	1
Average number of ratings per item	2.149	1.996	1.406
Minimum popularity of an artist	1	1	1
Average popularity of an artist	84.317	62.421	13.768
Maximum popularity of an artist	1670	1337	244
Sparsity	0.999971	0.999970	0.999922

We computed an accuracy value by means of the classical information retrieval measures: *precision* and *recall*. The corresponding formulas are as follows:

$$\text{precision} = \frac{|\{\text{relevant items}\} \cap \{\text{recommended items}\}|}{|\{\text{recommended items}\}|}, \tag{1}$$

$$\text{recall} = \frac{|\{\text{relevant items}\} \cap \{\text{recommended items}\}|}{|\{\text{relevant items}\}|}. \tag{2}$$

To measure serendipity of recommendations we use a simple metric presented in [34] that we called *expectedness* and showed in (3).

$$\text{expectedness} = \frac{1}{N} \sum_{i=1}^{N} \text{popularity}(\text{item}_i). \tag{3}$$

According to the meaning of the serendipity, the lower the value of the formula (3) is, the bigger the serendipity of the top-N recommendations. In contrast, for (1) and (2) higher values mean better precision accuracy.

We had to choose some existing recommendation technique to evaluate our approach since it is designed to work with existing algorithm. We used six algorithms from the LibRec library appropriate for the ranking task, i.e. Item kNN, User kNN, BPR [28], FISM [19], Latent Dirichlet Allocation (LDA) [14] and WRMF [17,26]. We applied each algorithm on both subsets of the Concert-Tweets data set twice: once as is, and the second time on the data generated by our ontology-based contextual pre-filtering technique. We were unable to finish computations for two algorithms, i.e. Item kNN and User kNN, on the subset with a descriptive rating scale, because of their computational complexity and the amount of data. We computed values of the accuracy and serendipity measures on two lists, i.e. top 5 and top 10 recommendations. The results are collected in Tables 2 and 3.

Table 2. Results obtained for ConcertTweets subset with a numerical rating scale. The prefix *onto-* denotes an algorithm applied combined with our approach.

Algorithm	Top 5			Top 10		
	Precision	Recall	Expectedness	Precision	Recall	Expectedness
BPR	0.00135	0.00676	38.37	0.00135	0.01353	36.40
ontoBPR	**0.03876**	**0.19382**	**32.36**	**0.03006**	**0.30056**	**28.84**
FISM	0.00000	0.00000	75.68	0.00045	0.00225	118.52
ontoFISM	**0.04103**	**0.20513**	**54.39**	**0.03615**	**0.36154**	**44.33**
ItemKNN	0.00068	0.00338	**26.22**	0.00034	0.00338	**18.32**
ontoItemKNN	**0.01729**	**0.08644**	30.63	**0.01370**	**0.13697**	29.90
LDA	0.00000	0.00000	**65.40**	0.00045	0.00338	118.58
ontoLDA	**0.02456**	**0.12281**	76.72	**0.02281**	**0.22807**	**69.49**
UserKNN	0.00068	0.00113	**26.43**	0.00034	0.00113	**18.51**
ontoUserKNN	**0.01655**	**0.08275**	33.20	**0.01259**	**0.12587**	30.41
WRMF	0.00113	0.00338	58.55	0.00101	0.00564	47.66
ontoWRMF	**0.03520**	**0.17598**	**24.56**	**0.02486**	**0.24860**	**23.46**

As expected, adding contextual information into the recommendation process increases the precision and recall values for all algorithms, subsets, and ranking lists. Moreover, the results show that our approach also improves the serendipity of the selected algorithms. Serendipity increases for all algorithms not based on kNN with one exception for the LDA algorithm in the case of the top 5 recommendation list on the subset with a numerical rating scale. Nonetheless, the same algorithm gives almost two times better serendipity value in the top 10 list on the same data set. The situation with kNN algorithms is quite different. We observed the deterioration of serendipity for all of the cases, for which we receive results. This may be due to the fact that the kNN algorithms are based on similarity. Thus, popular items will be similar to each other with higher probability than less popular items, and when we decrease the recommendation space by constraining it to some particular context, the probability that the less popular item would be considered is even smaller. The same applies to users.

We should further investigate whether we could measure the serendipity of recommendations made by traditional algorithms and context-aware ones in the same way, or if we should incorporate the context also in the serendipity formula. It is impossible to rely on an offline study to address this issue since it is hard to distinguish unexpected items from others without knowing the user's opinion. To this aim, some online experiments are necessary.

Table 3. Results obtained for ConcertTweets subset with a descriptive rating scale. The prefix *onto-* denotes an algorithm applied combined with our approach.

Algorithm	Top 5			Top 10		
	Precision	Recall	Expectedness	Precision	Recall	Expectedness
BPR	0.00058	0.00208	200.90	0.00054	0.00376	203.06
ontoBPR	**0.01800**	**0.09000**	**157.05**	**0.01588**	**0.15875**	**143.26**
FISM	0.00022	0.00040	480.00	0.00030	0.00168	648.38
ontoFISM	**0.01507**	**0.07537**	**206.84**	**0.01326**	**0.13257**	**198.00**
ItemKNN	-	-	-	-	-	-
OntoItemKNN	**0.01160**	**0.05801**	**123.23**	**0.01078**	**0.10779**	**105.17**
LDA	0.00021	0.00040	475.27	0.00020	0.00088	482.24
ontoLDA	**0.02151**	**0.10755**	**372.20**	**0.01644**	**0.16442**	**305.28**
UserKNN	-	-	-	-	-	-
ontoUserKNN	**0.00521**	**0.02606**	**158.48**	**0.00484**	**0.04839**	**133.78**
WRMF	0.00102	0.00264	255.38	0.00075	0.00368	214.09
ontoWRMF	**0.02029**	**0.10150**	**161.16**	**0.01583**	**0.15834**	**139.00**

5   Conclusions and Further Work

In this paper, we presented some experiments on the use of the ontology-based contextual pre-filtering technique together with existing algorithms for the ranking task on the ConcertTweets data set. We showed that incorporating contextual information in the recommendation process can significantly increase the precision and recall values for all the algorithms used for testing, i.e. Item kNN, User kNN, BPR, FISM, Latent Dirichlet Allocation and WRMF. Moreover, we observed improvement in serendipity for all the algorithms not based on the kNN approach. This suggests that the use of context in the recommendation process increases (desired) unexpectedness of recommended items.

Undoubtedly, some further research is needed. Firstly, we need to investigate how serendipity should be measured in context-aware recommendation systems. This requires a series of online experiments performed with trusted users. Secondly, we lack of a similarity measure which could be used to compare two contextual ontologies (built according to Structured Interpretation Model). As a result, the pre-filtering process is slow and cannot be used in real-life systems. These issues will be addressed in further research.

References

1. Adamopoulos, P., Tuzhilin, A.: Estimating the value of multi-dimensional data sets in context-based recommender systems. In: 8th ACM Conference on Recommender Systems (RecSys 2014) (2014)
2. Adamopoulos, P., Tuzhilin, A.: On unexpectedness in recommender systems: or how to better expect the unexpected. ACM Trans. Intell. Syst. Technol. 5(4), 54:1–54:32 (2014). http://doi.acm.org/10.1145/2559952

3. Adomavicius, G., Tuzhilin, A.: Context-aware recommender systems. In: Ricci, F., et al. (eds.) Recommender Systems Handbook, pp. 217–253. Springer, US (2011). doi:10.1007/978-0-387-85820-3_7

4. Anand, S.S., Kearney, P., Shapcott, M.: Generating semantically enriched user profiles for web personalization. ACM Trans. Internet Technol. **7**(4), 22:1–22:26 (2007). http://doi.acm.org/10.1145/1278366.1278371

5. Cantador, I., Bellogín, A., Castells, P.: Ontology-based personalised and context-aware recommendations of news items. In: Proceedings of the 2008 IEEE/WIC/ACM International Conference on Web Intelligence and Intelligent Agent Technology, WI-IAT 2008, vol. 1, pp. 562–565. IEEE Computer Society, Washington (2008). http://dx.doi.org/10.1109/WIIAT.2008.204

6. Costabello, L.: A declarative model for mobile context. In: Context-Aware Access Control and Presentation for Linked Data, pp. 21–32 (2013)

7. Dey, A.K.: Understanding and using context. Pers. Ubiquit. Comput. **5**(1), 4–7 (2001). http://dx.doi.org/10.1007/s007790170019

8. Di Noia, T., Ostuni, V.C.: Recommender systems and linked open data. In: Faber, W., Paschke, A. (eds.) Reasoning Web 2015. LNCS, vol. 9203, pp. 88–113. Springer, Cham (2015). doi:10.1007/978-3-319-21768-0_4

9. Goczyła, K., Waloszek, A., Waloszek, W.: Hierarchiczny podział przestrzeni ontologii na kontekst. In: Bazy Danych: Nowe Technologie, pp. 247–260. WKŁ, Gliwice (2007)

10. Goczyła, K., Waloszek, A., Waloszek, W.: Techniki modularyzacji ontologi. In: Bazy Danych: Rozwój metod i technologii, pp. 311–324. WKŁ, Gliwice (2008)

11. Goczyla, K., Waloszek, A., Waloszek, W.: Towards context-semantic knowledge bases. In: Proceedings of the Federated Conference on Computer Science and Information Systems - FedCSIS 2012, Wroclaw, Poland, pp. 475–482, 9–12 September 2012. https://fedcsis.org/proceedings/2012/pliks/388.pdf

12. Goczyła, K., Waloszek, A., Waloszek, W., Zawadzka, T.: Modularized knowledge bases using contexts, conglomerates and a query language. In: Bembenik, R., et al. (eds.) Intelligent Tools for Building a Scientific Information Platform. SCI, vol. 390, pp. 179–201. Springer, Heidelberg (2012). doi·10.1007/978-3-642-24809-2_11

13. Goczyla, K., Waloszek, W., Waloszek, A.: Contextualization of a DL knowledge base. In: Proceedings of the 2007 International Workshop on Description Logics (DL 2007) (2007). http://ceur-ws.org/Vol-250/paper_55.pdf

14. Griffiths, T.: Gibbs sampling in the generative model of Latent Dirichlet Allocation. Technical report, Stanford University (2002). www-psych.stanford.edu/~gruffydd/cogsci02/lda.ps

15. Hawalah, A., Fasli, M.: Utilizing contextual ontological user profiles for personalized recommendations. Expert Syst. Appl. **41**(10), 4777–4797 (2014). http://www.sciencedirect.com/science/article/pii/S0957417414000633

16. Herlocker, J., Konstan, J., Terveen, L., Riedl, J.: Evaluating collaborative filtering recommender systems. ACM Trans. Inf. Syst. **22**(1), 5–53 (2004)

17. Hu, Y., Koren, Y., Volinsky, C.: Collaborative filtering for implicit feedback datasets. In: Proceedings of the 2008 Eighth IEEE International Conference on Data Mining, ICDM 2008, pp. 263–272. IEEE Computer Society, Washington (2008). http://dx.doi.org/10.1109/ICDM.2008.22

18. Iaquinta, L., de Gemmis, M., Lops, P., Semeraro, G., Molino, P.: Can a recommender system induce serendipitous encounters? In: E-Commerce, pp. 1–17. IN-TECH, Vienna (2009)

19. Kabbur, S., Ning, X., Karypis, G.: Fism: factored item similarity models for top-n recommender systems. In: Proceedings of the 19th ACM SIGKDD International Conference on Knowledge Discovery and Data Mining, KDD 2013, pp. 659–667. ACM, New York (2013). http://doi.acm.org/10.1145/2487575.2487589

20. Karpus, A., Vagliano, I., Goczyła, K., Morisio, M.: An ontology-based contextual pre-filtering technique for recommender systems. In: 2016 Federated Conference on Computer Science and Information Systems (FedCSIS), pp. 411–420, September 2016

21. Kotkov, D., Veijalainen, J., Wang, S.: Challenges of serendipity in recommender systems. In: Proceedings of the 12th International Conference on Web Information Systems and Technologies, WEBIST, vol. 2, pp. 251–256 (2016)

22. Lops, P., de Gemmis, M., Semeraro, G.: Content-based recommender systems: state of the art and trends. In: Ricci, F., et al. (eds.) Recommender Systems Handbook, pp. 73–105. Springer, US (2011). doi:10.1007/978-0-387-85820-3_3

23. Maksai, A., Garcin, F., Faltings, B.: Predicting online performance of news recommender systems through richer evaluation metrics. In: Proceedings of the 9th ACM Conference on Recommender Systems, RecSys 2015, pp. 179–186. ACM, New York (2015). http://doi.acm.org/10.1145/2792838.2800184

24. Middleton, S.E., Shadbolt, N.R., De Roure, D.C.: Ontological user profiling in recommender systems. ACM Trans. Inf. Syst. **22**(1), 54–88 (2004). http://doi.acm.org/10.1145/963770.963773

25. Mobasher, B., Jin, X., Zhou, Y.: Semantically enhanced collaborative filtering on the web. In: Berendt, B., Hotho, A., Mladenič, D., Someren, M., Spiliopoulou, M., Stumme, G. (eds.) EWMF 2003. LNCS (LNAI), vol. 3209, pp. 57–76. Springer, Heidelberg (2004). doi:10.1007/978-3-540-30123-3_4

26. Pan, R., Zhou, Y., Cao, B., Liu, N.N., Lukose, R., Scholz, M., Yang, Q.: One-class collaborative filtering. In: Proceedings of the 2008 Eighth IEEE International Conference on Data Mining, ICDM 2008, pp. 502–511. IEEE Computer Society, Washington (2008). http://dx.doi.org/10.1109/ICDM.2008.16

27. Rack, C., Arbanowski, S., Steglich, S.: Context-aware, ontology-based recommendations. In: Proceedings of the International Symposium on Applications on Internet Workshops, SAINT-W 2006, pp. 98–104. IEEE Computer Society, Washington (2006). http://dx.doi.org/10.1109/saint-w.2006.13

28. Rendle, S., Freudenthaler, C., Gantner, Z., Schmidt-Thieme, L.: BPR: Bayesian personalized ranking from implicit feedback. In: Proceedings of the Twenty-Fifth Conference on Uncertainty in Artificial Intelligence, UAI 2009, pp. 452–461. AUAI Press, Arlington (2009). http://dl.acm.org/citation.cfm?id=1795114.1795167

29. Ricci, F., Rokach, L., Shapira, B.: Introduction to recommender systems handbook. In: Ricci, F., Rokach, L., Shapira, B., Kantor, P.B. (eds.) Recommender Systems Handbook, pp. 1–35. Springer, US (2011). doi:10.1007/978-0-387-85820-3_1

30. Rodríguez, J., Bravo, M., Guzmán, R.: Multidimensional ontology model to support context-aware systems (2013). http://www.aaai.org/ocs/index.php/WS/AAAIW13/paper/view/7187

31. Su, Z., Yan, J., Ling, H., Chen, H.: Research on personalized recommendation algorithm based on ontological user interest model. J. Comput. Inf. Syst. **8**(1), 169–181 (2012)

32. Weijnen, M., Drinkenburg, A. (eds.): Precision Process Technology: Perspectives for Pollution Prevention. Springer Science & Business Media, Dordrecht (2012)

33. Zhang, Y.C., Séaghdha, D.O., Quercia, D., Jambor, T.: Auralist: introducing serendipity into music recommendation. In: Proceedings of the Fifth ACM International Conference on Web Search and Data Mining, WSDM 2012, pp. 13–22. ACM, New York (2012). http://doi.acm.org/10.1145/2124295.2124300
34. Ziegler, C.N., McNee, S.M., Konstan, J.A., Lausen, G.: Improving recommendation lists through topic diversification. In: Proceedings of the 14th International Conference on World Wide Web, WWW 2005, pp. 22–32. ACM, New York (2005). http://doi.acm.org/10.1145/1060745.1060754

Extending Expressiveness of Knowledge Description with Contextual Approach

Aleksander Waloszek and Wojciech Waloszek[(✉)]

Gdańsk University of Technology,
ul. Gabriela Narutowicza 11/12, 80-233 Gdańsk, Poland
{alwal,wowal}@eti.pg.gda.pl

Abstract. In the paper we show how imposing the contextual structure of a knowledge base can lead to extending its expressiveness without changing the underlying language. We show this using the example of Description Logics, which constitutes a base for a range of dialects for expressing knowledge in ontologies (including state-of-the-art OWL).

While the contextual frameworks have been used in knowledge bases, they have been perceived as a tool for merging different viewpoints and domains, or a tool for simplifying reasoning by constraining the range of the statements being considered. We show, how it may also be used as a way of expressing more complicated interrelationships between terms, and discuss the import of this fact for authoring ontologies.

Keywords: Ontologies · Knowledge bases · Contexts · Knowledge base structure

1 Introduction

Contextual representation of knowledge is not a new topic and it receives raising attention within AI researchers community for many years. The scientists perceived them as a valuable way of structuring large ontologies[1]. For example CYC [13], a very well known project with the ambitious goal of covering understanding of natural language, used so-called micro-theories to organize modules of their knowledge base.

More recent projects, especially those undertaken within so called Semantic Web, were targeted towards broader audience, and had the ambitions of creating more general frameworks for various knowledge bases. The most significant early methods proposed for managing contextual ontologies were: DDL [4] and E-Connections [12]. However, the stress there was placed mainly on merging existing ontologies and knowledge bases, often encoded with different languages.

Only in the last few years, we have seen the advent of more mature contextual frameworks, which focus on the description of interrelationships between

---

[1] With a bit of abuse of the notions, for the sake of simplicity throughout this paper we use the terms "ontology" and "knowledge base" interchangeably.

© Springer International Publishing AG 2017
S. Kozielski et al. (Eds.): BDAS 2017, CCIS 716, pp. 260–272, 2017.
DOI: 10.1007/978-3-319-58274-0_22

modules. The methods underline the necessity of careful design of the structure of modules from the very beginning of the design any contextual knowledge base, and largely facilitate crucial tasks like dividing knowledge into small parts and ensuring proper flow of conclusions between them. Among those very interesting contextual frame-works are *Contextual Knowledge Repository* (CKR) [5] (partially based on earlier work [9]) and *Description Logics of Contexts* (DLC) [11]. Among them is also SIM (Structured Interpretation Framework) proposed in [7].

Dividing knowledge into small parts requires very careful decisions about the perspective from which the knowledge is presented. For example, in the context concerning present situation in the USA we can say *President*(*obama*). But in the context expanded to other countries, we should say *President*(*obama, usa*). If we change context taking into consideration a longer period of time our predicate becomes of arity 3: *President*(*obama, usa, 2008−2016*). It means that bigger contexts imply more complex languages and more complex structures of interpretations, however, the good and mature structural framework of a knowledge base should support transitions between larger and smaller contexts.

This paper is devoted to the observation that by exploiting the aforementioned ability of contextual framework of supporting context changes we can also achieve a new, perhaps partially unexpected, effect of increasing the expressiveness of a knowledge base. This effect can be particularly well observed within the family of Description Logics (DLs), commonly used in Semantic Web.

While Description Logics used to encode modern ontologies are very expressive, they, by their very nature, are not able to express knowledge about some kind of cyclic interrelationships between various individuals. One of the most frequently given example is when someone wants to select people who work for their parents (i.e. whose parent and employer is this same person).

The problem is important and well-known to the Semantic Web community and now is being solved by combining a DL language with a rule language. The Horn rules [10] are popular manner of encoding knowledge; moreover Horn rules focus on aspects of interrelationships between various individuals, but these two languages are significantly different in their nature. In our work we show how to engulf some of expressivity of the rules within a contextual framework, without the necessity of extending DLs with new syntactic constructs.

The rest of the paper is organized as follows. Section 2 introduces one of the basic approaches for contextual approach to reasoning, a box metaphor. Section 3 discusses SIM, Structured Interpretation Model for creating contextual knowledge bases. Section 4 is the main section of the paper. It introduces the reader to problem of reasoning with cyclic relationships and presents its contextual solution. The results are discussed in Sect. 5. Section 6 concludes the paper.

2   Basics of Contextual Approach

One of the most intuitive approaches to depict contextual approach to describing knowledge has been proposed by Bar-Hillel in 1964 [2] and has been developed in [3]. The approach is called context-as-a-box, as it depicts the knowledge being expressed as a set of sentences contained within a rectangle symbolizing *a context* (see Fig. 1a). Above the rectangle, there is a list of *contextual parameters*. According to the approach, no knowledge is context-free, so the parameters are used to describe the contextual factors that may influence the interpretation of sentences within the box. Each of contextual parameters is assigned a value, which reveals the specific contextual circumstances under which the sentences should be interpreted.

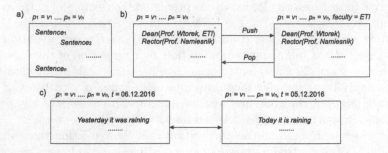

Fig. 1. Illustration of context-as-a-box metaphor (a), push/pop (b) and shift (c) operations

The main role of contextual parameters is to establish relationships between different boxes/contexts and to describe methods of transitioning between contexts. Usually three such methods are distinguished: *push, pop* and *shift*.

Push and *pop* are mutually inverse operations. *Push* corresponds to a transition from more general to more specific context, while *pop* corresponds to a transition from more specific to more general context. In Fig. 1b, there is shown the example of a sentence "Prof. Wtorek is the dean" (right box), whose more formal form can be expressed as $Dean(Prof.Wtorek)$. Contextual parameters reveal that the place where the sentence is uttered is the Faculty of Electronics, Telecommunication, and Informatics (ETI) of Gdansk University of Technology, and the time of utterance is presence. Basing on this information we can concur that Prof. Wtorek is currently the dean of this faculty. However, if we broaden our context a bit (by *popping* the information about faculty) and transit to the context of the whole University, it would be much more natural to formulate the sentence (conveying the same information) as "Prof. Wtorek is the dean of ETI", whose more formal form is $Dean(Prof.Wtorek, ETI)$.

The third and the last of the operations is *shift*. *Shift* consists in changing a value of one (or more) of the contextual parameters. It is most frequently illustrated with use of pronouns and with change of their meaning between contexts (boxes). Figure 1c depicts the shift in time, where in the left context we assume that the current date is 6.12, and in the right 5.12. Thus the sentence "Today it is raining" in the left context naturally implies the truth of the sentence "Yesterday it was raining" in the right context. Words "today" and "yesterday" may be treated as pronouns (technically they are adverbs) which are used instead of explicit dates. Their meaning changes when the value of the appropriate contextual parameter (*current date*) shifts.

For the purposes of the paper, *shift* operation can be perceived as a combination of *pop* and *push*, and those two former operations will be the most important for further discussion. As it can be seen from the above example, *push* involves enriching the set of contextual parameters and possibly decreasing the arity of predicates used to express sentences, and, conversely, *pop* simplifies the set of contextual parameters, but possibly increases the arity of predicates that have to be used to express the intended meaning. The work "possibly" is naturally very important here, as not all of the parameters have influence on all predicates. The example of such predicate is *Rector*, (Fig. 1b) which does not change its meaning, since we still operate within the context of Gdansk University of Technology.

3   SIM Framework

SIM [6,7] is a method of organizing contexts which takes into account some specifics of Description Logic (though is not necessarily used with DLs only). SIM is an acronym of Structured Interpretation Model.

Knowledge base in SIM is divided into modules. Modules resemble the boxes mentioned in the earlier Section, because sentences can be placed in the modules by a knowledge base user. SIM distinguishes two types of modules (for different kinds of sentences): first, called *context types*, for general, terminological knowledge, and second, called *context instances*, for knowledge about individual objects. This follows the general rule for Description Logics of dividing knowledge base into two parts: terminological TBoxes, and assertional ABoxes.

Contextual parameters are not explicit in SIM. Instead there are three types of relationships connecting knowledge modules, and governing the truth condition for sentences in appropriate boxes. *Inheritance*, the first relationship, binds context types. One context type can inherit form another. All the sentences true in the predecessor are also true in the ancestor. One context type can inherit from many other types. The relationship is acyclic, cyclic inheritances are not allowed.

Because of the character of inheritance relationship, the types lower in the hierarchy (assuming that the most general context type is on the top) reuse the vocabulary defined on higher levels and introduce new concepts. Therefore, with going down the hierarchy we are going towards more specialized contexts and vocabularies.

The second relationship is *instantiation*. It binds a context instance with exactly one context type, the type of this instance. Its semantics is very similar to the former relation, as the instance "inherits" all the sentences from its type (they must be true in the scope of the instance).

Similarly like with context types, instances may be placed on different levels of context inheritance hierarchy, depending on their types. The general rule here is that lower context instances describe smaller fragments of knowledge but in more details (with use of more specific vocabulary).

The third relationship is a bit more involved. It is called *aggregation* and it connects two context instances. The *aggregating* context instance has to be of more general type, and the *aggregated* context instance needs to be of a more specific inheriting type. The knowledge flows in both directions of the relationships, however with constraints. Sentences from the aggregated context instance need to be translated to the (smaller) vocabulary of the aggregating context instance, so they are transferred to the aggregating context but on higher level of abstraction.

The second direction of the flow, from aggregating context instance to aggregated one, does not require translation of sentences (as all the terms must be present in inherited vocabulary). However, this flow is also constrained, as only sentences concerning the smaller domain of aggregated context instance are transferred.

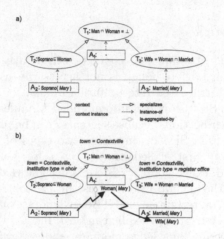

Fig. 2. An exemplary SIM knowledge base

The example of knowledge flow along different relationships is shown in Fig. 2. Context types are depicted as ovals, context instances as rectangles. The sentences contained within the modules are being translated (Fig. 2b, the flash-like arrow shows the translation/inference) with accordance to aforementioned rules, resulting in drawing the final conclusion that Mary is a mother.

SIM generally conforms to Bar-Hillel model. The conformance may not be apparent at the first sight, because contextual parameters are not explicit in SIM. However, we need to recall that the contexts which are lower in the hierarchy are more specific. More specific contexts mean more comprehensive list of contextual parameters. Therefore, while the contextual parameters themselves need not to be revealed, we may assume that they are present and arranged according to SIM hierarchy. SIM modules may be then perceived as boxes with explicit sentences and implicit contextual parameters whose list increases in size when going down towards inherited types and instantiated and aggregated context instances. Figure 2b shows an exemplary assignment of contextual parameters within a knowledge base.

Support for SIM has been implemented in [8]. CongloS is a system developed at Gdansk University of Technology for designing and reasoning over contextual knowledge bases. It has been created in the form of a set of plug-ins for a very well known Protégé ontology editor [15], and is available for download at http://www.conglos.org.

4 Contextual Approach to Cyclic Relationships

In this Section we will present the problem of cyclic relationships in DL language, and discuss the current methods of alleviating it. Then we turn to presenting the contextual approach which solves the problem, without the necessity of extending language used and with preserving decidability.

4.1 Cyclic Relationships and Description Logics

Description Logics constitute a family of logics very well suited for expressing advanced vocabularies called ontologies. They are a base for state-of-the-art OWL language [16], standardized by W3C (World Wide Web Consortium) and extensively used in knowledge management.

All Description Logics share common characteristics. They assume that a knowledge base is two-fold and consists of terminology (TBox) and description of world (ABox). Terminology introduces *concepts* (unary predicates) and *roles* (binary predicates), and defines relationships between them. Description of world introduces individuals (individual objects) and describes them in the terms of concepts and roles. Both of the parts of a knowledge base consist of sentences. Sentences used in TBoxes are called *axioms* and sentences in ABoxes *assertions*.

One of the distinguishing features of Description Logics is its special variable-less notation. Instead of using first order logic (FOL) form of predicates, in Description Logics we build *descriptions*, i.e. possibly complex concepts built from simpler concepts with use of special constructors. The following is the example of equivalence axiom:

$$Student \sqcap Foreigner \equiv ForeignStudent \qquad (1)$$

The equivalence axiom relates two descriptions by saying they are equivalent ($\equiv$). The right-hand description $ForeignStudent$ is a simple concept, while the left-hand description $Student \sqcap Foreigner$ is an intersection of two concepts, and a complex concept itself. With use of constructors, like the intersection constructor ($\sqcap$), one can easily introduce new concepts like $Student \sqcap Foreigner$ into a knowledge base. Concepts like this are in fact a form of new, quickly defined, unary predicates. Easiness of introducing such predicates into a knowledge base makes Description Logics a very good tool for modeling complicated structures of concepts.

Such new unary predicates can also involve roles. Roles are counterparts of binary predicates. They connect pairs of objects and can be used in description with use of special operators, among which most common are $\exists$ and $\forall$. They are existential and universal quantification which in Description Logics are used similarly to standard use in FOL. The example of use of $\exists$ is shown below:

$$Student \sqcap \exists comesFrom.ForeignCountry \equiv ForeignStudent \qquad (2)$$

In the above axiom the concept $Foreigner$ was replaced by $\exists comesFrom.ForeignCountry$ which denotes objects (most possibly people) related to members of the concept $ForeignCountry$ with the role $comesFrom$.

The use of descriptions in Description Logics has many advantages. The notation used is readable even to non-logicians. It is easy to introduce new concepts without complicated definitions. Focus on unary predicates is the reason why the structure of knowledge being defined resembles object models used in programming, as the terminology may basically be perceived as a hierarchy of concepts (classes).

However, all this features also introduce some difficulties in reasoning with specific concepts in Description Logics. Specifically these are concepts which refer to cyclic relationships.

By cyclic relationships we understand here a system of individuals connected by roles into a cycle. One of the simplest examples of a cycle is the concept of someone being employed by his relative, say uncle. For the sake of the example we assume that there are two roles in the knowledge base: $hasUncle$ and $isEmployedBy$. Individuals of interest are those who are related with $hasUncle$ and $isEmployedBy$ role with the same individual.

It is relatively easy to express the extension of such a concept in FOL, with the following formula (we are treating roles as binary predicates here):

$$EmployedByUncle(x) \leftrightarrow \exists y : hasUncle(x,y) \wedge isEmployedBy(x,y) \qquad (3)$$

In Description Logics we are facing a major difficulty here. The method of creating descriptions does not allow us to use the auxiliary variable y. One of the closest concepts we may get is $(\exists hasUncle.\top \sqcap \exists isEmployedBy.\top)$, where $\top$ denotes whole domain of individuals, but this concept only refers to objects (people) who have uncle and are employed, but the uncle and employer are not necessarily the same object (person).

This problem has been identified by scientists involved in the development of Description Logics [1], commencing the new area of research for methods of integrating DL with other methods of expressing knowledge, most importantly Horn rules.

Horn rules are used in languages like Prolog, and allows for defining new predicates with use of many variables. The solution to the presented problem could therefore be quite straightforward and consists in introducing the new Horn rule to the system (implication form):

$$hasUncle(x,y) \wedge isEmployedBy(x,y) \rightarrow EmployedByUncle(x) \qquad (4)$$

While the solution seems simple and efficient, there are major difficulties with its application. The reasoning methods for Description Logics rely on the

structure of descriptions and their "acyclic" character. Straightforward extension of DL by rules leads to undecidability, and as such cannot be applied
directly [14].

The solution is to interpret rules under a bit different regime. Description
Logics strictly follow the Open World Assumption (OWA). This feature results in
possibility of drawing a broad range of conclusion, even in the absence of detailed
knowledge about all the individuals involved in the process of reasoning. For
example, if we state that every Person has a parent ($Person \sqsubseteq \exists hasParent.\top$),
a simple assertion $Person(adam)$ can lead us to the conclusion that $adam$ has
at least one parent, even if we do not know the appropriate individual.

In contrast, Horn rules reasoners usually operate under Closed World
Assumption. This means that all the individuals involved in the system we are
reasoning about need to be known. For example for the rule:

$$hasParent(x, y) \land hasParent(y, z) \rightarrow GrandParent(z) \qquad (5)$$

we need to know at least one child and at least one grandchild of z to state
that z is a grandparent.

The discrepancies described above lead to necessity of assuming two ways of
interpreting sentences. In order to maintain satisfiability, DL sentences (axioms
and assertion) may still be understood under OWA, but Horn rules have to be
interpreted under CWA. It means that all the individuals that are assigned to
variables need to be known, which is sometimes denoted with special predicate
$\mathcal{O}$ (in the case of rule (5) we would write $\mathcal{O}(x), \mathcal{O}(y), \mathcal{O}(z)$).

As we could see, the traditional approach of combining DL and rules, forces
authors of ontologies to use in fact two very different languages and two very
different assumptions about the world (OWA and CWA). In the next subsection
we present an alternative approach to the problem, where we use the structure
of contexts to perform the same inference, without the necessity of employing
new language.

4.2   Cyclic Relationships and SIM

The alternative approach to reasoning with cycles employs the structure of contexts to enable the type of inference. No use of additional language for ontology
is necessary.

The scheme presented here is based on observation that Description Logics
is very well suited for reasoning over even very complicated and complex unary
predicates (which can be build with use of constructors and descriptions), but
offers limited support for higher level predicates (roles, which can be used within
specific constructors). Therefore, we will try to employ the idea of *push* and *pop*
operations for Bar-Hillel boxes and use them to reduce the arity of predicates in
ontology.

First of all, we need to notice that large majority of terms concerning people
(but not only people) can be used within different scopes-more general and more
specific. It is most clearly observable for titles from government and politics,

like *president* or *prime minister*. It is perfectly acceptable, and even polite, to address a former chief of government (like for instance in Poland Leszek Miller) with the title of *prime minister* (in general meaning). However, most people would immediately answer the question about Polish *prime minister* by giving the name of Beata Szydło, implicitly assuming the more specific scope of the term, constrained to the *current prime minister*.

To use this observation for the cyclic relationships, let us consider the relations from the previous subsection: *being an uncle* and *being an employer*. The relations were introduced as binary (*hasUncle* and *isEmployedBy*) but are associated with appropriate unary terms of *Uncle* and *Employer*.

In Sect. 2 we discussed the change of meaning and the arity of the *Dean* predicate with changing the context to more specific/more general. We need to notice, however, that the two phenomena (change of arity and change of scope), while very closely related, are to some extent independent. Clearly, we can speak about *Employers* in general sense without specifying whose employers they are (e.g. in sentences like *Employers express their concern about recent governmental announcements*).

Since SIM does not directly support change of arity of predicates, the main task to carry out was to introduce both unary and binary terms and to relate them to each other in different scopes with proper use of axioms (contents of modules) and contexts (contextual structure of SIM relationships).

We achieved this by introducing a special family of contexts. Each of the contexts represents a point of view of a single person. On the level of those "personal" contexts, the terms like *Uncle* and *Employer* (if introduced) should be understood specifically, locally, i.e. as uncles and employers of a specific person, which is called the *context owner*.

Above the level of those personal contexts, there is the aggregating context with original roles *hasUncle* and *isEmployedBy*. At this level the terms like *Uncle* and *Employer* (if introduced) should be understood generally, as uncles and employers in general.

Because of inability of SIM to reinterpret once introduced term at lower level of context hierarchy we decided about introducing the pair of concepts: *G-Uncle* and *G-Employer* (where *G* stands for general) at the aggregating level, and concepts *S-Uncle*, and *S-Employer* (where *S* stands for specific) at the aggregated ("personal") level.

This arrangement allows us to use SIM structure for reasoning over cyclic relationships. The shape of the knowledge base is presented in Fig. 3. There are two terminologies defined. Terminology T_1 introduces the roles *hasUncle* and *isEmployedBy* and general terms *G-Uncle* and *G-Employer*. It also introduces a general concept for people employed by their uncle (*EmployedByUncle*) without defining it. Terminology T_2 introduces specific terms *S-Uncle* and *S-Employer*.

Due to specificity of SIM, all the terms from terminology T_1 are also within the scope of all instances of T_2. This allows us to formulate bridging axioms which state that every specific uncle (*S-Uncle*) is also an uncle in general (*G-Uncle*),

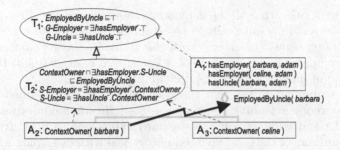

Fig. 3. Structure of contextual SIM ontology for reasoning with cyclic relationships

analogically for employers. Those bridging axiom allows for flow of the conclusions from lower contexts to the higher one.

Terminology T_2 also introduces a crucial concept of *ContextOwner*. A context owner is a special individual from whose point of view the knowledge in the context is presented. Existence of such concept allows us to formulate bridging axioms that allow for flow of the conclusions from more general context to the more specific ones. Namely, these axioms state that the local uncles are uncles of context owners (the same goes for the employers):

$$S\text{-}Uncle \equiv \exists hasUncle^-.ContextOwner \tag{6}$$

$$S\text{-}Employer \equiv \exists hasEmployer^-.ContextOwner \tag{7}$$

where r^- denotes the role inverse to r.

While defining the concept *EmployedByUncle* is not possible at the general level, it is extremely easy to define it at the level of aggregated ("personal") contexts. It suffices to formulate the following axiom:

$$ContextOwner \sqcap \exists hasEmployer.SUncle \sqsubseteq EmployedByUncle \tag{8}$$

The reason behind this easiness is that we operate now on the level of unary operators, where Description Logics is very expressive. The presence of the axiom (8) in the knowledge base ensures that each context owner who is employed by their (local) uncle is recognized by the base as such person.

The rules of flow of the conclusions in SIM guarantee that at the level of aggregating context instance A_1 all the knowledge about context owners employed by uncles is collected (flash-like arrow in Fig. 3 shows the flow of conclusions between contexts) Thus, from the point of view of A_1, we obtain an ontology with binary relations *hasUncle* and *isEmployedBy* and with the concept *EmployedByUncle* describing members of cyclic relationship about which we can reason (which would be impossible in a non-contextual DL knowledge base).

The condition under which the system works is creation of "personal" context instances for all the individuals about which we would like to reason. It is worth noting that this condition is very similar to the requirement following the integration with rules, that all the individuals should be known. Here, the set of individuals we reason about needs to be established, and for each of such individuals a context should be created with this individual as its *ContextOwner*. While this complicates the structure of a knowledge base, it is also the reason why this kind of inference maintains decidability.

It needs to be underlined, that the knowledge base we described here, can be created without extending the language used (inside modules we use only standard DL) and without any non-standard extensions to SIM structure. It means that authors of the ontology need not to deal with discrepant approaches like CWA and OWA in case of rules. It also indicates that SIM knowledge bases (and contextual bases in general) bear a potential for increasing expressiveness and for describing and dealing with a broader range of problems than standard non-contextual ontologies.

5   Discussion of the Results

In Sect. 4 we showed a novel approach to solving the problem of reasoning with cyclic relationships in Description Logics with use of contextual structure of a knowledge base.

The creators of contextual frameworks for ontologies tend to underline such features like natural support for modularization, easiness of introducing new modules etc. While they are certainly very important, it should not prevent us from studying more in-depth the additional possibilities of reasoning which open with introducing structured contextual knowledge bases.

Each kind of reasoning could then be referred to a specific contextual structure which supports it, and, consequently, a catalogue of such structures may be created. It could be a very valuable tool for revealing the promising directions of development of context-related methods and for assessing them and comparing to each other.

Our study demonstrates that SIM in general allows for creating knowledge bases of structure shown in Fig. 3, and, therefore, for reasoning with cyclic relationships. However, during the process we could identify that some of the actions were not sufficiently supported by tools or SIM itself.

One of the problems found during our research was little support for changing the scope of terms like *Uncle* and *Employer* while moving up and down the hierarchy of contexts. A consequence of this was the necessity of introducing duplicating concepts for "general" and "personal" uncle (*GUncle* and *SUncle*) and employer. This problem should be solved in next versions of SIM, and one of the possible improvements may consist in introducing new relationship between contexts which takes into account the changes in meaning of selected concepts (stemming from changing the scope of discussion).

Another problem encountered was more of technical nature and lied in difficulties with creating large numbers of contexts (one context for each individual). This operation should be automated to maximal possible extent or even performed in background, in the way transparent to user, during the reasoning.

6 Conclusions

In the paper we presented the method for performing inferences over cyclic relationships in Description Logics knowledge bases. While the standard approach to this problem consists in extending DLs of rule language, we exploited a contextual model for structuring a knowledge base and devised a type of structure which allowed for the required kind of reasoning.

Our study revealed a potential for performing non-standard inference tasks without changing the basic language for expressing knowledge, but by giving it a proper structure. We found this direction of work very promising, as it allows for expanding our expertise about contextual knowledge bases and to draw direction of their future development.

It is also worth mentioning that the knowledge base created during the study (and depicted in Fig. 3) is available for download from http://www.conglos.org along with CongloS system used for its design.

Acknowledgements. This work was partially supported by the Polish National Centre for Research and Development (NCBiR) under Grant No. PBS3/B3/35/2015, project "Structuring and classification of Internet contents with prediction of its dynamics" (Polish title: "Strukturyzacja i klasyfikacja treści internetowych wraz z predykcja ich dynamiki").

References

1. Baader, F., Calvanese, D., McGuiness, D.L., Nardi, D., Patel-Schneider, P.F. (eds.): The Description Logic Handbook: Theory, Implementation, and Applications, 2nd edn. Cambridge University Press, Cambridge (2007)
2. Bar-Hillel, Y.: Indexical expressions. Mind **63**, 359–379 (1954)
3. Benerecetti, M., Bouquet, P., Ghidini, C.: On the dimensions of context dependence: partiality, approximation, and perspective. In: Akman, V., Bouquet, P., Thomason, R., Young, R. (eds.) CONTEXT 2001. LNCS (LNAI), vol. 2116, pp. 59–72. Springer, Heidelberg (2001). doi:10.1007/3-540-44607-9_5
4. Borgida, A., Serafini, L.: Distributed description logics: assimilating information from peer sources. J. Data Semant. **1**, 153–184 (2003)
5. Bozzato, L., Serafini, L.: Materialization calculus for contexts in the semantic web. In: Proceedings of DL 2013. CEUR-WP, vol. 1014 (2013)
6. Goczyła, K., Waloszek, A., Waloszek, W., Zawadzka, T.: Towards context-semantic knowledge bases. In: Proceedings of the KEOD 2011 (2011)
7. Goczyła, K., Waloszek, A., Waloszek, W.: Contextualization of a DL knowledge base. In: Proceedings of the 20th International Workshop on Description Logics (DL 2007), pp. 291–299 (2007)
8. Goczyła, K., Waloszek, A., Waloszek, W., Zawadzka, T.: Theoretical and architectural framework for contextual knowledge bases. In: Bembenik, R., Skonieczny, L., Rybiński, H., Niezgódka, M. (eds.) Intelligent Tools for Building a Scientific Information Platform: Advanced Architectures and Solutions, pp. 257–280. Springer, Heidelberg (2013)
9. Homola, M., Serafini, L., Tamilin, A.: Modeling contextualized knowledge. In: Proceedings of the 2nd Workshop on Context, Information and Ontologies (2010)

10. Horn, A.: On sentences which are true of direct unions of algebras. J. Sym. Logic **16**(1), 14–21 (1951)
11. Klarman, S. (ed.): Reasoning with contexts in description logics. Ph.D. thesis, Free University of Amsterdam (2013)
12. Kutz, O., Lutz, C., Wolter, F., Zakharyaschev, M.: Distributed description logics: assimilating information from peer sources. Artif. Intell. **156**(1), 1–73 (2004)
13. Lenat, D.B.: CYC: a large-scale investment in knowledge infrastructure. Commun. ACM **38**(11), 33–38 (1995)
14. Motik, B., Sattler, U., Studer, R.: Query answering for OWL-DL with rules. In: McIlraith, S.A., Plexousakis, D., Harmelen, F. (eds.) ISWC 2004. LNCS, vol. 3298, pp. 549–563. Springer, Heidelberg (2004). doi:10.1007/978-3-540-30475-3_38
15. Musen, M.A.: The protégé project: a look back and a look forward. AI Matters **1**(4), 4–12 (2015)
16. W3C: OWL 2 Web Ontology Language Primer. W3C Recommendation (2012)

Ontology Reuse as a Means for Fast Prototyping of New Concepts

Igor Postanogov[1] and Tomasz Jastrząb[2(✉)]

[1] Perm State Universtity, Perm, Russia
ipostanogov@outlook.com
[2] Institute of Informatics, Silesian University of Technology, Gliwice, Poland
Tomasz.Jastrzab@polsl.pl

Abstract. In the paper we discuss the idea of ontology reuse as a way of fast prototyping of new concepts. In particular, we propose that instead of building a complete ontology describing certain concepts from the very beginning, it is possible and advisable to reuse existing resources. We claim that the available online resources such as Wikidata or wordnets can be used to provide some hints or even complete parts of ontologies aiding new concepts definition. As a proof of concept, we present the implementation of an extension to the Ontolis ontology editor. With this extension we are able to reuse the ontologies provided by Wikidata to define the concepts that have not been previously defined. As a preliminary evaluation of the extension, we compare the amount of work required to define selected concepts with and without the proposed ontology reuse method.

Keywords: Ontologies · Ontology-based data access · Data extraction · Data integration

1 Introduction

Ontologies, which can be used to describe concepts and relations pertaining to various domains have been a field of intensive study for many years. They can be applied in many areas such as Semantic Web [13], eCommerce and eBusiness [11], analysis of gene functions [21,22], autonomous mobile platforms control [10] and many others. As mentioned in [4], the ontologies can be classified into upper ontologies, an example of which is SUMO [27], describing the most general concepts, and domain ontologies, describing the concepts related to a particular domain. Moreover, the ontologies can be also classified as [4,31]:

- *informal* ontologies, in which the types are undefined or defined in natural language,
- *formal* ontologies, in which the concepts and relations are named and partially ordered,
- *axiomatized* ontologies, which are special forms of formal ontologies, in which the subtypes are defined in a formal language,

S. Kozielski et al. (Eds.): BDAS 2017, CCIS 716, pp. 273–287, 2017.
DOI: 10.1007/978-3-319-58274-0_23

- *prototype-based* ontologies, which are special types of formal ontologies, in which the subtypes are defined by means of comparison with typical members,
- *terminonological* ontologies, in which the concepts are described by labels without an axiomatic foundations. An example of such an ontology is a wordnet.

Various methods of ontologies creation are known, being typically divided into manual and automatic or semi-automatic methods. Among the existing automatic and semi-automatic methods of ontologies creation one may find methods based on data extraction from text, as described in [4,20]. It is also quite frequent that the corpora from which the ontologies can be built or enriched originate from the web resources [1,26]. Furthermore, the learning methods can be divided into clustering methods [5,9] and pattern matching methods [16]. For a broader overview of available (semi-)automatic methods of ontologies creation we refer the reader to [14].

There are also numerous papers dealing with ontology reuse methods, which typically use Wikipedia or WordNet to acquire some existing knowledge, e.g. [2,17,29,30]. Arnold and Rahm [2] propose a method, which extracts semantic relations from Wikipedia definitions using the semantic patterns matching. The semantic patterns application phase is preceded by text preprocessing, which is also similar to the approach in [17], where Robust Minimal Recursion Semantics (RMRS) representation is used for data preprocessing. The works by Ruiz-Cascado et al. [29,30] describe an approach in which the internal connections within Wikipedia (the hyperlinks), combined with WordNet relations were used to extract only relevant elements of the definitions. Then a set of generalized patterns is devised, which allows for identification of new relations.

Apart from the aforementioned approaches which mainly focus on the analysis of unstructured text, there are also solutions which try to investigate the already existing structures of the reused resources, such as Wikipedia info boxes used for the creation of DBpedia [3]. Finally, there are also approaches which aim at explicit modularisation of existing ontologies with the purpose of partial reuse. For examples of these and similar approaches see [12,33].

The main focus of the current paper is the manual creation of ontologies with the use of ontology editors. In particular, we focus on the Ontolis ontology editor described in [8]. It is a multi-platform, highly adaptable tool for visual definition of ontologies, supporting the standard ontology data formats such as OWL and RDF, as well as custom formats [6]. There are some alternative ontology editors, out of which one of the most popular ones is Protégé, which is an open source ontology editor available as a desktop as well as web application [15]. There are also similar solutions supporting the edition of terminological ontologies, i.e. wordnets, for instance the Dictionary Editor and Browser (DEB) platform described in [18].

The main contribution of the paper is the proposal of an idea of ontology reuse for the purpose of new concepts prototyping. We consider an approach in which instead of building the complete concept definition manually, it is possible to obtain partial or even full information on the concept's meaning from existing and widely available resources. The main motivation behind the proposed idea of ontology reuse is to ease the process of manual ontologies creation, which can be time-consuming and troublesome. However, the key difference between our approach and the ones described by Arnold and Rahm [2], or Ruiz-Cascado et al. [29,30] is that we do not aim at large-scale automatic data extraction. Instead we propose to reuse existing resources as a guidance for the user of the ontology editor.

In the context of manual ontology edition, our approach is similar to the work of Xiang et al. [33]. However, the Ontofox system they propose is targeted at the life sciences domain, providing the support for specialized ontologies only, while our approach does not impose such restrictions. In particular, we focus on using generic wordnets, which are lexical databases [24,25] or terminological fundamental ontologies [4], and Wikidata, which in turn is a collaborative knowledge base [32], as potential sources of information. As for the wordnets, we focus our attention on one of two Polish wordnets, namely plWordNet [23,28]. The reason for choosing this particular resource is that its contents have already been investigated by one of the authors in [19] and they were found suitable for the purposes of the current paper.

The main drawbacks of Ontofox system [33] are that it only allows for reuse of a single ontology at a time, and that it provides a limited set of configuration settings (such as to only include all, computed or no intermediates). Furthermore, some of the settings require additional knowledge related to the proper format of the parameters (see the Annotation/Axiom Specification part). The approach we propose is free from these restrictions as we allow for full and user-friendly customization of the data retrieval process. Although currently we support only the Wikidata resource, the extension to use plWordNet should not require much additional effort.

The paper is divided into four sections. Section 2 describes our proposal of ontology reuse for the purpose of new concepts definition. We also present an example of how the existing resources, including Wikidata and plWordNet can be reused. In Sect. 3 we discuss the implemented extension to the Ontolis ontology editor and present a preliminary evaluation of this extension. Finally, Sect. 4 contains the concluding remarks and future research perspectives.

2   The Idea of Ontology Reuse

The proposed idea of ontology reuse is primarily aimed at the simplification of the process of new concepts definition. We assume here that the new concept can be defined by means of an ontology, which may become a means of communication and collaborative learning as suggested in [6,7]. Let us then describe the steps necessary to perform the ontology reuse process:

1. A concept to be defined is selected. The concept can be represented by a single word or a compound phrase describing certain information. Let us denote the concept to be defined by c_0. We assume that the concept is given in a certain language, but one may consider the following two extensions which can increase the probability of ontology reuse:

(a) the user provides the concept in a multi-lingual form, i.e. the translations of the concept are provided manually, or

(b) the user provides the concept in a single language, but the translations are automatically searched for. For this purpose a previously prepared dictionary can be used or some online resource may be consulted for possible translations.

The increase of probability of ontology reuse mentioned above follows from the fact, that the concept may not be defined in the original language it has been provided. But it is possible that it has been defined in some other language, thus a chance of ontology reuse still exists.

2. A set of resources to be consulted is defined. This set can contain local or online resources, or a mix of both types of resources. Let us denote the set by $R = \{R_1, R_2, \ldots, R_k\}$, where k is the number of available resources. The resources may be consulted in a sequential manner, according to some predefined order or they may also be accessed simultaneously (in parallel) to make the process faster.

3. Given the concept c_0 and the set R, the resources are being asked for the definition of the concept c_0. The outcome of this step can be three-fold:

(a) none of the resources is able to locate a definition of the searched concept. If this is the case the user has to provide an additional part of the ontology defining the concept. This means that an element c_1 related to the original concept c_0 has to be defined, together with the name of the relation connecting the two elements. Element c_1 becomes then the new concept to be defined and the procedure defined above is repeated again.

(b) only one resource is able to locate a definition of the searched concept. If this is the case, the definition can be reused completely, allowing also for some modifications and extensions if needed. Such a result ends the concept definition process.

(c) mutliple resources are able to locate a definition of the searched concept. If this is the case, the definitions can be either compared with each other, leaving the user the chance to decide on the ontology to be reused, or they can be integrated with each other by means of relations and concepts comparison. Regardless of the approach taken, the reused ontology can be further edited or modified according to the needs. Such a result also ends the concept definition process.

4. The concept definition process ends with either complete or partial reuse of existing resources or, if none of the concept parts can be found in the resources belonging to set R, it ends with the ontology being built manually by the user.

Let us observe, based on the description provided above, that in the best case the concept to be defined can be retrieved in its entirety from the external resources. The worst-case scenario assumes that no part of the concept definition can be reused, but we conjecture that such a case should not be very frequent. We think so, because the concepts constituting the parts of the ontology will probably become more general and more common, thus the chance of finding their definitions will also increase.

Let us also consider the possible resources that may be contained in set R. We claim that both wordnets as well as resources such as Wikidata can be used for the purpose of aiding of ontologies construction. Although both types of resources provide some information on the words or phrases (i.e. concepts) and the relations between them, the type of information differs, for example in terms of naming.

In case of wordnets, the primary relation is a synonymy of words, since wordnet contents are typically organized in the form of synsets (i.e. **synonym sets**). Apart from the synonymy we often consider hyponymy as well as hypernymy, which can be respectively considered as specializations and generalizations. Furthermore, certain words may have similar meaning to some other words. In this case we may say that words can be inexact synonyms, as opposed to exact synonyms, which can be used interchangeably without changing the meaning of the sentence. The drawback of using wordnets for the purpose of ontology reuse is that they do not map directly to the ontological relations, or the mapping is not unambiguous. For instance a hyponym can represent the *subclass of* relation, but in some cases it will also represent the *part of* or *instance of* relations. Nevertheless, the information provided by the hypernymy/hyponymy hierarchy may still be helpful and, what is more it can be at least partially reused. Furthermore, although the synonyms (both exact and inexact) do not have their corresponding relations in the constructed ontology, they can serve the purpose of disambiguating terms, as some words can have multiple meanings and the user has to decide which ontology to reuse.

As for the Wikidata-like resources, the relations can be considered in direct ontological terms, including the aforementioned *subclass of, part of,* and *instance of* relations, as well as some other relations. The synonymy relation, although not represented directly can be also partially seen in the form of *also known as* descriptor available in Wikidata, which contains alternative names for a particular term or concept. What is more, Wikidata provides also some specific relations appearing only in certain contexts. Thus we may assume that the information provided by Wikidata can be much broader than the information provided by wordnets. Whether the additional amount of information provided by Wikidata is considered as an advantage or a drawback depends probably on the particular user's needs.

2.1   An Example

To illustrate the ideas presented above, and in particular to show the different types of information that can be provided by Wikidata and a wordnet, we will consider the concept of **doctor**. In the example we use plWordNet as the wordnet resource, but this should not affect the general conclusions drawn from the example. Following the procedure described above, the concept c_0 is thus defined as $c_0 = $ **doctor** and the set of resources is given as $R = \{$Wikidata, plWordNet$\}$.

The search for the concept in Wikidata shows that the word is considered a synonym of word **physician**. The information is obtained by observing the *also known as* descriptor. Focusing on the most basic relations it can be noticed that

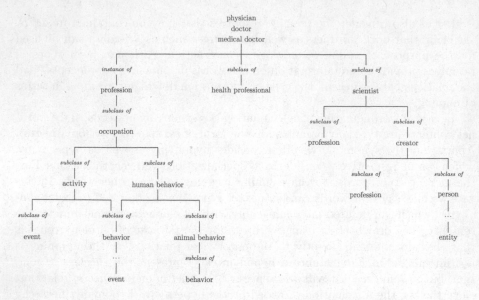

Fig. 1. The partial view of the ontology obtained from Wikidata for the concept **doctor**

a **physician** is an *instance of* **profession**, and also a *subclass of* two classes, namely **health professional** and **scientist**. Following these relations leads to the discovery that **profession** is a *subclass of* **occupation**, which in turn is a *subclass of* **activity** and **human behavior**. Since both the **activity** and the **human behavior** concepts are *subclasses of* **behavior**, and there also exists a path of *subclass of* relations of the form **human behavior** → **animal behavior** → **behavior**, we can conclude that the analysis of the **profession** concept leads to the **behavior** concept appearing on different levels of the hierarchy. Similarly, an **activity** is a *subclass of* an **event**, which also appears in the tree of **behavior** definition. Finally, the analysis of a **health professional** and **scientist** concepts will lead to the concept of a **person**, which, through some intermediate relations leads to the concept of an **entity**. Thus, from the analysis of the ontology, which is also partially shown in Fig. 1, we may conjecture that a **doctor** can be very generally described as an entity with certain behavior.

The search for the **doctor** concept in the plWordNet leads to the following discoveries[1]. The wordnet provides three synonyms, namely *lek.* (**dr.**), *doktor* (**doctor**) and *konsyliarz* (old name for a **doctor**), which can be considered as equivalents of *also known as* descriptor available in Wikidata. Furthermore, there are two hypernym paths, i.e. paths representing a sequence of *subclass of* relations, which "meet" at the level of *człowiek* (**human**) concept. By "meeting" of the hypernym paths we mean that at one point of the hypernymy hierarchy,

---

[1] As plWordNet contains primarily Polish terms, the actual search has been conducted using the Polish term *lekarz* which through the interlingual synonymy relation leads to the **doctor** concept.

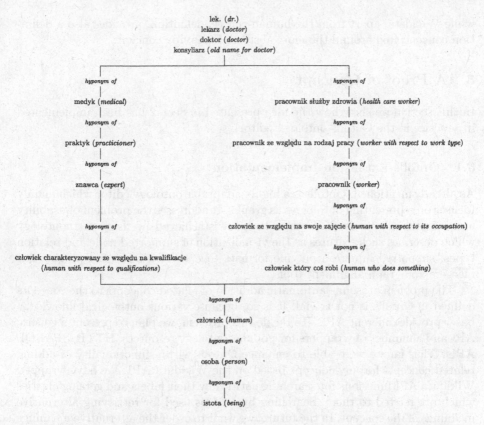

lek. (*dr.*)
lekarz (*doctor*)
doktor (*doctor*)
konsyliarz (*old name for doctor*)

hyponym of

medyk (*medical*)

hyponym of

praktyk (*practicioner*)

hyponym of

znawca (*expert*)

hyponym of

człowiek charakteryzowany ze względu na kwalifikacje
(*human with respect to qualifications*)

hyponym of

pracownik służby zdrowia (*health care worker*)

hyponym of

pracownik ze względu na rodzaj pracy (*worker with respect to work type*)

hyponym of

pracownik (*worker*)

hyponym of

człowiek ze względu na swoje zajęcie (*human with respect to its occupation*)

hyponym of

człowiek który coś robi (*human who does something*)

hyponym of

człowiek (*human*)

hyponym of

osoba (*person*)

hyponym of

istota (*being*)

Fig. 2. The view of the ontology obtained from plWordNet for the concept **doctor**

two different concepts become *subclasses of* a **human**. The complete information on the ontology available in plWordNet is presented in Fig. 2 (the English
translations of respective concepts were added for clarity).

Comparing the results presented in Figs. 1 and 2 as well as their respective
descriptions, the following conclusions can be drawn. Firstly, both resources provide a ready-to-use ontology describing the **doctor** concept. Thus it is possible to
reuse the existing data, without the need for the provision of own definition. Secondly, although we have restricted our attention to only selected relations available in Wikidata, the amount of information available from this resource is much
greater than the data obtained from plWordNet. Thirdly, in case of both resources
we can observe that initially different relations lead to some common concepts
appearing at different points of the hierarchy. In particular, the **behavior** and
event concepts appear in Wikidata ontology a couple of times, while the **human**
concept is the common hypernym for the two paths in the wordnet. Finally, comparing the actual ontologies, it can be observed that the wordnet-based ontology
is more human-centered, i.e. it focuses on the meaning of a **doctor** as a person,

while Wikidata, apart from the human-centric definition, provides also a definition concentrated around the more abstract **behavior** concept.

3   A Proof of Concept

In this section we show how the idea presented in Sect. 2 has been implemented in practice in the Ontolis ontology editor.

3.1   Ontolis Extension Implementation

As already mentioned, Ontolis is a highly adaptable ontology editor which mainly focuses on representing ontologies as graphs. It addresses the problem of usability and adaptability to user's needs. This goal is achieved by using metaontology which describes such features as the visualization of supported node and relation types, supported input/output file formats (e.g. OWL), advanced plugins like mergers and graph alignment tools.

The problem of (semi-)automatic addition of related concepts to the concepts defined in Ontolis is not trivial. It is so, because various ontological knowledge bases provide different APIs. To tackle this problem, we plan to present a generic API and a number of wrappers for popular ontology resources' HTTP/SPARQL APIs. Thus far we were able to enhance Ontolis with a functionality of adding related concepts for any concept based on the Wikidata API. We have wrapped Wikidata API functions for searching entities by their labels and getting entities which are related to them. Searching by label is used for retrieving alternative meanings of the concept. In the future, we want to order the alternative meanings using the context of the concept in an ontology.

The process of data retrieval is as follows. At first, the concept for which the ontology is being built is specified. Then it is searched for in Wikidata and after being found a set of its meanings is presented to the user. Selection of a particular meaning triggers the retrieval of entities related to this particular meaning, as shown in Fig. 3. As mentioned before the retrieved ontology can be also freely edited, taking into consideration the following possible situations:

– The user wants to add a relation to the concept not yet defined in the ontology. If this is the case, we automatically add new concept. Moreover, we also save the information on the authorship of the concept in concept's metadata and the Internationalized Resource Identifier (IRI) of the concept (and relation) in external repository.
– The user wants to add a relation to the concept with an existing external IRI. If this is the case, we add new relation only.
– The user wants to add a relation to the concept with a non-existing IRI, but with a matching label. If this is the case, we let the user decide whether the new concept should be created or the existing one should be used.

The important and interesting aspect of the use of both IRIs and labels is that we can model ontologies in various languages, an example of which is shown

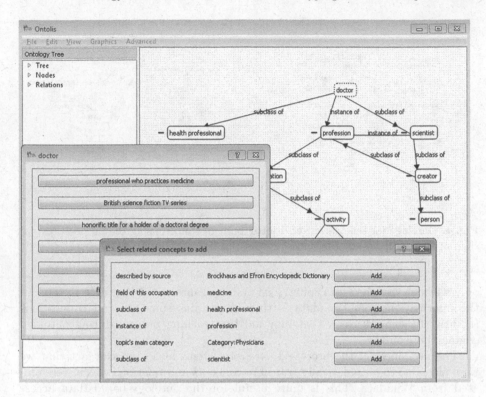

Fig. 3. The user interface for Wikidata's data retrieval in Ontolis

in Fig. 4. Both ontologies present the partial ontology for the concept **doctor**, giving the possibility of finding the English and Russian counterparts describing the various concepts and relations.

To implement the extension we have also used the fact that Ontolis uses a metaontology, in which plugins can be described. We created a new plugin node named #ExternalOntologyResources where we have described how this external plugin could be run. Plugin configuration includes program name, working directory, and general program arguments for the search functions. Concept's label to be searched and chosen identifier are passed as additional arguments. Although currently the interface for external ontological resources plugins is hard-coded into Ontolis, new implementations can be added in the future without source code modification and even without program restart.

3.2 Evaluation

To preliminarily evaluate our approach we have created a new ontology with the root **doctor** and checked whether it is difficult or not to create the ontology presented in Figs. 1 and 4. We found that the whole ontology could be created by successively extending the data with the information from Wikidata, using mouse

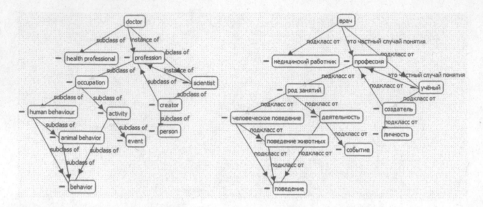

Fig. 4. Ontology for the concept of **doctor** built in Ontolis using the proposed extension, in English (left) and Russian

only without any need for typing. The main advantage of being able to retrieve the information from Wikidata is that it contains some verified connections (at least to some degree), that may aid the teaching process during ontology engineering courses.

The usefulness of the proposed extension stems also from the fact that we were also able to automatically translate concepts and relations which were created from Wikidata. This is quite useful for the ontology-based data access system Reply, described in [7], where user can formulate queries using natural language. Such queries may be used e.g. to manually control the autonomous mobile platforms described in [10, 34]. With automatic translation we were able to add support for new language for basic queries without any source code modification. We have tested only basic queries in Polish, because advanced queries require deep language-specific analysis which is not yet available for Polish.

To evaluate our approach more extensively we decided to compare the Wikidata's coverage of Data Science Ontology (http://www.datascienceontology. com) and bilingual (Russian, English) Ria News Ontology. The version of the Data Science Ontology we used, contains 322 concepts and 329 relations. It is a tree-structured ontology, so the only type of relation is *subclass of*. Furthermore each concept is associated with a single label. The Ria News Ontology in turn, includes 76 concepts and 129 relations, used mainly in news tagging, describing for instance places, main characters, etc.

The experiments were conducted according to the following scheme:

1. Wikidata was queried for concepts with labels taken from the analyzed ontologies.
2. For each found concept, concept's relations were retrieved.
3. The number of concepts related in initial ontologies, which were also connected in Wikidata, was found.

In the course of the experiment we have noticed, that typically the concepts with corresponding meanings can be found in the first two results of Wikidata search. Thus, in the analysis performed in steps 2–3 we have included at most 2 results per ontology concept. In the sequel we refer to this reduction as a disambiguation step.

The statistics gathered for the three analyzed ontologies (Data Science, Russian Ria News and English Ria News) are shown in Table 1. From the results in Table 1 we can observe that none of the ontologies could be completely covered by Wikidata. However, we have determined that in case of Data Science Ontology, the concepts that were not found corresponded usually to tree leaves denoting algorithm or technology names. Thus, such a behaviour can be justified.

We can also notice that the average number of Wikidata concepts per ontology concept is greater than 1 for all three ontologies, even after the meaning disambiguation step (see $\overline{N_{W/o}}$ and $\overline{N_{W/o}^d}$ statistics). Furthermore, looking at the median number of relations in Wikidata (N_r statistic), we may conclude that all three ontologies could benefit from the reuse of Wikidata, acquiring new relations.

Table 1. Statistical comparison of Data Science and Ria News ontologies with respect to Wikidata. N_o, N_W – number of found ontology and Wikidata concepts, N_W^d – number of Wikidata concepts after meanings disambiguation, $\overline{N_{W/o}}$, $\overline{N_{W/o}^d}$ – average number of Wikidata concepts per ontology concept (before and after disambiguation), N_r – number of relations

Statistic	Data Science Ont.	Ria News Ont. (ru)	Ria News Ont. (en)
N_o	187	51	58
N_W	737	250	318
$\overline{N_{W/o}}$	2.29	3.32	4.22
N_W^d	302	88	104
$\overline{N_{W/o}^d}$	1.61	1.73	1.79
Median of N_r	3	9	8

Apart from collecting the aforementioned statistics we have also investigated the most frequently appearing relations. Regardless of the analyzed ontology, the most common relations were *instance of, subclass of, category's main topic, topic's main category* and *part of*. The only difference was related to the fifth most common relation, which in case of Data Science Ontology was *official website*, while in case of both Ria News Ontologies was *is a list of*. The counts of relation occurrences are shown in Table 2. The table contains the 5 relations appearing as the most frequent in all analyzed ontologies.

Finally, we investigated the correspondence of relations between Wikidata and the analyzed ontologies. At first, we rejected all the relations for which either the domain (the subject) or range (the object) was not found in Wikidata. This way 136, 42 and 77 relations were left in Data Science, Russian Ria

Table 2. The summary of most common Wikidata relations appearing in concepts found for Data Science and Ria News ontologies

Relation	Data Science Ont.	Ria News Ont. (ru)	Ria News Ont. (en)
instance of	213	87	94
subclass of	72	57	68
category's main topic	56	55	61
topic's main category	55	52	59
part of	47	36	38

News and English Ria News ontologies, respectively. However, out of these relations only 13, 3 and 3 relations were found both in Wikidata and in the analyzed ontologies. It can be concluded that although Wikidata provides many new relations, the existing relations are rarely preserved between the analyzed ontologies and Wikidata.

4   Conclusions

The paper discusses the idea of ontology reuse for the purpose of new concepts definition. As a potential sources of information to be reused, we consider Wikidata knowledge base as well as plWordNet semantic dictionary – one of two Polish wordnets. We present an example of ontology reuse based on these two resources. An extension to one of the popular ontology editors, i.e. Ontolis is also being discussed and preliminarily evaluated. The evaluation proves that the proposed idea can be applied in practice.

The conducted experiments have shown that for the concepts that were found in Wikidata and analyzed ontologies, the information provided by Wikidata is typically richer. This observation is true both from the point of view of the number of concepts as well as the number of relations. However, as discussed in Sect. 3.2, the relations existing in Wikidata and in the analyzed ontologies, very rarely share common relations. It may indicate that Wikidata contains more intermediate nodes than Data Science or Ria News ontologies.

In the future we plan to extend the Ontolis system even further, to enable the wordnets data reuse. We will probably begin with plWordNet, although the idea presented in the paper can be easily applied to other wordnets, such as the Princeton Wordnet or other resources participating in the EuroWordNet project. We think that the ability to obtain the word translations automatically should be also considered an interesting approach, which we plan to pursue.

Acknowledgments. The reported study was partially supported by the European Union from the FP7-PEOPLE-2013-IAPP AutoUniMo project Automotive Production Engineering Unified Perspective based on Data Mining Methods and Virtual Factory Model (grant agreement no. 612207) and research work financed from funds for science in years 2016–2017 allocated to an international co-financed project (grant agreement no. 3491/7.PR/15/2016/2). It was also partially supported by the Institute of Informatics research grant no. BKM/507/RAU2/2016, at the Silesian Univeristy of Technology, Poland and the Government of Perm Krai, research project No. C-26/004.08 and by the Foundation of Assistance for Small Innovative Enterprises, Russia.

References

1. Agirre, E., Ansa, O., Hovy, E., Martínez, D.: Enriching very large ontologies using the WWW. In: Proceedings ECAI Workshop on Ontology Learning (2000)
2. Arnold, P., Rahm, E.: Extracting semantic concept relations from Wikipedia. In: Proceedings of the 4th International Conference on Web Intelligence, Mining and Semantics (WIMS 2014), pp. 1–11. ACM, New York (2014)
3. Auer, S., Bizer, C., Kobilarov, G., Lehmann, J., Cyganiak, R., Ives, Z.: DBpedia: a nucleus for a web of open data. In: Aberer, K., et al. (eds.) ASWC/ISWC - 2007. LNCS, vol. 4825, pp. 722–735. Springer, Heidelberg (2007). doi:10.1007/978-3-540-76298-0_52
4. Biemann, C.: Ontology learning from text: a survey of methods. LDV Forum **20**(2), 75–93 (2005)
5. Bisson, G., Nédellec, C., Namero, D.C.: Designing clustering methods for ontology building: the Mo'K workbench. In: Proceedings of the ECAI 2000 Workshop on Ontology Learning (2000)
6. Chuprina, S.: Steps towards bridging the HPC and computational science talent gap based on ontology engineering methods. Procedia Comput. Sci. **51**, 1705–1713 (2015)
7. Chuprina, S., Postanogov, I., Nasraoui, O.: Ontology based data access methods to teach students to transform traditional information systems and simplify decision making process. Procedia Comput. Sci. **80**, 1801–1811 (2016)
8. Chuprina, S., Zinenko, D.V.: Adaptable visual ontology editor Ontolis. Vestnik Perm State Univ. **3**(22), 106–110 (2013). (in Russian)
9. Cimiano, P., Staab, S.: Learning concept hierarchies from text with a guided agglomerative clustering algorithm. In: Proceedings ICML 2005 Workshop on Learning and Extending Lexical Ontologies with Machine Learning Methods (OntoML 2005) (2005)
10. Cupek, R., Ziebinski, A., Fojcik, M.: An ontology model for communicating with an autonomous mobile platform. In: Kozielski, S., Kasprowski, P., Mrozek, D., Małysiak-Mrozek, B., Kostrzewa, D. (eds.) BDAS 2017. CCIS, vol. 716, pp. 480–493. Springer, Cham (2017)
11. De Nicola, A., Missikoff, M., Navigli, R.: A software engineering approach to ontology building. Inf. Syst. **34**, 258–275 (2009)
12. Doran, P., Tamma, V., Iannone, L.: Ontology module extraction for ontology reuse: an ontology engineering perspective. In: Proceedings of the 16th ACM Conference on Information and Knowledge Management (CIKM 2007), pp. 61–70. ACM, New York (2007)
13. Gómez-Pérez, A., Fernández-López, M., Corcho, O.: Ontological Engineering: With Examples from the Areas of Knowledge Management. E-Commerce and the Semantic Web. Springer, London (2004)
14. Gómez-Pérez, A., Manzano-Macho, D.: Deliverable 1.5: a survey of ontology learning methods and techniques (2003)
15. Gonalves, R., Mardi, J., Horridge, M., Musen, M., Nyulas, C., Tu, S., Tudorache, T.: http://protege.stanford.edu/. Accessed 11 June 2016

16. Hearst, M.A.: Automatic acquisition of hyponyms from large text corpora. In: Proceedings of the 14th International Conference on Computational Linguistics (COLING 1992), vol. 2, pp. 539–545 (1992)

17. Herbelot, A., Copestake, A.: Acquiring ontological relationships from Wikipedia using RMRS. In: Proceedings of ISWC 2006 Workshop on Web Content Mining with Human Language Technologies, pp. 1–10 (2006)

18. Horák, A., Pala, K., Rambousek, A., Povolný, M.: DEBVisDic - first version of new client-server wordnet browsing and editing tool. In: Proceedings of the Third International WordNet Conference (GWC 2006), pp. 325–328 (2006)

19. Jastrząb, T., Kwiatkowski, G., Sadowski, P.: Mapping of selected synsets to semantic features. In: Kozielski, S., Mrozek, D., Kasprowski, P., Małysiak-Mrozek, B., Kostrzewa, D. (eds.) BDAS 2016. CCIS, vol. 613, pp. 357–367. Springer, Cham (2016). doi:10.1007/978-3-319-34099-9_28

20. Kostareva, T., Chuprina, S., Nam, A.: Using ontology-driven methods to develop frameworks for tackling NLP problems. In: Supplementary Proceedings of the 5th International Data Science Conference: Analysis of Images, Social Networks and Texts (AIST 2016), CEUR-WS, vol. 1710 (2016)

21. Kozielski, M., Gruca, A.: Visual comparison of clustering gene ontology with different similarity measures. Studia Informatica $32(2A(96))$, 169–180 (2011). Presented at BDAS 2011

22. Kozielski, M., Gruca, A.: Application of binary similarity measures to analysis of genes represented in gene ontology domain. Studia Informatica $33(2A(105))$, 543–554 (2012). Presented at BDAS 2012

23. Maziarz, M., Piasecki, M., Rudnicka, E., Szpakowicz, S., Kędzia, P.: plWordNet 3.0 - a comprehensive lexical-semantic resource. In: Proceedings of COLING 2016, The 26th International Conference on Computational Linguistics: Technical Papers, pp. 2259–2268 (2016)

24. Miller, G.A.: Nouns in wordnet: a lexical inheritance system. Int. J. Lexicography $3(4)$, 245–264 (1990)

25. Miller, G.A., Beckwith, R., Fellbaum, C., Gross, D., Miller, K.: Introduction to wordnet: an on-line lexical database. Int. J. Lexicography $3(4)$, 235–244 (1990)

26. Navigli, R., Velardi, P.: Learning domain ontologies from document warehouses and dedicated web sites. Comput. Linguist. $30(2)$, 151–179 (2004)

27. Pease, A., Niles, I.: IEEE standard upper ontology: a progress report. Knowl. Eng. Rev. Spec. Issue Ontol. Agents $17(1)$, 65–70 (2002)

28. Piasecki, M., Szpakowicz, S., Broda, B.: Toward plWordNet 2.0. In: Bhattacharyya, P., Fellbaum, C., Vossen, P. (eds.) Proceedings of the 5th Global Wordnet Conference on Principles, Construction and Application of Multilingual Wordnets, pp. 263–270. Narosa Publishing House (2010)

29. Ruiz-Cascado, M., Alfonseca, E., Castells, P.: Automatic extraction of semantic relationships for WordNet by means of pattern learning from Wikipedia. In: Montoyo, A., Muñoz, R., Métais, E. (eds.) NLDB 2005. LNCS, vol. 3513, pp. 67–79. Springer, Heidelberg (2005). doi:10.1007/11428817_7

30. Ruiz-Cascado, M., Alfonseca, E., Castells, P.: Automatising the learning of lexical patterns: an application to the enrichment of WordNet by extracting semantic relationships from Wikipedia. Data Knowl. Eng. $61(3)$, 484–499 (2007)

31. Sowa, J.F.: Ontology (2010). http://www.jfsowa.com/ontology/. Accessed 11 June 2016

32. Vrandečić, D., Krötsch, M.: Wikidata: a free collaborative knowledgebase. Commun. ACM $57(10)$, 78–85 (2014)

33. Xiang, Z., Courtot, M., Brinkman, R., Ruttenberg, A., He, Y.: OntoFox: web-based support for ontology reuse. BMC Res. Notes **3**(175), 1–12 (2010)
34. Ziebinski, A., Cupek, R., Erdogan, H., Waechter, S.: A survey of ADAS technologies for the future perspective of sensor fusion. In: Nguyen, N.-T., Manolopoulos, Y., Iliadis, L., Trawiński, B. (eds.) ICCCI 2016. LNCS (LNAI), vol. 9876, pp. 135–146. Springer, Cham (2016). doi:10.1007/978-3-319-45246-3_13

Reading Comprehension of Natural Language Instructions by Robots

Irena Markievicz[1(✉)], Minija Tamosiunaite[1,2], Daiva Vitkute-Adzgauskiene[1],
Jurgita Kapociute-Dzikiene[1], Rita Valteryte[1], and Tomas Krilavicius[1]

[1] Faculty of Informatics, Vytautas Magnus University, Kaunas, Lithuania
irena.markievicz@gmail.com
[2] Bernstein Center for Computational Neuroscience,
University of Goettingen, Goettingen, Germany

Abstract. We address the problem of robots executing instructions written for humans. The goal is to simplify and speed-up the process of robot adaptation to certain tasks, which are described in human language. We propose an approach, where semantic roles are attached to the components of instructions which lead to robotic execution. However, extraction of such roles from the sentence is not trivial due to the prevalent non determinism of human language. We propose algorithms for extracting actions and object names with roles and explain, how it leads to the robotic execution via attached sub-symbolic information of previous execution examples for rotor assembly and bio(technology) laboratory scenarios. The precision for the main action extraction is 0.977, for the main, primary and secondary objects is 0.828, 0.943 and 0.954, respectively.

Keywords: NLP · Robotics · Ontology

1 Introduction

"Take a cup from the cupboard and prepare some coffee, please" - this or more complicated instructions are easily comprehendible and executable by a human. However robots are still far away from the understanding and executing instructions written for humans. The first problem is semantic analysis at the world level. Robot needs to perform sentence analysis and to determine the names of the actions which require to be executed and at the same time to indicate all the objects with their roles necessary for these actions. Thus, as we can see from the given compound sentence example, the main action of the first clause is "take", the main object which needs to be taken and placed is "a cup", the object performing location function is "the cupboard". The second problem is grounding: robot needs to know (at the sub-symbolic level!) how to identify a cupboard in the room, how to reach it (possibly how to open it); how to identify a cup, how to grasp and fetch it without hitting any obstacles or toppling down other dishes from the cupboard. The scope of the paper involves tabletop

S. Kozielski et al. (Eds.): BDAS 2017, CCIS 716, pp. 288–301, 2017.
DOI: 10.1007/978-3-319-58274-0_24

manipulation instructions, where we are addressing the first problem in such a way that the solution could make possible to approach the second (grounding) problem, too. For this problem we define an appropriate closed-set of actions that robot can potentially execute and map those actions to the action names that appear in the instruction. We define object roles of each action in a rather general way, based on object touching and un-touching events observed in the video sequence [1]. This allows re-using object roles through groups of actions. Alternative approaches concentrate on more specific applications, e.g. action cores in [10] are defined for very specific actions like neutralization using a highly specific object roles, like "acid" and "base". Tellex et al. [11] analyze driving with landmarks domain handling entities like "event", "object", "place" and "path". Misra et al. [9] are considering a wider set of actions, however they do not pre-structure object roles but acquire language- to-robot-action mapping by means of learning from examples. Yang et al. [12] are using action grammar for manipulation action interpretation at the generality level similar to ours, however they have not attempted to interpret natural language instructions from this perspective.

2   Transforming HL Instructions to Robotic Instruction

Our approach is based on the definition of instruction semantics in such a way that it directly matches to the observable entities of robotic execution (Fig. 1). First, we define robotic action at a fixed granularity in temporal domain: the action starts by a robot manipulator approaching an object for a grasp and ends when the manipulator has released the object and retracted. E.g. "Pick a test tube, shake it and put it back on the table" would be an instruction corresponding to a full robotic action by our definition. This, however, raises a problem of correct identification of the "main" action word, where in the example it is "shake", while "pick" and "put back" we call "supportive" action words. The main action word is always present in the instruction, while supportive action words may be either given or omitted. Identification of the "main" action in the instruction is important for correct retrieval of the sub-symbolic level action grounding information described further.

Second, we define the roles of objects in the action based on touching or un-touching relations of objects in the scene which can be identified by vision or force sensing in the robotic setup. The following object roles are identified: - Main object: the object that is first touched by a robot manipulator; - Primary object: the object from which the main object first un-touches in the course of a manipulation - Secondary objects: the object with which the main object establishes a touch relation in the course of a manipulation.

In the given example the main object is the "test tube", the primary and the secondary objects coincide and both are the "table". The main object is easy to extract from a sentence as it is linked with the main root action by the object relation (e.g. "rotor cap" is main object of "pick up" action in Fig. 3).

To avoid some dependency parsing errors, which may be crucial in the further system compilation steps (when linking with the ontology information, creating sequences of action categories, etc.), dependency parsing is complemented with the following capabilities: complex object mapping to pseudo simple and anaphora resolution problem solving.

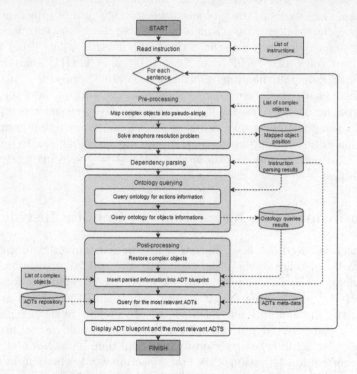

Fig. 1. General workflow on instruction parsing

In order to treat complex objects (e.g. "rotor cap", "robot platform") as indivisible units, they are replaced with the appropriate pseudo simple objects (e.g. "rotor cap" = "cap", "robot platform" = "platform"). This replacement protects sentences from redundant and often erroneous dependency relations.

POS analysis, as well as dependency parsing is used for the anaphora resolution problem solving. If the part-of-speech of a word indicated a personal, possessive, interrogative or relative pronouns, then it has to be replaced with the appropriate noun or noun phrase. The noun or noun phrase is determined by searching back in the sentence for the first dependent with the dependency label, indicating direct/indirect/of preposition object or (passive) nominal/clausal subject. E.g. instruction "take a rotor cap from conveyor and place it on the fixture" is replaced with "take a rotor cap from conveyor and place rotor cap on the fixture" (Fig. 4).

The primary and secondary objects can either be determined from the same sentence (e.g. "conveyor" in Fig. 3; "conveyor", "fixture" in Fig. 4) or by more specific approaches present further in this chapter.

Our supportive data structure is WordNet type domain-specific action and object ontology, where in addition to WordNet hierarchical structures (synonyms, hypernyms, troponyms, hyponyms, holonyms, meronyms) (Fig. 2), actions are connected to objects with which they can be performed, including indication of the object roles (main, primary, secondary). Actions are given features "Can be main action" (yes/no), "Can be supportive action" (yes/no), used for identifying the main action word in the instruction sentence.

For action grounding, sub-symbolic forms called Action Data Tables (ADTs) are attached. The ADTs contain object and manipulator relations, trajectories and poses that were used for an instruction with given action and object names previously and allow action re-execution. We will not expand more about ADT structure as here we are emphasizing symbolic analysis allowing to introduce the sub-symbolic link through ADTs.

The details on ontology building, defining the main action in the sentence as well as defining object roles are provided in the sub-sections next.

2.1 Action Ontology

The conceptual structure of the ACAT ontology is presented in Fig. 2. Two main ACAT ontology classes (ACTION and OBJECT) determine the hierarchical structure of action hypernyms/troponyms and object hypernyms/hyponyms (property subClassOf). Each action and object synset contains a subset of synonymous instances (property instanceOf) having the same definition as their parent class. The subclasses of ACTION class are described by the following properties: main_action, robotic_action, supportive_action.

The subclasses of OBJECT class are defined by the property part_of to describe object holonym and meronym relations. Also, each action and object can be described by the values of annotation properties from WordNet and corresponding instruction sheets (label, gloss and example). Relation between an action and an object is determined by the following restriction properties: with_tool, with_main_object, with_primary_object and with_secondary_object.

Action ontology is formed by assigning an appropriate action or object synset for each action and, also, all action details required for action execution. Here, a synset is a set of synonyms, which groups semantically equivalent data elements. An action synset contains verbs, prepositional verbs, phrasal verbs and other multiword verbs, having the same sense. E.g., the following verbs can be marked as synonyms: "put in = put into", "put out = put away". An object synset contains nouns or multiword expressions (usually consisting of adjectives and nouns) having the same sense. For example, the synset of the "conveyor" object consists of the following members: "conveyor", "conveyor belt", "transporter". Such an approach allows the textual instruction compiler to handle situations of synonym object names used in the instruction text (Fig. 1, ontology querying step in general instruction parsing workflow).

The relations between each ontology concept are identified depending on ontology knowledge source. E.g., when the source of ontology objects is the relational database, concept relation types can be described with the table links and additional semantic meta-model [5] or with Fuzzy-Syllogistic reasoning [6]. In our approach, when we are building ontology from texts, objects in action environment synsets are grouped by their semantic roles, e.g. main, primary, and secondary objects. This is implemented by inserting corresponding object properties linking actions and objects in the ACAT ontology. These properties include: "with main object", "with primary object", "with secondary object" and "with tool". Such a linking approach allows to establish different types of links between actions and objects - e.g., object "conveyor" can have the "primary object" or the "secondary object" role, depending on the instruction sentence.

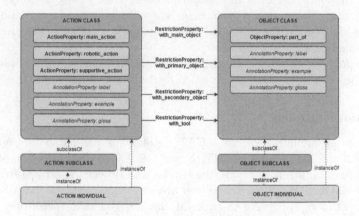

Fig. 2. The structure of the action ontology

Complex objects for the ACAT ontology are identified by extracting possible collocations using corresponding textual search patterns and calculating logDice statistical coefficient for each candidate [3,8]. Two types of collocations are used: "adjective + noun" and "noun + noun" (collective noun) - both are identified by using corresponding patterns for matching node configurations the semantic graph structure of the sentence. A collective noun always identifies a complex object, but "adjective + noun" collocation often describes the object and its feature. E.g. "black bottle" should be interpreted as the object "bottle" with the color feature "black". Calculating the logDice statistical coefficient helps in classifying the collocation candidates to complex objects and object-feature combinations. logDice coefficient is based on comparing the frequency of two or more words co-occurrence in text to the frequency of these words occurring separately. It gives better results with the high number word occurencies in the text. In this case, first we are preparing the list of possible complex objects by extracting them from a corpus.

Table 1 presents the most common collocations from the focused biotechnology corpus. Extracted glossary of biotechnology environment elements contains not just domain terms (e.g. "periodic table"), but also named entities, such as chemical elements (e.g. "carbon dioxide", etc.), measurement data (e.g. "room temperature") and names of tools (e.g. "water bath").

Table 1. Most common collocations in the biotechnology focused corpus

Collocation	logDice	Freq.	Collocation	logDice	Freq.
Reductive amination	13,740	287	Aqueous layer	11,851	290
Baking soda	13,709	206	Science fair project	11,79	213
Science fair	13,319	459	Diethyl ether	11,784	232
Carbon dioxide	13,098	361	Reflux condenser	11,715	188
Essential oil	12,891	296	Acetic acid	11,696	426
Periodic table	12,838	244	Hydrochloric acid	11,579	359
Copper sulfate	12,798	220	Organic layer	11,482	202
Hydrogen peroxide	12,705	270	Small amount	11,404	207
Methylene chloride	12,639	487	Reaction mixture	11,337	787
Sodium hydroxide	12,474	661	Sassafras oil	11,222	284
Reduced pressure	12,371	239	Chemical abstracts	11,085	205
Room temperature	12,359	551	Sodium borohydride	10,872	196
Alkali metal	12,285	200	Formic acid	10,783	202
Ammonium chloride	12,018	309	Sodium acetate	10,766	213
Sulfuric acid	11,856	504	Sodium chloride	10,664	219

By comparing each collocation candidate from the instruction against the collocations from the ontology, we can identify complex objects in the instruction sheet.

In case of a complex object, ontology queries return not only the hierarchical structure of the object, but also their relations to each other. If the ontology contains a complex object, it is often classified as a holonym/meronym of the other object from the ontology. E.g., class "rotor" and its hierarchical structure is described by the relation part_of: "rotor axle" and "rotor core" are parts of a "rotor shaft", and "rotor shaft" together with "rotor cap" creates whole "rotor".

Also, ontology information about the physical state of an object (e.g., liquid or solid) allows to identify any additional tool, which is needed to grasp object by the robot platform. E.g., the object "rotor cap" is identified as a solid one, so it can be grasped by a gripper, but the object "essential oil" is described as a liquid one, so it must be provided in some container.

The main source of the ontology data are the texts from transcribed videos of related experiments: 14 videos on rotor assembly and 11 videos on DNA extraction. We hand label small fragments of the targeted instruction sheets then propagate on WordNet type hierarchical structure. Manual text labeling gave the 110 actions and 98 related objects. Propagated with WordNet data, the ontology contains 426 action and object classes and 692 individuals (the instances of synset classes).

2.2 Main Action Identification

Dependency parser is the main tool for identifying actions due to the structure of dependency grammars (DG), where the verb is the structural center of the clause and all other syntactic units are connect by directed links (directly or indirectly) to it. The part-of-speech tags and the dependency relations are determined with the Stanford NLP parser [4]. E.g., for a sentence "Pick up rotor cap from conveyor" parses into the structure depicted in Fig. 3. It is easy to see, that the action is identified.

Fig. 3. Dependency tree for the single instruction

In the more complex instructions the correct identification of the main action word raises a problem. E.g. instruction "take a rotor cap from conveyor and place it on fixture" contains two action verbs ("take" and "place") - there is no possibility, to identify, which of them is "main" action.

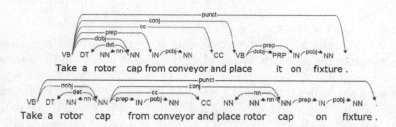

Fig. 4. Dependency trees for complex instruction sentence with anaphora resolution problem solving

For action robotic roles recognition the constituency analysis is used - identified verb phrases indicate possible main actions. Then the action ontology is queried to extract action structure for known actions. With SPARQL queries we define the action synsets for each action and their possible robotic roles. E.g. querying ontology with the "take" and "place" verbs results with the sets of action synonyms: "choose, pick up, take" and "place, pose, put, identify" (Fig. 5). Both of parsed actions can be executed by robots - property "*robotic_action* - 1".

In the case of several verbs ontology allows determining the action, i.e. "pick-and-place in this" case.

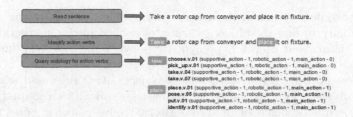

Fig. 5. Main action identification with action ontology queries information

2.3 Object Role Identification

Knowing the structure of the action background, the next step of instruction sheet knowledge processing is to fill-in information for each action background object.

The same dependency tree (e.g., Figs. 2 and 3), show how dobj relation directs to the main object ("rotor cap" in this case), and then primary and secondary objects are identified by prep/pobj relations, i.e. conveyor and fixture, respectively. All supporting objects information is gained from the sentence substitution, expansions and pro forms (e.g. "put rotor cap [on the conveyor/on the top of the conveyor/there]").

Additional action object properties, which are not mentioned explicitly in the instruction sheets, can be obtained by querying the action ontology using SPARQL queries (Fig. 6).

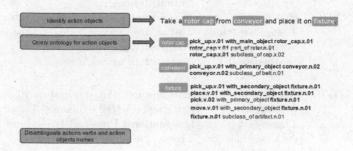

Fig. 6. Object roles identification with action ontology queries information

The action ontology allows more detailed action environment descriptions by establishing links between an ontology data collection consisting of an action and a set of environment objects on one side and a so-called ADT table on the other side. E.g. the instruction "Take a rotor cap and place it on fixture" contains two action verbs: "take" and "place". By querying the ontology, the system gains knowledge, which of these two verbs takes the "main action" role. Also, parsed instruction does not include information about primary object for the action (the position of main object) - ontology data allows to resolve this issue and gives two possible positions of the main object: "table" and "conveyor". If there is no possibility to find the object, which was defined in the instruction sheet, additional data about object hypernyms/hyponyms and holonyms/meronyms can be used.

When categorizing action environment objects using rules and search patterns, one source of information is the VerbNet lexicon with structured description of the syntactic behavior of verbs. Alternatively, syntactic parse trees for instruction sentences are used. When applying automated extraction of rules from VerbNet lexicon database, mapping of VerbNet thematic roles to the elements of the predefined conceptual action background model is done. Rules are extracted from VerbNet syntactic and semantic frames for corresponding verbs.

E.g. with "place" verb (Table 2), we obtain two possible search patterns, which are then used in action environment classification: NP V NP PP. DESTINATION, NP V NP ADVP. Defined patterns by their semantics means, that an Agent places a Theme at a Destination - the Theme is under the control of the Agent/Cause at the time of its arrival at the Destination. These patterns are then applied to morphologically annotated domain specific corpus for filling the action ontology with classified action environment elements and the definition of action main (semantic role Theme) and secondary (semantic role Destination) objects.

Table 2. VerbNet syntactic and semantic frames for verb "place" (Source: VerbNet)

Description	Syntax	Semantics	Example
NP V NP PP.DESTINATION	Agent-NP (putter) V Theme-NP (thing put) [+*loc*] Destination-PP (where put)	motion(during(E), Theme) not(Prep(start(E), Theme, Destination)) Prep(end(E), Theme, Destination) cause(Agent, E)	Place the rotor cap on the fixture
NP V NP ADVP	Agent-NP (putter) V Theme-NP (thing put) Destination (where put) [+*advloc*]	motion(during(E), Theme) not(Prep(start(E), Theme, Destination)) Prep(end(E), Theme, Destination) cause(Agent, E)	Place the rotor cap here

Then, using predefined rules corresponding ADTs are selected from the knowledge base. We do not discuss the rules in the detail in this paper, however the basic idea is that actions should coincide (be a part of a synset), and then, the main, primary and secondary objects (their synsets) are less important, correspondingly.

3   Experimental Evaluation

In our experiments we parse 87 instructions of DNA extraction and rotor assembly. These texts were collected by video transcription. All used sentences are in imperative mode and contains maximum two verb phrases. The precision for the main action extraction for both scenarios is 0.977, for the main, primary and secondary objects is 0.828, 0.943 and 0.954 respectively.

As our dataset was small, we could not allow ourselves having both training and testing datasets. Thus we were testing our system without action-object links, described in the section "Action Ontology". Thus, we were, eventually, testing a heuristic rule set, independent of any training data. This way we could use all the available data as the test set.

Table 3 includes sentences from robot assembly and DNA extraction scenarios instructions sheets (column "Instruction" - e.g. "Take a rotor cap from the table and place it on a fixture"). Next columns contains information parsed from an instruction sentence: main action, main object, primary object and secondary object (e.g. main action: put, main object: ring, primary object: table, secondary object: fixture).

Analysis of the instruction completion experiment, pointing to inaccuracies, results with the following classes of compiler errors.

- Too general, non-robotic action - some of symbolic level actions are too general and describe whole actions process, but not concrete sequence of actions steps. E.g. action "centrifuge the suspension of e-coli" can be described as sequence of action: open the cover of centrifuge, pick up bottle of e-coli and put into centrifuge, close the cover of centrifuge, press the start button.
- Wrong complex object recognition - object features like "top", "bottom", "side", "surface" mistakenly are taken instead of the object. Some of the objects from the instructions (especially biotechnology texts) are even more complex and are described by the more features or identifies things, that cannot be grasped by the robotic system. E.g., "the suspension of e-coli from the flask PB1" is interpreted by the compiler as main object "suspension". Robotic platform cannot grasp the liquid without using any container, so the information about the "flask PB1" is more useful. Complex object features are interpreted as primary object.
- Not clear whole action background context - compiler could interpret only single instruction sentence it is not able to identify the correct possible position of main object: primary and secondary object information in all cases. E.g., the instruction "Place the bottle on the table. Pick up the bottle and mix it gently." is interpreted as two independent actions, so the position of main object from the second action is not clear.
- Not relevant instruction parsing - compiler is able to recognize sentence semantics, but could not evaluate whether the parsed instruction is relevant to the domain-specific scenario. E.g. instruction "Take me out of here." is parsed correctly, but could not be applied to the rotor assembly or bio(technology) laboratory scenarios.

Table 3. The excerpt of the instruction parsing result table from robot assembly and DNA extraction scenarios

Instruction	Main action	Main object	Primary object	Secondary object
Take a rotor cap from conveyor and place it on fixture	place	rotor cap	conveyor	fixture
Insert a shaft into a hole in a magnet	insert	rotor shaft	-	magnet hole
Take the rotor shaft from the fixture and insert it into the cylindrical holder	insert	rotor shaft	Fixture	cylindrical holder
Pick a magnet from the magnet dispenser and insert it into the magnet hole that is directly in front of the robot	insert	magnet	magnet dispenser	magnet hole
Pick a rotor cap from the fixture on the robot platform	pick	rotor cap	fixture	robot platform (wrong rec.)
Put the cap on the rotor shaft	put	rotor cap	-	rotor shaft
Open the centrifuge	open	centrifuge	-	-
Pour the suspension of e-coli from a flask into a bottle	pour	suspension (not complex rec.)	flask	bottle
Turn the lid to cover the bottle PB1	cover	bottle pb1	-	-
Close the lid of centrifuge	close	lid	-	-
Start centrifuge by pressing a button But1	pressing	button but1	centrifuge	-
Open the lid of the centrifuge	open	lid	-	-
Take out the bottle PB1 from centrifuge and put it on fixture	put	bottle pb1	centrifuge	fixture
Open the plastic bottle PB1	open	bottle pb1	-	-
Pour the liquid from the bottle into a flask	pour	liquid	bottle	flask
Close the bottle PB1 again	close	bottle pb1	-	-
Open the bottle with lysis buffer PB2	open	bottle	-	-
Close the bottle PB1 by turning the lid	turning	lid	-	-
Put the bottle PB1 into centrifuge	put	bottle pb1	-	centrifuge

To correct those error sources bigger dataset and more elaborate ontology structure is required. Thus we state attribute correction of those errors to future work.

4   Discussion and Future Work

The proposed technique including parsing and action ontology querying is able to define the actions and syntactic roles for action background elements required in robotic execution.

The designed instruction completion schema and the corresponding instruction compilation application are in principle able to parse instruction sheet information and define the actions and syntactic roles for action background elements. The approach of combining parsed instruction sheet information with ontology query results seems to give adequate information for instruction completion.

In our work we are using the Stanford parser - i.e. statistical dependency parser, which is reported as one of the fastest (1000 sentences per second) and the most accurate (92.2% unlabelled attachment score) currently known approaches for parsing English sentences [2]. Due to the proposed novel technique (based on learning neural network classifier for use in greedy, transition-based dependency parser) it outperforms baselines of arc-eager and arc-standard parsers and achieves 2% improvement on both labeled and unlabeled attachment scores, while running about 20 times faster. Despite all these advantages and superiority over the other dependency parsing techniques, the Stanford parser still has shortcomings that emerge mostly because it is not adjusted to any specific domain or solving task. The Stanford parser is trained on the English Penn Treebank [7], composed of 40,000 sentences, taken from the Wall Street Journal. This newspaper domain is indeed very different from the robotics and chemistry domains that we are dealing in ACAT project. However, major accuracy problems (in both, indicating part-of-speech tags and dependency labels) cause the sentence structure itself: i.e. we need to process sentences written in imperative mood, but the Stanford parser is trained on the sentences in indicative.

There is always a possibility, that POS and dependency parser will return incorrect annotation results due to disambiguities in the sentence. This can happened, when instructions are very domain specific (e.g., chemical or mechanical texts) or have unusual grammatical and syntactic sentence structure (e.g., imperative mood used for verbs). E.g. the instruction "Open centrifuge" is annotated as Open-NNP centrifuge-VBP (POS annotation) and nsubj(centrifuge-2, Open-1) (dependency parsing results). In this case, centrifuge is interpreted as verb form - the interpretation of sentence can be similar to Object "Open" is centrifuged. Wrong parser results affect the overall quality of the compiler. In order to reduce the probability of parser errors, an additional dictionary in the form of an XML file with "always true" statements can be used (e.g., ¡statement¿centrifuge-NN¡/statement¿ means, that centrifuge should always be interpreted as a noun, as centrifuge always takes object role).

The results of instruction completion using ontology queries depend, in addition to size, on the focus of the ontology and the domain-specific corpus focus w.r.t. the specific tasks. Future work on instruction completion should include the improvement of algorithm precision and accuracy by expanding the ontology from additional domain-specific corpus data, and also by including additional weight information, obtained from the feedback of ADT validation by the human and execution. The technique can be further improved by extracting more details on action execution (e.g. size, quantity, material, etc.) from instruction sentences.

Knowledge of actions and action background objects can be further improved by querying the action ontology, including more details on action execution parameters (properties, size, quantity, location, etc.).

We are going to implement the function for filling in missing action or object information with the most probable action or object instances - in this case, an SVM classification algorithm can be used.

Information about non-robotic actions is included in the ACAT action ontology - each action class contains a property *robotic_action*, which allows to describe whether the action can be executed by the robot system, or it is too general for execution. By querying the ontology with each action recognized in an instruction sentence, the symbolic information compiler will identify such complex instructions and will return a request to the user to provide a more specified instruction with all process details described.

Complex sentences can include the explanation of too general instruction. E.g., the instruction "Mix the liquid by inverting the bottle 4 times" explains, how to execute general mixing process. These kind of instructions will be recognized using patterns for matching node configurations in one of the text semantic graph structures. We have identified two main action explanation patterns: - with minimum two actions and subordinating conjunction "by" in relation with one gerund (verbal noun) - e.g. "Wash the flask by placing it in the washing machine."; - with a list of minimum two actions, which is introduced by a colon mark - e.g. "Wash the flask: open the washing machine, place the dirty flask in it, close the washing machine and press the START button".

Action background learning can be accomplished by identifying all background components - main object and its pre and post positions: primary object and secondary object, all this being done in each action step. When the compiler will able to parse not only one single sentence, but the whole complex instruction, there will a possibility to fix the action background information from each previous action steps. E.g., the complex two-sentence instruction "Place the bottle from the cupboard on the table. Pick up the bottle and mix it gently." contains two actions, changing the states of the main object "bottle". In the first step, the compiler identifies, that the bottle is moved from the cupboard (primary object), to the table (secondary objects). In the second step, there is no information about the primary position of the main object. However, from the context of the previous steps, the compiler will able to identify the table as the last position of the main object.

Acknowledgments. The research leading to these results has received funding from the European Community's Seventh Framework Programme FP7/2007-2013 (Programme and Theme: ICT-2011.2.1, Cognitive Systems and Robotics) under grant agreement No. 600578, ACAT.

References

1. Aksoy, E.E., Abramov, A., Dörr, J., Ning, K., Dellen, B., Wörgötter, F.: Learning the semantics of object-action relations by observation. The Int. J. Robot. Res. **30**(10), 1229–1249 (2011)
2. Chen, D., Manning, C.D.: A fast and accurate dependency parser using neural networks. In: Proceedings of Empirical Methods in Natural Language Processing (EMNLP), pp. 740–750 (2014)
3. Daudaravičius, V., Marcinkevičienė, R.: Gravity counts for the boundaries of collocations. Int. J. Corpus Linguist. **9**(2), 321–348 (2004)
4. De Marneffe, M.C., Manning, C.D.: Stanford typed dependencies manual. Technical report, Stanford University (2008)
5. El Idrissi, B., Baïna, S., Baïna, K.: Ontology learning from relational database: how to label the relationships between concepts? In: Publishing, S.I. (ed.) International Conference: Beyond Databases, Architectures and Structures, pp. 235–244 (2015)
6. Kumova, B.İ.: Generating ontologies from relational data with fuzzy-syllogistic reasoning. In: Kozielski, S., Mrozek, D., Kasprowski, P., Małysiak-Mrozek, B., Kostrzewa, D. (eds.) BDAS 2015. CCIS, vol. 521, pp. 21–32. Springer, Cham (2015). doi:10.1007/978-3-319-18422-7_2
7. Marcus, M.P., Marcinkiewicz, M.A., Santorini, B.: Building a large annotated corpus of english: the penn treebank. Computat. Linguist. **19**(2), 313–330 (1993)
8. Markievicz, I., Vitkute-Adzgauskiene, D., Tamosiunaite, M.: Semi-supervised learning of action ontology from domain-specific corpora. In: Skersys, T., Butleris, R., Butkiene, R. (eds.) ICIST 2013. CCIS, vol. 403, pp. 173–185. Springer, Heidelberg (2013). doi:10.1007/978-3-642-41947-8_16
9. Misra, D.K., Sung, J., Lee, K., Saxena, A.: Tell me dave: context-sensitive grounding of natural language to manipulation instructions. Int. J. Robot. Res. **35**(1–3), 281–300 (2016)
10. Nyga, D., Beetz, M.: Cloud-based probabilistic knowledge services for instruction interpretation. In: ISRR, Genoa, Italy (2015)
11. Tellex, S., Kollar, T., Dickerson, S., Walter, M.R., Banerjee, A.G., Teller, S., Roy, N.: Approaching the symbol grounding problem with probabilistic graphical models. AI Mag. **32**(4), 64–76 (2011)
12. Yang, Y., Guha, A., Fermuller, C., Aloimonos, Y.: A cognitive system for understanding human manipulation actions. Adv. Cogn. Syst. **3**, 67–86 (2014)

A New Method of XML-Based Wordnets' Data Integration

Daniel Krasnokucki, Grzegorz Kwiatkowski, and Tomasz Jastrząb[✉]

Institute of Informatics, Silesian University of Technology, Gliwice, Poland
{Daniel.Krasnokucki,Grzegorz.Wojciech.Kwiatkowski,
Tomasz.Jastrzab}@polsl.pl

Abstract. In the paper we present a novel method of wordnets' data integration. The proposed method is based on the XML representation of wordnets content. In particular, we focus on the integration of VisDic-based documents representing the data of two Polish wordnets, i.e. plWordNet and Polnet. One of the key features of the method is that it is able to automatically identify and handle the discrepancies existing in the structure of the integrated documents. Apart from the method itself, we briefly discuss a C#-based implementation of the method. Finally, we present some statistical measures related to the data available before and after the integration process. The statistical comparison allows us to determine, among other things, the impact of particular wordnets on the integrated set of data.

Keywords: Natural language processing · Ontologies · Wordnets · XML data integration

1 Introduction

Data integration has been a field of intensive study of computer scientists for many years. It plays the crucial role in various practical applications, including for instance the integration of data from sensors [5,28]. In the paper we deal with data integration problem related to the area of natural language processing (NLP). To be more precise, we consider the integration of resources which can aid the process of natural language processing, namely wordnets. The data contained in wordnets can be used to support the semantic analysis of text, which is the third phase of natural language text processing, following the morphological and syntactic analysis [21]. The semantic knowledge contained in wordnets has been pointed out in [13] as one of the important elements for the task of semantic features recognition and assignment. Furthermore, the semantics has been also mentioned in [2,3] as a useful element of different database-related tasks, such as querying databases in natural language.

Wordnets can be perceived as graphs, connecting words with semantic relations [8], or as terminological fundamental ontologies, as suggested in [4]. The most common relations forming the wordnet's graph of words include synonymy,

© Springer International Publishing AG 2017
S. Kozielski et al. (Eds.): BDAS 2017, CCIS 716, pp. 302–315, 2017.
DOI: 10.1007/978-3-319-58274-0_25

hypernymy and hyponymy. The synonymy relation is particularly crucial, since the sets of synonyms (called synsets) form the most basic building blocks of the wordnet's structure [19,20]. The hypernymy relation can be in turn considered as a generalization of the synsets, while the hyponymy represents specialization of the synsets. The wordnets that we analyze are available in VisDic format, which is an XML representation of wordnet data developed by Pavelek, Horák and others [9]. According to the examples presented in [10] the main requirements related to the XML VisDic format are that:

- each synset is a mini-XML document,
- there is no enclosing (root) tag in the document,
- there should be no attributes in the tags, but they may be rewritten as nested tags.

The contribution of the paper is a novel method of wordnets integration. The integration is based on the XML representation of wordnets. We evaluate the method using the data of two Polish wordnets, i.e. plWordNet (also known as Słowosieć) [18,22] and Polnet [25,26]. The motivation behind the integration of wordnets' data is three-fold:

- to be able to compare and contrast the contents of both wordnets. This will allow us to evaluate the quality of integrated wordnets, but also to verify their correctness and to observe potential discrepancies.
- to create a single resource, which enables easier and faster access to the data available in multiple resources.
- to unify the format according to which the data is stored. Although both resources considered in the paper are available in VisDic format, there are certain differences in their representation. The integration removes these discrepancies.

The paper is divided into five sections. In Sect. 2 we review some of the existing solutions related to data integration. In Sect. 3 we present our proposal and an exemplary run of the integration method. Section 4 contains the description of the C#-based implementation of the method as well as short discussion of the conducted experiments. The final section summarizes the contents of the paper and the most important conclusions.

2   Related Works

Before we present our wordnet integration method, let us first consider some of the already existing solutions. We begin with a brief review of the methods and algorithms that can be used for the integration of ontologies.

Goczyła and Zawadzka [7] discuss an approach in which a global ontology describes the contents of the local ontologies to be integrated. Similarly, Świderski [24], who tackles the integration of geospatial XML data, being also implicitly indicated as a potential challenge for the sensor fusion in [28], discusses an approach which requires the creation of a global conceptual schema, which is later related to the local views. The main weakness of these approaches is that, the global ontologies or schemas only describe the contents of the local ontologies, instead of actually integrating them. In contrary to these approaches, we do not define any additional descriptive ontologies, but we try to perform the integration directly, based on the analysis of the integrated wordnets.

The approach that is similar to our approach was presented in [16]. Magnini and Speranza [16] present a plug-in approach in which a set of basic synsets is built and, based on the integration of these basic synsets the remaining elements are connected. They also discuss a way to disambiguate merged synsets basing on the hypernymy and hyponymy relations. The similarity between our approach and theirs is that we also use the aforementioned relations as a means of disambiguation. However, in case of [16], the integration was related to generic and specialized wordnets, while in our approach we deal with two generic wordnets. The main disadvantage of the approach discussed in [16] is that it requires experts' knowledge to select the set of basic synsets, and to make decisions in the disambiguation steps. The method we propose minimizes the manual labour, since it is able to perform both structural as well as semantic matching automatically. The other disadvantage of the approach proposed by Magnini and Speranza is that it requires definition of certain precedence rules, which can make integration of multiple resources hard. In contrast, our method allows for integration of multiple resources without additional effort.

Amaro and Mendes [1] also deal with matching of technical (specialized) and common (generic) wordnets, although their approach does not seem to take into consideration any structural similarities. On the other hand, Xiang et al. [27] focus only on structural analysis of data representation, using deep neural networks. Contrarily to these two approaches, we conjecture that the structural and semantic analysis can be combined to achieve proper matching. An approach that utilizes WordNet relations to perform ontology matching has been proposed in [14]. However, in their approach WordNet is used as an external resource facilitating the ontology matching process, while in our case, the wordnets are being actually integrated. Besides, their approach takes into account a broader set of relations than our approach, which may affect performance. Finally, the experimental results shown in [14] indicate that the proposed similarity measure can vary significantly, thus the parameters of the method probably need to be tuned depending on the input data sets.

Comparing our method to various ontology matching techniques, described extensively in [6], we can find certain similarities. For instance, the structural analysis of wordnets' data formats described in Sect. 3, applies name-based methods. In particular, we deal with string-based methods, e.g. substring tests. We also use the structure-based techniques, including the internal structure methods related to datatypes and multiplicities of respective elements, as well as relational structure methods, involving the taxonomic structure following from hypernym/hyponym relations. For a broader overview of existing methods and tools the reader is referred to [6,14] and their references.

Let us also describe the similarities and differences between the wordnet data integration method presented in this paper and selected approaches related to database schema integration. For an overview of the available schema integration methods see e.g. [23]. Hossain et al. [11] and Ibrahim et al. [12] point out the importance of both structural and semantic aspects of database schema matching. As described in Sect. 3 our method also involves these two phases, proposing some ways of dealing with both structural as well as semantic inconsistencies. Some authors propose also to use external resources to resolve the problems with mapping of the schema semantics, for examples see [15,17]. Contrarily, we try to explore the information included in the integrated resources only, without using any additional knowledge.

3 Method Description

The proposed method is based on a two-phase approach involving the structural content matching and semantic content matching. Both phases can be performed automatically. The aim of both phases is first, to identify the similarities and differences among the data to be integrated, and second, to define a means of data integration which successfully deals with any differences or discrepancies occurring in the input data. In the sequel we assume that only two documents are being integrated, although the method can be easily applied to more than two documents. Whenever needed, we will refer to these documents as D_1 and D_2.

Let us begin with the description of the structural matching phase. It is based on the analysis of the structure of integrated documents, which to some extent follows the general contract of VisDic format. However, the considered XML documents can also contain certain differences regarding the overall structure of the document elements or nodes. In particular we primarily have to deal with the following problems:

- different element names – this issue can be related to both the top-level (root) nodes of the document as well as their subelements. The problem arising from the unmatched element names is that they cannot be directly mapped to each other without additional knowledge, related for instance to the semantics of the nodes.
- different ways of data presentation – this issue can be related to the use of node attributes as well as subnodes as a means of providing certain information. Given that the same information can be represented in different forms in the integrated documents, a simple direct structural match between the data is not possible. However, as we explain later this problem can be solved quite easily.

Taking into account the problems mentioned above we may describe the structural matching phase as follows.

At first, the XML documents are parsed to identify a unique set of structural elements (i.e. nodes and attributes) that exist. For each element, the name and type are registered, where by type we mean here either *a node* or *an attribute*. As an outcome of this step, we obtain two sets of unique name-type pairs, denoted by T_{D_1} and T_{D_2}.

Next, given the sets T_{D_1} and T_{D_2}, we build the partial structure of the integrated document by matching the elements with equal (or very similar) names. We begin the name-matching approach by considering the elements type-wisely, meaning that we try to separately match nodes and attributes. The matching process can be also aided with the analysis of the node levels, to ensure that the elements are matched properly, and not only because of accidental name similarity. Finally, to tackle the problem of different data representation, we consider the attributes as subnodes of a particular node, and we repeat the name-matching procedure once again, but this time only with respect to the nodes that have not been previously matched.

Let us shortly comment on the complexity of the structural matching phase. Let n_1 and n_2 denote the number of top-level nodes in D_1 and D_2 respectively, and let k_1 and k_2 denote the maximum number of subelements. To get the complete view of documents structure we need to parse each individual node, therefore the worst-case complexity of the first part is given by $O(n_1 k_1 + n_2 k_2)$. Next, assuming that the sizes of sets T_{D_1} and T_{D_2} are given by t_1 and t_2, the complexity of the second step is given by $O(t_1 t_2)$. However, since $t_1 t_2$ will typically be much smaller than either $n_1 k_1$ or $n_2 k_2$ the complexity of this phase may be estimated as $O(n_1 k_1 + n_2 k_2)$.

The semantic matching phase, used if the structural matching phase does not manage to completely resolve the differences in the input documents, proceeds as follows:

1. Both input documents are parsed and the data contained in respective nodes is extracted. In particular, we extract the textual data, that describes the synonym sets. To identify the data we need, i.e. the words representing synonyms, we consider the type of node content. For instance we observe whether the node contains numbers or characters, or whether the content is usually a short phrase, or a sentence.

2. Given that the synonyms are properly identified in the previous step we loop through the synsets related to one of the input documents, say D_1 and for each synonym we search for its equivalent in the other document. If an equivalent is not found then the complete synset is added to the integrated document. Otherwise, we have to deal with the cases of a single match or multiple matches occurring between the documents. To handle these situations, regardless of the cardinality of the match, we always perform a disambiguation of the matched words, according to the following procedure:

   (a) the sets of synonyms, understood as the sets of words within given synset that share the same structural similarity (e.g. have the same enclosing node name), are compared to each other. If the intersection of the two sets produces a set containing at least two elements (the originally found synonym and at least one other common element), we assume that these two synsets should be integrated with each other.

(b) if the previous step fails to disambiguate the matches, we compare the sense identifiers to identify which synsets contain the synonyms with the same sense and we may try to reject synonyms with a different sense. However, as the sense identifiers are most probably not cross-document compliant, it only helps to analyze the data within single document.

(c) if the previous step fails to disambiguate the matches, we perform a comparison of the relations between synsets. In this step we take advantage of the fact that the set of possible relations is limited, hence it is relatively easy to identify the nodes describing relations and to follow these relations. Furthermore, we mainly focus on the hypernymy and hyponymy relations, since we assume that they may provide the best similarity measure between matched synsets. If as a result of the comparison we obtain some non-empty intersections of hypernym and hyponym sets, we decide to integrate the synsets.

3. After the disambiguation step is finished we either add the synsets as separate nodes of the integrated document, or we merge them together by combining the sets of synonyms and the sets of relations. This step requires also to handle the potentially mismatched node identifiers. To deal with this problem, we decided to generate new identifiers for the integrated nodes, but keeping track of the previously assigned identifiers. The storage of previous identifier values facilitates the identification of the sources of elements in the integrated document. Finally, the unmatched nodes can be either copied to the final document, or they can be ignored – the decision may be left to the user performing the integration.

The disambiguation method allows us to deal with polysemy, since we require not only to match a single word but we also follow the existing relations. Words with different meanings will not share the same relations.

As for the complexity of the presented approach it requires to perform $n_1 l_1 + n_2 l_2$ data extraction steps, where l_1, l_2 define the maximum number of synonyms in a particular synset of the respective input documents. Then the disambiguation procedure requires to perform $n_1 n_2 l_1 l_2$ synonym comparisons and presumably a similar number of relation comparisons. Finally, the merging phase's complexity can be estimated as $O(l_1 l_2)$.

3.1 An Example

To illustrate how the proposed method works in practice, let us consider the following example. Let us assume that the synset we want to integrate is related to the word advice (*rada*). An excerpt of synset definition taken from the plWordNet XML document is shown in Fig. 1. The listing contains the information on the synset identifier (ID node), the set of synonyms (SYNONYM and LITERAL elements, combined with SENSE identifiers denoting different meanings), as well as hyponymy and hypernymy relations (ILR elements). The English translations written in italics were added for clarity and are not part of the original XML document.

```
<SYNSET>
  <ID>PLWN-00002845-n</ID>
  <SYNONYM>
    <LITERAL>porada (advice)<SENSE>1</SENSE></LITERAL>
    <LITERAL>rada (advice)<SENSE>2</SENSE></LITERAL>
    <LITERAL>wskazówka (tip)<SENSE>4</SENSE></LITERAL>
    <LITERAL>wskazanie (indication)<SENSE>6</SENSE></LITERAL>
    <LITERAL>podpowiedź (hint)<SENSE>1</SENSE></LITERAL>
    <LITERAL>instrukcja (instruction)<SENSE>1</SENSE></LITERAL>
    <LITERAL>sugestia (suggestion)<SENSE>2</SENSE></LITERAL>
  </SYNONYM>
  <ILR>PLWN-00010900-n<TYPE>hiponimia (hyponymy)</TYPE></ILR>
  <ILR>PLWN-00007128-n<TYPE>hiponimia (hyponymy)</TYPE></ILR>
  <ILR>PLWN-00007164-n<TYPE>hiponimia (hyponymy)</TYPE></ILR>
  <ILR>PLWN-00010812-n<TYPE>hiponimia (hyponymy)</TYPE></ILR>
  <ILR>PLWN-00036794-n<TYPE>hiponimia (hyponymy)</TYPE></ILR>
  <ILR>PLWN-00236322-n<TYPE>hiponimia (hyponymy)</TYPE></ILR>
  <ILR>PLWN-00398457-n<TYPE>hiponimia (hyponymy)</TYPE></ILR>
  <ILR>PLWN-00416547-n<TYPE>hypernym (hypernymy)</TYPE></ILR>
  ...
</SYNSET>
```

Fig. 1. An excerpt from plWordNet XML document describing word rada (*advice*)

The search for the rada (*advice*) word in the Polnet XML document produces three synsets, whose excerpts are presented in Figs. 2, 3 and 4. The XML fragments again contain the information on the synset identifier (ID element), the set of synonyms (SYNONYM, WORD and LITERAL elements) as well as the hypernymy relations descriptors (ILR element). Let us note that the WORD elements appearing within the SYNONYM node are in fact repeated twice, since they are first mentioned consecutively, and then they are also placed next to the corresponding LITERAL element (omitted in the excerpts).

It is easy to observe that the structural matching phase will be able to match most of the elements presented in Fig. 1 with the corresponding elements in Figs. 2, 3 and 4. In particular, the SYNSET, ID, SYNONYM, LITERAL and ILR can be matched according to the name-matching approach described above. Additionally, the SENSE and TYPE nodes can be matched with the sense and type attributes, by applying the second step of structural matching. The only unmatched elements remaining are lnote attribute and WORD element appearing in the Polnet documents.

The semantic matching phase will be used to disambiguate between the three proposed matches for the word rada (*advice*). In the presented example the disambiguation will be finished after the comparison of synonym sets, since the intersection of LITERAL elements shown in Figs. 1 and 4 provides a set consisting of six common synonyms. The size of the intersection suggests that the match is indeed correct, even though the comparison of available relations will certainly not lead to synsets matching.

```
<SYNSET>
  <ID>POL-2141573557</ID>
  <SYNONYM>
    <WORD>rada (council)</WORD>
    <LITERAL lnote="U2" sense="2">rada (council)</LITERAL>
  </SYNONYM>
  <ILR type="hypernym">POL-2141574560</ILR>
  ...
</SYNSET>
```

Fig. 2. An excerpt from Polnet XML document describing word rada (*council*)

```
<SYNSET>
  <ID>POL-2141573558</ID>
  <SYNONYM>
    <WORD>rada (council)</WORD>
    <LITERAL lnote="U3" sense="3">rada (council)</LITERAL>
  </SYNONYM>
  <ILR type="hypernym">POL-2141574214</ILR>
  ...
</SYNSET>
```

Fig. 3. An excerpt from Polnet XML document describing word rada (*council*)

```
<SYNSET>
  <ID>Prz-519335304</ID>
  <SYNONYM>
    <WORD>rada (advice)</WORD>
    <WORD>instrukcja (instruction)</WORD>
    <WORD>podpowiedź (hint)</WORD>
    <WORD>porada (advice)</WORD>
    <WORD>pouczenie (instruction)</WORD>
    <WORD>wskazanie (indication)</WORD>
    <WORD>wskazówka (tip)</WORD>
    <LITERAL lnote="U1" sense="1">rada (advice)</LITERAL>
    <LITERAL lnote="U2" sense="3">instrukcja (instruction)</LITERAL>
    <LITERAL lnote="U2" sense="2">podpowiedź (hint)</LITERAL>
    <LITERAL lnote="U1" sense="1">porada (advice)</LITERAL>
    <LITERAL lnote="U1b" sense="2">pouczenie (instruction)</LITERAL>
    <LITERAL lnote="U1b" sense="2">wskazanie (indication)</LITERAL>
    <LITERAL lnote="U3" sense="3">wskazówka (tip)</LITERAL>
  </SYNONYM>
  <ILR type="hypernym">POL-2141574336</ILR>
  ...
</SYNSET>
```

Fig. 4. An excerpt from Polnet XML document describing word rada (*advice*)

```
<SYNSET>
  <ID plwn="PLWN-00002845-n" pn="Prz-519335304">...</ID>
  <SYNONYM>
    <LITERAL plwnsense="2" pnsense="1">rada (advice)</LITERAL>
    <LITERAL plwnsense="1" pnsense="3">instrukcja (instruction)</LITERAL>
    <LITERAL plwnsense="1" pnsense="2">podpowiedź (hint)</LITERAL>
    <LITERAL plwnsense="1" pnsense="1">porada (advice)</LITERAL>
    <LITERAL pnsense="2">pouczenie (instruction)</LITERAL>
    <LITERAL plwnsense="6" pnsense="2">wskazanie (indication)</LITERAL>
    <LITERAL plwnsense="4" pnsense="3">wskazówka (tip)</LITERAL>
  </SYNONYM>
  <ILR type="hiponimia" source="plwn">PLWN-00010900-n</ILR>
  <ILR type="hiponimia" source="plwn">PLWN-00007128-n</ILR>
  <ILR type="hiponimia" source="plwn">PLWN-00007164-n</ILR>
  <ILR type="hiponimia" source="plwn">PLWN-00010812-n</ILR>
  <ILR type="hiponimia" source="plwn">PLWN-00036794-n</ILR>
  <ILR type="hiponimia" source="plwn">PLWN-00236322-n</ILR>
  <ILR type="hiponimia" source="plwn">PLWN-00398457-n</ILR>
  <ILR type="hypernym" source="plwn">PLWN-00416547-n</ILR>
  <ILR type="hypernym" source="pn">POL-2141574336</ILR>
</SYNSET>
```

Fig. 5. An example of the integrated XML document describing word rada (*advice*)

Since the `lnote` element cannot be disambiguated in any way it may be copied into the integrated document or skipped, depending on the approach towards dealing with the unmatched elements. As for the `WORD` nodes, we may observe that they do not carry any additional information with respect to the `LITERAL` elements so they could possibly be safely omitted. Finally, the integrated synset can take the form presented in Fig. 5. The additional attributes that appear in the document presented in Fig. 5 allow to track the sources of particular pieces of information, such as the origin of the `LITERAL` elements and their senses, as well as the identifiers appearing in the `ID` and `ILR` elements.

4   Experiments

4.1   Implementation

The method was implemented using .Net framework 4.6.1 using the C# 6.0 features. The full implementation with additional features (i.e. searching through the XML files by words or numbers) is available at https://goo.gl/WbdzgY.

The solution consists of two main parts – Data Transfer Object (DTO) and the engine responsible for performing the merging part. The DTO is used for reflecting the structure of the output file. In the current version it contains all the fields and attributes we wanted to include in the output document. However, the structure can be changed according to the needs of the experiment or user, since the implementation has been prepared in easily extendable architecture.

What is more the changes or additions can be made on-the-fly and they have no impact on the efficiency of the program. The engine (enclosed in Engine.cs file) is responsible for matching proper fields and finding correspondences between them. After loading two XML files it iterates through all unique words (in our case it takes the node LITERAL as a reference), searches for synonyms in second file and merges all information about this words – synonyms, relations, meanings and their referenced IDs from both sets.

4.2   Results

As already mentioned, in the experimental evaluation we attempted to integrate the data from plWordNet and Polnet. However, taking into consideration the size of both wordnets, the experiments were conducted for a subset of all the synsets gathered in the XML documents, because the whole process turned out to be very time-consuming (the experiments took around 200 min to complete). The summary of obtained results for two different sizes of the D_2 file (Polnet), is presented in Table 1.

As can be observed from the results presented in Table 1, in both experiments there were some synsets that could not be merged together and had to be simply copied. The level of mismatch was equal to approximately 4% and 12% of the size of the D_2 document, respectively, for the first and second experiment. Let us also note that both the set of literals and the set of relations was enriched in both experiments as a result of the integration process (as compared to D_2).

As a general conclusion we may say that mapping and merging of those two sets can be beneficial in many ways. Starting from the wider range of words available in the output file, through many established relations and connections, ending with one logical and consistent file to be used.

Let us also present an exemplary output of merging the synsets for word *pizzeria*, as shown in Fig. 6. The example is particularly interesting, because it proves the correctness of our method, since:

Table 1. The summary of the input and output data characteristics

	File D_1	File D_2	Output file
Size (kb)	135627	9	34
Number of relations (ILR)	1120482	23	431
Number of words (LITERAL)	307533	48	91
Unmatched words	2	0	-
Size (kb)	135627	1393	5656
Number of relations (ILR)	1120482	5090	88552
Number of words (LITERAL)	307533	1716	2984
Unmatched words	203	0	-

```
<SYNSET>
  <ID>ID_4
    <P1WN>PLWN-00004853-n</P1WN>
    <PN>POL-1060151321</PN>
  </ID>
  <SYNONYM>
    <LITERAL pnsense="2">lokal (place)</LITERAL>
    <LITERAL plwnsense="1" pnsense="1">zakład gastronomiczny
                          (cateringfacility)</LITERAL>
  </SYNONYM>
</SYNSET>
...
<SYNSET>
  <ID>ID_10
    <P1WN>PLWN-00004853-n</P1WN>
    <P1WN>PLWN-00306652-n_pwn</P1WN>
    <PN>1163</PN>
  </ID>
  <SYNONYM>
    <LITERAL plwnsense="1" pnsense="1">pizzeria</LITERAL>
    <LITERAL plwnsense="1">pizza shop</LITERAL>
    <LITERAL plwnsense="1">pizza parlor</LITERAL>
  </SYNONYM>
  <ILR type="hypernym" source="pnid" merged="ID_4">POL-1060151321</ILR>
  <ILR type="category_domain" source="pnid">1356</ILR>
  <ILR type="hypernym" source="plwnid" merged="ID_4">PLWN-00004853-n</ILR>
  <ILR type="Hypernym" source="plwnid">PLWN-00308114-n_pwn</ILR>
</SYNSET>
```

Fig. 6. An exemplary output for the integration of word *pizzeria*

– the synonyms shown in Fig. 6 result from the integration of plWordNet and Polnet data, but also from the interlingual synonymy within the plWordNet itself (thus we have multiple P1WN nodes within the ID node),
– the merged attribute in ILR nodes corresponds to the new ID of a particular synset, which was also previously included in the output. Let us observe that the same value of merged attribute can appear in many ILR nodes, which results from the relation integration between plWordNet and Polnet. In case of Fig. 6 it results from hypernymy relations pointing to the merged synset with identifier ID_4 that has been also shown in Fig. 6.

5   Conclusions

The paper discusses a new method of integration of XML-based wordnets' data. The method and its practical implementation has been experimentally evaluated on the basis of two Polish wordnets. As a result we were able to obtain a coherent, easy-readable and useful resource providing the information contained in both input resources. The corresponding original identifiers were always added to particular output nodes, to have transparent and easy access to the original source of the literals.

The experiments have shown, that the biggest problem in merging the two resources, being very similar and very different at the same time, was the size of data we had to work with. The number of elements that can be corresponding or related to each other, together with lexical ambiguity, gives us a challenging problem of creating the proper output. Although the proposed method was successful in performing the data integration process, we think that there is no common solution, which can be always used. Therefore, we consider our method as a data preprocessing step for the course of further experiments.

In the future we plan to parametrize the prepared software to have the ability of easy configuration of the output format. Furthermore, we reckon that there exists a possibility to combine wordnets based on different languages and to create a smart dictionary, covering the whole grammatical and semantic complexity.

Acknowledgments. The reported study was partially supported by the European Union from the FP7-PEOPLE-2013-IAPP AutoUniMo project Automotive Production Engineering Unified Perspective based on Data Mining Methods and Virtual Factory Model (grant agreement no. 612207) and research work financed from funds for science in years 2016–2017 allocated to an international co-financed project (grant agreement no: 3491/7.PR/15/2016/2). It was also partially supported by Institute of Informatics research grant no. BKM/507/RAU2/2016.

References

1. Amaro, R., Mendes, S.: Towards merging common and technical lexicon wordnets. In: Proceedings of the 3rd Workshop on Cognitive Aspects of the Lexicon (CogALex-III): 24th International Conference on Computational Linguistics, COLING 2012, pp. 147–160 (2012)
2. Arfaoui, N., Akaichi, J.: Automating schema integration technique case study: generating data warehouse schema from data mart schemas. In: Kozielski, S., Mrozek, D., Kasprowski, P., Małysiak-Mrozek, B., Kostrzewa, D. (eds.) BDAS 2015. CCIS, vol. 521, pp. 200–209. Springer, Cham (2015). doi:10.1007/978-3-319-18422-7_18
3. Bach, M., Kozielski, S., Świderski, M.: Zastosowanie ontologii do opisu semantyki relacyjnej bazy danych na potrzeby analizy zapytań w języku naturalnym. Studia Informatica **30**(2A(83)), 187–199 (2009). Presented at BDAS 2009
4. Biemann, C.: Ontology learning from text: a survey of methods. LDV Forum **20**(2), 75–93 (2005)
5. Cupek, R., Ziebinski, A., Fojcik, M.: An ontology model for communicating with an autonomous mobile platform. In: Kozielski, S., Kasprowski, P., Mrozek, D., Małysiak-Mrozek, B., Kostrzewa, D. (eds.) BDAS 2017. CCIS, vol. 716, pp. 480–493. Springer, Cham (2017)
6. Euzenat, J., Schvaiko, P.: Ontology Matching. Springer, Heidelberg (2013)
7. Goczyła, K., Zawadzka, T.: Zależności między ontologiami i ich wpływ na problem integracji ontologii. In: Kozielski, S., Małysiak, B., Kasprowski, P., Mrozek, D. (eds.) Bazy Danych: Struktury, Algorytmy, Metody, pp. 331–340. WKŁ, Warsaw (2006)

8. Hajnicz, E.: Automatyczne tworzenie semantycznych słowników walencyjnych. Akademicka Oficyna Wydawnicza EXIT, Warsaw (2011)
9. Horák, A., Smrž, P.: VisDic - wordnet browsing and editing tool. In: Sojka, P., Pala, K., Smrž, P., Fellbaum, C., Vossen, P. (eds.) Proceedings of the 2nd International WordNet Conference, pp. 136–141 (2003)
10. Horák, A., Smrž, P.: New features of wordnet editor VisDic. Rom. J. Inf. Sci. Technol. **7**(1–2), 1–13 (2004)
11. Hossain, J., Sani, F., Affendey, L.S., Ishak, I., Kasmiran, K.A.: Semantic schema matching approaches: a review. J. Theor. Appl. Inf. Technol. **62**(1), 139–147 (2014)
12. Ibrahim, H., Karasneh, Y., Mirabi, M., Yaakob, R., Othman, M.: An automatic domain independent schema matching in integrating schemas of heterogeneous relational databases. J. Inf. Sci. Eng. **30**, 1505–1536 (2014)
13. Jastrząb, T., Kwiatkowski, G., Sadowski, P.: Mapping of selected synsets to semantic features. In: Kozielski, S., Mrozek, D., Kasprowski, P., Małysiak-Mrozek, B., Kostrzewa, D. (eds.) BDAS 2015-2016. CCIS, vol. 613, pp. 357–367. Springer, Cham (2016). doi:10.1007/978-3-319-34099-9_28
14. Kwak, J., Yong, H.S.: Ontology matching based on hypernym, hyponym, holonym, and meronym sets in wordnet. Int. J. Semant. Technol. **1**(2), 1–14 (2010)
15. Lawrence, R., Barker, K.: Integrating relational database schemas using a standardized dictionary. In: Proceedings of the 2001 ACM Symposium on Applied Computing (SAC 2001), pp. 225–230. ACM (2001)
16. Magnini, B., Speranza, M.: Integrating generic and specialized wordnets. In: Proceedings of the 2nd Conference on Recent Advances in Natural Language Processing (RANLP 2001) (2001)
17. Mahdi, A.M., Tiun, S.: Utilizing wordnet for instance-based schema matching. In: Proceedings of the International Conference on Advances in Computer Science and Electronics Engineering (CSEE 2014), pp. 59–63. Institute of Research Engineers and Doctors (2014)
18. Maziarz, M., Piasecki, M., Rudnicka, E., Szpakowicz, S., Kędzia, P.: plWordNet 3.0 - a comprehensive lexical-semantic resource. In: Proceedings of the 26th International Conference on Computational Linguistics: Technical Papers, COLING 2016, pp. 2259–2268 (2016)
19. Miller, G.A.: Nouns in wordnet: a lexical inheritance system. Int. J. Lexicogr. **3**(4), 245–264 (1990)
20. Miller, G.A., Beckwith, R., Fellbaum, C., Gross, D., Miller, K.: Introduction to wordnet: an on-line lexical database. Int. J. Lexicogr. **3**(4), 235–244 (1990)
21. Mykowiecka, A.: Inżynieria lingwistyczna: komputerowe przetwarzanie tekstów w języku naturalnym. Wydawnictwo PJWSTK, Warsaw (2007)
22. Piasecki, M., Szpakowicz, S., Broda, B.: Toward plWordNet 2.0. In: Bhattacharyya, P., Fellbaum, C., Vossen, P. (eds.) Proceedings of the 5th Global Wordnet Conference on Principles, Construction and Application of Multilingual Wordnets, pp. 263–270. Narosa Publishing House (2010)
23. Rahm, E., Bernstein, P.: A survey of approaches to automatic schema matching. VLDB J. **10**, 334–350 (2001)
24. Świderski, M.: Metodologia LAV w systemie semantycznej integracji geoprzestrzennych źródeł danych. In: Kozielski, S., Małysiak, B., Kasprowski, P., Mrozek, D. (eds.) Bazy Danych: Modele, Technologie, Narzędzia, pp. 213–220. WKŁ, Warsaw (2005)
25. Vetulani, Z.: Komunikacja człowieka z maszyną. Akademicka Oficyna Wydawnicza EXIT, Warsaw (2014)

26. Vetulani, Z., Vetulani, G., Kochanowski, B.: Recent advances in development of a lexicon-grammar of polish: Polnet 3.0. In: Calzolari, N., Choukri, K., et al. (eds.) Proceedings of the 10th International Conference on Language Resources and Evaluation (LREC 2016), pp. 2851–2854. European Language Resources Association (ELRA) (2016)
27. Xiang, C., Jiang, T., Chang, B., Sui, Z.: ERSOM: A structural ontology matching approach using automatically learned entity representation. In: Proceedings of the 2015 Conference on Empirical Methods in Natural Language Processing, pp. 2419–2429 (2015)
28. Ziebinski, A., Cupek, R., Erdogan, H., Waechter, S.: A survey of ADAS technologies for the future perspective of sensor fusion. In: Nguyen, N.-T., Manolopoulos, Y., Iliadis, L., Trawiński, B. (eds.) ICCCI 2016. LNCS (LNAI), vol. 9876, pp. 135–146. Springer, Cham (2016). doi:10.1007/978-3-319-45246-3_13

Authorship Attribution for Polish Texts Based on Part of Speech Tagging

Piotr Szwed[(✉)]

AGH University of Science and Technology, Kraków, Poland
pszwed@agh.edu.pl

Abstract. Authorship attribution aims at identifying the author of an unseen text document based on text samples originating from different authors. In this paper we focus on authorship attribution of Polish texts using stylometric features based on part of speech (POS) tags. Polish language is characterized by high inflection level and in consequence over 1000 POS tags can be distinguished. This allows building a sufficiently large feature space by extracting POS information from documents and performing their classification with use of machine learning methods. We report results of experiments conducted with Weka workbench using combinations of the following features: POS tags, an approximation of their bigrams and simple document statistics.

Keywords: Authorship attribution · Polish texts · Part of speech tagging

1 Introduction

Authorship attribution aims at identifying the author of an unseen text document based on text samples originating from different authors. In contrast to the text categorization problem, whose goal is to assign a topic or a list of topics to a text based on its content, authorship attribution abstracts away from a particular domain and attempts to grasp content independent text traits that are "linguistic expressions" of particular authors [3]. Such content independent text properties are commonly referred as *stylometric features*. Various their types has been proposed and applied in authorship attribution tasks, including: word frequencies, character n-grams, function words and Part of Speech (POS) tags [14].

In this paper we report experiments aimed at authorship attribution of Polish text based mainly on POS features. The idea of using tagging information is not new and it was successfully applied to a number of style classification tasks, where in particular English texts were processed, see for example [4,14]. Typically, based on part of speech tagging, POS n-grams were extracted: (unigrams, bigrams and/or trigrams). The described approach is computationally feasible for English texts, as the number of POS tags for English is small. For example in [7] 38 unigrams and 1000 bigrams were used, whereas in [3] only 8 POS tags and their trigrams were considered. In contrast to English, Polish is a highly

© Springer International Publishing AG 2017
S. Kozielski et al. (Eds.): BDAS 2017, CCIS 716, pp. 316–328, 2017.
DOI: 10.1007/978-3-319-58274-0_26

inflected language and the number of POS tags that are used to describe words ranges at 1000. Moreover, attribution of tags to words is ambiguous, i.e. a given inflected form of a word can be linked to a number of lemmas (basic word forms) and a set o POS tags.

Such situation can be considered both an opportunity and an obstacle. The number of Polish POS tags seems to be high enough to cover style variations and build accurate classification models using various Machine Learning techniques. Thus, a hypothesis can be made that, in opposition to the English language, where the feature space based solely on POS tags occurrences is not sufficient to efficiently perform authorship attribution, there is no need for Polish texts to extend the prospective set of over 1000 features related to POS tags by including extra n-gram information.

On the other side, POS n-grams are commonly seen as good representations of stylistic figures characteristic for authors and their use might improve the classification results. Unfortunately, with over 1000 POS tags in use in Polish, even calculation of bigrams would consume a vast amount of resources and result in extremely high dimensional feature space having negative impact on classifier learning speed and performance.

To circumvent this problem we decided to build an approximation of POS n-grams having the form of a closed set of syntactic patterns to be detected in the processed texts. Actually, we reused previously developed software that aims at extracting occurrences of compound terms describing objects and actions from Polish texts and normalizes them by translating to nominal form [16,17]. The translation is done with rules, hence, executing a rule set over a document and counting, how many times a particular rule fired would produce a value of a corresponding attribute.

In the rest of the paper we investigate application of Machine Learning techniques for various combinations of three sets of features: POS tags, features obtained by counting rule firing (referred by BPOS) and a dozen of simple statistical features related to sentence length and presence of punctuation marks.

The paper is organized as follows: next Sect. 2 provides a short review of authorship attribution problems. It is followed by Sect. 3, which describes the datasets used. Section 4 presents sets of features used in classification. It is followed by Sect. 5 reporting results of conducted experiments. Section 6 provides concluding remarks.

2 Related Works

Authorship attribution is an important problem with many practical application in various fields, falling beyond the scope of traditional literary studies: intelligence, criminal law, computer forensic, plagiarism detection or author profiling [14]. A common approach to authorship attribution consists in extracting *stylometric features* from documents. Then documents represented as vectors in the feature space are assessed according to their similarity or the original problem is transformed into a typical classification task suitable for machine learning techniques.

Sets of proposed features vary, depending on available data or the intended generality of their extraction method and applicability to various languages.

The simplest features describe *statistical properties* of documents: word length, sentence length and vocabulary richness. *Function words* are features based on word frequencies. In contrast to text categorization problems, where the most frequent words are regarded useless or even harmful for classification, in authorship attribution problems they are often used as personal style markers. However, not all the most frequent words are good candidates to be included to that set of features: an important characteristic is *instability* [6], i.e. the possibility to be replaced by another word from the dictionary. Another problem with function words is that their selection may be biased for a particular topic, e.g. the word "set" would probably be promoted as function word in texts on mathematics: there are 124 occurrences in Wikipedia article on functions[1].

Another word-based features are *word sequences* (n-grams). An example of this approach can be found in [1], where classification using word sequences was tested on 350 poems in Spanish by 5 authors giving about 83% accuracy.

Features, which usually give very high accuracy measures are *character n-grams*, e.g. sequences of four characters extracted from words appearing in documents. They are considered language independent, i.e. they can be extracted from texts in various languages regardless of character sets used. See for example [5] for reports on authorship attribution of English, Greek and Chinese texts.

In our opinion very good results of their application should be treated with caution: there is an obvious functional dependence between document content and character n-grams, so they may constitute and alternative representation of function words (what is probably good) or they may just render document content (what seems to be worse).

Feature that apparently abstract away from topic related vocabulary are *part of speech tags* and their n-grams. They were used in mentioned earlier works [3,7] in experiments, where POS n-grams were used in authorship attribution of English documents. Another types are *syntactic features*: rewrite rules and chunks [14].

An issue related to authorship attribution is the difficulty of the particular case [7]. Generally, the problems, where the number of authors is small and large amounts of data samples are available are considered easy and high accuracy is expected. On the other side, the difficulty increases with the growing number of authors and smaller data sizes [10], what results in inferior accuracy measures.

The test cases reported in this paper, where authorship attribution was conducted for three authors, can be considered easy. However, a frequent approach appearing in many studies is to conduct experiments on a simpler case with artificially generated data [3,15], especially, if they aim at confirming a hypothesis or validating a certain method.

Finally, we would like to mention works tackling authorship attribution of Polish texts. Research related to stylometry in various languages, including

---

[1] https://en.wikipedia.org/wiki/Function_(mathematics).

Polish, was conducted by members of Computational Stylistics Groups[2]. In their works they used frequent words [12], and character n-grams [2].

Typical Machine Learning approach for authorship attribution of Polish newspaper articles was reported in [8]. Authors compared classifier performance for various feature sets including character n-grams, word n-grams, lemmas, POS tags and function words. This publication is particularly interesting in the context of this work, as it is probably the single source of information on usage of POS tags in classification of Polish texts. The conducted experiments showed that the best accuracy was achieved with use of character n-grams and lemmas. This result is not surprising, as in our opinion, the mentioned features tend to describe document content, rather than stylistic features.

3   Datasets

Datasets used in the experiments were extracted from three classic Polish novels written at the turn of twentieth century: *Ziemia obiecana* (Eng. Promised Land) by Władysław Reymont, *Rodzina Połanieckich* (Eng. Połaniecki family) by Henryk Sienkiewicz and *Syzyfowe prace* (Eng. Sisyphus works) by Stefan Żeromski. The texts were selected to have the same genre and similar time and place of action. The content of the books were broken into sentences, then documents to be classified were formed by concatenating consecutive n ($n = 10, 20, ..., 100$) sentences attributed to one author. Resulting datasets are collections of equal length documents being disjoint excerpts from three books.

While splitting the books content, we marked the sentences as either narrative or dialog (the distinction was based on dash character appearing at the sentence start, which traditionally is used in Polish typography.) Hence, datasets can be formed to represent either narrative parts (N* group), dialogs (D* group) or their mixture (A* all) sentences. Such division was made to verify a hypothesis that author's style manifests itself primarily in narrative sentences, whereas dialogs rather mimic the way of speaking of portrayed persons, often they employ expressions observed in the real life and used in less formal communication.

The statistics for the source data is given in Table 1. Based on these values it can be stated that for example A10 (10-sentence) dataset contains 1365 documents by Reymont, 1571 by Sienkiewicz and 433 by Żeromski and average document length from N10 is 136.9 words. For other An, Nn and Dn datasets, where n is the number of sentences indicated in the name, the values should be correspondingly scaled.

4   Features

Based on the documents' content three sets of features were computed: *POS* – part of speech tags, *BPOS* – an approximation of POS n-grams (mostly bigrams) and *Stat* – 10 features capturing such document properties, as numbers of words and punctuation marks (including mean values, variance and standard deviation).

---

[2] https://sites.google.com/site/computationalstylistics/.

Table 1. Source data summary

Selection	Words	Sentences	Avg. $\frac{\#words}{sentence}$	#sentences Reymont	#sentences Sienkiewicz	#sentences Żeromski
A* (all)	33680	461140	13.69	13649 (40.53%)	15708 (46.64%)	4323 (12.84%)
N* (narrative)	23569	363923	15.44	8244 (34.98%)	11725 (49.75%)	3600 (15.27%)
D* (dialog)	10111	97217	9.61	5405 (53.46%)	3983 (39.39%)	723 (7.15%)

4.1 Part of Speech Tags

As POS tagger we used Morfologik [11], which is both a comprehensive dictionary of Polish inflected forms based on PoliMorf [18], and a software library written in Java accompanied by a number of utility tools.

Morfologik dictionary can be seen as a relation $D \subset IF \times L \times \mathcal{P}$, where IF is a set of inflected forms, L is a set of lemmas, $L \subset IF$, and $\mathcal{P}$ is a set of POS tags defining properties of inflected forms (part of speech, gender, singular vs. plural, declination case, etc.)

The basic function offered by the library is $stem: IF \rightarrow 2^{L \times \mathcal{P}}$. It takes as input an inflected form and returns a set of lemmas with accompanying tags. Figure 1 gives an example of tags attributed by Morfologik to two terms: *rewolucyjny* (Eng. *revolutionary*) and *idei* (Eng. *ideas*). In this case only one lemma (indicated in parentheses) was returned to each word, however, multiple POS tags.

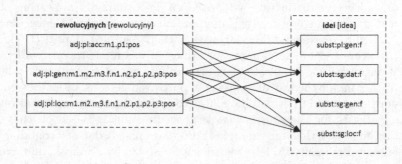

Fig. 1. Example of POS tagging for two words: *rewolucyjny* (Eng. revolutionary) and *idei* (Eng. ideas)

Let us define the function $pos(w)$ for $w \in IF$ that returns the union of POS sets for a given word (inflected form) w:

$$pos(w) = \bigcup_{(l,p) \in stem(w)} p$$

The set of POS features for a given document comprising words W is computed in two steps:

- At first the union $POS(W) = \bigcup_{w \in W} pos(w)$ is calculated.
- Then the large string of multiple tags representing $POS(W)$ is converted with the Weka filter StringToWordVector configured to apply IDF transform.

In the datasets discussed in Sect. 3, about 800 distinct POS tags were found.

4.2 Approximating POS n-grams

The specificity of Polish language makes the problem of finding adequate representation of POS n-grams hard. It is illustrated in Fig. 1: as each word can be attributed with a set of tags, a POS n-gram corresponds to a sequence of sets. Such representation can be flattened to a Cartesian product of n sets of tags $\mathcal{P}$. Considering bigrams only, for the example shown in Fig. 1, a set of 12 bigrams would be obtained: $POS(\{rewolucyjnych\}) \times POS(\{idei\})$. For the analyzed datasets the total number of POS tags appearing in documents was about 800; it can be noticed that, even after flattening, processing bigram information can be very resource demanding, e.g. it would require storage for 800×800 matrices, and result in a huge number of features.

As it was mentioned in the Sect. 1, to avoid such extensive computational effort, we decided to build a partial POS n-gram representation. For this purpose we used a software tool described in [16,17], which aims at extracting concepts form documents looking for compound terms matching specific syntactic pattern and then performing their normalization, usually consisting in changing nouns with complements to Nominative and singular form and verbs to verbal nouns. Specification of text transformations has the form of rules, each of them configured to match a specific pattern of POS tags. Actually, rule premises are sequences of POS sets: $(pos_1, \ldots, pos_n)$.

The rule execution engine analyzes POS tags of words forming a sentence: $(pos(w_1), \ldots, pos(w_k), \ldots, pos(w_{k+n}), \ldots, pos(w_m))$. If there is at least a partial match between a rule premise and tags attributed to words for a given k, i.e. $\forall_{j=1,n} . pos_j \cap pos(w_{k+j}) \neq \emptyset$, the rule fires and produces output strings.

Hence the idea of building a partial representation of POS n-grams by processing an input document with the concept extraction tool and creating a set of additional features based on frequencies of rule firings. The disadvantage of this method is that previously defined rules were specified to extract mostly nouns and verbal nouns. We extended the basic set by specifying a few dozen of extra rules related to actions (verbs) and their objects.

To give an example of the coverage reached with this approach, we gathered in Table 2 sequences of words that triggered the rules for the first sentence from the novel *Ziemia obiecana* (Eng. *Promissed land*) by W. Reymont (parts of the sentences marked with boldface):
*"**Pierwszy wrzaskliwy świst fabryczny** rozdarł ciszę **wczesnego poranku**, a za nim we wszystkich **stronach miasta zaczęły się zrywać** coraz zgiełkliwiej inne i darty się chrapliwymi, **niesfornymi głosami** niby chór **potwornych kogutów**, piejących **metalowymi gardzielami** hasło do pracy."*

Table 2. Word n-grams that triggered the rules

1.	Pierwszy wrzaskliwy	6.	zaczęły się zrywać
2.	wrzaskliwy świst	7.	niesfornymi głosami
3.	świst fabryczny	8.	potwornych kogutów
4.	wczesnego poranku	9.	metalowymi gardzielami
5.	stronach miasta		

The extended ruleset comprised 419 rules, however, a part of it was pruned with a mechanism preventing rule redundancy described in [17]. Finally, for the datasets described in Sect. 3, 276 rules at least once fired, what allowed to build the set of BPOS features counting 276 attributes.

It can be observed, that the described above method of feature extraction is somehow similar to detecting syntactic phrases (chunks) described in [13,14], however, the current coverage assured by the defined ruleset is smaller.

5   Experiments

We performed numerous experiments using Weka classifiers for datasets from A*, N* and D* document groups varying in document length. To assess classifier performance we used weighted precision (Pr), recall (Rc) and $F1 = \frac{2 \cdot Pr \cdot Rc}{Pr + Rc}$ measures returned by Weka. In all cases 10-fold cross-validation was preformed. The obtained values of these three measures are shown in tables, however, for clarity in plots only values of $F1$, which aggregates both precision and recall, are displayed. It should be mentioned that, basically, due to datasets imbalance visible in Table 1 classification results for excerpts of Żeromski book were worse.

At the beginning it is worth to note that categorization of documents from the datasets, i.e. classification based on their *content* represented as bag of words, in general, gives great results. For example for the dataset N10, the values of precision, recall and F1 measure determined with Naive Bayes Multinomial classifier were respectively $P = 0.998$, $R = 0.998$, $F1 = 0.998$. For D10, which was assumed to be a harder case, the results are similar: $P = 0.992$, $R = 0.992$, $F1 = 0.992$. In experiments aiming at authorship attribution such values were never reached, they were at most approached for longer documents. This is due to the fact that the documents contents were ignored, and only features expected to represent authors' style (POS, BPOS and Stat) were considered.

Figure 2 shows, how F1 measure values determined with Naïve Bayes Multinomial classifier vary with document length. Not surprisingly, classification results for N* (narrative) documents are usually better than for A* and much better than for D* (dialogs).

It can be also observed that authorship attribution is more accurate for longer documents. The best values $(Pr = 0.934, Rc = 0.933$ and $F1 = 0.931)$ were obtained for N100 dataset, where the average number of words in a document

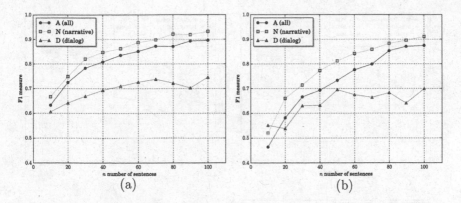

Fig. 2. F1 measure values determined with Naïve Bayes Multinomial classifier for various document lengths (n-number of sentences). Features used: (a) POS + BPOS + Stat (b) BPOS + Stat

Table 3. Comparison of classifier performance. NB - Naïve Bayes, J48-decision tree, RF - random forest, SMO - support vector machines, IBk - k nearest neighbors ($k = 5$). All calculations made using the three sets of features: POS + BPOS + Stat

Data	NB			J48			RF			SMO			IBk		
	Pr	Rc	F1	Pr	Rc	F1	Pr	Rc	F1	Pr	Rc	F1	Pr	Rc	F1
A10	0.705	0.671	0.673	0.718	0.722	0.720	0.776	0.749	0.719	0.792	0.793	**0.792**	0.611	0.615	0.600
A20	0.769	0.750	0.751	0.774	0.778	0.775	0.819	0.798	0.774	0.815	0.816	**0.815**	0.666	0.668	0.650
A30	0.798	0.792	0.790	0.777	0.781	0.779	0.825	0.804	0.777	0.840	0.839	**0.838**	0.729	0.725	0.703
A50	0.852	0.851	0.850	0.773	0.768	0.770	0.855	0.841	0.823	0.885	0.885	**0.884**	0.764	0.759	0.740
A80	0.875	0.875	0.874	0.826	0.825	0.825	0.884	0.875	0.864	0.909	0.908	**0.907**	0.809	0.797	0.786
A100	0.899	0.900	0.899	0.858	0.858	0.855	0.877	0.867	0.856	0.905	0.906	**0.905**	0.787	0.782	0.757
N10	0.734	0.710	0.706	0.635	0.638	0.637	0.746	0.715	0.685	0.768	0.767	**0.767**	0.648	0.659	0.635
N20	0.786	0.782	0.778	0.686	0.689	0.687	0.784	0.756	0.728	0.819	0.820	**0.820**	0.702	0.703	0.681
N30	0.836	0.836	0.835	0.741	0.745	0.742	0.812	0.789	0.769	0.881	0.882	**0.881**	0.723	0.723	0.699
N50	0.878	0.879	**0.878**	0.788	0.793	0.790	0.858	0.843	0.828	0.873	0.873	0.871	0.812	0.795	0.775
N80	0.889	0.889	0.888	0.869	0.872	0.869	0.875	0.865	0.856	0.909	0.909	**0.909**	0.834	0.828	0.817
N100	0.907	0.903	0.904	0.895	0.891	0.892	0.883	0.870	0.864	0.942	0.941	**0.941**	0.830	0.819	0.808
D10	0.683	0.658	0.664	0.878	0.887	**0.882**	0.742	0.786	0.752	0.693	0.695	0.694	0.554	0.566	0.554
D20	0.736	0.732	0.732	0.861	0.860	**0.860**	0.735	0.780	0.746	0.671	0.677	0.673	0.582	0.596	0.581
D30	0.783	0.782	0.778	0.858	0.864	**0.861**	0.713	0.758	0.724	0.743	0.746	0.741	0.616	0.617	0.606
D50	0.837	0.838	0.830	0.913	0.917	**0.914**	0.766	0.804	0.772	0.723	0.735	0.721	0.622	0.632	0.618
D80	0.798	0.859	0.827	0.963	0.961	**0.958**	0.724	0.773	0.736	0.800	0.789	0.784	0.593	0.625	0.601
D100	0.746	0.806	0.772	0.939	0.942	**0.940**	0.745	0.786	0.750	0.810	0.816	0.804	0.723	0.650	0.634

was 1544. In this case all features: POS, BPOS and Stat were used. However slightly worse results were observed for the reduced set of attributes consisting of BPOS and Stat groups ($Pr = 0.914$, $Rc = 0.912$ and $F1 = 0.911$).

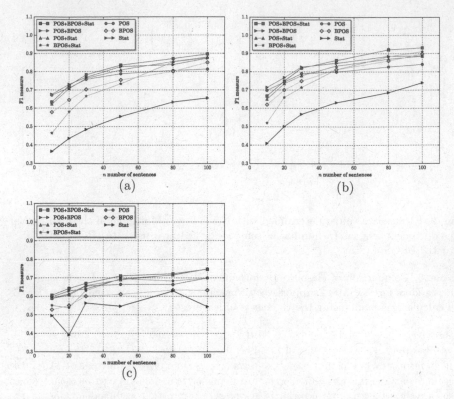

Fig. 3. F1 measure values (Naïve Bayes Multinomial classifier used) for various various document lengths (*n*-number of sentences) and subsets of features. (a) document group A* (all), (b) document group N* (narrative), (c) document group D* (dialog)

Naïve Bayes Multinomial classifier was the our first choice due to its known efficiency in text classification. We performed additional tests aiming at assessing performance of various classifier: naïve Bayes, J48 (Weka implementation of C4.5), random forest, support vector machines and k-NN. In all cases default values were used. Their results are summarized in Table 3. Based on their examination it can be stated that for A* and D* groups of documents SMO (support vector machines, polynomial kernel as a default) outperformed other classifiers. For documents extracted from dialogs, J48 decision tree occurred the best in all cases.

5.1 Feature Selection: Sets of Attributes

We analyzed, how selection of sets of features (POS, BPOS and Stat) affects the performance of Naïve Bayes Multinomial (NBM) classifier. NBM was used due to its learning speed (usually an order of magnitude faster than tree models). As there are three sets of features, their seven nonempty combinations were tested. The results for particular types of documents are shown in Fig. 3a–c. It can

Table 4. Feature selection: BF-CFS: BestFirst-CfsSubsetEval, Rank-Corr: Ranker-CorrelationAttributeEval, Rank-IG: Ranker-InfoGainAttributeEval

Data	BF-CFS			Rank-Corr			Rank-IG		
	Pr	Rc	F1	Pr	Rc	F1	Pr	Rc	F1
A10	0.777	0.777	0.774	0.792	0.793	**0.792**	0.792	0.793	**0.792**
A20	0.852	0.851	**0.849**	0.815	0.816	0.815	0.815	0.816	0.815
A30	0.885	0.884	**0.883**	0.840	0.839	0.838	0.840	0.839	0.838
A50	0.917	0.916	**0.916**	0.885	0.885	0.884	0.885	0.885	0.884
A80	0.920	0.920	**0.919**	0.909	0.908	0.907	0.909	0.908	0.907
A100	0.956	0.956	**0.956**	0.905	0.906	0.905	0.905	0.906	0.905
N10	0.775	0.777	0.772	0.768	0.767	**0.767**	0.768	0.767	**0.767**
N20	0.847	0.849	**0.846**	0.819	0.820	0.820	0.819	0.820	0.820
N30	0.873	0.872	0.871	0.881	0.882	0.881	0.881	0.882	**0.881**
N50	0.899	0.898	**0.898**	0.873	0.873	0.871	0.873	0.873	0.871
N80	0.977	0.976	**0.976**	0.909	0.909	0.909	0.909	0.909	0.909
N100	0.975	0.975	**0.975**	0.942	0.941	0.941	0.942	0.941	0.941
D10	0.857	0.847	**0.820**	0.693	0.695	0.694	0.693	0.695	0.694
D20	0.827	0.833	**0.821**	0.671	0.677	0.673	0.671	0.677	0.673
D30	0.918	0.920	**0.907**	0.743	0.746	0.741	0.743	0.746	0.741
D50	0.944	0.941	**0.935**	0.723	0.735	0.721	0.723	0.735	0.721
D80	0.925	0.930	**0.925**	0.800	0.789	0.784	0.807	0.797	0.791
D100	0.919	0.922	**0.918**	0.810	0.816	0.804	0.810	0.816	0.804

be observed, that usually the best classification results were obtained, when all features were used, however, in many cases the subsets POS + Stat and BPOS + Stat gave comparable results. Apparently, in the presence of bias including more features into a model may increase classifier performance, nevertheless it comes with higher computational cost related to feature extraction and more expensive classifier learning.

5.2 Feature Selection: Attributes

We performed feature selection for individual attributes applying three combinations of search and evaluation methods available in Weka: BF-CFS (BestFirst, CfsSubsetEval), Rank-Corr (Ranker, CorrelationAttributeEval) and Rank-IG (Ranker, InfoGainAttributeEval). Results for other are not reported, because either they are partial due to unacceptable processing time or inferior performance, as in the case of PCA.

The dataset with reduced set of features was further submitted to SMO (support vector machines) classifier. The results are gathered in Table 4. It can be observed that in most cases BF-CFS was the most effective and its application resulted in increased classifier performance by 2–3% (c.f. Table 3).

Analysis of selections made by BF-CFS indicated that attributes of all feature sets were selected with a dominance of POS tags, which were the most numerous. For example, when applied to A20 dataset BF-CFS pointed out 37 attributes (out of 1092) including: 27 POS tags (verbs, nouns, adjectives and pronouns), 7 BPOS features (nouns + adjectives, verbal nouns + nouns and one 3-gram rule) and 3 statistic features (average number of punctuation, quotations and exclamation).

6   Conclusions

At the beginning of this paper we put forward a hypothesis that the high variability of inflected forms appearing in the Polish language can be employed to perform authorship attribution. More than 1000 POS tags present in Polish dictionaries create a feature space, whose dimension is high enough to be used to discriminate authors' styles. Such approach is probably not sufficient for less inflected languages, e.g. English, where about 40 tags returned by commonly used Brill tagger.

To verify our hypothesis, we conducted experiments for combinations of three sets of features: simple statistics related to sentence length and frequencies of punctuation marks, POS tags and BPOS, which as described in Sect. 4.2 partially covers POS bigrams.

The obtained results show that the employed *bag of POS* model provides quite a good accuracy, especially for narrative texts and longer documents, which better reflect distributions of inflected form of nouns, verbs, pronouns, etc. used by particular authors. Although the coverage of BPOS features (approximation POS n-grams) was only partial, interestingly, in many experiments their use returned results comparable to POS (see Fig. 3).

Extraction of features based of POS tags in highly inflected languages, including Polish, seems a promising direction in authorship attribution. Classification accuracy for POS features is lower than for character n-grams or function words, see for example [8], but their advantage lies in apparent abstraction of the document domain.

In the future we plan to focus on further development of BPOS features. This can be done with extending the set of patterns (rules). During the experiments we reused the ruleset, which was developed with an intent to extract candidate concept names from texts, hence, it was less sensitive for more specific stylistic constructs. Another interesting direction is investigating various feature selection algorithms for POS based features, e.g. applying feature maximization method [9], which turned out to be very efficient in text categorization.

References

1. Coyotl-Morales, R.M., Villaseñor-Pineda, L., Montes-y-Gómez, M., Rosso, P.: Authorship attribution using word sequences. In: Martínez-Trinidad, J.F., Carrasco Ochoa, J.A., Kittler, J. (eds.) CIARP 2006. LNCS, vol. 4225, pp. 844–853. Springer, Heidelberg (2006). doi:10.1007/11892755_87
2. Eder, M.: Style-markers in authorship attribution a cross-language study of the authorial fingerprint. Stud. Pol. Linguist. **6**(1), 99–114 (2011)
3. Gamon, M.: Linguistic correlates of style: authorship classification with deep linguistic analysis features. In: Proceedings of the 20th International Conference on Computational Linguistics. COLING 2004, Stroudsburg. Association for Computational Linguistics (2004). http://dx.doi.org/10.3115/1220355.1220443
4. Juola, P.: Authorship attribution. Found. Trends Inf. Retr. **1**(3), 233–334 (2006)
5. Kešelj, V., Peng, F., Cercone, N., Thomas, C.: N-gram-based author profiles for authorship attribution. In: Proceedings of the Conference Pacific Association for Computational Linguistics, PACLING, vol. 3, pp. 255–264 (2003)
6. Koppel, M., Akiva, N., Dagan, I.: Feature instability as a criterion for selecting potential style markers. J. Am. Soc. Inf. Sci. Technol. **57**(11), 1519–1525 (2006)
7. Koppel, M., Schler, J., Argamon, S.: Authorship attribution: what's easy and what's hard? J. Law Policy **21**, 317–331 (2013)
8. Kuta, M., Puto, B., Kitowski, J.: Authorship attribution of Polish newspaper articles. In: Rutkowski, L., Korytkowski, M., Scherer, R., Tadeusiewicz, R., Zadeh, L.A., Zurada, J.M. (eds.) ICAISC 2016. LNCS (LNAI), vol. 9693, pp. 474–483. Springer, Cham (2016). doi:10.1007/978-3-319-39384-1_41
9. Lamirel, J.-C.: New metrics and related statistical approaches for efficient mining in very large and highly multidimensional databases. In: Kozielski, S., Mrozck, D., Kasprowski, P., Małysiak-Mrozek, B., Kostrzewa, D. (eds.) BDAS 2015. CCIS, vol. 521, pp. 3–20. Springer, Cham (2015). doi:10.1007/978-3-319-18422-7_1
10. Luyckx, K., Daelemans, W.: The effect of author set size and data size in authorship attribution. Lit. Linguist. Comput. **26**(1), 35–55 (2011)
11. Miłkowski, M.: Morfologik (2016). http://morfologik.blogspot.com/. Accessed Dec 2016
12. Rybicki, J.: Success rates in most-frequent-word-based authorship attribution: a case study of 1000 Polish novels from Ignacy Krasicki to Jerzy Pilch. Stud. Pol. Linguist. **10**(2), 87–104 (2015). http://www.ejournals.eu/SPL/2015/Issue-2/art/5409/
13. Stamatatos, E., Fakotakis, N., Kokkinakis, G.: Computer-based authorship attribution without lexical measures. Comput. Humanit. **35**(2), 193–214 (2001). http://dx.doi.org/10.1023/A: 1002681919510
14. Stamatatos, E.: A survey of modern authorship attribution methods. J. Am. Soc. Inf. Sci. Technol. **60**(3), 538–556 (2009)
15. Stańczyk, U.: The class imbalance problem in construction of training datasets for authorship attribution. In: Gruca, A., Brachman, A., Kozielski, S., Czachórski, T. (eds.) Man–Machine Interactions 4. AISC, vol. 391, pp. 535–547. Springer, Cham (2016). doi:10.1007/978-3-319-23437-3_46
16. Szwed, P.: Concepts extraction from unstructured Polish texts: a rule based approach. In: 2015 Federated Conference on Computer Science and Information Systems (FedCSIS), pp. 355–364, September 2015

17. Szwed, P.: Enhancing concept extraction from Polish texts with rule management. In: Kozielski, S., Mrozek, D., Kasprowski, P., Małysiak-Mrozek, B., Kostrzewa, D. (eds.) BDAS 2015-2016. CCIS, vol. 613, pp. 341–356. Springer, Cham (2016). doi:10.1007/978-3-319-34099-9_27

18. Wolinski, M., Milkowski, M., Ogrodniczuk, M., Przepiórkowski, A.: PoliMorf: a (not so) new open morphological dictionary for Polish. In: LREC, pp. 860–864 (2012)

Fast Plagiarism Detection in Large-Scale Data

Radosław Szmit(⊠)

Institute of Control and Industrial Electronics, Warsaw University of Technology,
ul. Koszykowa 75, 00-662 Warsaw, Poland
radoslaw.szmit@ee.pw.edu.pl

Abstract. This paper presents some research results involved in building Polish semantic Internet search engine called the Natively Enhanced Knowledge Sharing Technologies (NEKST) and its plagiarism detection module. The main goal is to describe tools and algorithms of the engine and its usage within the Open System for Antiplagiarism (OSA).

Keywords: Plagiarism detection · Sentence hashing · Cloud computing · Semantic comparison · Big Data · NEKST · OSA

1 Introduction

Efficient large-scale text comparison is a common problem in data processing. It occurs in any plagiarism detection system [1]. Usually we have to find similarities between many textual documents and at the same time new coming document must be check for plagiarism. Very often the granularity of comparisons has to be kept at the level of sentences. In such situations linear comparison to each reference document is not suitable and more efficient approach is required. Dramatic performance gains are obtained using hashing methods which significantly reduce the time required for searching. In this paper we present a complete, high scalable system for plagiarism detection which works on sentences level granularity.

The semantic search engine NEKST has been built for Polish language at the Institute of Computer Science of Polish Academy of Sciences[1] to analyze huge amount of data, estimated to one billion of Polish Internet documents.

The novelty of this paper is in showing composition of known ideas of pursuing similarities between documents in large-scale databases [3,36,42,45,46] with special emphasis on natural language processing (NLP) adapted to Polish. Many methods described in this paper are widely used in other fields or smaller data sets, but the main goal of this research was to tailor and implement them in a fast and efficient, Polish-oriented, plagiarism detection software. The implemented systems (NEKST and OSA) are based on the Apache Hadoop platform [17,21] and the MapReduce paradigm [15] which leads to highly scalable and constant-time response solution.

---

[1] http://ipipan.waw.pl/.

© Springer International Publishing AG 2017
S. Kozielski et al. (Eds.): BDAS 2017, CCIS 716, pp. 329–343, 2017.
DOI: 10.1007/978-3-319-58274-0_27

2   Plagiarism Detection Algorithm

2.1   Brief History

Document copy detection has been extensively researched since the ninetys. The simplest and well-known method to deal with this issue of matching the longest common subsequence [22]. The most popular and naive implementation has a high $O(n^2)$ time complexity. Sophisticated implementations reduce it to $O(n^2/log(n))$ [7,23,24,26,34].

Another approach is based on comparison of sentences being natural and relatively small pieces of documents' messages. This technique is very popular and used in many systems [6,20,28,29,38,52]. Some of them search only for identical sentences, whereas others pursue close enough ones using similarity measures.

Moving from all possible subsequences to sentences has many advantages. Sentences and/or subsentences are natural pieces in which people divide a text. Copying some part of a document ending in the middle of a such a piece usually has no sense for potential readers. Most of algorithms matching the longest common subsequence require to define frame size. Choosing too short frame results in too many indications and slows down execution, on the other hand too long frame may cause ignoring some phrases, [45]. In order to overcome this difficulty one may rely on the documents' natural pieces, and their natural lengths.

Common subsequence approaches are mostly used in pairwise comparison of documents, which leads to good results at the expense of processing speed. This makes the approaches unusable when pursuing similarities in large collections of documents.

A popular approach used in large collections is to divide all documents into small pieces that can be indexed and then rapidly compared to other documents pieces to identify similarities, [2,33]. This technique is called "fingerprinting" and is very often used in similarity search in large document collections due to high search speed.

The simples and well-known method of generating fingerprints was introduced in [33]. This technique is called "shingling" and it is based on dividing texts into all possible k characters' subsequences without repetitions.

Example 1 (nosign). Shingling For the string *"Warsaw is the capital of Poland."* with $k = 7$ we obtain the following set of the k–shingles: *warsawi, arsawis, rsawist, sawisth, awisthe, wisthec, istheca, sthecap, thecapi, hecapit, ecapita, capital, apitalo, pitalof, italofp, talofpo, alofpol, lofpola, ofpolan, ofpolan, fpoland.*

As we can notice, we obtain at most $n - k$ shingles for text of n characters because any repeated shingle will be ignored. White spaces can be treated in a special way, but in the example above they are simply ignored.

Shingling algorithm can be adopted from characters to words. The change produce less shingles and allows ignoring stop words like "and", "to", "you", etc., [45].

Example 2 (nosign). Shingling on words For the string *"Warsaw is the capital of Poland."* with $k = 3$ we get the following set of k−shingles: *(Warsaw is the), (is the capital), (the capital of), (capital of Poland).*

There is a relation between shingling and algorithms matching the longest common subsequence. Comparing shingles between two documents allow us to find all common subsequences with length greater than or equal to k. However, in contrast to those algorithms, the shingling approach allows to hash and map all shingles into buckets of documents with the same shingle, and then to find all documents with common subsequences easily, without comparing all possible pairs.

2.2 Basic Concept

Modern anti-plagiarism systems are usually hybrid solutions, combining quality and high speed of detection. That is why we propose a hybrid algorithm based on natural language processing, see [3, 36, 42, 45, 46] and sentence hashing concept firstly introduced in [6] for English, and in [11] for Polish.

By combining the technique of hashing with comparing sentences we profit from both methods. The usage of sentences to compare documents lead to reducing costs of the longest common subsequence algorithms. Of course a plagiarist could change all sentences in copied text, but it is very hard for long documents because it requires a lot of additional work. This situation usually appears when short fragments are copied or when the plagiarist is very determined. The solution to this problem is proposed in Sect. 3.

In addition, hashing sentences allows to divide them into buckets fast (one bucket for each unique sentence) and easily find documents with the same sentences without checking all possible pairs of documents. We also produce less fingerprints than shingling algorithm which reduces both disk usage and time of search. For shingling algorithm we must to calculate hash function for every k word sequence in every document producing $n - k$ fingerprints, where n is the document length measured by the number of its words. Replacing words with sentences we usually produce approximately n/s hashes, where s is the average sentence length.

2.3 Text Preprocessing

NEKST uses web crawlers to monitor and collect Polish Internet sites and store them in a distributed Apache Hadoop database. Their usage as references in plagiarism detection requires multi-step preprocessing.

First of all, every document file is parsed to text by special parser according to the file extension (HTML, PDF, DOC, etc.). The resulting text is then tokenized, see [3]. Next the paragraphs, sentences, subsentences and speech forms are identified for deeper natural language processing leading to lemmatization (see the *Morfeusz* project[2]) in which each word is brought to its basic dictionary form called lemma. This is a complicated task for languages like Polish since lemmas of many words depend òn their context.

In the sequel, for the sake of simplicity, we make no difference between subsentences and sentences and simple call them sentences.

2.4 Sentence Hashing

The main idea of sentence hashing is to map each extracted sentence from the documents to its shortened representation being either a string or a number with predefined length. The latest is the case in this paper.

First of all, after extracting sentences from a document, we replace each lemma with a number being either its index in a dictionary of the lemmas or dynamically generated, unique positive integer. In the first case we have good control over transforming words into numbers and we easily assign each word a unique value. On the other hand, in the second case we have better hash distribution in the given range and we avoid problems with transforming out-of-dictionary words into numbers by dynamic extension of the dictionary. This requires synchronization of dictionaries if hashing is performed at different locations (see Sect. 2.12). In the NEKST dynamic word hashing is used with detection of eventual collisions.

It is difficult to tell the number of words in Polish language. *Wielki słownik ortograficzny PWN* (the biggest Polish orthographic dictionary) has about 140 000 lemmas [43]. This dictionary contains the biggest collection of Polish words in their base form inclouding c.a. 10 000 proper nouns. However, these are other words which occur in written texts. It is estimated there are more than one million names of chemical compounds. The biggest online encyclopedia *Wikipedia*[3] has over 1 134 000 articles in Polish and over 4 966 000 in English. The biggest Polish corpus *National Corpus of Polish* has about 5 million of unique words [44]. Creating hash functions we take into account that the language has a huge variety of words and therefore we use large number range, for example 32 bits (corresponding to over 4 billion possible hashes).

After mapping lemmas to their hashes we can simply create representations of sentences by summing up the corresponding hashes:

$$h(\boldsymbol{x}) = \sum_{i=1}^{u} x_i. \tag{1}$$

This operation has $O(1)$ time cost and it is easy to implement. Unfortunately using arithmetic sum of hashes often leads to bad hash distribution [10]. Better hash function for vector $\boldsymbol{x}$ of words was presented in [16]:

$$h(\boldsymbol{x}) = (\sum_{i=1}^{n} x_i a^i) \bmod p, \tag{2}$$

---

[2] http://sgjp.pl/morfeusz/index.html.en.
[3] https://www.wikipedia.org/.

where p is a large enough prime and a is uniformly random. This is known as *multiplicative hash function*. In practice, one uses modular arithmetic and the mod operator and the parameter p can be omitted, [19]. Time cost of this method is $O(\log n)$.

Hashing can also be applied to whole sentences, however, when used word-by-word it gives robustness to changes of words order. In the case of 2, this requires sorting the hashes. This is a supremacy over the shingling algorithm. By lemmatization we also get additional resistance to changing inflexion and/or grammatical time.

Pseudo code of the sentence hashing algorithm is presented below:

```
hashList = []
for all sentence in document do
    sentenceHash = 0
    for all word in sentence do
        lemma = extract(word)
        wordHash = hashWord(lemma)
        sentenceHash += wordHash * a^i
    end for
    hashList += [sentenceHash]
end for
return hashList
```

2.5   Hash Functions

Creating good hash functions is not a simple task. Implementation of such functions must run fast and always return the same result for the same input.

In large data sets function may produce the same results for different input, and collisions may occur with high probability. There are some methods to overcome this difficulty be creating so-called perfect hashes [13] but they have special requirements which are difficult to meet in efficient big data processing [5,18]. In practice one uses well-distributed hashes with small probability of collision. Such a solution is implemented in the NEKST system [51]. More complex collision solvers are presented in [39].

2.6   Search Indices

In [27] the author propose creating server containing repository for a copyright recording and document registration system. The paper [6] describes implementation of such a system based on a hash table. The biggest advantage of this data structure is that it enables searching at the expected cost $O(1)$. However the worst case cost is $O(n)$, since many elements might be hashed into the same key. To overcome this disadvantage we can use the Cuckoo hashing [39] which guarantees the wost case cost $O(1)$ lookups or we may choose hashing with small collision probability and decide that eventual collisions are resolved in detailed comparisons (see Sect. 2.7).

Searching at the expected cost $O(1)$ requires adequate implementation. Most of nowadays databases use B-Tree indices which can be used both for ranged access and ordering but cannot be used to immediate access to a record since one needs to traverse the child records. Such implementation leads to the expected cost $O(\log n)$ [30]. Therefore to reduce the cost to $O(1)$ we build a specialized index to find all common sentences. The NEKST search index is based on inverted lists [14], and all words occurring in documents are mapped to lists of documents containing this word. We use the same idea in plagiarism detection, by mapping the sentence hashes values to the corresponding documents. The expected cost is $O(1)$ for a single sentence and $O(n)$ for n sentences [9]. It is also possible to use some cache structures like Bloom filter to improve search time [4].

The search algorithm is presented below:

Step 1. Parse document to check and extract all words and sentences.

Step 2. Run document sentence hashing algorithm.

Step 3. Search all documents with similar sentences in hash table (it maps sentence hash to document identity).

Step 4. Skip documents with the number of common sentences below the threshold (optional).

Step 5. Collect found documents from database.

Step 6. For each document do detailed comparison with document to check.

Step 7. Generate report for end user about plagiarism detection results.

2.7   Detailed Comparison

The search algorithm gives all similar documents and counts the number of common sentences. The average number of those documents is much, much smaller than the size of the reference database. The documents can be additionally processed to bring more information to users. We can also use less efficient but more accurate algorithms in order to perform a deeper check of resemblance.

Most anti-plagiarism systems determine and return long common phrases (longer than a given threshold), their length and numbers of appearance. We can calculate this by counting the number of consecutive sentences and their lengths or we may use a longest common subsequence algorithm, e.g., by Smith and Waterman [47]. The first leads to faster algorithms but the second provides deeper insight into the resemblance. We can also use other much more complex algorithms to improve the similarity detection at the expense of efficiency. Since the whole process in performed on a very small part of the reference database we may relax on small costs.

The next thing we can do is sentence validation. As mentioned in Sect. 2.5, if two different sentences have the same hash value, we must to resolve it in a special way, as described in [6].

The last, but not least thing we have to do in detailed comparison is generating presentation layer for the end user. In Poland advisers decide whether theses are ready for defense. Therefore, one makes their task easy by endowing then with a tool for analysis and comparison of common passages. Relying solely on the percentage of similarity is often inadequate and much more careful analysis is required. One exposes the similarities and indicates the fragments copied from reference documents.

2.8 Spell Checking

Another way to improve similarity detection is to use spelling correction. In NEKST the correction is based on the Levenshtein distance [32]. It uses a list build upon the *Morfeusz*[4] project and words collected from Polish websites (chosen with some frequency threshold). The system is capable of correcting authors mistakes and using corrected texts for plagiarism detection and detailed comparison.

2.9 Distributed Computing

The NEKST project is based on the Apache Hadoop framework implemented on the MapReduce paradigm. It was firstly presented in [15] and nowadays is the most used in processing large scale distributed data.

In the MapReduce we have to define the following signatures:

map: $(k_1, v_1) \rightarrow [(k_2, v_2)]$
reduce: $(k_2, [v_2]) \rightarrow [(k_3, v_3)]$

called mapper and reducer respectively.

In the first signature we map a pair of key value input data (k_1, v_1) to another pair (k_2, v_2). In the second we obtain their groups assigned to the key k_2, which forces the programmer to reduce all values to pairs of output data (k_3, v_3).

We input a list of documents, where k_1 is the identity number of a document and v_1 is the document body with metadata. This is to obtain information about sentences (hash values) and a list of document identities containing the hashes.

The mapper is used to process all documents crawled from the Internet and to extract pairs of sentence hash values and document identities in which these sentences occur. The Apache Hadoop groups the mapper output and passes the resulting pairs of sentence hash plus document's identities where the hash occurs. Next we loop over the list of identities, store them in single collection, and return a pair consisting of sentence hash value and the corresponding set of document identities, as illustrated in the signature below.

map: $(ID, Document) \rightarrow [(Hash, ID)]$
reduce: $(Hash, [ID]) \rightarrow [(Hash, IDs)]$

The output data can be exported to a database or a specialized search index and used to perform validation on single document (see Sect. 2.6).

---

[4] http://sgjp.pl/morfeusz/index.html.en.

2.10   Internal Plagiarism Detection in Collections

Sometimes there is a need to find all pairs of similar texts in large collections (sets) of documents. Determining the pairs with the longest common subsequence search leads to quadratic computational cost. By implementing the specialized search indices and the algorithm presented in Sect. 2.6, we reduce the cost estimate to Cn, where n is the number of documents in the set, and $C > 0$ is independent of n.

Using the MapReduce we simplify the process and take a good control over the size of C. We begin with hashing each sentence in the collection documents, as described in Sect. 2.9. Next we need to get information about sentences and all documents containing the same sentence. To this end we invert the relation document—sentence to produce pairs of similar documents and map them to the list of all common sentences. This can be achieved with the following MapReduce process:

map: $(Hash, IDs) \rightarrow [(Pair, Hash)]$
reduce: $(Pair, [Hash]) \rightarrow [(Pair, Hashes)]$

This results with pairs of similar documents and number of their common sentences. We can do additional filtration to ignore the pairs with number of similar sentences below some threshold. Finally, we may compare relevant documents, as described in Sect. 2.7, to give additional information to the end user.

This techniques can be also used to compare one collection to another simply by ignoring all pairs of documents belonging to the same collection.

2.11   Preventing Camouflage

One of the most popular technique used to camouflage illegal borrowings involves manipulating with entity codes from different scripts, for instance by replacing all instances of Roman uppercase A with looking the same Cyrillic uppercase A or Greek uppercase alpha. The simplest way of decrypting such replacements is based on lists of entity codes of look-alike characters from different scrips.

It is also important to take into account that determined plagiarist may use white characters instead of white spaces. Such characters are invisible when printed. They may jeopardise similarity detection when unattended. White dots and commas are especially dangerous for sentence-based systems.

To detect deceptions mentioned above one statistically analyzes out-of-dictionary words and usage of uppercase letters as well as occurrence of basic sentence elements like subjects, finite verbs and objects. If the results are essentially different from the reference average, a parsed version of the examined document is presented to the end-user.

2.12   Data Protection

Working on confidential documents require high security of the data. Sending original documents to an external system need good protection of network communication and additional safeguards. In the hash-based approach there is no actual need to send originals, since we may pursue similarities in the reference database using irreversible hases as the sole knowledge about the documents. This observation has been successfully adopted and implemented in the Open System for Antiplagiarism[5] (OSA) for communication with the Natively Enhanced Knowledge Sharing Technologies (NEKST) module handling the Internet. The document to be check is transformed into a sequence of hashes on the client side and then the sequence is sent to NEKST. This requires detailed comparisons described in Sect. 2.7 to be performed on the client side.

Sometimes sending original documents is not allowed by law. This is the case for texts in the All-Polish Repository of Diploma Writings[6]. Only partial and irreversible information about its documents can be transfer to OSA or any other antiplagiarism systems. And the hashes do the job. Even with very sophisticated techniques like rainbow tables [37], from sequences of document's hashes it is practically impossible to construct a text which is even though close to the original.

3   Semantic Similarity

3.1   Similar Documents

In most cases detecting plagiarism on sentence level is very effective. We can easily find copied sentences, paragraphs and whole documents. Usually plagiarists change only some passages and leave the rest. In such situations unchanged parts of the text are usually enough to reveal the source.

At the expense of extensive effort, a determined plagiarist can make serious modifications to texts without changing their meaning. Detecting such cases is much more difficult.

The most common method of finding similar texts use some similarity coefficients, see [30].

In [8] the author initiated a technique called MinHash. It is based on the Jaccard coefficient (or index) defined as:

$$J(A,B) = \frac{|A \cap B|}{|A \cup B|} \tag{3}$$

where A and B are two lemmatized texts (multisets) and $|\cdot|$ stand for the size, i.e. the number of lemmas counted wit multiplicities. The closer to 1, the more similar. If a hash function h maps the elements of A and B to distinct integers, then for any multiset S one defines $h_m(S)$ to be the member x of S such that

$$h_m(x) = \min\{h(w) : w \in S\}. \tag{4}$$

There is simple relation between minhashes and Jaccard coefficients. If A and B are randomly chosen, then $h_m(A) = h_m(B)$ holds with probability $J(A, B)$, In

---

[5] http://osaweb.pl/.
[6] http://polon.nauka.gov.pl/repozytorium-orpd.

[45] the authors compare minhashing with shingling described in Sect. 2.1. Since minhashing consumes less memory we decided to combine it with the Locality Sensitive Hashing [25] and implement it the NEKST engine. This allowes us to eliminate duplicated or nearly the same webpages from the search [50]. Unfortunately, this leads to very good results when comparing texts of more or less similar size.

Another approach to comparing texts is presented in [31]. It is based on frequency analysis and measuring the similarities with a cosine-like distance of weighted term frequency vectors. This leads to very good results, especially when applied to overlapping fragments, say up to 10, of the input texts. Since term frequency vectors can be easily computed during text preprocessing (see Sect. 2.3), the additional measuring can be applied in pairwise comparisons of input texts to the reference ones preselected in the hash-based search. This leads to a hybrid algorithm which gives very good plagiarism detection.

In [12] the authors propose to use Semantic Compression introduced in [41] as a process of transforming documents to more concise forms with the same meaning as the originals. They propose to replace some word with their representatives using synonyms and ontology.

In this work we introduce an idea of using some research results on natural language processing such as advanced text parsers and sentence decomposition (see [3, 40]).

The main idea is to enlarge the input document with some additional sentences corresponding to each sentence present in the document, while maintaining the condition that each sentence added has essentially the same meaning as its origin. Leaving the original sentences in search structures allows saying whether the copied text is identical or similar. This leads to very effective plagiarism detection at the expense of the increased search index.

3.2 Synonym Replacement

The most common way to confuse automatic plagiarism detection is to replace some words with their synonyms. There are many computer programs that can perform this process automatically.

The indicated drawback can be easily resolved by generating additional sentences based on words in synonym databases like Thesaurus or WordNet [35]. To prevent generating too many sentences, in each group of synonyms one selects a single representative to be used as the replacement of any other word in the group.

3.3 Acronyms

The NEKST system has its own module to detect and recognize acronyms on the Polish websites. The biggest advantage of this module is the ability to find new acronyms. When some phrase and its acronym occurs on many pages, say above a given threshold, then the phrase and the acronym are treated as synonyms.

4   How It Works

The NEKST database consists of over 0.7 billion Polish language web documents gathered and monitored by specialized web crawlers. The system is based on the MapReduce framework and the Hadoop Distributed File System (HDFS) which can run on thousands commodities (see [17,21] and Sect. 2.9).

In [49] the author presented initial ideas of the Open System for Antiplagiarism (OSA) and its operation in the Polish academic community. In [48] authors described OSA architecture before integration with NEKST. Actually, OSA and NEKST are joint together to prevent copyright infringement. Both systems communicate using the HTTPS protocol. After a thesis is added to OSA, the system starts plagiarism detection. On the client side the document is parsed and transformed into a bunch of its sentences' hashes, as described in Sects. 2.3 and 2.4. Then the bunch is sent to the NEKST REST API. On the server side NEKST searches for documents containing the sentences encrypted in the bunch (see Sect. 2.6) and returns the results to the client. This completes the initial stage of processing. After that OSA performs the final stage on the end user server which results in eliminating false positive indications and preparing a visual report for the user. The separation of the stages is to ensure data protection (Sect. 2.12).

At at the time of submitting this paper, each reindexing of 716,407,819 Polish web documents took about six hours on 70 machines (24 CPU cores, 128GB RAM, 6×3T hard drive). In this process all documents are divided into sentences and the corresponding hash table is generated. This involves MapReduce and Apache Ooozie Workflow Scheduler[7].

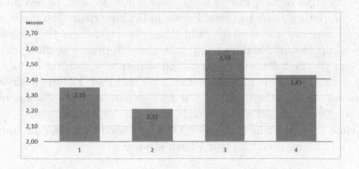

Fig. 1. Preselection tests in NEKST database

Figures 1 and 2 show performance tests of the preselection (see Sect. 2.6). The first one shows response times corresponding to four groups of 100 documents randomly chosen from a set consisting of 20,000 real (but anonymous) academic theses. The horizontal line shows the average time of response, which is 2,40 s per thesis. The differences are caused by different sizes of theses (more sentences

---

[7] http://oozie.apache.org/.

340   R. Szmit

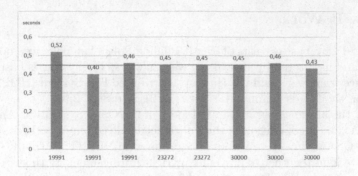

Fig. 2. Preselection tests in local database

gives more hashes to find) and technical reasons like communication with the
NEKST database. The second figure refers to preselection tests performed on
8 varied-size subsets of 30,000 anonymous theses stored on ThinkPad T440p
(Intel Core i7-4710MQ, CPU 2.50 GHz 4 cores, 8 threads, Hynix/Hyundai DDR3
16 GB 1600 MH, Crucial SSD 512 GB). The sizes are marked on the x-axis and
the horizontal line indicates the average time per thesis, which is 0,45 s. The
processing time is shorter because OSA and the subsets ware installed on the
same machine (no network communication) and a solid-state drive (SSD) was
used, which was much faster than hard drive used in NEKST.

In the tests described above the final stage of processing was ignored, since
its time depends on number and length of documents found in the initial stage.
The final stage time usually ranges from several milliseconds to several seconds.

So far, comparing OSA to other Polish anti-plagiarism systems is a hassle,
since their owners are not open for publishing descriptions of the methods used
nor for live comparisons. Needless to say, the naive approach to plagiarism detection based on the best longest common subsequence algorithm totally fails on
large-scale databases. This algorithm run on ThinkPad T440p (with specifications as above) would take about a year to compare a single document to 700
million reference texts, and about two days on the NEKST cluster.

At the time of submitting this paper, OSA was used by 30 leading universities
in Poland[8], educating about 35% of Polish students.

Acknowledgement. The author acknowledges his contribution in the NEKST
project (http://nekst.ipipan.waw.pl) founded by the Innovative Economy Operational
Programme (POIG.01.01.02-14-013/09) and in the OSA project founded by the
Interuniversity Centre for IT (MUCI – Miedzyuniwersyteckie Centrum Informatyzacji,
http://muci.edu.pl).

---

[8] http://osaweb.pl/node/22?language=en.

References

1. Alzahrani, S.M., Salim, N., Abraham, A.: Understanding plagiarism linguistic patterns, textual features, and detection methods. IEEE Trans. Syst. Man Cybern. Part C (Appl. Rev.) **42**(2), 133–149 (2012)
2. Barrón-Cedeño, A., Rosso, P.: On automatic plagiarism detection based on n-grams comparison. In: Boughanem, M., Berrut, C., Mothe, J., Soule-Dupuy, C. (eds.) ECIR 2009. LNCS, vol. 5478, pp. 696–700. Springer, Heidelberg (2009). doi:10. 1007/978-3-642-00958-7_69
3. Becker, M., Drożdżyński, W., Krieger, H.U., Piskorski, J., Schäfer, U., Xu, F.: Sprout - shallow processing with typed feature structures and unification. In: Proceedings of ICON 2002 - International Conference on NLP, Mumbai, India (2002)
4. Bloom, B.H.: Space/time trade-offs in hash coding with allowable errors. Commun. ACM **13**(7), 422–426 (1970)
5. Botelho, F.C., Ziviani, N.: External perfect hashing for very large key sets. In: Proceedings of the Sixteenth ACM Conference on Information and Knowledge Management, pp. 653–662. ACM (2007)
6. Brin, S., Davis, J., Garcia-Molina, H.: Copy detection mechanisms for digital documents. In: ACM SIGMOD Record, vol. 24, pp. 398–409. ACM (1995)
7. Brodal, G.S., Kaligosi, K., Katriel, I., Kutz, M.: Faster algorithms for computing longest common increasing subsequences. In: Lewenstein, M., Valiente, G. (eds.) CPM 2006. LNCS, vol. 4009, pp. 330–341. Springer, Heidelberg (2006). doi:10. 1007/11780441_30
8. Broder, A.Z.: On the resemblance and containment of documents. In: Compression and Complexity of Sequences 1997, Proceedings, pp. 21–29. IEEE (1997)
9. Cárdenas, A.F.: Analysis and performance of inverted data base structures. Commun. ACM **18**(5), 253–263 (1975)
10. Carter, J.L., Wegman, M.N.: Universal classes of hash functions. In: Proceedings of the Ninth Annual ACM Symposium on Theory of Computing, pp. 106–112. ACM (1977)
11. Ceglarek, D., Haniewicz, K.: Fast plagiarism detection by sentence hashing. In: Rutkowski, L., Korytkowski, M., Scherer, R., Tadeusiewicz, R., Zadeh, L.A., Zurada, J.M. (eds.) ICAISC 2012. LNCS (LNAI), vol. 7268, pp. 30–37. Springer, Heidelberg (2012). doi:10.1007/978-3-642-29350-4_4
12. Ceglarek, D., Haniewicz, K., Rutkowski, W.: Robust plagiary detection using semantic compression augmented SHAPD. In: Nguyen, N.-T., Hoang, K., Jędrzejowicz, P. (eds.) ICCCI 2012. LNCS (LNAI), vol. 7653, pp. 308–317. Springer, Heidelberg (2012). doi:10.1007/978-3-642-34630-9_32
13. Cichelli, R.J.: Minimal perfect hash functions made simple. Commun. ACM **23**(1), 17–19 (1980)
14. Czerski, D., Ciesielski, K., Dramiński, M., Kłopotek, M.A., Wierzchoń, S.T.: Inverted lists compression using contextual information. In: Pejaś, J., Saeed, K. (eds.) Advances in Information Processing and Protection, pp. 55–66. Springer, Boston (2007)
15. Dean, J., Ghemawat, S.: MapReduce: simplified data processing on large clusters. Commun. ACM **51**(1), 107–113 (2008)
16. Dietzfelbinger, M., Gil, J., Matias, Y., Pippenger, N.: Polynomial hash functions are reliable. In: Kuich, W. (ed.) ICALP 1992. LNCS, vol. 623, pp. 235–246. Springer, Heidelberg (1992). doi:10.1007/3-540-55719-9_77
17. Foundation, A.S.: Apache Hadoop (2015). http://hadoop.apache.org/

18. Fox, E.A., Heath, L.S., Chen, Q.F., Daoud, A.M.: Practical minimal perfect hash functions for large databases. Commun. ACM **35**(1), 105–121 (1992)
19. Gauss, C.F.: Disquisitiones Arithmeticae, vol. 157. Yale University Press, New Haven (1966)
20. Gustafson, N., Pera, M.S., Ng, Y.K.: Nowhere to hide: finding plagiarized documents based on sentence similarity. In: Proceedings of the 2008 IEEE/WIC/ACM International Conference on Web Intelligence and Intelligent Agent Technology, vol. 01, pp. 690–696. IEEE Computer Society (2008)
21. Hameurlain, A., Morvan, F.: Big Data management in the cloud: evolution or crossroad? In: Kozielski, S., Mrozek, D., Kasprowski, P., Małysiak-Mrozek, B., Kostrzewa, D. (eds.) BDAS 2015-2016. CCIS, vol. 613, pp. 23–38. Springer, Cham (2016). doi:10.1007/978-3-319-34099-9_2
22. Hirschberg, D.S.: Algorithms for the longest common subsequence problem. J. ACM (JACM) **24**(4), 664–675 (1977)
23. Hunt, J.W., Szymanski, T.G.: A fast algorithm for computing longest common subsequences. Commun. ACM **20**(5), 350–353 (1977)
24. Iliopoulos, C.S., Rahman, M.S.: A new efficient algorithm for computing the longest common subsequence. Theor. Comput. Syst. **45**(2), 355–371 (2009)
25. Indyk, P., Motwani, R.: Approximate nearest neighbors: towards removing the curse of dimensionality. In: Proceedings of the Thirtieth Annual ACM Symposium on Theory of Computing, pp. 604–613. ACM (1998)
26. Irving, R.W.: Plagiarism and collusion detection using the smith-waterman algorithm. DCS Technical report, University of Glasgow, pp. 1–24 (2004)
27. Kahn, R.E.: Deposit, registration and recordation in an electronic copyright management system. Technical report, Corporation for National Research Initiatives, Reston, Virginia (1992)
28. Kang, N.O., Gelbukh, A., Han, S.Y.: PPChecker: plagiarism pattern checker in document copy detection. In: Sojka, P., Kopeček, I., Pala, K. (eds.) TSD 2006. LNCS (LNAI), vol. 4188, pp. 661–667. Springer, Heidelberg (2006). doi:10.1007/11846406_83
29. Kang, N.O., Han, S.Y.: Document copy detection system based on plagiarism patterns. In: Gelbukh, A. (ed.) CICLing 2006. LNCS, vol. 3878, pp. 571–574. Springer, Heidelberg (2006). doi:10.1007/11671299_60
30. Knuth, D.E.: The Art of Computer Programming: Sorting and Searching, vol. 3. Pearson Education (1998)
31. Kowalski, M., Szczepański, M.: Identity of academic theses. In: Dobrzyńska, T., Kuncheva, R. (eds.) Resemblance and Difference. The problem of identity, pp. 259–278. Instytut Badań Literackich Polskiej Akademii Nauk (2015)
32. Levenshtein, V.I.: Binary codes capable of correcting deletions, insertions, and reversals. Soviet physics doklady **10**, 707–710 (1966)
33. Manber, U., et al.: Finding similar files in a large file system. Usenix Winter **94**, 1–10 (1994)
34. Masek, W.J., Paterson, M.S.: A faster algorithm computing string edit distances. J. Comput. Syst. Sci. **20**(1), 18–31 (1980)
35. Maziarz, M., Piasecki, M., Szpakowicz, S.: Approaching plWordNet 2.0. In: Proceedings of the 6th Global WordNet Conference, Matsue, Japan, pp. 50–62 (2012)
36. Miłkowski, M., Lipski, J.: Using SRX standard for sentence segmentation. In: Vetulani, Z. (ed.) LTC 2009. LNCS (LNAI), vol. 6562, pp. 172–182. Springer, Heidelberg (2011). doi:10.1007/978-3-642-20095-3_16

37. Oechslin, P.: Making a faster cryptanalytic time-memory trade-off. In: Boneh, D. (ed.) CRYPTO 2003. LNCS, vol. 2729, pp. 617–630. Springer, Heidelberg (2003). doi:10.1007/978-3-540-45146-4_36
38. Osman, A.H., Salim, N., Kumar, Y.J., Abuobieda, A.: Fuzzy semantic plagiarism detection. In: Hassanien, A.E., Salem, A.-B.M., Ramadan, R., Kim, T. (eds.) AMLTA 2012. CCIS, vol. 322, pp. 543–553. Springer, Heidelberg (2012). doi:10.1007/978-3-642-35326-0_54
39. Pagh, R., Rodler, F.F.: Cuckoo hashing. J. Algorithms **51**, 122–144 (2004)
40. Paik, W., Liddy, E.D., Liddy, J.H., Niles, I.H., Allen, E.E.: Information extraction system and method using concept-relation-concept (CRC) triples, 17 July 2001. US Patent 6,263,335
41. Percova, N.N.: On the types of semantic compression of text. In: COLING, pp. 229–231 (1982)
42. Piskorski, J.: Rule-based named-entity recognition for polish. In: Proceedings of the Workshop on Named-Entity Recognition for NLP Applications held in Conjunction with the 1st International Joint Conference on NLP (2004)
43. Polaski, E.: Wielki sownik ortograficzny PWN. Wydawnictwo Naukowe PWN (2003)
44. Przepirkowski, A., Bako, M., Grski, R., Lewandowska-Tomaszczyk, B.: Narodowy korpus jezyka polskiego. PWN, Warszawa (2012)
45. Rajaraman, A., Ullman, J.D.: Mining of Massive Datasets, vol. 77. Cambridge University Press, Cambridge (2012)
46. Saloni, Z., Gruszczyński, W., Woliński, M., Wołosz, R., Skowrońska, D.: Website of the morphological analyser morfeusz (2011). http://sgjp.pl/morfeusz/index.html.en
47. Smith, T.F., Waterman, M.S.: Identification of common molecular subsequences. J. Mol. Biol. **147**(1), 195–197 (1981)
48. Sobieski, Ś., Kowalski, M.A., Kruszyński, P., Sysak, M., Zieliński, B., Maślanka, P.: OSA architecture. In: Kozielski, S., Mrozek, D., Kasprowski, P., Małysiak-Mrozek, B., Kostrzewa, D. (eds.) BDAS 2015-2016. CCIS, vol. 613, pp. 571–584. Springer, Cham (2016). doi:10.1007/978-3-319-34099-9_44
49. Szczepański, M.: Algorytmy klasyfikacji tekstów i ich wykorzystanie w systemie wykrywania plagiatów. Oficyna Wydawnicza Politechniki Warszawskiej (2014)
50. Szmit, R.: Locality sensitive hashing for similarity search using MapReduce on large scale data. In: Kłopotek, M.A., Koronacki, J., Marciniak, M., Mykowiecka, A., Wierzchoń, S.T. (eds.) IIS 2013. LNCS, vol. 7912, pp. 171–178. Springer, Heidelberg (2013). doi:10.1007/978-3-642-38634-3_19
51. Wang, T.: Integer hash function. (2007). http://www.concentric.net/~ttwang/tech/inthash.htm
52. White, D.R., Joy, M.S.: Sentence-based natural language plagiarism detection. J. Educ. Res. Comput. (JERIC) **4**(4), 2 (2004)

RDF Validation: A Brief Survey

Dominik Tomaszuk[(✉)]

Institute of Informatics, University of Bialystok,
ul. Konstantego Ciołkowskiego 1M, 15-245 Białystok, Poland
d.tomaszuk@uwb.edu.pl

Abstract. The last few years have brought a lot of changes in the RDF
validation and integrity constraints in the Semantic Web environment,
offering more and more options. This paper analyses the current state
of knowledge on RDF validation and proposes requirementsL for RDF
validation languages. It overviews and compares the previous approaches
and development directions in RDF validation. It also points at the pros
and cons of particular implementation scenarios.

Keywords: Validation · Semantic Web · RDF · Integrity constraints

1 Introduction and Motivation

Schemes pay a key role in databases. They organize data and determine the way
in which the database is constructed and what integrity constraints it is affected
by. Databases allow for checking the conformity of their instances with the given
scheme. Schemes in the RDF context describe integrity constraints imposed by
the application on documents and/or RDF graph stores. Integrity constraints
may account for the multitude of relations, restrictions in the allowed property
values, the presence of properties, certain types of data values or default values.
Current solutions [6,17,18,20,26] are trying to answer the needs of defining the
graph's structure for validation and the way of matching the data with the
description. The main purpose of this paper is to support discussions of those
solutions and to help working groups in the development of suitable approaches.

In this paper, we perform an overview and comparison of current options for
RDF validation. Section 2 describes the basic notions of RDF. Section 3 presents
requirements for RDF validation languages. In Sect. 4 we perform an overview of
solutions and we compare their characteristics and expressiveness. This section
is also devoted to related work. Section 5 presents experiments performed to
evaluate presented approaches. The last section is a conclusion.

2 RDF Background

An RDF constitutes a universal method of the description and information
modeling accessible in Web resources. RDF is a very common data model for

© Springer International Publishing AG 2017
S. Kozielski et al. (Eds.): BDAS 2017, CCIS 716, pp. 344–355, 2017.
DOI: 10.1007/978-3-319-58274-0_28

resources and relationship description between them. In other words, tt provides the crucial foundation and infrastructure to support the management and description of data.

An assumption in RDF [12] is to describe resources by means of the statement consisting of three elements (the so-called RDF *triple*): *subject*, *predicate* and *object*. RDF borrows often from natural languages. An RDF *triple* may then be seen as an expression with *subject* corresponding to the subject of a sentence, *predicate* corresponding to its verb and *object* corresponding to its object. The RDF data model depends on the concept of creating web-resource expressions in the form of subject-predicate-object statements, which in the RDF terminology, are referred to as *triples*.

An RDF triple comprises a subject, a predicate, and an object. In [12], the meaning of subject, predicate and object is explained. The *subject* denotes a resource, the *object* fills the value of the relation, the *predicate* refers to the aspects or features of resource and expresses a subject – object relationship. In other words, the predicate (also known as a property) denotes a binary relation.

The elemental constituents of the RDF data model are RDF terms that can be used in reference to resources: anything with identity. The set of RDF terms is divided into three disjoint subsets:

- IRIs,
- literals,
- blank nodes.

Definition 1 (IRIs). IRIs *serve as global identifiers that can be used to identify any resource. For example,*

Example 1 (IRIs). <http://dbpedia.org/page/Car> is used to identify the car in DBpedia [1].

Definition 2 (Literals). Literals *are a set of lexical values.*

Example 2 (Literals). Literals comprise a lexical string and a datatype, such as "1"^^http://www.w3.org/2001/XMLSchema#int. Datatypes are identified by IRIs, where RDF borrows many of the datatypes defined in XML Schema 1.1 [29].

Definition 3 (Blank nodes). Blank nodes *are defined as existential variables used to denote the existence of some resource for which an IRI or literal is not given. They are inconstant for blank nodes and are in all cases locally scoped to the RDF space.*

RDF triple is composed of the above terms. Following [12], we provide definitions of RDF triples below.

Definition 4 (RDF triple). *Assume that $\mathcal{I}$ is the set of all Internationalized Resource Identifier (IRI) references, $\mathcal{B}$ (an infinite) set of blank nodes, $\mathcal{L}$ the set of literals. An* RDF triple *is defined as a triple $t = \langle s, p, o \rangle$ where $s \in \mathcal{I} \cup \mathcal{B}$ is called the* subject, *$p \in \mathcal{I}$ is called the* predicate *and $o \in \mathcal{I} \cup \mathcal{B} \cup \mathcal{L}$ is called the* object.

Example 3 (RDF triple). The example presents an RDF triple consisting of subject, predicate and object.

```
<http://example.net/spec#d> rdfs:label "specification".
```

A set of RDF triples intrinsically represents a labeled directed multigraph. The nodes are the subjects and objects of their triples. RDF is often referred to as being *graph structured data* where each $\langle s, p, o \rangle$ triple can be interpreted as an edge $s \xrightarrow{p} o$.

Definition 5 (RDF graph). *Let* $\mathcal{O} = \mathcal{I} \cup \mathcal{B} \cup \mathcal{L}$ *and* $\mathcal{S} = \mathcal{I} \cup \mathcal{B}$. $G \subset \mathcal{S} \times \mathcal{I} \times \mathcal{O}$ *is a finite subset of* RDF *triples and called an* RDF *graph.*

Example 4 (RDF graph). The example presents an RDF graph of a FOAF [8] profile in Turtle [2] syntax. This graph includes two RDF triples:

```
<#d>      rdf:type      foaf:Document.
<#d>      rdfs:label    "specification".
```

3   RDF Validation Language Requirements

In this section, we propose requirements that an RDF validation language should fulfil. From presented approaches [6,17,18,20,26] and our experience in using RDF validation language we have composed a list of key requirements:

1. representatable in RDF and concise language,
2. expressive power,
3. shortcuts to recurring patterns,
4. self-describability,
5. provide a standard semantics.

3.1   Representatable in RDF

Any non-RDF syntax will lose a key advantage of a triple-based notation. RDF is well-implemented so RDF validation language shall use RDF as its syntax. On the other hand, RDF constraints should be specifiable in a compact form. Table 1 presents key approaches and their syntaxes. A language shall be designed to easily integrate into deployed systems that use RDF (i.e. graph stores), and provides a smooth upgrade. The use of RDF in a validation language makes constraints accessible to developers without the obligation to install additional parsers, software libraries or other programs. This requirement is satisfied by all RDF validation languages, but an RDF syntax is the primary format for SHACL, ReSh and SPIN.

Table 1. Summary of RDF validation approaches

Name	Syntax	Related work
ShEx	Compact, RDF, JSON	[4,20,24]
SHACL	RDF	[17]
ReSh	RDF	[25,26]
DSP	XML, RDF[a]	[6,11]
SPIN	RDF, SPARQL	[14,18]
OWL, RDFS	RDF (Turtle), XML, Manchester, functional	[7,21,22]

[a]The main syntax is XML.

3.2 Expressive Power

An RDF validation language should express:

- RDF terms restrictions:
  - value restrictions,
  - allowed values and not allowed values,
  - default values,
  - literal values,
  - datatypes comparison,
  - IRI pattern matching,
- cardinality constraints:
  - minimum cardinality,
  - maximum cardinality,
  - exact cardinality,
- predicate constraints:
  - class-specific property range,
  - OR operator (including groups),
  - required predicates,
  - optional predicates,
  - repeatable predicates,
  - negative predicate constraints.

Table 2 presents features of the below-mentioned approaches, namely: RDF terms constraints, cardinality and predicates constraints compared to RDFS and OWL. Note that OWL and RDFS are developed for inference and do not provide the features strictly for validation. This requirement is not satisfied by any languages, but SHACL supports most of the features.

3.3 Shortcuts to Recurring Patterns

Another important requirement is a macro supporting. An RDF validation language shall enable the definition of shortcuts to recurring patterns that improve overall readability and maintainability. Macros also can separate various parts of schema and enable users define rich constraints. It should provide a way to define high level reusable components in SPARQL [15], JavaScript or other languages. This requirement is satisfied by ShEx, SHACL and SPIN.

Table 2. Validation features

	ShEx	SHACL	ReSh	DSP	SPIN	RDFS, OWL
RDF terms						
Value restrictions	☑	☑	☑	☑	☑	☑
Allowed values	☑	☑	☑	☑	☑	☑
Not allowed values	☑	☑	☒	☒	☑	☑
Default values	☒	☑	☑	☒	☑	☒
Literal values	☑	☑	☒	☒	☑	☒
Datatypes comparison	☑	☑	☑	☑	☑	☑
IRI pattern matching	☑	☑	☒	☒	☑	☒
Cardinality						
Min. cardinality	☑	☑	☐	☑	☑	☑
Max. cardinality	☑	☑	☐	☑	☑	☑
Exact cardinality	☑	☑	☐	☑	☑	☑
Predicates						
Class-specyfic property range	☑	☑	☑	☑	☑	☑
OR operator	☑	☑	☒	☒	☑	☑
OR operator (groups)	☑	☐	☑	☒	☑	☐
Required predicates	☑	☑	☑	☑	☑	☑
Optional predicates	☑	☑	☑	☑	☑	☑
Repeatable predicates	☑	☑	☑	☑	☑	☑
Negative predicate constraints	☑	☑	☒	☒	☑	☑

All features are marked as ☑ (yes), ☐ (partial) or ☒ (no).

3.4   Self-describability

An RDF validation language shall represent the schema specification using the schema itself being defined. This makes creating meta scheme easier than in a language without this property, since constraints can be treated as data. It means that examining the schema's entities depends on a homogeneous structure, and it does not have to handle several different structures that would appear in a complex syntax. Moreover, this meta schema can valuable in bootstrapping the implementation of the constraint language. This requirement is satisfied by all RDF validation languages.

3.5   Standard Semantics

An RDF validation language shall provide a standard semantics for describing RDF constraints. It should be defined in a commonly agreed-upon formal specification, which describe a model-theoretic semantics for RDF constraints, providing an exact formal specification of when truth is preserved by various validation operations. This requirement is satisfied only by SHACL.

4   RDF Validation

In this section we discuss main approaches, which are strictly designed for validation (Sect. 4.1) and other approaches, which can be used for some scenarios (Sect. 4.2).

4.1   Main Approaches

Shape Expressions. The first approach is Shape Expressions [4, 20, 24], which is a language for describing constraints on RDF triples. In addition to validating RDF, ShEx also allows to inform expected graph patterns for interfaces and create graphical user interface forms. It is a domain specific language used to define shapes, which describe conditions that handle a given node. It can be transformed into SPARQL queries. The most common syntax of ShEx compact, which is similar to RELAX NG Compact [9].

Example 5 (ShEx). The example presents a Shape Expressions schema in compact syntax according to Example 3.

```
<Shape1> {
  (rdfs:label xsd:string+)
}
```

Shapes Constraint Language. The next approach is the Shapes Constraint Language [17], which is a language constraining the contents of graphs. SHACL, similarly to ShEx, organises these constraints into shapes, which provide a high-level vocabulary to distinguish RDF predicates and their constraints. Moreover, constraints can be linked with shapes using SPARQL queries and JavaScript. In addition to validating RDF, it also allows to describe information about data structures, generate RDF data, and build GUIs.

Example 6 (SHACL). The example presents a Shapes Constraint Language schema in Turtle according to Example 3.

```
<Shape1> a sh:Shape;
  sh:property [
    sh:predicate rdfs:label;
    sh:datatype xsd:string;
    sh:minCount 1.
  ].
```

Resource Shapes. Another approach is Resource Shapes [25, 26] which is a vocabulary for specifying the RDF shapes. ReSh authors assume that RDF terms come from many vocabularies. The ReSh shape is a description of the RDF graphs to integrity constraints those data are required to satisfy.

Example 7 (ReSh). The example presents a Resource Shapes schema in Turtle according to Example 3.

```
<Shape1> a oslc:ResourceShape;
  oslc:property [
    a oslc:Property;
    oslc:propertyDefinition rdfs:label;
    oslc:valueType xsd:string;
    oslc:occurs oslc:One-or-many
  ].
```

Description Set Profiles and SPARQL Inferencing Notation. ShEx, SHACL and ReSh are based on declarative and a high-level descriptions of RDF graph contents. These three approaches have similar features. Description Set Profiles [6,11] and SPARQL Inferencing Notation [18] are different approaches. The first is used to define structural constraints on data in a Dublin Core Application Profile. That approach allows to declare the metadata record contents in terms of validatable constraints. In addition to validating RDF, it can be used as configuration for databases and configuration for metadata editing software.

Example 8 (DSP). The example presents a Description Set Profiles schema in XML according to Example 3.

```
<dsp:StatementTemplate minOccurs="1"
 maxOccurs="infinity" type="literal">
  <dsp:Property>
http://www.w3.org/2000/01/rdf-schema#label
  </Property>
</dsp:StatementTemplate>
```

The latter one is a constraint and SPARQL-based rule language for RDF. It can link class with queries to capture constraints and rules which describe the behavior of those classes. SPIN is also a method to represent queries as templates. It can represent SPARQL statement as RDF triples. That proposal allows to declare new SPARQL functions.

Example 9 (SPIN). The example presents a SPARQL Inferencing Notation schema in Turtle according to Example 3.

```
<#c> spin:constraint [
  a spl:Attribute;
  spl:predicate rdfs:label;
  spl:minCount 1;
  spl:valueType xsd:string.
].
```

4.2   Other Approaches

In addition to the approaches presented in Sect. 4.1 there are OWL/RDFS [7,22] based approaches [10,21,23,28,30] and SPARQL [15] based approaches [13,19, 27]. The first group uses of RDF and OWL expressions with a CWA (Closed World Assumption) to express integrity constraints. The main drawback of those approaches is that inference engines cannot be used for checking the constraints.

Kontokostas *et al.* [19] focus on the data quality test patterns and data quality integrity constraints, which are represented in SPARQL query templates. Fischer *et al.* present RDD, language for expressing RDF constraints, which can be easily transformed into SPARQL queries. In the [27] authors propose approach based on SPARQL property paths. This proposal is similar to Schematron [16] in XML documents. The main drawback of those approaches is the need to know SPARQL. Another approach to the requirement is presented in [5].

5   Evaluation

In this Section we evaluate the most promising constraint languages. We evaluate all shape approaches (ShEx, SHACL and ReSh) that have community support and advanced features (see Table 2). In this Section we analyze whether languages are concise and how fast we can process them.

All experiments have been executed on a Intel Core i7-4770K CPU @ 3.50GHz (4 cores, 8 thread), 8GB of RAM, and a HDD with reading speed rated at ∼160 MB/s[1]. We have been used Linux Mint 17.3 Rosa (kernel version 3.13.0) and Python 3.4.3 with RDFLib 4.2.1 and Virtuoso Open-Source Edition 7.2.4.

We prepare constraint rules for RDF data that was generated by Berlin SPARQL Benchmark (BSBM) [3]. According to [3], the BSBM benchmark is settled in an e-commerce scenarios in which a set of products (denotated P) is offered by different vendors and consumers who have posted reviews on various sites. The benchmark defines an abstract data model for this scenarios together with data production rules that allow benchmark datasets to be scaled to arbitrary sizes using the number of products as a scale factor. We enrich BSBM to fake datatypes (denotated E). Our implementation is available at https://github.com/domel/bsbm_validation. The data description which was used in the experiment, is presented in Table 3.

All RDF validation languages were created in RDF syntax. We choose N-Triples[2] because this serialization is normalized. In ShEx, SHACL and ReSh we declared appropriate datatypes for predicates such as: `foaf:name`, `dc:publisher`, `dc:title`, `bsbm:price`, `bsbm:validFrom`, `bsbm:validTo` and `bsbm:deliveryDays`. Incorrect values refer to `dc:date` predicate. We choose that scenario, because al validation languages support tested features, such as value and datatype restrictions. We also prepare constraint rules for all BSBM datasets.

---

[1] We test it in `hdparm -t`.
[2] https://www.w3.org/TR/n-triples/#canonical-ntriples.

Table 3. Datasets description

$|P|$	$|G|$	$|E|$	$|P|$	$|G|$	$|E|$
10	4987	614	60	26143	2913
20	8458	928	70	29707	3230
30	11962	1244	80	33194	3544
40	16901	1845	90	36699	3859
50	22629	2598	100	40177	4174

$|P|$ - cardinality of BSBM products.
$|G|$ - cardinality of RDF triples.
$|E|$ - cardinality of fake datatypes.

To test our schemes for different datasets, we built an RDF data validation system. The prototype implementation of the testbed has been fully implemented in the Python programming language with RDFLib. For storage, a Virtuoso Open-Source Edition 7.2.4 RDF graph store has been used. Figure 1 shows architecture of our testbed. Our data validation system checks datatypes on-the-fly before loading data.

Definition 6 (RDF data validation system). *An* RDF data validation system *is a tuple* $\langle \mathcal{D}, C, \Sigma \rangle$ *where* $\mathcal{D}$ *is the source data,* C *is the constraints and* Σ *is the state of the system.*

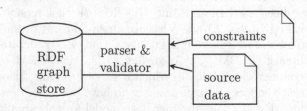

Fig. 1. Testbed architecture

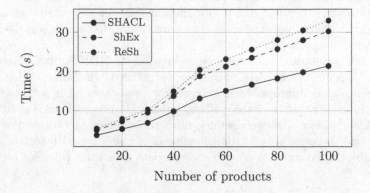

Fig. 2. Evaluation times of validation and loading valid data

Figure 2 shows that the number of predicates needed to describe validation rules has the greatest influence on the differences in the validation times. The fastest processing time belongs to SHACL and the slowest belongs to ReSh.

6   Conclusions

RDF validation is an important issue in the Semantic Web stack. There is no standard way to validate RDF data conforming to RDF constraints like DDL for relational databases or DTD for XML documents. Integrity constraints may be necessary in editing documents and actions performed on RDF graph stores. In this article, we conduct an overview of approaches to RDF validation and we show the differences in these approaches, drawing attention to the advantages and disadvantages of particular solutions.

We argue that RDF validation on the Semantic Web nowadays faces some of the challenges we were facing in the past, when databases were at their infancy. However, this area evolves very fast and attracts the attention of many researchers, the resulting in the vast scope of works we showed in this paper. We hope that this survey contributes to a better understanding of RDF validation. As part of future work, we will continuously analyze solutions within the RDF validation area.

Acknowledgment. We thank David Wood, co-chair of the RDF Working Group, for comments that greatly improved the paper.

A   Used Prefixes

In Table 4 we enumerate the prefixes used throughout this paper to abbreviate IRIs.

Table 4. Used prefixes

Prefix	IRI
rdf	http://www.w3.org/1999/02/22-rdf-syntax-ns#
rdfs	http://www.w3.org/2000/01/rdf-schema#
foaf	http://xmlns.com/foaf/0.1/
xsd	http://www.w3.org/2001/XMLSchema#
sh	http://www.w3.org/ns/shacl#
oslc	http://open-services.net/ns/core#
dsp	http://purl.org/metainfo/terms/dsp#
spin	http://spinrdf.org/spin#
spl	http://spinrdf.org/spl#
dc	http://purl.org/dc/elements/1.1/
bsbm	http://www4.wiwiss.fu-berlin.de/bizer/bsbm/v01/vocabulary/

References

1. Auer, S., Bizer, C., Kobilarov, G., Lehmann, J., Cyganiak, R., Ives, Z.: DBpedia: a nucleus for a web of open data. In: Aberer, K., et al. (eds.) ASWC/ISWC-2007. LNCS, vol. 4825, pp. 722–735. Springer, Heidelberg (2007). doi:10.1007/978-3-540-76298-0_52

2. Beckett, D., Berners-Lee, T., Prud'hommeaux, E., Carothers, G.: Turtle - Terse RDF Triple Language. W3C recommendation, World Wide Web Consortium. https://www.w3.org/TR/2014/REC-turtle-20140225/

3. Bizer, C., Schultz, A.: The Berlin SPARQL benchmark. Int. J. Semant. Web Inf. Syst. (IJSWIS) 5(2), 1–24 (2009)

4. Boneva, I., Labra Gayo, J.E., Hym, S., Prud'hommeau, E.G., Solbrig, H.R., Staworko, S.: Validating RDF with shape expressions. CoRR abs/1404.1270 (2014)

5. Bosch, T., Eckert, K.: Requirements on RDF constraint formulation and validation. In: Proceedings of the 2014 International Conference on Dublin Core and Metadata Applications, pp. 95–108. Citeseer (2014)

6. Bosch, T., Eckert, K.: Towards description set profiles for RDF using SPARQL as intermediate language. In: Proceedings of the 2014 International Conference on Dublin Core and Metadata Applications (DCMI 2014), pp. 129–137. Dublin Core Metadata Initiative (2014)

7. Brickley, D., Guha, R.: RDF Schema 1.1. W3C recommendation, World Wide Web Consortium, February 2014. http://www.w3.org/TR/2014/REC-rdf-schema-20140225/

8. Brickley, D., Miller, L.: FOAF vocabulary specification 0.99. Technical report FOAF Project, January 2014. http://xmlns.com/foaf/spec/20140114.html

9. Clark, J.: RELAX NG compact syntax. Committee specification, The Organization for the Advancement of Structured Information Standards (2002). http://www.oasis-open.org/committees/relax-ng/compact-20021121.html

10. Clark, K., Sirin, E.: On RDF validation, stardog ICV, and assorted remarks. In: RDF Validation Workshop. Practical Assurances for Quality RDF Data, Cambridge, MA, Boston (2013)

11. Coyle, K., Baker, T.: Dublin core application profiles. Separating validation from semantics. In: RDF Validation Workshop. Practical Assurances for Quality RDF Data, Cambridge, MA, Boston (2013)

12. Cyganiak, R., Lanthaler, M., Wood, D.: RDF 1.1 concepts and abstract syntax. W3C recommendation, World Wide Web Consortium, February 2014. http://www.w3.org/TR/2014/REC-rdf11-concepts-20140225/

13. Fischer, P.M., Lausen, G., Schätzle, A., Schmidt, M.: RDF constraint checking. In: EDBT/ICDT Workshops, pp. 205–212 (2015)

14. Fürber, C., Hepp, M.: Using SPARQL and SPIN for data quality management on the semantic web. In: Abramowicz, W., Tolksdorf, R. (eds.) BIS 2010. LNBIP, vol. 47, pp. 35–46. Springer, Heidelberg (2010). doi:10.1007/978-3-642-12814-1_4

15. Harris, S., Seaborne, A.: SPARQL 1.1 Query Language. W3C recommendation, World Wide Web Consortium. https://www.w3.org/TR/2013/REC-sparql11-query-20130321/

16. Jelliffe, R.: The schematron assertion language 1.5. Academia Sinica Computing Center (2000)

17. Knublauch, H.: Shapes constraint language (SHACL). W3C editor's draft, World Wide Web Consortium, September 2014. http://w3c.github.io/data-shapes/shacl/

18. Knublauch, H., Hendle, J.A., Idehen, K.: SPIN - overview and motivation. W3C member submission, World Wide Web Consortium, February 2011. http://www.w3.org/Submission/2011/SUBM-spin-overview-20110222/
19. Kontokostas, D., Westphal, P., Auer, S., Hellmann, S., Lehmann, J., Cornelissen, R., Zaveri, A.: Test-driven evaluation of linked data quality. In: Proceedings of the 23rd International Conference on World Wide Web, pp. 747–758. ACM (2014)
20. Labra Gayo, J., Prud'hommeaux, E., Solbrig, H., Rodríguez, J.: Validating and describing linked data portals using RDF shape expressions. In: Workshop on Linked Data Quality (2015)
21. Motik, B., Horrocks, I., Sattler, U.: Adding integrity constraints to OWL. In: Third International Workshop on OWL: Experiences and Directions 2007 (OWLED 2007), Innsbruck, Austria (2007)
22. Patel-Schneider, P., Hayes, P.: RDF 1.1 semantics. W3C recommendation, World Wide Web Consortium, February 2014. http://www.w3.org/TR/2014/REC-rdf11-mt-20140225/
23. Pérez-Urbina, H., Sirin, E., Clark, K.: Validating RDF with OWL integrity constraints (2012). http://docs.stardog.com/icv/icv-specification.html
24. Prud'hommeaux, E., Labra Gayo, J.E., Solbrig, H.: Shape expressions: an RDF validation and transformation language. In: Proceedings of the 10th International Conference on Semantic Systems (SEM 2014), NY, USA, pp. 32–40. ACM, New York (2014)
25. Ryman, A.: Resource shape 2.0. W3C member submission, World Wide Web Consortium, February 2014. http://www.w3.org/Submission/2014/SUBM-shapes-20140211/
26. Ryman, A.G., Hors, A.L., Speicher, S.: OSLC resource shape: a language for defining constraints on linked data. In: LDOW, CEUR Workshop Proceedings, vol. 996. CEUR-WS.org (2013)
27. Simister, S., Brickley, D.: Simple application-specific constraints for RDF models. In: RDF Validation Workshop. Practical Assurances for Quality RDF Data (2013)
28. Sirin, E., Tao, J.: Towards integrity constraints in OWL. In: OWLED, vol. 529 (2009)
29. Sperberg-McQueen, M., Thompson, H., Peterson, D., Malhotra, A., Biron, P.V., Gao, S.: W3C XML schema definition language (XSD) 1.1 Part 2: datatypes. W3C recommendation, World Wide Web Consortium, April 2012. http://www.w3.org/TR/2012/REC-xmlschema11-2-20120405/
30. Tao, J., Sirin, E., Bao, J., McGuinness, D.L.: Integrity constraints in OWL. In: AAAI (2010)

Bioinformatics and Biological Data Analysis

Objective Clustering Inductive Technology of Gene Expression Sequences Features

Sergii Babichev[1(✉)], Volodymyr Lytvynenko[2], Maxim Korobchynskyi[3], and Mochamed Ali Taiff[2]

[1] Jan Evangelista Purkyne University, Usti nad Labem, Czech Republic 8, Ceske Mladeze Street, 400 96 Usti nad Labem, Czech Republic
sergii.babichev@ujep.cz
[2] Kherson National Technical University, Kherson, Ukraine
immun56@gmail.com
[3] Military-Diplomatic Academy Named Eugene Bereznyak, Kiev, Ukraine
http://www.sci.ujep.cz

Abstract. Technology of high dimensional data features objective clustering based on the methods of complex systems inductive modeling is presented in the paper. Architecture of the objective clustering inductive technology as a block diagram of step-by-step implementation of the objects clustering procedure was developed. Method of criterial evaluation of complex data clustering results using two equal power data subsets is proposed. Degree of clustering objectivity evaluates on the basis of complex use of internal and external criteria. Researches on the simulation results of the proposed technology based on the SOTA self-organizing clustering algorithm using the gene expression data obtained by DNA microarray analysis of patients with lung cancer GEOD-68571 Array Express database, the datasets "Compound" and "Aggregation" of the Computing School of the Eastern Finland University and the data "soods" are presented.

Keywords: Clustering · Inductive modeling · Gene expression · High dimensional data

1 Introduction

The process of the gene regulatory networks creation based on the gene expression sequences suggests the steps to group genes using different proximity metrics at the stage of data preprocessing. Currently this problem is solved by various methods. Using the component or factor analysis partially solves the problem of the feature space dimension reducing, however, the partial loss and distortion of the initial information occurs during data transformation that has a direct influence to the problem solution accuracy. Technology of the bicluster analysis preserves the object-feature structure of data, but the feature space dimension of the objects subsets derived much smaller than the dimension of the original data that allows us to construct the gene regulatory networks in real-time with

© Springer International Publishing AG 2017
S. Kozielski et al. (Eds.): BDAS 2017, CCIS 716, pp. 359–372, 2017.
DOI: 10.1007/978-3-319-58274-0_29

the preservation of the information about the influence specifics of the individual genes to the target node. The bicluster analysis questions for gene expression sequences processing are considered in [10,17]. The authors analyzed various biclustering algorithms and extracted their advantages and disadvantages. In [6] was conducted a comparative analysis of different biclustering algorithms for the analysis of gene expression profiles. The [11] presents a study on the use of the spectral biclustering technique for the analysis of the gene expression data on the example of the simulated data. The distribution diagram of objects and the specifics of their grouping in different biclusters are showed. It should be noted that this technology has high actuality in the context of feature extraction for the construction of the gene regulatory networks nowadays. However it should be noted, that in spite of archived progress in this subject area, there are some problems associated with: the choice of the biclusters quantity and the degree of detailing of the objects and features in corresponding biclusters; the choice of the metrics to estimate the proximity of the objects and features vectors concurrently. The use of traditional clustering algorithms to group the feature vectors according to their degree of similarity is an alternative to the bicluster analysis. A lot of clustering algorithms exist nowadays. Each of them has its advantages and disadvantages and is focused on a specific type of data. One of the essential disadvantages of the existing clustering algorithms is the nonreproductivity error, i.e., high clustering quality on a single dataset does not guarantee the same results on another similar dataset. To raise the clustering objectivity is possible by developing the hybrid models based on the inductive methods of complex systems modeling, which is a logical continuation of the group method of data handling (GMDH) [9]. The questions of inductive methods of complex systems objective self-organizing models creation are presented in [14] and further developed in [18]. The authors have presented the researches concerning the implementation of the inductive modeling principles for creating the systems of objects complex nature self-organizing based on the group method of data handling. Researches concerning the use of inductive modeling methods to create the inductive technologies of informational and analytical researches for different nature information analysis are presented in [16]. However, it should be noted that the authors' studies are focused primarily on the low dimensional data, at the same time insufficient attention is paid to the inductive models based on the clustering enumeration for the purpose of their self-organizing with the use of the external balance criteria of clustering quality assessing by equal power subsets.

The aim of the paper is working out the technology of creation the objective clustering inductive model of complex nature high dimensional data and its practical implementation by DNA microarray experiments use.

2   Problem Statement

Let the initial dataset of the objects is a matrix: $A = \{x_{ij}\}, i = 1, \ldots, n;\ j = 1, \ldots, m$, where n – is the number of objects observed, m – is the number of features characterizing the objects. The aim of the clustering process is a partition of the objects into nonempty subsets of pairwise nonintersecting clusters, herewith a surface which divides the clusters can take any shape [9,18]:

$$K = \{K_s\}, s = 1, \ldots, k; K_1 \bigcup K_2 \bigcup \cdots \bigcup K_k = A; K_i \bigcap K_j = \emptyset, i \neq j,$$

where k – is the number of clusters, $i, j = 1, \ldots, k$. Inductive model of objective clustering assumes a sequential enumeration of clustering in order to select from them the best variants. Let W – is the set of all admissible clustering for given set A. The best objective on quality criteria QC(K) is the clustering for which is:

$$K_{opt} = \arg \min_{K \subseteq W} CQ(K) or \ K_{opt} = \arg \max_{K \subseteq W} CQ(K)$$

Clustering $K_{opt} \subseteq W$ is an objective if it has the least difference from an expert by the number of objects, the character of the objects distribution in the appropriate clusters and the number of discrepancies [9,18].

The technology of the objective clustering inductive model creation assumes the following stages:

1. Assignment an affinity function of studied objects, i.e., finding the metric to determine the degree of objects similarity in m-dimensional feature space.
2. Development of the algorithm to partition the initial set of the objects into two equal power subsets. The equal power subsets are the subsets which contain the same number of pairwise similar objects.
3. Assignment a method of clusters formation (sorting, regrouping, grouping, division, etc.).
4. Assignment the criterion QC of quality clustering estimation as a measure of the clusters similarity in various clustering.
5. Organization of motion to max, min or optimal value of the criteria QC of quality clustering estimation.
6. Assignment an objective clustering fixation method corresponding to the extremum of the criteria value of quality clustering estimation.

Figure 1 shows the chart of the modules interaction in the objective clustering inductive model. The choice of affinity functions to assess the degree of proximity from objects to clusters is determined by the nature of the studied objects features. The method of clusters formation in inductive model is determined by clustering algorithm used for parallel grouping the objects in equal power subsets. The character of equal power subsets formation is determined by the choice of the objects similarity measure which depends on the objects feature space properties. To choose the objective clustering it is necessary at the early stage to define the internal and the external criteria, extremum value of which will allows to fix an objective clustering for the studied data subsets during clustering enumeration.

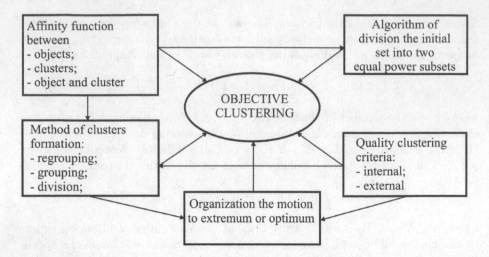

Fig. 1. Charts of the modules interaction in the objective clustering inductive model

3   Principles of the Objective Clustering Inductive Technology

Three fundamental principles borrowed from different scientific fields allowed to create the complete, organic and interconnected theory, are the basis of the methodology of the complex systems inductive modeling [9, 16, 18]:

- the principle of heuristic self-organizing, i.e., enumeration of the models set aiming to select from them the best on the basis of the external balance criteria;
- the principle of external addition, i.e., the necessity to use several equal power data subsets with the purpose of objective verification of models;
- the principle of solution inconclusive, i.e., generation of certain sets of intermediate results in order to select from them the best variants.

Implementation of these principles in the adapted version provides conditions to create the methodology of inductive model of complex data objective clustering.

3.1   Principle of Heuristic Self-Organizing

Inductive model of objective clustering assumes a sequential enumeration of the clustering by using the two equal power subsets; herewith the result of clustering is estimated at each step by calculating the external balance criterion, which determines the difference of the objects clustering results on the two subsets. The model self organizes so that the best clustering correspond to an extremum value of this criterion depending on the type of the algorithm and the measures of the objects and clusters similarity. During the process of clustering enumeration it is possible that the value of the external criterion has several local extremums corresponding to different objects clustering. This phenomenon is occurred in case of a hierarchical clustering, when clustering on the two subsets are sufficiently similar during the sequential process of objects grouping or separation that leads to the appearing of the local minimum at a given level of the hierarchy. In this case the choice of an objective clustering is determined by the goals of the task whereas each of the clustering that corresponds to the extremum value of the external balance criteria may be considered as the objective and the choice of the final clustering is determined by the required objects partition or grouping detailed elaboration level.

3.2 Principle of the External Edition

The principle of the external addition in the model of the group method of data handling (GMDH) assumes the use of "fresh information" for an objective verification of the model and selection of the best model during the process of multiserial inductive procedures of optimal model synthesis. The implementation of this principle in the framework of the objective clustering inductive model supposes the existence of the two equal power subsets, which contain the same number of pairwise similar objects in terms of their attributes values of objects. Clustering is carried out on the two equal power subsets concurrently during the algorithm operation with the sequential comparison of clustering results by chosen external balance criteria. The idea of the algorithm to divide the initial dataset of the objects Ω into two equal power subsets Ω^A and Ω^B is stated in [9] and further developed in [18]. Implementation of this algorithm assumes the following steps:

1. Calculation of $\dfrac{n \times (n - 1)}{2}$ pairwise distances between the objects in the original sample of data. The result of this step is a triangular matrix of the distances.
2. Allocation of the pairs of objects X_s and X_p, the distance between which is minimal:
$$d(X_s, X_p) = \min_{i,j} d(X_i, X_j);$$
3. Distribution of the object X_s to subset Ω^A, and the object X_p to subset Ω^B.
4. Repetition of the steps 2 and 3 for the remaining objects. If the number of objects is odd, the last object is distributed to the both subsets.

3.3 Principle of the Solution Inconclusive

The implementation of this principle, which is relative to the inductive model of objective clustering, assumes a fixation of clustering which correspond to the local minimum or maximum of external balance criterion for different levels of the hierarchical tree. Each local extremum corresponds to an objective clustering with a certain degree of detailing. The final choice and therefore the fixation of the obtained clustering is determined by the goals of the task at this stage of its solving.

4   Criteria in the Objective Clustering Inductive Technology

The necessity of the clustering quality estimation on the several equal power subsets occurs during the process of implementation of the objective clustering inductive technology, herewith separate estimations may not coincide with each other while using different algorithms and different evaluation functions for the same data. Thus, there is a necessity of the estimation of the correspondence of the modeling results to the purposes of the task in view.

4.1   Internal Criteria in the Objective Clustering Inductive Technology

Usually in most cases the number of clusters is unknown, therefor the best solutions which correspond to the extremums of the internal criteria are allocated during the process of clustering algorithm operation. High level of the clustering, obviously, corresponds to a high separating capability of various clusters and high density of objects distribution within clusters. Therefore, an internal criterion of clustering quality evaluation should include two components: the sum of squared deviations of objects relative to the corresponding centroid within clusters QC_W and the sum of the squared deviations of clusters centroids relative to a general mass center of all clusters QC_B. The formulas to calculate these internal criterion components can be presented as follows:

$$QC_W = \sum_{j=1}^{K} \sum_{i=1}^{N_j} d(x_i^j, C_j)^2$$

$$QC_B = \sum_{j=1}^{K} N_j d(C_j, \bar{C})^2$$

where K – is the quantity of clusters, N_j – is the quantity of the objects in j cluster, C_j – is the centroid of cluster j: $C_j = \frac{1}{N_j} \sum_{i=1}^{N_j} x_i^j$, x_i^j – is the object i in cluster j, $\bar{C}$ – is the general centroid of studied objects, $d(X_a, X_b)$ – is the similarity distance between vectors X_a and X_b. The correlation distance was used as similarity distance in the case of gene expression sequences analysis:

$$d(X_a, X_b) = 1 - \frac{\sum_{i=1}^{m}(x_{ai} - \bar{x}_a)(x_{bi} - \bar{x}_b)}{\sqrt{\sum_{i=1}^{m}(x_{ai} - \bar{x}_a)^2} \times \sqrt{\sum_{i=1}^{m}(x_{bi} - \bar{x}_b)^2}}$$

where m – is the number of sequences features, x_{ai} and x_{bi} – are the i-th features of the X_a and X_b sequences respectively, $\bar{x}$ – is the mean value of the correspond sequence features.

Comparative analysis of the quality clustering internal criteria estimation by the use of various types and combinations of the presented measures are showed in [7,8,12,15,19,20]. Structural block diagram of the process of the clusters quantity determination on the basis of the internal criteria is shown in Fig. 2. Implementation of this process assumes the following steps:

1. Application of the selected clustering algorithm for clustering K within the limits of allowable range $K = [K_{min}, K_{max}]$.
2. Fixation of the obtained clustering, calculation of the clusters centroids.
3. Calculation of the internal criteria for obtained clustering.
4. Repetition of the steps 1–3 to obtain the required number of clusters within the given range.
5. Construction of the graphs of internal criteria versus the number of clusters. Analysis of the graphs, selection of the optimal clustering.

As can be seen from Fig. 2, an objective clustering corresponds to a local minimum values of the internal criterion, herewith several extremums can be observed within a clustering process. Each of the local minimum corresponds to an adequate grouping of objects with various degree of the process detailing. However, it should be noted that it is not possible to evaluate the clustering objectivity based on the internal criteria because this evaluation is possible if there is a "fresh" information based on an external criteria of evaluation of the corresponding clustering difference by using the two equal power subsets.

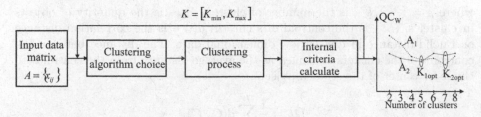

Fig. 2. Flowchart to choose the optimal clustering based on the internal criterion for the data A1 and A2

4.2   External Criteria to Estimate the Quality of the Objects Grouping

As noted hereinbefore, an adequate selection of the criteria to estimate the clustering quality on the different stages of the model operation is one of the major factors which promotes to the high efficiency of the objective clustering inductive model. These criteria should take into account both the location of the objects in the respective clusters relative to the corresponding mass center and the centroid position of the respective clusters according to the relation of clusters to each other in different clustering. An example of a possible location of the objects in a three-cluster objective clustering inductive model is shown in Fig. 3. The position of the k-th cluster's centroid is defined as the average of the objects features in this cluster:

$$C_k = \frac{1}{n_k} \sum_{i=1}^{n_k} x_{ij},$$

where n_k – is the quantity of objects in k cluster, $j = 1, \ldots, m$ – is the quantity of features what characterize the objects.

The first component of the external criterion based on the assumption that the average value of the total displacement of corresponding clusters mass centers at the different clustering in the case of the objective clustering should be minimal. In case of normalization of criterion value formula takes the form:

$$QC_1(A, B) = \sqrt{\sum_{j=1}^{m} (\frac{\sum_{i=1}^{k}(C_i(A) - C_i(B)^2}{\sum_{i=1}^{k}(C_i(A) + C_i(B)^2})^2} \longrightarrow \min$$

The second component of the external criterion takes into account the difference of the character clusters and the objects distribution in the respective clusters in different clustering. The average distance from the objects to the corresponding clusters centroids can be calculated as follows:

$$D_W = \frac{1}{k} \sum_{s=1}^{k} (\frac{1}{n_s} \sum_{i=1}^{n_s} d(X_i, C_s))$$

where $s = 1, \ldots, k$ – is the number of clusters, n_s – is the quantity of objects in cluster s, C_s – is the centroid of s cluster, $d(\cdot)$ – is the correlation distance or Euclid distance in case of low dimensional data. The distance between the centroids of the clusters is defined as the average distance from the centroids to the mass center of the studied objects:

$$D_B = \frac{1}{k} \sum_{s=1}^{k} d(C_s, \bar{C}).$$

It is obviously that the clustering will be more qualitative when the density of the objects distribution within clusters is higher and the distance from the centroids of the clusters to the total mass center of objects are more: $D_W \longrightarrow \min$, $D_B \longrightarrow \max$. The complex internal criterion was calculated using Calinski-Harabasz criterion [20]:

$$D = \frac{D_W(K - 1)}{D_B(N - K)}.$$

The second component of the external criterion can be represented as an absolute value of the internal criterion difference for various clustering. A normalized form of this formula takes the form:

$$QC_2(A, B) = \frac{|D(A) - D(B)|}{|D(A) + D(B)|}.$$

The objective clustering is selected based on the local minimum analysis of the both external criteria during the enumeration of all accessible clustering.

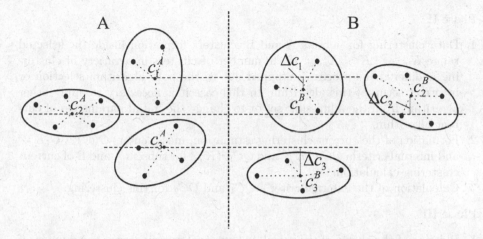

Fig. 3. An example of the objects and clusters location in objective clustering inductive technology

5   Architecture of the Step by Step Procedure of the Objective Clustering Inductive Technology

Figure 4 shows the general architecture of the objective clustering inductive technology implementation. There is a data matrix where the studied objects are given in rows and the features, defining the properties of the given objects are presented in the columns, are applied to the input of the system. The set of clusters, each of which includes a group of objects features which have a high affinity for these objects is the output of the system. Implementation of this technology supposes the following steps:

Phase I

1. Problem statement. Formation of the clustering aims.
2. Analysis of the studied data, definition of the studied objects feature character, brining of the data to a matrix view: $A = \{x_{ij}\}$, $i = 1, \ldots, n$; $y = 1, \ldots, m$, where n – is the quantity of the studied objects, m – is the quantity of the features characterizing the objects.
3. Data preprocessing that includes the data filtration, data normalization, missing value restoration, dimension of the feature space reducing.
4. Determination of the affinity function for the further degree of objects' affinity estimation.
5. Formation of the equal power subsets A and B in accordance with hereinbefore algorithm.
6. Choosing and setup of the clustering algorithm. Initialization of the input parameters of this algorithm.

Phase II

1. Data clustering for subsets A and B, clusters formation inside the selected range $K_{min} \le K \le K_{max}$. If the number of clusters in a variety of clustering is differed the process is stopped due to the poor algorithm selection or incorrect setup of this algorithm. In this case it is necessary to apply other algorithm from the admissible set or to change the initial parameters of current algorithm.
2. Formation of the current clustering estimation, mass centers $C(K)^A$, $C(K)^B$ and internal criteria $QC_W(K)^A$ and $QC_W(K)^B$ for subsets A and B of current clustering calculation.
3. Calculation of the external criteria QC_1 and QC_2 for this clustering.

Phase III

1. Plotting of the charts of the calculated external criteria versus the number of the obtained clusters within a given range $K_{min} \le K \le K_{max}$.
2. Analysis of the obtained results. In case of external criteria local minimum absence or if the values of these criteria are more than permissible norms (Fig. 4 sign "–") selection of another clustering algorithm or reinitialization of the current algorithm initial parameters. Repetition of the steps 2–5 of the Phase 1 of this procedure.
3. In case of the local minimum presence under a condition of enumerating all clustering within given range fixation of the objective clustering corresponding to the minimum of the external criteria.

6   Experiment, Results and Discussion

Approbation of the proposed model was carried out by using the patients' data with lung cancer E-GEOD-68571 of the database Array Express [5], which includes the gene-expression profiles of 96 patients, 10 of which were healthy and 86 patients were divided by the degree of the disease into three groups. The size of the initial data matrix was (96×7129). The researches on the optimization of the gene expression data preprocessing for the purpose of features space informativity increasing and quantity of the genes reducing are presented in the [2,3]. Sample of 400 genes was used at the present stage of the simulation to simplify a computing, herewith the initial dataset was divided into two equal power subsets using hereinbefore algorithm. To compare the simulation results during inductive model operation also the datasets "Compound" and "Aggregation" of the Computing School of the Eastern Finland University [1] and the dataset "seeds", representing the researches of kernels of different kinds of wheat [13] were used. Each kernel was characterized by seven attributes. In work [4] authors used the agglomerative hierarchical clustering algorithm to data clustering within the framework of presented model. In this work the SOTA self-organizing clustering algorithm was used as a base within the framework of the proposed model. The simulation of the clustering was carried out by software R. The charts of

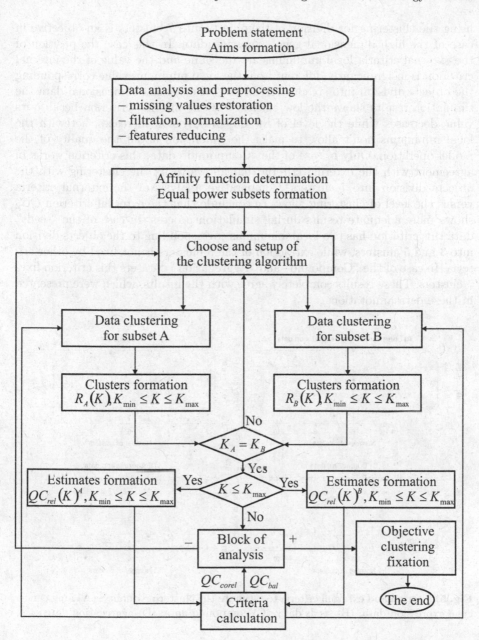

Fig. 4. Architecture of the objective clustering inductive technology

the used criteria versus the obtained clusters quantity for studied datasets are shown in Fig. 5. The number of the obtained clusters was changed from 2 to 11. The analysis of the charts allows to conclude that in terms of all external criteria using the clustering for division of the objects into 7 clusters is an objective in case of the high dimensional gene expression data. In this case the position of the external criteria local minimums are the same and the value of the internal criterion is insignificantly different from the local minimum value corresponding the objects division into six clusters. In case of other low dimensional data the simulation results shows the low efficiency of the internal criterion because its value decreases while the level of the objects division increases, herewith the local minimums don't allow to make the conclusion about the quality of the model operation. Only in case of the "Compound" data, this criterion works in agreement with the external criteria that allows to fix the clustering with the objects division into 7 clusters. Comparative analysis of the external criteria versus the level of clustering allows to conclude that the external criterion QC_2 shows more adequate results during simulation process. In case of the "seeds" data this criterion has two local minimums corresponding to the objects division into 3 and 5 clusters, while extraction of the 3 clusters is not fixed by other criteria. In case of the "Compound" and "Aggregation" datasets this criterion fixes 7 clusters. These results completely agree with the results, which were presented in these data annotations.

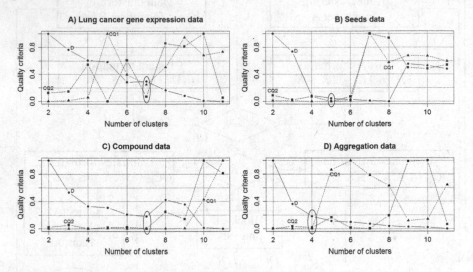

Fig. 5. Internal and external criteria to estimate the clustering quality: (A) lung cancer gene expression data; (B) seeds data; (C) compound data; (D) aggregation data

7   Conclusions

The researches aiming to create the technology of the objective clustering inductive model of the complex nature objects are presented in the paper. To improve the objectivity of the objects grouping the original data set is divided into two

equal power subsets, which contained the same number of the pairwise similar objects in terms of the correlation distance of their attributes profiles. The
architecture of the objective clustering inductive model has been developed and
practically implemented on the basis of the self-organizing SOTA clustering algorithm, while the evaluation of the partition objects into clusters quality at each
step was estimated using an external criteria, which take into account the difference of the objects and the clusters distribution in different clustering. The
sample of the lung cancer patients' gene expression profiles which contains 400
profile genes of 96 patients, "Compound", "Aggregation" and "Seeds" data were
used to approbate the proposed model. The simulation results proved high efficiency of the proposed model operation. The local minimums values of the internal and external criteria allow to take more adequate solution about the choice
of the studied data objective clustering. The further authors' researches will
be focused on a more detailed study of the proposal model operation based on
various clustering algorithms with the use of different nature data.

References

1. Machine learning school of computing university of eastern finland. Clustering datasets. https://cs.joensuu.fi/sipu/datasets/
2. Babichev, S.A., Kornelyuk, A.I., Lytvynenko, V.I., Osypenko, V.: Computational analysis of microarray gene expression profiles of lung cancer. Biopolymers Cell **32**(1), 70–79 (2016). http://biopolymers.org.ua/content/32/1/070/
3. Babichev, S., Taif, M.A., Lytvynenko, V.: Filtration of dna nucleotide gene expression profiles in the systems of biological objects clustering. Int. Front. Sci. Lett. **8**, 1–8 (2016). https://www.scipress.com/IFSL.8.1
4. Babichev, S., Taif, M.A., Lytvynenko, V.: Inductive model of data clustering based on the agglomerative hierarchical algorithm. In: Proceeding of the 2016 IEEE First International Conference on Data Stream Mining and Processing (DSMP), pp. 19–22 (2016). http://ieeexplore.ieee.org/document/7583499/
5. Beer, D.G., Kardia, S.L., et al.: Gene-expression profiles predict survival of patients with lung adenocarcinoma. Nat. Med. **8**(8), 816–824 (2002). http://www.nature.com/nm/journal/v8/n8/full/nm733.html
6. Eren, K., Deveci, M., Kucuktunc, O., Catalyurek, U.V.: A comparative analysis of biclustering algorithms for gene expression data. Briefings Bioinform. **14**(3), 279–292 (2012). https://doi.org/10.1093/bib/bbs032
7. Halkidi, M., Batistakis, Y., Vazirgiannis, M.: Clustering validity checking methods: Part 2. ACM SIGMOD Rec. **31**(3), 19–27 (2002). https://www.researchgate.net/publication/2533655_Clustering_Validity_Checking_Methods_Part_II
8. Halkidi, M., Vazirgiannis, M.: Clustering validity assessment: finding the optimal partitioning of a data set, pp. 187–194 (2001). http://ieeexplore.ieee.org/document/989517/?reload=true&arnumber=989517
9. Ivakhnenko, A.: Group method of data handling as competitor to the method of stochastic approximation. Sov. Autom. Control **3**, 64–78 (1968)
10. Kaiser, S.: Biclustering: methods, software and application (2011). https://edoc.ub.uni-muenchen.de/13073/
11. Kluger, Y., Basry, R., Chang, J., Gerstein, M.: Spectral biclustering of microarray data: coclustering genes and conditions. Genome Res. **13**(4), 703–716 (1985). http://genome.cshlp.org/content/13/4/703.abstract

12. Krzanowski, W., Lai, Y.: A criterion for determining the number of groups in a data set using sum of squares clustering. Biometrics **44**(1), 23–34 (1985). https://www.jstor.org/stable/2531893?seq=1#page_scan_tab_contents

13. Kulczycki, P., Kowalski, P.A., Lukasik, S., Zak, S.: Seeds data set. http://archive.ics.uci.edu/ml/datasets/seeds

14. Madala, H., Ivakhnenko, A.: Inductive Learning Algorithms for Complex Systems Modeling, pp. 26–51. CRC Press (1994). http://www.gmdh.net/articles/theory/ch2.pdf

15. Milligan, G., Cooper, M.: An examination of procedures for determining the number of clusters in a data set. Psychometrika **50**(2), 159–179 (1985). http://link.springer.com/article/10.1007/BF02294245

16. Osypenko, V.V., Reshetjuk, V.M.: The methodology of inductive system analysis as a tool of engineering researches analytical planning. Agric. Forest Eng. **58**, 67–71 (2011). http://annals-wuls.sggw.pl/?q=node/234

17. Pontes, B., Giraldez, R., Aguilar-Ruiz, J.S.: Biclustering on expression data: a review. J. Biomed. Inf. **57**, 163–180 (2015). https://www.ncbi.nlm.nih.gov/pubmed/26160444

18. Sarycheva, L.: Objective cluster analysis of data based on the group method of data handling. Probl. Control Automatics **2**, 86–104 (2008)

19. Still, S., Bialek, W.: How many clusters? An information theoretic perspective. Neural Comput. **16**(12), 2483–2506 (2004). http://www.mitpressjournals.org/doi/abs/10.1162/0899766042321751#.WJst02_hCUl

20. Xie, X., Beni, G.: A validity measure for fuzzy clustering. IEEE Trans. Pattern Anal. Mach. Intell. **13**(8), 841–847 (1991). http://dl.acm.org/citation.cfm?id=117682

Novel Computational Techniques for Thin-Layer Chromatography (TLC) Profiling and TLC Profile Similarity Scoring

Florian Heinke[1,2(✉)], Rico Beier[1,2], Tommy Bergmann[1], Heiko Mixtacki[3], and Dirk Labudde[1]

[1] Bioinformatics Group Mittweida, University of Applied Sciences Mittweida, Technikumplatz 17, 09648 Mittweida, Germany
heinke@hs-mittweida.de
[2] TU Bergakademie Freiberg, Akademiestrasse 6, 09599 Freiberg, Germany
[3] biostep GmbH, Innere Gewerbestrasse 7, 09235 Burkhardtsdorf, Germany
http://www.hs-mittweida.de/bigm

Abstract. Thin-layer chromatography (TLC) is an experimental separation technique for multi-compound mixtures widely applied in various fields of industry and research. In contrast to comparable techniques, TLC is straightforward, cost- and time-efficient, and well-applicable in field operations due to its flexibility. In TLC, after applying a mixture sample to the adsorbent layer on the TLC plate, the compounds ascent the plate at different rates due to their individual physicochemical characteristics, whereas separation is eventually achieved.

In this paper, we present novel computational techniques for automated TLC plate photograph profiling and fast TLC profile similarity scoring that allow advanced database accessibility for experimental TLC data. The presented methodology thus provides a toolset for automated comparison of plate profiles with gold standard or baseline profile databases. Impurities or sub-standard deviations can be readily identified. Hence, these techniques can be of great value by supporting the pharmaceutical quality assessment process.

Keywords: Thin-layer chromatography · TLC · Quality assessment · TLC plate profiling · Profile similarity scoring · Profile database querying

1 Introduction

Thin-layer chromatography (TLC) is an experimental method for qualitative analyses of non-volatile organic compound mixtures with a wide range in application. In fact, even fifty years after its introduction, it is one of the most applied analytical methods in todays organic chemistry laboratories [2,6,9]. Areas of application are in various fields of industry and research, such as food chemistry, quality assessment in pharmaceutical research and development, forensic diagnostics as well as basic biochemical research. In contrast to comparable experimental techniques (such as high-performance liquid chromatography and high

© Springer International Publishing AG 2017
S. Kozielski et al. (Eds.): BDAS 2017, CCIS 716, pp. 373–385, 2017.
DOI: 10.1007/978-3-319-58274-0_30

performance TLC), conventional TLC is straightforward, cost- and time-efficient, and well-applicable in field operations due to its flexibility and portability [6]. In addition, multiple samples can be analyzed in one single experimental run simultaneously and separately [11].

The general aim of TLC is to separate individual compounds (the so-called analytes) present in a mixture sample. The two most common applications are qualitative tests for the presence or absence of mixture impurities and the isolation of analytes. In its core, the experimental principle (see Fig. 1) is strikingly simple: a mixture sample is applied to a layer of adsorbent material. Silica gel is used mostly [6] as such. The layer of adsorbent material is referred to as TLC plate in the following text. In the process a solvent is applied to the TLC plate. Capillary action now causes the solvent and the analytes to migrate through the adsorbent. Individual physicochemical properties of the analytes however lead to different molecular interactions with the moving solvent, whereas the rate of migration differs between analytes. Eventually, separation of analytes is achieved over time. The fraction of migration length between solvent (known as solvent front) and the migration length of a separated analyte is commonly referred to as retardation factor R_f. To allow additional scaling of analyte migration lengths and characterizing separated analytes, a mixture of marker species with known R_f values is applied to the plate as well.

In this work, we introduce and discuss a novel knowledge-based technique for TLC plate lane and spot detection, and demonstrate its performance on photographs with varying light sources and image quality. We further elucidate how this technique can be employed to obtain TLC lane profiles and databases of such profiles valuable for in-house applications. We show computational techniques for TLC lane database searching as well as automated impurity identification and separation evaluation. In this respect we address the problem of scoring lane similarity, and present a dynamic programming-based algorithm with scoring capabilities.

2   Methods

Over the last decade, numerous computational techniques have been proposed that address the problem of TLC image analyses. This issue can be divided into the problems of lane detection and spot detection from raw TLC plate images. Lane detection methods presented in [1,8,12] perform reliably (F_1 measures of about 0.95 are reported in all three studies). However, subsequent analyses of extracted profiles retrieved from identified lanes have to be conducted manually. The second problem has shown itself to be difficult to address. Most notably, straightforward spot recognition by means of extracting and tracing edges is not applicable, since analyte spots can "overlap" and often show no distinct edges (see Fig. 1c). Methods, such as presented by Tie-xin and Hong [14] or Zhang and Lin [16], utilize image grayscaling and image transformations to detect individual spots in lanes. In a study presented by Vovk and Prosek [15] a technique is demonstrated which shows reliable in this respect. Nonetheless, additional plate

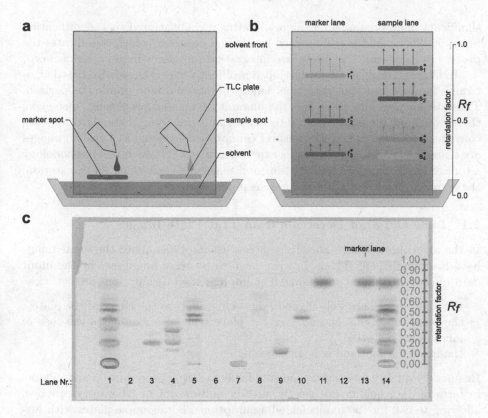

Fig. 1. The experimental principle of TLC. (a) Spots of mixture samples and mixtures of marker species are applied to a TLC plate covered by adsorbent material. A solvent is applied next, which starts to migrate through the adsorbent due to capillarity action. (b) Marker substances and analytes comprising the sample start to migrate with the solvent flow; however migration rates differ between analytes due to individual molecular properties. Thus, separation of analytes and marker species is eventually achieved (shown by spots labeled s_1^* through s_4^* and r_1^* through r_3^*, respectively). The normalized analyte-specific migration length relative to the solvent is referred to as retardation factor R_f. Marker substances with known R_f values provide statistics valuable for analyte characterization. (c) Photograph of a TLC plate in lab application. Note that four marker species have been applied to lanes 9, 10, 11 and 12 separately, and to lane 13 as a mixture. The R_f scale has been provided by the experimentalist.

scanning hardware and post-processing software is required. Hemmateenejad et al. [5] demonstrated a multivariate image analyses technique that integrates feasibly into conventional image acquisition and processing pipelines. Although this technique shows to perform reliably (error in analyte concentration prediction is reported to be less than 10%), the extend of required manual image preprocessing is unclear. In a more recent study, Kerr et al. [7] demonstrated that straightforward RGB value decomposition of spots can be valuable in estimating analyte concentrations. However the extend of manual preprocessing is significant in all presented techniques. To this day automated spot identification remains difficult and is still assisted by manual curation, which complicates the process of large-scale TLC database integration and accessing in consequence.

To address these issues, we developed problem-tailored knowledge-based algorithms for lane and spot detection, lane profiling and lane profile comparison. These techniques aim to reduce the amount of manual processing, increasing the throughput for profile data integration, and providing a standard for data consistency. Note that the latter aspect cannot be considered as a given if images are manually processed by multiple experimentalists. Through this methodology a toolset for profile database accessing and searching is provided. In this section, the core algorithms used in our analysis pipeline are outlined.

2.1   Lane and Spot Detection from TLC Plate Images

In the algorithmic design knowledge-based assumptions about the input image have been implemented that, in turn, have to be accordingly met by the input data in order to enable processing. The implemented assumptions are:

- the image edges are well-aligned (almost parallel) to the edges of the plate,
- the lanes are evenly spaced over the plate (this also eliminates the case of lanes being distorted),
- the image background is homogeneous.

Ensured by quality assessment and standard operating procedures, requirements in term of data quality are mostly already quite strict in laboratory practice, allowing to call for the implemented assumptions. For example plates with distorted and deformed lanes, which would pose problematic for processing, are usually discarded beforehand as they are not in agreement with laboratory standards in the first place. Thus, although these aspects require the data to be almost ideal, meeting these quality demands is rarely problematic in laboratory practice.

With these requirements met, as the first step of lane detection, statistics about the background are obtained based on the HSV color space. The image outline is now traversed in order to detect plate edges. As the next step, a background-corrected average saturation profile is computed along the x-axis of the image (see Fig. 2a). Peaks and plateaus in this profile correspond to lanes. Using this profile as input, lane x-coordinates are extracted by progressively inferring lane separation. A least-square error scheme for measuring the correlation of observed peak spaces and hypothetical evenly spaced *ad hoc* coordinates has been developed for this purpose. According to this scheme, the extracted coordinates correspond to the set of coordinates with lowest cumulative error of smallest common lane separation. With the lane x-coordinates being now deduced, the spots are identified during the next steps. Here, per-pixel RGB values are background-corrected and transformed to HSV color space. This transformation allows to trace hue and saturation values along the y-axis, whereas lane-specific saturation and hue profiles are obtained. Spot detection is conducted next based on the saturation profile. Information on overlapping spots

Algorithm 1. Image analysis

```
1: function PROCESS(image)
2:     detect background hue and saturation ranges
3:     eliminate non-backround border pixels
4:     compute profile of hue and saturation differences to the background
5:     progressively detect all profile peaks
6:     lanes ← select subset of evenly spaced peaks by error minimization
7:     bg ← extract background stripe of lane width
8:     for each lane in lanes do
9:         subtract bg from lane in rgb-mode and convert to averaged hsv
10:        apply hue changes with sigmoid function to saturation profile
11:        smoothing lane saturation profile with moving average
12:        spot[lane] ← detecting peaks by gaussian fingerprinting
13:     export spots
```

is included by adapting saturation values according to hue changes. If abrupt hue changes are observed, a spot edge or overlap is inferred. In this case, the saturation is reduced according to a sigmoid activation function (see Fig. 2b). Note that, in an effort to increase sensitivity, the saturation value is set to 1. Hence hue-based saturation adaption introduces additional peaks to the saturation profile which are eventually identified as spots during the next step. Finally, the spots are iteratively retrieved from the obtained saturation profile. During each iteration, a Gaussian function is fitted to the maximum saturation peak, and subtracted from the profile until all remaining peak maxima are below a given threshold. For each fit a spot is now defined as the set of assigned pixels including their color values, and a representative centroid coordinate. This data can now be stored to an in-house database for further analyses. A succinct summary of the method is given in Algorithm 1.

Experimental testing has shown that not only spots are identified reliably (see Sect. 3) but minor separation characteristics of the analyte mixture, such as inconspicuous lubrication effects, are also identified. Thus, the algorithm retrieves an entire profile of major and minor intensities from a lane in addition to individual spots, giving it a fingerprint-like nature. Henceforth, the obtained set of spots is referred to as lane profile in the process. In addition the term 'spot' is treated synonymous to actual spots and said detected minor signals in the following sections for textual clarity.

2.2   TLC Lane Profile Similarity Scoring and Database Searching

The problem of TLC lane profile similarity scoring, which is the underlying principle for database-wide searching, is similar to the classic global sequence alignment problem found in bioinformatics. Similarly to the bioinformatics solution to the problem [10,13], we propose a dynamic programming-based approach for generating profile alignments and subsequent alignment scoring. An alignment of two given profiles can be understood as the optimal sequential arrangement of

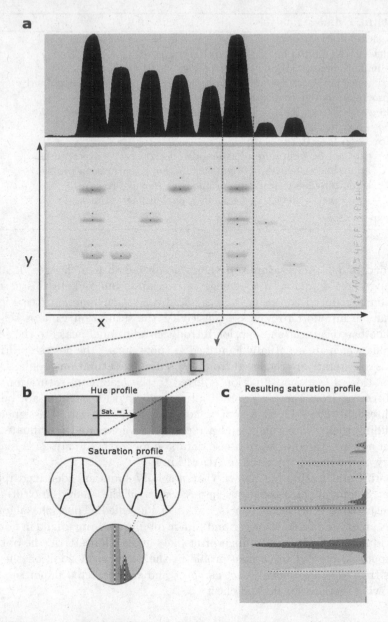

Fig. 2. Schematic outline of the lane and spot detection process. (a) First, a mean HSV saturation profile is deduced along the x-axis from which lane x-coordinates are computed subsequently. (b) Mean hue and saturation profiles are determined for each lane. With the saturation set to 1, abrupt hue changes indicating spot edges or overlapping spots can be readily identified, as shown exemplarily for the rectangular detail. The saturation profile is adapted accordingly. (c) The resulting saturation profile for the example lane. Gaussian functions are iteratively fit to the profile to obtain locations of spots (indicated by black dots in a). (Color figure online)

detected spots. Suboptimally matching spots and introducing gaps are allowed editing operations in alignment computation. The score of the alignment indicates the quality of the arrangement. The more gaps and suboptimal matches are present, the smaller the score. Reference points for spot matching are corresponding y-coordinates. The aim of the algorithm is to find the best arrangement of two profile spot y-coordinate sets S^1 and S^2, denoted as $S^1 = \{s^1_1, \ldots, s^1_m\}$ and $S^2 = \{s^2_1, \ldots, s^2_n\}$ in the following text. The algorithms further requires the y-coordinates in both sets to be arranged in ascending order.

Before introducing the algorithmic core of the alignment technique, it needs to be noted that the y-coordinates are not adequate for comparing lane profiles obtained from different plates. The spot y-coordinates of an analyte can vary greatly if different adsorbent materials are used or experimental conditions are present. Even using different experimental running times can impact the profile. Hence, the y-coordinates need to be normalized. The y-coordinates of the marker species can be applied for this purpose, as experimental migration length variances are canceled out. Let s^*_i and r^*_j be the y-coordinate of an analyte spot and marker spot retrieved from the same TLC plate, respectively. The normalized analyte spot y-coordinate s_i is simply obtained by the following expression:

$$s_i = \frac{s^*_i - r^*_m}{r^*_1 - r^*_m},$$

where r^*_1 and r^*_m are the y-coordinates of the marker species with longest and shortest migration length, respectively (see Fig. 1b for an illustration of spot indexing). Thus, two normalized profiles S^1 and S^2 become comparable between different TLC plates. Although the R_f of i can be considered as an appropriate s_i as well, this would however require additional manual annotation of the plate prior to processing by providing a solvent front marking on either the actual plate or the image. Therefore, normalization is proposed by means of marker spots. In a standardized TLC workflow, the marker species can be applied to a single predefined lane as a mixture (or to multiple predefined lanes if marker species have to be regarded individually) whereas the algorithm extracts marker coordinates automatically from said specified lanes without impeding or reducing data throughput.

Generally speaking, by utilizing dynamic programming, an optimal spot alignment is determined by computing an optimization score for each pair of spot y-coordinates s^1_i and s^2_j and storing it in a so-called optimization matrix O, whereas optimization score calculation considers the distance between normalized spot y-coordinates, optimization scores computed in previous iteration steps and a constant gap penalty. The optimization score calculation further yields information whether two spots are accepted as a match or a gap is introduced in S^1 or S^2. This information is stored in edit traceback matrix T. After computing O, the alignment raw score x_{raw} is read from O and the alignment traceback is inferred from T. A more formal description of the algorithmic core is listed in Algorithm 2.

The function $dist(s^1_i, s^2_j)$ (see Algorithm 2, line 11) yields the distance between normalized spot y-coordinates, and the constant ε acts as a gap penalty.

Although any gap penalty value could be considered for ε, a reasonable choice can be deduced from the pairwise distance density distribution function obtained from random y-coordinates initialized in $[0, 1]$. From this distribution a distance cut-off can be estimated according to a given level of statistical significance α. In our study an α of 0.05 was chosen, which results to an ε of 0.03. In this case, the interpretation of ε is as follows: given two randomly chosen y-coordinates, the probability to observe an inter-spot distance $\leq \varepsilon$ is 0.05. A measured inter-spot distance $\leq \varepsilon$ is therefore considered as significantly small, leading to a positive δ (see Algorithm 2, line 11) and increasing the chance of the spot pair to be matched in the alignment. Whether this is finally the case however is dependent on the final outcome of optimization matrix O and the underlying alignment traceback path encoded in T.

The obtained raw alignment score x_{raw} shows to be unsuited for proper similarity scoring, as its order can vary by the size of the optimization matrix O and ε. For example, given a third profile S^3 with $|S^3| > |S^2|$, the obtained x_{raw} of alignment $S^1 - S^3$ tends to be greater than x_{raw} of alignment $S^1 - S^2$ although the degree of dissimilarity between S^1 and S^3 could theoretically be smaller. Additional standardization of x_{raw} is thus needed to infer profile dissimilarity independent of m and n. In the scenario of profile database searching, this further allows alignment ranking and reporting of best profile matches.

A technique for standardizing x_{raw} has been presented in [3, 4] and requires a large number of alignment raw scores obtained from randomly initialized profiles of the same length as the two profiles in question. This set of raw scores provides an estimate on baseline profile dissimilarity observed by chance. For this purpose, profiles of lengths m and n are generated by initializing m and n random numbers in $[s_m^1, s_1^1]$ and $[s_n^2, s_1^2]$, respectively. This gives random normalized spot y-coordinates which finally allow to compute the standardized alignment score x_A according to:

$$x_A = -\log\left(\frac{x_{\text{raw}} - \overline{x}_p}{x_{\text{opt}} - \overline{x}_p}\right),$$

where

$$x_{\text{opt}} = \frac{1}{2}\varepsilon(m + n).$$

As the average of raw scores obtained from random profile alignments, $\overline{x}_p$ is a point estimate of the expected x_{raw} for any randomly selected profile pair. x_A grows with increasing profile dissimilarity. Impurities in an analyte mixture are indicated by increased x_A when compared to a standard lane (which is for example retrieved from a profile database). To perform a database search, a query TLC lane profile S is run against the set of database profiles, which can be ranked subsequently according to lane dissimilarity expressed by x_A.

Finally, background-corrected spot color can be scored as well. At the moment matched spots are scored individually, which provides an indication when a spot is abnormally colored due to impurity. Let P and Q be the RGB color vector set of assigned pixels in two matched spots. Color dissimilarity $C(P, Q)$ is then defined as:

$$C(P,Q) = \frac{\sqrt{3\varrho^2}}{|P| \cdot |Q|} \sum_{p \in P} \sum_{q \in Q} \text{dist}_{\text{RGB}}(p, q),$$

where ϱ corresponds to the number color levels in each RGB color channel[1]. The function $\text{dist}_{\text{RGB}}(p, q)$ simply yields the euclidean distance between two RGB color vectors. The proposed color dissimilarity function can be interpreted as the mean RGB color vector distance normalized by the maximum possible RGB color distance $\sqrt{3\varrho^2}$.

Algorithm 2. TLC lane profile alignment

1: **function** ALIGN(S^1, S^2)
2:     $m \leftarrow |S^1|, n \leftarrow |S^2|$
3:     initialize optimization matrix $O_{(m+1) \times (n+1)}$
4:     initialize edit traceback matrix $T_{(m+1) \times (n+1)}$
5:     fill $\forall i \in [1, ..., m+1] : O_{i,1} \leftarrow -\varepsilon \cdot (i - 1)$
6:     fill $\forall j \in [1, ..., n+1] : O_{1,j} \leftarrow -\varepsilon \cdot (j - 1)$
7:     **for** $i \leftarrow 2, ..., m+1$ **do**
8:         **for** $j \leftarrow 2, ..., n+1$ **do**
9:             $u \leftarrow O_{i-1,j} - \varepsilon$
10:             $l \leftarrow O_{i,j-1} - \varepsilon$
11:             $d \leftarrow O_{i-1,j-1} - \underbrace{\left[\text{dist}(s_{i-1}^1, s_{j-1}^2) - \varepsilon \right]}_{\delta}$
12:             $c \leftarrow \max(u, l, d)$
13:             $O_{i,j} \leftarrow c$
14:             $T_{i,j} \leftarrow \begin{cases} \text{'up'}, & u = c \\ \text{'left'}, & l = c \\ \text{'diag'}, & \text{else} \end{cases}$
15:         raw score $x_{\text{raw}} \leftarrow O_{m+1,n+1}$
16:         infer alignment path P from traceback on T with $T_{m+1,n+1}$ as initial element
17:         introduce gaps to S^1 and S^2 according to P
18:         return aligned profiles S^1 and S^2, and raw score x_r

---

3   Results and Discussion

Lane and spot detection algorithm performance was tested on 41 TLC plate images with a total number of 2,714 spots and 556 lanes. It needs to be noted that in common lab practice TLC plates are often labeled by hand-written markings by the experimentalist for documentation purposes. These additional markings can be problematic in lane recognition and subsequent spot identification. In our experiments however the written labels were only problematic in one case.

With respect to lane recognition, the average F_1-measure was observed to be 0.98, with the smallest F_1-measure being as low as 0.53 for one of the images.

---

[1] In most software tools and programming languages, 256 color levels are implemented.

This image suffers from low image quality, low spot resolution, a low signal-to-noise ratio and image artifacts introduced by the camera flash. We refrained from removing this image from the dataset in order to test the robustness of the algorithms under suboptimal conditions. The average F_1-measure indicates performance superior to recent techniques (Moreira et al. [8]: $F_1 = 0.96$; Sousa et al. [12]: $F_1 = 0.95$; Akbari et al. [1]: $F_1 = 0.96$). However further testing is needed in an effort to further quantify lane recognition performance and its limitations. Furthermore a gold standard for benchmarking is yet to be defined.

With respect to spot recognition performance, it needs to be highlighted that a spot, as discussed in Sect. 2.1, could correspond to an actual analyte spot or a minor, yet potentially characteristic signal identified in a lane. Furthermore, an actual spot could be represented by two or more sequential signals in a profile, yielding normalized y-coordinates s_i through s_{i+k}, if the saturation profile is shaped accordingly. Hence, the underlying classification statistics (the classifier confusion matrix obtained from numbers of (in-)correctly identified or missed spots) can be deduced depended on the perspective on how (in-)correctly identified spots and additional signals are treated. These perspectives are:

1. Most conservative. If an actual spot is reported by two or more profile spots, only one profile spot is treated as a true-positive. The remainder is treated as false-positives. Spots obtained from peaks with low signal-to-noise ratio are treated as analyte-uncharacteristic and are accordingly reported as false-positives.
2. Semi-conservative. If an actual spot is reported by two or more profile spots, all profile spots are treated as true-positives. Spots obtained from peaks with low signal-to-noise ratio are treated as analyte-uncharacteristic and are accordingly reported as false-positives.
3. Least conservative. If an actual spot is reported by two or more profile spots, all profile spots are treated as true-positives. Spots obtained from peaks with low signal-to-noise ratio are treated as analyte-characteristic and are accordingly reported as true-positive.

Through perspective 1 to 3, the F_1-measures are 0.85, 0.89 and 0.95. Furthermore, incorrect spots obtained from written markings need to be considered separately. If such false-positive spots are excluded beforehand in this case, the F_1-measures increase to 0.87, 0.91 and 0.97 for perspective 1 to 3. Classification statistics were deduced manually by curating the recognition outputs produced by the proposed methods.

In general, spot recognition performs reasonably well on the test data, however hand-written image labels show to negatively affect performance in this respect as expected. This issue could be addressed by implementing a well defined masked area on the right or left side of the plate ignored by the lane detection algorithm. This would still allow the experimentalist to add labels to the plate without impeding subsequent lane detection. Further improvements could be made by considering a more complex lane shape. At the moment, a lane is assumed to be rectangular in all cases, which could lead to reduced outlier-sensitive averaging and canceling out signals valuable for lane identification.

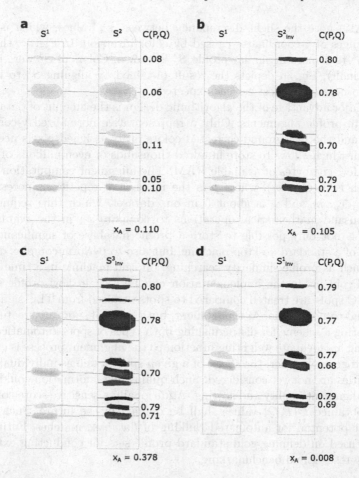

Fig. 3. Four cases for profile alignments obtained from TLC plate profile database searching. (a–d) Alignments of profiles with varying global profile similarity reported by x_A. x_A increases with increasing profile dissimilarity. Green lines indicate matching profile spots, red lines in profiles highlight gaps (spots with no corresponding match). Color dissimilarity values $C(P, Q)$ are reported for all matches. In b–d color inverted profiles were used to illustrate color dissimilarity descriptiveness. For a colored figure please refer to the online version of this article. (Color figure online)

In Fig. 3 four lane profile alignment outputs are shown exemplarily for matches obtained from database searching. The database consisted of profiles computed from our test set of 41 images. Profile S^1 was used as input. Green lines indicate matching profile spots, whereas spots highlighted by red lines indicate alignment gaps. As shown in Fig. 3a, minor dissimilarity to profile S^2 was identified, reported by a gap and a slightly increased x_A. Color dissimilarities of matched spots indicate good agreement. To illustrate color dissimilarity descriptiveness, the colors of S^2 were inverted and the alignment was recalculated. The result is shown in Fig. 3b. As expected, color dissimilarity scores are significantly greater. Also note the slight discrepancy between x_A values, which is due to minor changes of y-coordinates caused by color inversion. In Fig. 3c, the alignment of S^1 to a more dissimilar profile S^3 is shown. Colors of S^3 were inverted as well. Finally, Fig. 3d depicts the result obtained by aligning S^1 to its color inverted form. x_A is almost zero as expected.

One major advantage of the algorithmic design is the amount of data needed to compute profile alignments. Only a representative normalized y-coordinate of a spot and the background-corrected colors of its assigned pixels need to be considered, which allows to store hundred-thousands or even millions of profiles in only a few gigabytes of available RAM. The alignment computation itself is fast, which is in $O(pmn)$, where p is the number of repetitions necessary for computing $\overline{x}_p$. m and n is about 5 in our dataset, which thus requires only a few thousand mathematical operations to compute x_A in the average case. Hence, it is not only possible to store a profile database of significant size in the RAM of a standard desktop machine, but also to perform complex database queries, such as profile similarity searching, on said machine in a time efficient manner. Problems in the implementation can be seen in how profile spots of actual TLC spots are treated compared to spots deduced from TLC signals with lower signal-to-noise ratio. At the moment, both types of spots are treated equal. Implementing a scheme for distinguishing both types of spots automatically and introducing an adequate weighting function to the alignment process is currently work in progress. Further, the shape of a given spot as well as individual saturation densities are not yet considered. Such qualitative information could greatly enhance alignment quality and scoring. Additionally a sensitive score combining x_A and obtained $C(P,Q)$ values shall be defined in the future. Such a score is of great potential for automated ranking of database matches. Future work is also focused on defining gold standard profile sets for conducting exhaustive sensitivity testing and benchmarking.

Acknowledgments. We are very grateful to Pia Altenhofer, who provided TLC images, test sets and helpful ideas.

References

1. Akbari, A., Albregtsen, F., Jakobsen, K.S.: Automatic lane detection and separation in one dimensional gel images using continuous wavelet transform. Anal. Methods **2**(9), 1360 (2010). http://dx.doi.org/10.1039/c0ay00167h
2. Bernard-Savary, P., Poole, C.F.: Instrument platforms for thin-layer chromatography. J. Chromatogr. A **1421**, 184–202 (2015). http://dx.doi.org/10.1016/j.chroma.2015.08.002
3. Feng, D.F., Doolittle, R.F.: [21] progressive alignment of amino acid sequences and construction of phylogenetic trees from them. Methods Enzymol. **266**, 368–382 (1996). Elsevier BV. http://dx.doi.org/10.1016/S0076-6879(96)66023-6
4. Heinke, F., Schildbach, S., Stockmann, D., Labudde, D.: eProS-a database and toolbox for investigating protein sequence-structure-function relationships through energy profiles. Nucleic Acids Res. **41**(D1), D320–D326 (2012). http://dx.doi.org/10.1093/nar/gks1079

5. Hemmateenejad, B., Mobaraki, N., Shakerizadeh-Shirazi, F., Miri, R.: Multivariate image analysis-thin layer chromatography (MIA-TLC) for simultaneous determination of co-eluting components. Analyst 135(7), 1747 (2010). http://dx.doi.org/10.1039/C0AN00078G

6. Johnsson, R., Träff, G., Sundén, M., Ellervik, U.: Evaluation of quantitative thin layer chromatography using staining reagents. J. Chromatogr. A 1164(1–2), 298–305 (2007). http://dx.doi.org/10.1016/j.chroma.2007.07.029

7. Kerr, E., West, C., Hartwell, S.K.: Quantitative TLC-image analysis of urinary creatinine using iodine staining and RGB values. J. Chromatogr. Sci. 54(4), 639–646 (2015). http://dx.doi.org/10.1093/chromsci/bmv183

8. Moreira, B., Sousa, A., Mendonça, A.M., Campilho, A.: Automatic lane segmentation in TLC images using the continuous wavelet transform. Comput. Math. Methods Med. 2013, 1–19 (2013). http://dx.doi.org/10.1155/2013/218415

9. Mustoe, S.P., McCrossen, S.D.: A comparison between slit densitometry and video densitometry for quantitation in thin layer chromatography. Chromatogr. 53(1), S474–S477 (2001). http://dx.doi.org/10.1007/BF02490381

10. Needleman, S.B., Wunsch, C.D.: A general method applicable to the search for similarities in the amino acid sequence of two proteins. J. Mol. Biol. 48(3), 443–453 (1970). http://dx.doi.org/10.1016/0022-2836(70)90057-4

11. Poole, C.F.: Planar chromatography at the turn of the century. J. Chromatogr. A 856(1–2), 399–427 (1999)

12. Sousa, A.V., Aguiar, R., Mendonça, A.M., Campilho, A.: Automatic lane and band detection in images of thin layer chromatography. In: Campilho, A., Kamel, M. (eds.) ICIAR 2004. LNCS, vol. 3212, pp. 158–165. Springer, Heidelberg (2004). doi:10.1007/978-3-540-30126-4_20

13. Sung, W.K.: Algorithms in Bioinformatics: A Practical Introduction. CRC Press, Boca Raton (2009)

14. Tie-xin, T., Hong, W.: An image analysis system for thin-layer chromatography quantification and its validation. J. Chromatogr. Sci. 46(6), 560–564 (2008). http://dx.doi.org/10.1093/chromsci/46.6.560

15. Vovk, I., Prošek, M.: Quantitative evaluation of chromatograms from totally illuminated thin-layer chromatographic plates. J. Chromatogr. A 768(2), 329–333 (1997). http://dx.doi.org/10.1016/S0021-9673(96)01070-9

16. Zhang, L., Lin, X.: Quantitative evaluation of thin-layer chromatography with image background estimation based on charge-coupled device imaging. J. Chromatogr. A 1109(2), 273–278 (2006). http://dx.doi.org/10.1016/j.chroma.2006.01.018

Extending the Doctrine ORM Framework Towards Fuzzy Processing of Data
Exemplified by Ambulatory Data Analysis

Bożena Małysiak-Mrozek, Hanna Mazurkiewicz, and Dariusz Mrozek[✉] [ID]

Institute of Informatics, Silesian University of Technology,
ul. Akademicka 16, 44-100 Gliwice, Poland
{bozena.malysiak,dariusz.mrozek}@polsl.pl

Abstract. Extending standard data analysis with the possibility to formulate fuzzy search criteria and benefit from linguistic terms that are frequently used in real life, like *small*, *high*, *normal*, *around*, has many advantages. In some situations, it allows to extend the set of results by similar cases that would not be possible or difficult with precise search criteria. This is especially beneficial when analyzing biomedical data, where sets of important measurements or biomedical markers describing particular state of a patient or person have similar, but not the same values. In other situations, it allows to generalize the data and aggregate it, and thus, quickly reduce the volume of data from Big to small. Extensions that allow the fuzzy data analysis can be implemented in various layers of the database client-server architecture. In this paper, on the basis of the ambulatory data analysis, we show extensions to the Doctrine object-relational mapping (ORM) layer that allow for fuzzy querying and grouping of crisp data.

Keywords: Databases · Fuzzy sets · Fuzzy logic · Querying · Information retrieval · Biomedical data analysis · Object-relational mapping · ORM

1   Introduction

Nowadays people live faster and more dynamically. Stress affects almost everyone and sleep deprivation and neurosis occur even in children. Unfortunately, this way of life is directly related to the appearance of various civilization diseases, such as: heart disease, diabetes, nervousness, cancer, allergies. Fortunately, the human conscience in this regard and attention to health are growing. People more often report to doctors and often are sent for laboratory testing. Moreover, they often are in control of their health, performing basic laboratory tests, such as: morphology, erythrocyte sedimentation rate (ESR), blood sugar, lipid profile, or by measuring independently blood pressure at home. The majority of these tests are performed in medical laboratories, which services range from routine testing, such as basic blood counts and cholesterol tests, to highly complex methods that

© Springer International Publishing AG 2017
S. Kozielski et al. (Eds.): BDAS 2017, CCIS 716, pp. 386–402, 2017.
DOI: 10.1007/978-3-319-58274-0_31

assist in diagnosing genetic disorders, cancers, and other rare diseases. Almost 70% of health care decisions are guided by lab test results.

Laboratories must keep information about the examined people – their personal data and the results of laboratory tests together with ranges for particular test types. These results more often have numeric character.

Some examples of laboratory test are as follows:

– Blood sugar
  Normal blood sugar levels are as follows:
  - Between 4.0 to 6.0 mmol/L (72 to 108 mg/dL) when fasting
  - Up to 7.8 mmol/L (140 mg/dL) 2 h after eating
  But for people with diabetes, blood sugar level targets are as follows:
  - Before meals: 4 to 7 mmol/L for people with type 1 or type 2 diabetes
  - After meals: under 9 mmol/L for people with type 1 diabetes and under 8.5 mmol/L for people with type 2 diabetes

– Lipid profile
  The lipid profile is used as part of a cardiac risk assessment to help determine an individual's risk of heart disease and to help make decisions about what treatment may be best, if there is a borderline or high risk. A lipid profile typically includes the following tests:
  - LDL Cholesterol
    * Optimal: Less than 100 mg/dL (2.59 mmol/L)
    * Near/above optimal: 100–129 mg/dL (2.59–3.34 mmol/L)
    * Borderline high: 130–159 mg/dL (3.37–4.12 mmol/L)
    * High: 160–189 mg/dL (4.15–4.90 mmol/L)
    * Very high: Greater than 190 mg/dL (4.90 mmol/L)
  - Total Cholesterol
    * Desirable: Less than 200 mg/dL (5.18 mmol/L)
    * Borderline high: 200–239 mg/dL (5.18 to 6.18 mmol/L)
    * High: 240 mg/dL (6.22 mmol/L) or higher
  - HDL Cholesterol
    * Low level, increased risk: Less than 40 mg/dL (1.0 mmol/L) for men and less than 50 mg/dL (1.3 mmol/L) for women
    * Average level, average risk: 40–50 mg/dL (1.0–1.3 mmol/L) for men and between 50–59 mg/dl (1.3–1.5 mmol/L) for women
    * High level, less than average risk: 60 mg/dL (1.55 mmol/L) or higher for both men and women
  - Fasting Triglycerides
    * Desirable: Less than 150 mg/dL (1.70 mmol/L)
    * Borderline high: 150–199 mg/dL(1.7–2.2 mmol/L)
    * High: 200–499 mg/dL (2.3–5.6 mmol/L)
    * Very high: Greater than 500 mg/dL (5.6 mmol/L)

The volume of such ambulatory data obtained in laboratory tests is large, and the data must be stored in a repository. In biomedical laboratories this is usually a relational database that holds all the data. Similarly, when people make measurements in their homes, for example blood pressure, the amount of data related to measurements can be huge, especially, when the data are collected remotely by an external system. Frequently it happens that a doctor recommends measuring the patient blood pressure three times a day, and the patient must record all measurements and report them to the doctor, or those measurements are automatically sent to doctors though a telemedicine system to monitor patient and provide clinical health care from a distance. No matter where the data is stored, access to it by searching and retrieving appropriate patients' records should be quick.

Since results of various laboratory tests should fall into certain ranges or may exceed them, people, including medical doctors, usually use common terms, like *normal*, *above* or *below* to describe levels of particular biochemical markers. This provides a good motivation to apply such a logic in computer processing of the data, which would be able to appropriately assign particular values to a certain range, or mark them as going beyond the range: above or below. These conditions can be met by using fuzzy logic [20,21], which becomes particularly handy when dealing with Big Data sets containing results of various ambulatory tests. The fuzzy logic allows to generalize the data, which is of great importance in large medical screenings.

2   Fuzzy Logic in Data Processing and Analysis

Extending standard data analysis with the possibility to formulate fuzzy search criteria and benefit from imprecise and proximity-based linguistic terms that are frequently used in real life, like *small*, *high*, *normal*, *around*, *near* has many advantages. In some situations, it allows to extend the set of results by similar cases, which would not be possible, or at least, difficult with precise search criteria. This is especially beneficial when analyzing biomedical data [16], where sets of important measurements or biomedical markers, e.g., blood pressure, BMI, cholesterol, age, describing particular state of a patient or person may have similar, but not the same values. Enriching the set of results with similar cases can be then very helpful in drawing appropriate conclusions, reporting on important lesions, suggesting certain clinical actions, and preparing similar treatment scenarios for patients with similar symptoms. On the other hand, incorporating routines for fuzzy processing in the data analysis pipeline allows to generalize the data, group it and aggregate, or classify and assign to clusters or subgroups, and thus, change the granularity of information that we have to deal with. This provides a way to quickly reduce the volume of data from big to small, which is highly required in the era of Big Data.

2.1   Related Works

Extensions that allow fuzzy processing, querying and data analysis can be implemented in various layers of the database client-server architecture (Fig. 1). On the client side, various software tools and applications may incorporate fuzzy extensions as procedures and functions that are parts of the software, bound to particular controls of the Graphical User Interface (GUI) or invoked internally from other functions. This has several advantages, including adaptation of the fuzzy procedures to the object-oriented environment (which is frequently used in development of such applications), and adjustment to the specificity of the application and processed data. The fuzzy extensions seem to be tailored to data being processed. The huge disadvantage of such a solution is a tight coupling to a particular application and negligible re-usability of procedures for fuzzy processing for other applications and the analysis of other data. Examples of the implementation of fuzzy data processing on the client side include: risk assessment based on human factors [2], database flexible querying [3], modeling and control [6], historical monuments searching [7], damage assessment [8], decision support systems [11], searching candidates for a date, human profile analysis for missing and unidentified people, automatic news generation for stock exchange [13], decision making in business [19], and others.

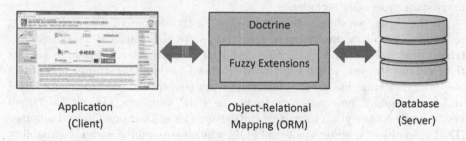

Fig. 1. Client-server architecture with object-relational mapping layer and location of fuzzy extensions proposed in the paper within the architecture.

On the other hand, procedures and functions that allow fuzzy processing of data can be implemented on the server side. This involves implementation of the procedures and functions in the programming language native for the particular database management system (DBMS). Examples of such implementations are: SQLf [5], FQUERY [10], Soft-SQL [4], fuzzy Generalised Logical Condition [9], FuzzyQ [13], fuzzy SQL extensions for relational databases [12,14,18], possibilistic databases [17], and for data warehouses [1,15]. Such an approach usually delivers universal routines that can be used for fuzzy processing of various data coming from different domains. This versatility is a great asset, but it binds users and software developers to particular database management system and its native language, which may also have some limitations.

2.2 Problem Formulation and Scope of the Work

In both mentioned approaches, the prevalent problem is mapping between classes of the client software application and database tables, which is necessary for applications that manipulate and persist data. In the past, each application created its own mapping layer in which the client application's specific classes were mapped to specific tables in the relational database using dedicated SQL statements. Object-relational mapping (ORM) brought evolution in software development by delivering programming technique that allows for automatic conversions of data between relational databases and object-oriented programming languages. ORM was introduced to programming practice in recent decade in response to the incompatibility between relational model of database systems and object-oriented model of client applications. This incompatibility, which is often referred to as the object-relational impedance mismatch, covers various difficulties while mapping objects or class definitions in the developed software application to database tables defined by relational schema. Object-relational mapping tools mitigate the problem of OR impedance mismatch and simplify building software applications that access data in relational database management systems (RDBMSs). Figure 1 shows the role and place of the ORM tools in a typical client-server architecture. However, ORM tools do not provide solutions for people involved in the development of applications that make use of fuzzy data processing techniques.

In the paper, we show extensions to the Doctrine object-relational mapping tool that allow fuzzy processing of crisp data stored in relational database. Doctrine ORM framework is one of several PHP libraries developed as a part of the Doctrine Project, which is primarily focused on database storage and object mapping. Doctrine has greatly benefited from concepts of the Hibernate ORM and has adapted these concepts to fit the PHP language. One of Doctrine's key features is the option to write database queries in Doctrine Query Language (DQL), an object-oriented dialect of SQL, which we extended with the capability of fuzzy data processing.

3   Extensions to Relational Algebra

For fuzzy exploration of crisp data stored in a relational database we have extended a collection of standard operations of the relational algebra by *fuzzy selection* operation.

Given a relation R with n attributes:

$$R = \{A_1 A_2 A_3 ... A_n\}, \tag{1}$$

a *fuzzy selection* $\tilde{\sigma}$ is a unary operation that denotes a subset of a relation R on the basis of fuzzy search condition:

$$\tilde{\sigma}_{A_i \overset{\lambda}{\approx} v}(R) = \{t : t \in R, t(A_i) \overset{\lambda}{\approx} v\}, \tag{2}$$

where: $A_i \overset{\lambda}{\approx} v$ is a fuzzy search condition, A_i is one of attributes of the relation R for $i = 1..n$, n is the number of attributes of the relation R, v is a fuzzy set (e.g., *young* person, *tall* man, *normal* blood pressure, *age near* 30), $\approx$ is a comparison operator used to compare crisp value of attribute A_i for each tuple

t from database with fuzzy set v, λ is a minimum membership degree for which the search condition is satisfied.

The selection $\tilde{\sigma}_{A_i \underset{\approx}{\lambda} v}(R)$ denotes all tuples in R for which $\approx$ holds between the attribute A_i and the fuzzy set v with the membership degree greater or equal to λ. Therefore,

$$\tilde{\sigma}_{A_i \underset{\approx}{\lambda} v}(R) = \{t : t \in R, \mu_v(t(A_i)) \geq \lambda\}, \tag{3}$$

where μ_v is a membership function of a fuzzy set v.

The fuzzy set v can be defined by various types of membership functions, including:

– triangular

$$\mu_v(t(A_i); l, m, n) = \begin{cases} 0, & \text{if } t(A_i) \leq l \\ \frac{t(A_i)-l}{m-l}, & \text{if } l < t(A_i) \leq m \\ \frac{n-t(A_i)}{n-m}, & \text{if } m < t(A_i) \leq n \\ 0, & \text{if } t(A_i) > n \end{cases}, \tag{4}$$

where m is the core of the fuzzy set v, $[l, m]$ determines the left spread (boundary) of the fuzzy set, and $[m, n]$ determines the right spread (boundary) of the fuzzy set.

– trapezoidal

$$\mu_v(t(A_i); l, m, n, p) = \begin{cases} 0, & \text{if } t(A_i) \leq l \\ \frac{t(A_i)-l}{m-l}, & \text{if } l \leq t(A_i) \leq m \\ 1, & \text{if } m \leq t(A_i) \leq n, \\ \frac{p-t(A_i)}{p-n}, & \text{if } n \leq t(A_i) \leq p \\ 0, & \text{if } t(A_i) > p \end{cases} \tag{5}$$

where $[m, n]$ is the core of the fuzzy set v, $[l, m]$ determines the left spread (boundary) of the fuzzy set, and $[n, p]$ determines the right spread (boundary) of the fuzzy set.

– Gaussian

$$\mu_v(t(A_i); c, s, m) = e^{-(\frac{t(A_i)-c}{2s})^m}, \tag{6}$$

where c is called a centre, s is a width, and m is a fuzzification factor (e.g., $m = 2$).

Figure 2 shows how filtering (selection) with fuzzy search conditions works for sample data stored in tables *Measurement* and *Measure* of a relational database (Tables 1 and 2).

Fuzzy selection can be performed on the basis of multiple fuzzy search conditions (or mixed with crisp search conditions), e.g.:

$$\tilde{\sigma}_{A_i \underset{\approx}{\lambda_i} v_i \, \Theta ... \Theta \, A_j \underset{\approx}{\lambda_j} v_j}(R) = \{t : t \in R, t(A_i) \overset{\lambda_i}{\approx} v_i \, \Theta ... \Theta \, t(A_j) \overset{\lambda_j}{\approx} v_j\}, \tag{7}$$

Table 1. Simplified structure of the *Measure* table

ID	Name
1	BMI
2	Systolic blood pressure
3	Diastolic blood pressure
4	PLT

Table 2. Simplified structure of the *Measure* table

MEASURE_ID	VALUE	USER_ID
1	27	1
2	105	1
1	31	2
2	136	2

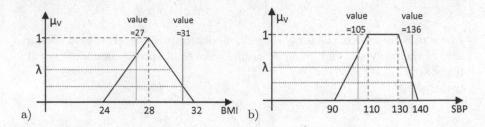

Fig. 2. Filtering crisp data (selection) with fuzzy search conditions for fuzzy sets: (a) body mass index (BMI) *around* 28 (b) *normal* systolic blood pressure (SBM is *normal*, In both cases $\lambda = 0.5$.

where: A_i, A_j are attributes of relation R, $i, j = 1...n, i \neq j$, v_i, v_j are fuzzy sets, λ_i, λ_j are minimum membership degrees for particular fuzzy search conditions, and Θ can be any of logical operators of conjunction or disjunction $\Theta = \{\wedge, \vee\}$. Therefore:

$$\tilde{\sigma}_{A_i \tilde{\approx} v_i}^{\lambda_i} \Theta ... \Theta \, _{A_j \tilde{\approx} v_j}^{\lambda_j} (R) = \{t : t \in R, \mu_{v_i}(t(A_i)) \geq \lambda_i \tag{8}$$

$$\Theta ... \Theta \, \mu_{v_j}(t(A_j)) \geq \lambda_j\},$$

where: μ_{v_i}, μ_{v_j} are membership functions of fuzzy sets v_i, v_j.

4   Extensions to Doctrine ORM

We have extended the Doctrine ORM library with a *Fuzzy* module that enables fuzzy processing of crisp data stored in a relational database. In this section, we describe the most important classes extending standard functionality of the

Doctrine ORM library with the possibility of fuzzy data processing. Additionally, we present a sample usage of the classes in a real application, while performing a simple fuzzy analysis of ambulatory data.

4.1   Class Model of the *Fuzzy* Module

Extensions to Doctrine object-relational mapping tool are gathered in a dedicated programming module, called *Fuzzy*. The module is available at https:// gitlab.com/auroree/fuzzy. The *Fuzzy* module is universal, i.e., independent of the domain of developed application and data stored in a database. Software developers can use the implemented functionality for fuzzy processing of any values stored in the database, e.g., atmospheric pressure, body temperature, person's age, or the number of hours spent watching television. It is necessary to select the type of membership function defining a fuzzy set and provide the relevant parameters.

The *Fuzzy* module provides implementations of functions extending Doctrine Query Language (DQL), which is query language of the Doctrine ORM library. Classes delivered by the *Fuzzy* module are presented in Fig. 3. They are marked in green, in contrast to native PHP classes and classes of the Doctrine ORM/DBAL library that are marked in white. In order to enable fuzzy data processing in the ORM layer we had to implement a set of classes and methods that lead to proper generation of SQL queries for particular functions of fuzzy data processing. To this purpose, we have extended the *Doctrine\ORM\Query\AST\Functions/FunctionNode* class provided by the Doctrine ORM library (Fig. 3). Classes that inherit from the *FunctionNode* class are divided into two groups placed in separate namespaces: membership functions (e.g., *InRange*, *Near*, *RangeUp*) and general-purpose functions (e.g., *Floor*, *Date*). They all implement two important methods: *parse* and *getSql*. The *parse* method detects function parameters in the DQL expression and stores them for future use, then the *getSql* method generates an expression representing a particular mathematical function in the native SQL for a database.

The *InRange* class represents classical (LR) trapezoidal membership function and is suitable to describe values of a domain, e.g., a fuzzy set of *normal* blood pressure, but also *near optimal* LDL Cholesterol (which are in certain ranges of values). The *RangeUp* and *RangeDown* classes represent special cases of trapezoidal membership functions - L-functions (with parameters $n = p = +\infty$) and R-functions (with parameters $l = m = -\infty$), respectively. They are both defined automatically with respect to the fuzzy sets of chosen values of a domain and are suitable to represent selection conditions, such as *HDL below the norm* or *slow heart rate* (R-functions) and *LDL above the norm* or *high blood pressure* (L-functions). The *Near* class represents triangular membership functions and is suitable, e.g., in formulating fuzzy search conditions, like *age about 35*. The *NearGaussian* class represents Gaussian membership function and has similar purpose to triangular membership function.

The *Fuzzy* module also provides a function factory (*FuzzyFunctionFactory* class), which creates instances of classes for the selected membership functions,

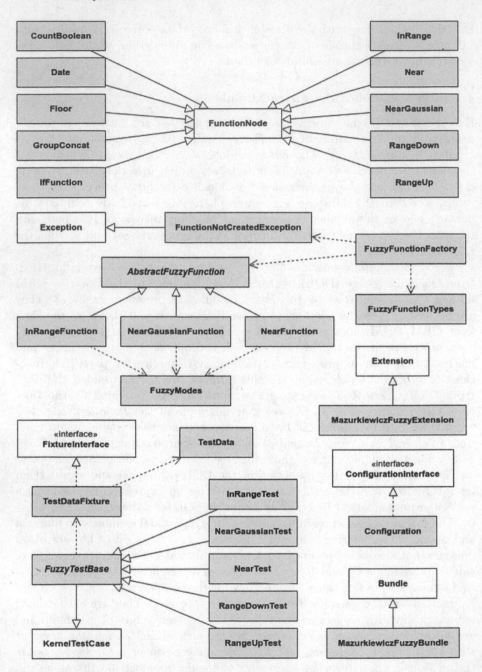

Fig. 3. Overview of classes provided by the *Fuzzy* module extending the Doctrine ORM library.

based on the specified type of the function (one of the values of *FuzzyFunction-Types*), e.g., instance of the *NearFunction* class for the *Near* characteristic function. The function factory class generates appropriate DQL expression together with fuzzy selection condition (as formally defined in Sect. 3) on the basis of declared query type (accepted types are constants of the class *FuzzyModes*). The *Fuzzy* module also contains classes for various tests, e.g., for testing SQL statements generated by the extended ORM library for particular membership functions (e.g., *InRangeTest*, *NearTest* that inherit from *FuzzyTestBase* class).

4.2 Sample Usage of the Doctrine ORM Library with Fuzzy Extensions

In this section, we present a sample usage of the Doctrine ORM library with developed *Fuzzy* module in the analysis of ambulatory data. We show how the fuzzy extensions are utilized in the PHP code of our software application that allows reporting on measurements stored in MySQL relational data repository by calling appropriate DQL queries of the Doctrine ORM library. Finally, we present the form of the SQL query executed in the relational database that corresponds to the DQL query.

Presented sample of the code refers to relational table *measurement* containing ambulatory measurements in the *value* attribute for particular measures identified by *measure_id* attribute (like in Table 2 in Sect. 3). In the ORM layer, this table is mapped to a class called *Measurement*, which attributes correspond to fields (columns) of the *measurement* table. Fuzzy search conditions, created by means of appropriate classes of the *Fuzzy* module, will be imposed on the *value* attribute - Fig. 4, Sect. 2 - for selected measures of systolic blood pressure and diastolic blood pressure (1).

Part of the PHP code was skipped for the sake of clarity of the presentation. In the presented example we assume that domains of both measures are divided into three fuzzy sets: *normal, low,* and *high* blood pressure, according to applicable standards for systolic and diastolic blood pressure. The starting point in this case is to define fuzzy sets for *normal* systolic and diastolic blood pressure with respect to which we define *low* and *high* fuzzy sets for both measures. To represent *normal* blood pressure we use trapezoidal membership functions (specified by *IN_RANGE* function type in the *Fuzzy* module) with $90, 110, 130, 135$ parameters for systolic blood pressure (3) and with $50, 65, 80, 90$ parameters for diastolic blood pressure (4). We define both membership functions by invocation of the *create* function of the *FuzzyFunctionFactory* class. We are interested in selecting patients, whose blood pressure (both types) is elevated (*high*), i.e., above the *normal* value, with the minimum membership degree equal to 0.5 (8). Therefore, we have to define fuzzy search conditions by using *getDql* method of the *InRangeFunction* class instance returned by the *FuzzyFunctionFactory*. Then, we have to use *ABOVE_SET* fuzzy mode in the *getDql* method in order to get values above the *normal*. In such a way, we obtain two fuzzy search conditions for DQL query that will be used in the *where* clause. To build the whole analytical report we formulate a query by using Doctrine Query Builder with

```
/** @var MeasurementRepository $repository */
$repository = ...;
$sysMeasureId = ...; // (1)
$diaMeasureId = ...; // (1)

$valueColumnName = 'mm.value'; // (2)

// (3)
$sysFunction = FuzzyFunctionFactory::create(
  FuzzyFunctionTypes::IN_RANGE,
  [90, 110, 130, 135]
);
// (8)
$sysDqlCondition = $sysFunction->getDql(
  FuzzyModes::ABOVE_SET,
  $valueColumnName,
  0.5
);

// (4)
$diaFunction = FuzzyFunctionFactory::create(
  FuzzyFunctionTypes::IN_RANGE,
  [50, 65, 80, 90]
);
// (8)
$diaDqlCondition = $diaFunction->getDql(
  FuzzyModes::ABOVE_SET,
  $valueColumnName,
  0.5
);

// (9)
$query = $repository->createQueryBuilder('mm')
  ->select('u.id, m.name, mm.value')
  ->join('mm.user', 'u') // (5)
  ->join('mm.measure', 'm') // (6)
  ->where("(mm.measure = {$sysMeasureId} AND {$sysDqlCondition})
    OR (mm.measure = {$diaMeasureId} AND {$diaDqlCondition})") // (7)
  ->getQuery();

$result = $query->getResult();
```

Fig. 4. Sample usage of the *Fuzzy* module in PHP code.

appropriate clauses of the query statement (9). We join *User* (5) and *Measure* (6) entities/classes to add data about patients and measure types. Finally, we add fuzzy search conditions in (7). The query will return only those rows for which values of both measures belong to *high* fuzzy set (i.e., *above the norm*) defined for the particular measure.

The PHP code presented in Fig. 4 produces the DQL query shown in Fig. 5, which will be translated to SQL query for relational database (Fig. 6) by the Doctrine library. Translation of the query built up with the PHP Query Builder (Sect. (9) in Fig. 4) to DQL query produces WHERE clause containing two invocations of the RANGE_UP functions with appropriate parameters of L-type membership functions representing *above the norm* fuzzy sets for particular measures.

```
SELECT u.id, m.name, mm.value
FROM Mazurkiewicz\TrackerBundle\Entity\Measurement mm
    INNER JOIN mm.user u
    INNER JOIN mm.measure m
WHERE (mm.measure = 2 AND RANGE_UP(mm.value, 130, 135) >= 0.5)
    OR (mm.measure = 3 AND RANGE_UP(mm.value, 80, 90) >= 0.5)
```

Fig. 5. DQL query with two fuzzy search conditions for PHP code presented in Fig. 4

These invocations are then translated to the CASE ... WHEN ... THEN statements in the WHERE clause of the SQL query command (Fig. 6).

```
SELECT u0_.id AS id_0, m1_.name AS name_1, m2_.value AS value_2
FROM measurement m2_
INNER JOIN user u0_ ON m2_.user_id = u0_.id
INNER JOIN measure m1_ ON m2_.measure_id = m1_.id
WHERE
    (m2_.measure id = 2 AND CASE
        WHEN m2_.value <= 130 THEN 0
        WHEN m2_.value <= 135 THEN (m2_.value-130)/(135-130)
        ELSE 1 END >= 0.5
    )
    OR (m2_.measure_id = 3 AND CASE
        WHEN m2_.value <= 80 THEN 0
        WHEN m2_.value <= 90 THEN (m2_.value-80)/(90-80)
        ELSE 1 END >= 0.5
    )
```

Fig. 6. SQL query translated by fuzzy extension of the Doctrine library from DQL query presented in Fig. 5

5   Experimental Results

We tested performance of the fuzzy extension for the Doctrine ORM library in several series of tests. We were primarily interested in verification of how the

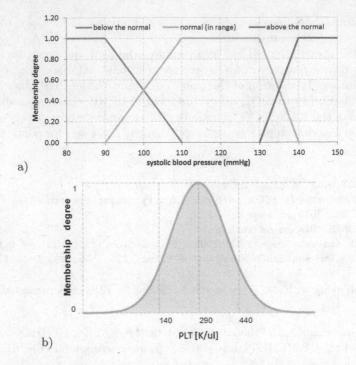

a)

b)

Fig. 7. Membership functions for *systolic blood pressure* (a). Membership function for the fuzzy set *normal PLT* (b).

necessity of calculation of value of a membership function influences the execution time of particular fuzzy queries with respect to classical queries that operate on given ranges of values (appropriately chosen intervals). For this purpose, we used a database containing 2,500,000 records in the *Measurement* table coming from laboratory tests.

Results of performance tests are presented in Table 3 for three chosen sample fuzzy queries (Q1–Q3) retrieving measurement data for patients having:

- Q1 - *normal* systolic blood pressure (Fig. 7a),
- Q2 - systolic blood pressure *above the normal* (Fig. 7a),
- Q3 - *normal* platelet count (PLT) (Fig. 7b),

with a minimum membership degree $\lambda = 0.5$. In a real implementation, we tested many more queries, but they all shown the same execution time tendency.

Queries Q1–Q3 contain fuzzy search conditions. Definitions of fuzzy sets used in these search conditions are presented in Fig. 7. Particular parameters of the membership functions were set on the basis of arbitrary expert's knowledge, and include some tolerance. These parameters can be changed in specific implementations, which leads to different results. Therefore, they must be assumed carefully, while consulting the shape of membership functions with domain experts.

Additionally, for queries Q1–Q3 we created their classical counterparts with precise search criteria based on intervals, where left and right boundaries of the intervals were calculated for the membership degree $\lambda = 0.5$. Particular fuzzy queries and their precise counterparts returned the same sets of results, but were parametrized in a different way - precise queries need exact values of left and right boundaries of intervals, while fuzzy queries need only the minimum membership degree λ, above which the search condition is satisfied.

Table 3. Results of performance tests for fuzzy queries Q1–Q3 and their precise counterparts for the minimum membership degree $\lambda = 0.5$.

Query	Average execution time (s)		Difference	Relative
	Precise query	Fuzzy query	(s)	difference (%)
Q1	0.491274	0.496286	0.005012	1.02
Q2	0.485716	0.497575	0.011859	2.44
Q3	0.610142	0.627314	0.017172	2.81

Results of performance tests presented in Table 3 proved that execution times of fuzzy queries were only slightly worse than execution times of precise queries that returned the same sets of results. Fuzzy queries were executed relatively 1–3% longer than their precise counterparts. This means that for users of the ORM library with fuzzy extensions the difference in execution time is almost imperceptible.

6   Discussion and Concluding Remarks

Our research on extending the Doctrine object-relational mapping framework toward fuzzy data processing show that it is possible to incorporate fuzzy logic in the ORM layer and enhance standard database querying with new capabilities of imprecise, proximity-based or similarity-based searching. The enhancement brings new power to the analysis of crisp, numerical data stored in databases, which is important when processing large volumes of biomedical or ambulatory data, and can be now performed in the ORM layer, which is important for software developers. As proved by our experiments, performance costs of such an enhancement are negligible compared to the additional analytic possibilities that are obtained by developers of database applications.

Fuzzy querying with fuzzy search conditions provides several benefits compared to precise queries. Synthetic comparison of precise and fuzzy queries in terms of flexibility of queries and corresponding requirements is presented in Table 4. First of all, fuzzy queries give the possibility to easily filter out uninteresting data on the basis of soft search conditions, while still keeping similar data in the final result set. Therefore, they narrow the result set to similar cases, which is very important while performing large-scale medical screenings based on the

Table 4. Comparison of fuzzy and precise queries in terms of flexibility, requirements, and performance.

Fuzzy queries	Precise queries
+ Soft filtering - including similar cases	− Hard filtering - similar data filtered out
+ Require value of the minimum membership degree λ in a fuzzy search condition	− Require crisp values in search conditions, e.g., left and right boundaries of an interval
± Require expert that arbitrary defines fuzzy sets and corresponding membership functions	− Require users to have knowledge of data domain
− Require additional calculations of membership degrees, which may affect performance	+ No additional calculations are required

+ advantage, − disadvantage

analysis of various types of biomedical data, including ambulatory data. Hard filtering with precise queries may cause that important cases leading to the same medical conclusions and therapeutic recommendations will be just skipped, as they not satisfy precise search conditions. Secondly, precise queries require specification of crisp values for their filtering conditions. These crisp values may not be known for the final users or may require further investigations to know them, especially in medical domain. On the other hand, fuzzy queries require specifying the minimum membership degrees λ for fuzzy search conditions, which may also be a problem, but we must remember that they decide about the similarity degree for data that is returned in the result set. Therefore, they can be chosen in several trials narrowing the final result set in several following steps. Consequently, precise queries require that users of the developed system have a specialized knowledge of the domain of analyzed data, which is sometimes very difficult to gain unless they are experts. For example, when analyzing results of laboratory tests users have to find out what are the normal ranges for specific ambulatory tests, if they are not doctors or laboratory staff, which is a weakness. This can be also a weakness of fuzzy queries, since for many domains the requirement for dividing the analyzed domain into proper ranges and defining proper membership functions for identified fuzzy sets is prevalent and can be a spark for discussion. However, for ambulatory data and many other types of biomedical data these values are usually arbitrary defined by experts and are indisputable for, at least, some period of time. Therefore, for such domains this does not constitute a problem and causes the use of fuzzy search conditions with their large flexibility a more natural solution. Finally, a weak point of fuzzy queries is the necessity to calculate membership degrees for each tuple from the database processed by the fuzzy query, which may negatively affect performance of the query. However, the fuzzy extensions to the ORM layer that we have developed proved to be only 1–3% slower for tested queries, which allows us to ignore this slight decrease in performance in the face of much better querying capabilities.

Our fuzzy extensions to the Doctrine library mitigate the problem of object-relational impedance mismatch for those software developers that want to perform fuzzy searches while working in the object-oriented model, regardless of the data domain being analyzed. It is limited to the PHP technology of building client software tools, but universal in terms of the type of analyzed data and built application. The *Fuzzy* module for Doctrine framework presented in the paper enables re-usability of procedures for fuzzy data processing for any client application that is developed and any data that is analyzed, which was a limitation of client-based solutions mentioned in Sect. 2.1. On the other hand, software developers are not bound to a particular database management system and its native query language, which was a weakness of server-side solutions presented in Sect. 2.1. This ensures broader portability of our fuzzy extension. In such a way, our solution complements a collection of existing solutions and, to the best of our knowledge, is first such an extension for the ORM layer.

References

1. Appelgren Lara, G., Delgado, M., Marín, N.: Fuzzy multidimensional modelling for flexible querying of learning object repositories. In: Larsen, H.L., Martin-Bautista, M.J., Vila, M.A., Andreasen, T., Christiansen, H. (eds.) FQAS 2013. LNCS (LNAI), vol. 8132, pp. 112–123. Springer, Heidelberg (2013). doi:10.1007/978-3-642-40769-7_10
2. Aras, F., Karaka, Y.: Fuzzy logic-based user interface design for risk assessment considering human factor: a case study for high-voltage cell. Saf. Sci. **70**, 387–396 (2014). http://www.sciencedirect.com/science/article/pii/S0925753514001726
3. Ben Hassine, M.A., Ounelli, H.: IDFQ: an interface for database flexible querying. In: Atzeni, P., Caplinskas, A., Jaakkola, H. (eds.) ADBIS 2008. LNCS, vol. 5207, pp. 112–126. Springer, Heidelberg (2008). doi:10.1007/978-3-540-85713-6_9
4. Bordogna, G., Psaila, G.: Customizable flexible querying in classical relational databases. In: Handbook of Research on Fuzzy Information Processing in Databases, pp. 191–217 (2008)
5. Bosc, P., Pivert, O.: SQLf query functionality on top of a regular relational database management system. In: Pons, O., Vila, A.M., Kacprzyk, J. (eds.) Knowledge Management in Fuzzy Databases, vol. 39, pp. 171–190. Physica-Verlag HD, Heidelberg (2000). doi:10.1007/978-3-7908-1865-9_11
6. Cheng, S., Dong, R., Pedrycz, W.: A framework of fuzzy hybrid systems for modelling and control. Int. J. Gen Syst **39**(2), 165–176 (2010). http://dx.doi.org/10.1080/03081070903427358
7. Czajkowski, K., Olczyk, P.: Fuzzy interface for historical monuments databases. In: Kozielski, S., Mrozek, D., Kasprowski, P., Małysiak-Mrozek, B., Kostrzewa, D. (eds.) BDAS 2014. CCIS, vol. 424, pp. 271–279. Springer, Cham (2014). doi:10.1007/978-3-319-06932-6_26
8. Furuta, H., Shiraishi, N.: Fuzzy data processing in damage assessment. In: Natke, H.G., Yao, J.T.P. (eds.) Structural Safety Evaluation Based on System Identification Approaches, pp. 381–392. Vieweg+Teubner Verlag, Wiesbaden (1988). doi:10.1007/978-3-663-05657-7_18

9. Hudec, M.: An approach to fuzzy database querying, analysis and realisation. Comput. Sci. Inf. Syst. **12**, 127–140 (2009)

10. Kacprzyk, J., Zadrożny, S.: Data mining via fuzzy querying over the internet. In: Pons, O., Vila, A.M., Kacprzyk, J. (eds.) Knowledge Management in Fuzzy Databases, vol. 39, pp. 211–233. Physica-Verlag HD, Heidelberg (2000). doi:10.1007/978-3-7908-1865-9_13

11. Macwan, N., Sajja, P.S.: Fuzzy logic: an effective user interface tool for decision support system. Int. J. Eng. Sci. Innov. Technol. **3**(3), 278–283 (2014). http://www.ijesit.com/Volume%203/Issue%203/IJESIT201403_35.pdf

12. Małysiak, B., Mrozek, D., Kozielski, S.: Processing fuzzy SQL queries with flat, context-dependent and multidimensional membership functions. In: IASTED International Conference on Computational Intelligence, Calgary, Alberta, Canada, 4–6 July 2005, pp. 36–41 (2005)

13. Małysiak-Mrozek, B., Kozielski, S., Mrozek, D.: Modern software tools for researching and teaching fuzzy logic incorporated into database systems. In: Proceedings of the iNEER International Conference on Engineering Education, Gliwice, Poland, pp. 1–8. iNEER, July 2010. http://www.ineer.org/Events/ICEE2010/papers/T11D/Paper_954_1141.pdf

14. Małysiak-Mrozek, B., Mrozek, D., Kozielski, S.: Data grouping process in extended SQL language containing fuzzy elements. In: Cyran, K.A., Kozielski, S., Peters, J.F., Stańczyk, U., Wakulicz-Deja, A. (eds.) Man-Machine Interactions, vol. 59, pp. 247–256. Springer, Heidelberg (2009). doi:10.1007/978-3-642-00563-3_25

15. Małysiak-Mrozek, B., Mrozek, D., Kozielski, S.: Processing of crisp and fuzzy measures in the fuzzy data warehouse for global natural resources. In: García-Pedrajas, N., Herrera, F., Fyfe, C., Benítez, J.M., Ali, M. (eds.) IEA/AIE 2010. LNCS (LNAI), vol. 6098, pp. 616–625. Springer, Heidelberg (2010). doi:10.1007/978-3-642-13033-5_63

16. Mrozek, D., Kasprowski, P., Małysiak-Mrozek, B., Kozielski, S.: Life sciences data analysis. Inf. Sci. **384**, 86–89 (2017)

17. Myszkorowski, K.: Inference rules for fuzzy functional dependencies in possibilistic databases. In: Kozielski, S., Mrozek, D., Kasprowski, P., Małysiak-Mrozek, B., Kostrzewa, D. (eds.) BDAS 2015-2016. CCIS, vol. 613, pp. 181–191. Springer, Cham (2016). doi:10.1007/978-3-319-34099-9_13

18. Portinale, L., Montani, S.: A fuzzy logic approach to case matching and retrieval suitable to SQL implementation. In: Proceedings of the 2008 20th IEEE International Conference on Tools with Artificial Intelligence, ICTAI 2008, vol. 02, pp. 241–245. IEEE Computer Society, Washington, DC (2008). http://dx.doi.org/10.1109/ICTAI.2008.88

19. Ribeiro, R.A., Moreira, A.M.: Fuzzy query interface for a business database. Int. J. Hum.-Comput. Stud. **58**(4), 363–391 (2003)

20. Zadeh, L.: Fuzzy sets. Inf. Control **8**, 338–353 (1965)

21. Zadeh, L.: Fuzzy logic. Computer **21**(4), 83–93 (1988)

Segmenting Lungs from Whole-Body CT Scans

Maksym Walczak[1,2](✉), Izabela Burda[1], Jakub Nalepa[1,2](✉),
and Michal Kawulok[1,2](✉)

[1] Future Processing, Gliwice, Poland
{mwalczak,iburda,jnalepa,mkawulok}@future-processing.com,
{maksym.walczak,jakub.nalepa,michal.kawulok}@polsl.pl
[2] Institute of Informatics, Silesian University of Technology, Gliwice, Poland

Abstract. Image segmentation is an initial, yet crucial procedure in
a number of medical imaging systems. Despite the existence of numerous generic solutions that address this problem, there is still a need for
developing fast and accurate techniques specialized at extracting particular organs from the CT scans. In this paper, we present an approach
based on simple operations, which is controlled with a few easy-to-adjust
parameters and works without any user interaction. The proposed approach, despite its simplicity, was shown to be reliable and efficient for a
dataset of over 50 studies, containing both healthy and pathologic lungs.

Keywords: Lung segmentation · Computed tomography · Medical
image processing

1 Introduction

Computed tomography (CT) was introduced over three decades ago and since
then it has become a crucial technique for diagnosing many diseases, including
cancer. As the imaging devices and diagnosis procedures evolved, the need for
automated image segmentation algorithms gradually increased. The amount of
generated medical data grows extraordinarily fast and its efficient analysis and
handling, e.g., in the field of medical imaging, play a pivotal role and attract
research attention [3]. The first attempts to the automated segmentation date
back to the beginning of the 80s' in the 20th century [18] and they are based on
relatively simple approaches, such as thresholding and elementary morphological operations. These techniques still offer building blocks for many successful
modern applications [2,7,21].

Gradually, many new categories of approaches emerged. A remarkable example are the knowledge-based methods, where a learned set of anatomical models is used to increase robustness of the segmentation [20]. Another group of
approaches relies on fuzzy logic—i.e., fuzzy connectedness [8], where the connectivity analysis of segmented regions is extended to grayscale instead of a
previously thresholded binary image. Fuzzy logic can also be used for reasoning
in the previous, model-based approach. There are also algorithms that rely on
pseudo-physical simulation of region [9] or contour growing, such as the active-contour model [19]. Here, a 2D or 3D growing element is iteratively modified by

S. Kozielski et al. (Eds.): BDAS 2017, CCIS 716, pp. 403–414, 2017.
DOI: 10.1007/978-3-319-58274-0_32

its *internal energy*, *image energy* and *constraint energy* which influence its shape change. It has been proved to be a very good technique for segmenting regions such as trachea and pulmonary vessels. However, for this method to be efficient, its parameters require difficult manual adjustments and such optimization can be computationally intensive.

Generally, all of the above-mentioned algorithms rely—at least in terms of preprocessing—on elementary operations which are widely used in computer vision [17]. Hence, finding optimal combinations of these elementary processing steps and introducing various improvements to them can contribute to many segmentation frameworks.

1.1 Contribution

The proposed algorithm is partly based on [6], however it puts special emphasis on untypical and pathologic lungs and offers important improvements. First, we introduce a procedure for removing non-body false positives. Next, we introduce an approach to distinguish between the trachea region and the airways above it to avoid faulty segmentation in whole-body CT scans. Finally, we improve the procedure to separate the lungs from each other, which was presented in [6]. We achieve this by introducing auxiliary steps and by modifying the original solution so that it relies solely on image morphology. Furthermore, we present a slightly faster, but similarly efficient method based on finding local minimum along the outcome of the distance transform. Thanks to these improvements, the proposed algorithm can process untypical and pathologic lungs.

1.2 Paper Structure

In Sect. 1, a brief overview of existing methods is given. Section 2 describes the related literature. Section 3 gives a detailed description of the proposed algorithm. In Sect. 4, the applied dataset and the experimental setup are described. The last section contains evaluation of the results and sketches further improvements we plan to introduce.

2 Related Literature

Costa and Carvalho [2] showed that using basic image processing algorithms such as image morphology and thresholding, it is possible to achieve very good segmentation results. Their *Simple Automatic Lung Segmentation Algorithm* (SALSA) achieved top level lung segmentation accuracy for a large dataset. Another example, where the above-mentioned algorithms proved to be quite successful, is the work by Hu and Hoffman [6], where they achieved a good coverage with human-segmented ground truth in a series of 24 CT scans. Certain improvements have been made by applying fuzzy logics [16] and fuzzy connectedness [8]. The main deficiency of lung segmentation using thresholding is that it is sensitive to noise and inconsistencies in image acquisition techniques. This problem is partly overcome by the active contour model. Annangi et al. [1] exploit this approach and combine it with descriptors of edge feature points and region-based statistics in order to obtain very precise segmentation results.

Recently, a remarkable progress was achieved in image segmentation using deep neural networks. So far, in terms of computer aided diagnosis, most of the proposed solutions use reinforced learning and manually selected features [8]. A paper by Shin et al. [15] offers an in-depth study of image segmentation for computer aided diagnosis applications. There are currently three major techniques employing convolutional neural networks to medical image classification: unsupervised training with supervised fine tuning [13], training the networks from scratch [14] and using pre-trained features [5]. The main challenges in image segmentation using deep neural networks are the dimensionality of the problem, time needed to train the networks and finally making sense of the data. The last can still be achieved using simpler and faster algorithms such as the one presented in this paper, which can be further enhanced if needed.

3   Proposed Algorithm

The flowchart of the proposed algorithm is rendered in Fig. 1. The first two steps are initially performed separately for each slice. Non-body regions removal and lung extraction rely on the previously found clusters of air-filled voxels, later called 3D connected components. Since the lungs are extracted together with the trachea and the upper airways, it is necessary to further segment and separate these regions using the trachea extraction procedure. The last step is the lung separation and it is necessary only when the optimal thresholding fails to detect the thin tissue wall between the lungs (as a result, both lungs appear as one). After this final step, regions of interest in the form of contours or binary masks covering the lung regions can be generated as the output. In the following subsections, each of these steps is described in details.

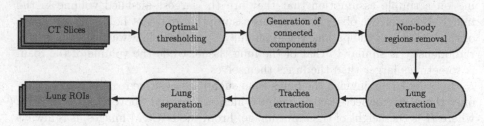

Fig. 1. Flowchart of the proposed algorithm.

3.1   Optimal Thresholding

Optimal thresholding is an iterative procedure aimed at finding the optimal threshold value, which segments the image into *radioopaque* and *radiotransparent* regions. *Radioopaque* regions have higher radiodensity (measured with

Hounsfield units—Hu), such as water-filled tissues and bones, while the *radio-transparent* regions are those with lower radiodensity, which includes the lung candidates and gas in the bowels. Before optimal thresholding, the air surrounding the patient is removed from each slice using flood fill algorithm. At the beginning of the procedure, an initial threshold value is set for the first iteration to $T_0 = 0$ Hu, which equals to the radiodensity of water that is present in the tissues. Then, in the subsequent iterations the threshold value is calculated by averaging mean of *radioopaque* and *radiodense* pixel radiodensities from each previous iteration. In each iteration, the threshold value is updated as $T_{i+1} = (\mu_b + \mu_{nb})/2$, where μ_b is the value of mean radiodensity for the radioopaque pixels and μ_{nb} is the mean value for the radiotransparent ones. The resulting binary image forms a mask, whose white and black pixels are the radioopaque and radiotransparent regions, respectively. This procedure is terminated when $T_{i+1} = T_i$.

3.2 3D Connected-Components Labeling

The goal of 3D connected-components labeling is to extract 3D volumes from a series of 2D slices, whose properties are exploited to segment the lungs. The labeling first occurs in each slice in 2D for the radiotransparent regions. Later, the 2D connected components are merged into layered 3D connected components by scanning subsequent slices and building an undirected graph $G(V,E)$. The vertices V are labelled with each component label and the identifier of the layer. Edges E connect the vertices whenever a connectivity between subsequent slices occurs. Relabeling the graph yields data about 3D connectivity and metadata containing the location of centroids and areas of each component are generated.

3.3 Non-body Region Removal and Lung Extraction

Once all the 3D connected components are labelled, the lungs are detected relying on a simple assumption that they are the largest air-filled volume in the image. However, we observed that this assumption may not hold, as some parts of the bed, which the patient is laid on, may be made of lightweight foam, whose radiodensity is similar to that of the lung tissue, while the volume of the foam is sometimes larger than the lungs themselves.

In order to distinguish the lungs from such false positive regions, we consider only those 3D-connected components which are located above $H_{max} = 0.25H$, where H is the height of the input image. From the DICOM images, it is always possible to identify the top of the image, so there is no risk of the image being flipped. There are usually no other large air-filled regions above the patient and the patient himself is positioned in the center of the CT scan. We have also considered extracting shape-related features [10], however the algorithm was tested using data derived from different patients and scanning equipment, and this simple verification procedure was correct in all cases.

3.4   Trachea Extraction

In order to extract the trachea, areas of 2D connected components belonging to the lungs are scanned from head to feet. When the area growth becomes too rapid, the scanning is aborted and all the previously scanned 2D connected components are separated from the lungs as trachea and mainstem bronchi. The growth is controlled with a parameter T_a, which refers to the maximum area accepted during the slice-by-slice scanning. The value of T_a was chosen experimentally to be 800 pixels—smaller values result in premature interruption of the procedure and larger values lead to confusion between the trachea and the lungs. The largest 3D connected component found using this method is classified as trachea and mainstem bronchi.

3.5   Lung Separation

In some cases, the optimal thresholding fails to correctly identify the radiotransparent regions and the thin tissue wall between the lungs is not detected. As a result, the lungs appear as connected together. Below, we discuss two methods that we propose to overcome this issue—distance transform splitting and morphological splitting. In order to determine whether the lungs need splitting, a centroid of each 2D connected component in each lungs mask is calculated and compared to the 3D centroid of the whole group of masks representing lungs. If the distance along the horizontal plane between the centroids is below a certain threshold T_c, then the connected component should be split. The value of T_c is expressed as a horizontal distance from the centroid of the current component to the centroid of the lungs—both divided by the width of the lungs.

In the distance transform splitting, distance transform [4] is calculated for the binary lungs mask A, from which a morphological skeleton is generated [22]. Next, all the side branches are removed, leaving only the "stem". In the final step, all pixels are scanned from the resulting skeleton, the value of the underlying distance transform is checked and the minimum value of it is found. The point minimizing the value lies in the thinnest section of the input lungs mask and the lungs are simply separated with a vertical line along that point.

Morphological splitting is based on conditional erosion and dilation. Conditional erosion consists of eroding the image as long as the connectivity is not affected. This can be expressed as follows:

$$S = A \ominus nB, \tag{1}$$

where S is the resulting conditionally eroded image, $\ominus$ denotes the erosion, A is the input mask representing lungs, B is a 3×3 diamond shaped structuring element and n is the smallest number of repetitions of the erosion resulting in S having more connected components than A. In order to prevent premature splitting of the lungs due to holes and irregularities, the holes are filled using flood fill algorithm. The whole procedure is presented in Algorithm 1.

If the lungs connect in more than one region, the algorithm is repeated as long as $A - D$ has the same number of 2D connected components as the original lungs

Algorithm 1. Lung separation algorithm.

1: $S \leftarrow$ conditionalErosion(A, B)
2: $S \leftarrow$ fillHoles(S)
3: $D, D_{prev} \leftarrow \emptyset$
4: leftLung, rightLung $\leftarrow$ getTwoLargestConnectedComponents(S)
5: **while** $D = \emptyset \vee D \neq D_{prev}$ **do**
6:      leftLung $\leftarrow [($leftLung $\oplus B) \cap A] \setminus D$ ▷ $\oplus$ denotes dilation
7:      rightLung $\leftarrow [($rightLung $\oplus B) \cap A] \setminus D$
8:      $D \leftarrow$ leftLung $\cap$ rightLung $\cup D_{prev}$
9:      swap(D, D_{prev})
10: **end while**
11: **return** $A - D$

---

mask A. After executing the algorithm, mask D contains common part of left and right lung accumulated across all the iterations. The exemplary execution of Algorithm 1 is rendered in Fig. 2.

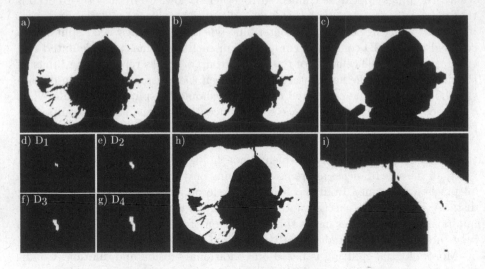

Fig. 2. Exemplary execution of the lung separation algorithm: (a) unprocessed lungs mask, (b) filled holes, (c) result of conditional erosion, (d–g) intersection growth in mediastinum, (h) separated lungs, and (i) magnified region of the separation.

4   Experimental Validation

4.1   Setup

The proposed algorithms were implemented in C++ using the OpenCV 3.1 library and compiled with Microsoft Visual C Compiler 2015. The test platform was a computer equipped with Intel Core i5-6500 CPU, 16 GB DDR4 RAM and 256

GB SSD. In order to see how the algorithms influence the quality of segmentation, they are run in 4 different variants, presented in Table 1. The parameters $T_c = 0.05$, $T_a = 800$, $H_{max} = 0.25H$ were adjusted experimentally.

The results were verified using the test dataset, which consists of 56 CT scans containing 28 healthy and 28 pathologic lungs—$2.2 \cdot 10^4$ frames in total. For each CT scan, 3 or 4 randomly chosen frames containing lungs were manually segmented to provide ground truth data. This resulted in 164 ground-truth slices. We quantify the segmentation quality using the following metrics:

- *Specificity* $q = TN/(TN + FP)$, where FP—false positives, TN—true negatives. It measures the ratio of negative pixels in the ground truth that are correctly identified as negatives by the algorithm being evaluated.
- *Sensitivity* $p = TP/(TP + FN)$, where FN—false negatives, TP—true positives. It measures the ratio of positive pixels in the ground truth that are correctly identified as positives by the algorithm being evaluated.
- *Precision* $PPV = TP/(TP + FP)$—positive predictive value.
- *C-Factor*—defined when $p > 1 - q$. It gives information whether segmentation being evaluated is an under-segmentation (negative values) or over-segmentation. The C-Factor (C) is given as:

$$C = \begin{cases} d, p \geq q \\ -d, p < q \end{cases}, \text{ and } d = \frac{2p(1-q)}{p+(1-q)} + \frac{2(1-p)q}{(1-p)+q}. \tag{2}$$

- DICE $= 2TP/(2TP+FP+FN)$. It is used for comparing the similarity of two samples. It determines how well automatically segmented values cover ground truth data.

Table 1. Investigated variants of the proposed algorithm.

Variant	Meaning
B	Baseline [6]
BB	Baseline + false positives removal
DB	Distance transform splitting + false positives removal
MB	Morphological splitting + false positives removal

4.2 Quantitative Results

Segmentation quality metrics for each variant of the algorithm are presented in Tables 2, 3 and 4, where μ—mean for each metrics with respect to ground truth slices, σ—standard deviation from the mean.

From the tables, it can be observed that each of the variants from BB, up to MB, improves segmentation quality at the cost of slightly increased processing time. Another advantage was a visible drop of false positive rate and standard deviation in the metrics due to successful detection and exclusion of the false positives and thanks to the upgraded lung splitting procedure.

Table 2. Statistical evaluation of segmentation for the left lung.

Left lung						
Measure	Variant	DICE	Sensitivity	Precision	C-Factor	Specificity
μ	B	0.92212	0.94101	0.92062	−0.03924	0.98808
	BB	0.94029	**0.96666**	0.93559	**−0.03598**	0.98847
	DB	0.97531	0.96634	0.98666	−0.06110	0.99917
	MB	**0.97660**	0.96652	**0.98890**	−0.06090	**0.99926**
σ	B	0.18234	0.16448	0.21045	0.11750	0.03697
	BB	0.11113	0.05136	0.16227	0.11923	0.03693
	DB	0.03389	0.05143	0.03044	0.08054	0.00191
	MB	**0.03294**	**0.05131**	**0.02512**	**0.08023**	**0.00179**

Table 3. Statistical evaluation of segmentation for the right lung.

Right lung						
Measure	Variant	DICE	Sensitivity	Precision	C-Factor	Specificity
μ	B	0.94315	0.93595	0.95301	−0.05421	0.99863
	BB	0.93471	0.93644	0.94279	**−0.05245**	0.99812
	DB	0.96806	0.96949	0.97437	−0.05431	0.99766
	MB	**0.97100**	**0.97088**	**0.97788**	−0.05409	**0.99869**
σ	B	0.18426	0.18357	0.18865	0.06951	0.00820
	BB	0.19613	0.18174	0.20665	0.07228	0.00922
	DB	0.07320	0.04027	0.09963	0.07633	0.01336
	MB	**0.06810**	**0.03853**	**0.09121**	**0.06783**	**0.00438**

Table 4. Statistical evaluation of segmentation for both the lungs.

Both lungs						
Measure	Variant	DICE	Sensitivity	Precision	C-Factor	Specificity
μ	B	0.95508	0.94723	0.96397	**−0.05093**	0.99681
	BB	0.97574	**0.97157**	0.98230	−0.05195	0.99750
	DB	0.97572	0.97150	0.98232	−0.05206	0.99751
	MB	**0.97618**	0.97145	**0.98322**	−0.05286	**0.99780**
σ	B	0.15623	0.15693	0.15794	0.05894	0.00894
	BB	0.03620	0.03221	0.05397	0.06067	0.00647
	DB	0.03617	0.03217	0.05390	0.06065	0.00647
	MB	**0.03595**	**0.03215**	**0.05304**	**0.05946**	**0.00549**

4.3   Qualitative Results

Figure 3 shows the only case among 56 analyzed studies, where $T_a = 800$ caused improper segmentation of trachea. In all the other cases, the experimentally chosen value works correctly. In the presented case, only a 2× larger value resulted in correct extraction of the trachea. This occurred partly because the section where the trachea "splits" into mainstem bronchi is really large and partly because the tissue has subpixel thickness and the initial segmentation using the optimal thresholding failed to distinguish it.

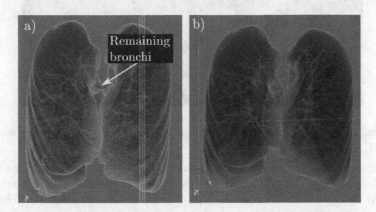

Fig. 3. Influence of the parameter T_a on the quality of trachea segmentation (a) $T_a = 800$, (b) $T_a = 1600$, correctly extracted trachea and the mainstem bronchi.

In Fig. 4, regions classified as the lungs are colored green. In Fig. 4a, the bed is mistaken with lungs because it was not removed from the pool of lung candidates and in fact it had the largest volume out of all the candidates. Figure 4b shows properly recognized lungs using the proposed false-positive removal algorithm.

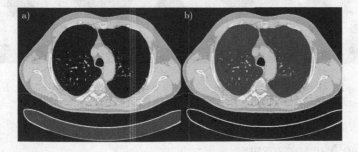

Fig. 4. Influence of the non-body false positive removal algorithm on the lung recognition. The algorithm is disabled (a) and enabled (b). (Color figure online)

In Fig. 5, it can be seen that without proper recognition of segmented regions after lung extraction, the upper airways remain to be recognized as lungs as in Fig. 5a. With the proper recognition, the problem no longer occurs as in Fig. 5b. The issue is present in the baseline version because it assumes that the scan contains only a section of the body from the trachea to the lowest part of the lungs, and therefore full body CT scans cannot be properly processed.

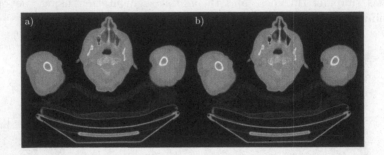

Fig. 5. Trachea segmentation (a) before and (b) after the improvements.

In Fig. 6, three different methods of lung separation are compared for a person with severe scoliosis. Red-colored sections in green circles depict regions which were split using each of the variants of the lung separation algorithm. The lungs were originally connected in both the frontal section (near sternum) and the rear section. Variant B, in Fig. 6a failed completely, since the presence of only one section at which the lungs connect is assumed. Variant BB failed in the same way as variant B, since the only improvement it introduces is the non-body false positives removal. Variant DB in Fig. 6b of the algorithm separated the lungs along a straight line, which partly failed because they are not symmetrical. Variant MB, rendered in Fig. 6c separated the lungs correctly.

Fig. 6. Results of three different lung separation algorithms for pathologic lungs—(a) variant B and BB, (b) variant DB, (c) variant MB. (Color figure online)

4.4   Time Measurements

In Table 5, running times for each variant of the algorithm are reported. The times measured exclude data source input/output operations in order to provide better consistency. All the measurements were obtained for the whole dataset (see Sect. 4.1). Processing time increased by 37.5% from the worst baseline variant B to the best variant MB. Average processing time per study is approximately 4.3 s for variant MB and 3.2 s for variant B. The most time-consuming process was the morphological analysis—it often required dozens of iterations in order to finish processing. In the DB variant, computing distance transform prolonged the processing time. Implementation of false positive removal does not seem to increase processing time noticeably. Overall, the processing times are very satisfactory compared to more complicated methods such as active contour, where the segmentation often takes several minutes or longer.

Table 5. Processing times for each variant of the algorithm given in milliseconds.

Variant	Total processing time (s)	Processing time/slice (ms)
B	175.30	7.89
BB	176.51	7.95
DB	236.68	10.66
MB	241.07	10.86

5 Conclusions and Outlook

Overall, the proposed algorithm yields highly accurate results, maintaining relatively short execution times. The already mentioned methods can be further improved. First, area threshold for trachea extraction T_a could be replaced with a gradient-based method. Secondly—even though the lung separation procedure implemented in the variant MB of the solution works well, the region where the splitting occurs is not being verified, which could lead to improper results. The algorithm could be tested with other thresholding methods such as [11,12].

The algorithm we propose does not explicitly segment such structures as the lung nodules, smaller airways or bronchial tubes—addressing these problems is the goal of our ongoing work. Moreover, the algorithms which were implemented will be adapted for segmenting other organs and they are not limited to only one modality such as CT.

Acknowledgments. This research was supported by the National Centre for Research and Development under the Innomed Research and Development Grant No. POIR.01.02.00-00-0030/15.

References

1. Annangi, P., Thiruvenkadam, S., et al.: A region based active contour method for X-Ray lung segmentation using prior shape and low level features. In: Biomedical Imaging From Nano to Macro. IEEE (2010)
2. Costa, A., Carvalho, B.: SALSA–A simple automatic lung segmentation algorithm. In: Progress in Pattern Recognition, Image Analysis, Computer Vision, and Applications, pp. 501–508 (2015)

3. Cyganek, B., Graña, M., Porwik, P., Wozniak, M.: Intelligent methods applied to health-care information systems. Appl. Artif. Intell. **30**(6), 495–496 (2016)
4. Felzenszwalb, P., Huttenlocher, D.: Distance transforms of sampled functions. Theory Comput. **8**, 415–428 (2012)
5. van Ginneken, B., et al.: Off-the-shelf convolutional neural network features for pulmonary nodule detection in computed tomography scans. In: Proceedings of IEEE ISBI, pp. 286–289 (2015)
6. Hu, S., Hoffmann, E.: Automatic lung segmentation for accurate quantitation of volumetric X-Ray CT images. IEEE Trans. Med. Imaging **20**(6), 490–498 (2001)
7. Kawulok, M., Kawulok, J., Nalepa, J., Smolka, B.: Self-adaptive algorithm for segmenting skin regions. EURASIP J. Adv. Sig. Proc. **2014**, 170 (2014)
8. Mansoor, A., et al.: A generic approach to pathological lung segmentation. IEEE Trans. Med. Imaging **33**(12), 2293–2310 (2014)
9. Mostafa, A., Elfattah, M.A., Fouad, A., Hassanien, A.E., Hefny, H.: Enhanced region growing segmentation for CT liver images. Adv. Intell. Syst. Comput. **407**, 115–127 (2015)
10. Nalepa, J., Kawulok, M.: Fast and accurate hand shape classification. In: Kozielski, S., Mrozek, D., Kasprowski, P., Małysiak-Mrozek, B., Kostrzewa, D. (eds.) BDAS 2014. CCIS, vol. 424, pp. 364–373. Springer, Cham (2014). doi:10. 1007/978-3-319-06932-6_35
11. Otsu, N.: A threshold selection method from gray level histograms. IEEE Trans. Syst. Man Cybern. **SMC–9**(1), 62–66 (1979)
12. Perez, M.G., et al.: A multi-level thresholding method based on histogram derivatives for accurate brain MRI segmentation. Rev. Politcnica **35**, 82 (2015)
13. Schlegl, T., Ofner, J., Langs, G.: Unsupervised pre-training across image domains improves lung tissue classification. In: Menze, B., Langs, G., Montillo, A., Kelm, M., Müller, H., Zhang, S., Cai, W.T., Metaxas, D. (eds.) MCV 2014. LNCS, vol. 8848, pp. 82–93. Springer, Cham (2014). doi:10.1007/978-3-319-13972-2_8
14. Shen, W., Zhou, M., Yang, F., Yang, C., Tian, J.: Multi-scale convolutional neural networks for lung nodule classification. In: Ourselin, S., Alexander, D.C., Westin, C.-F., Cardoso, M.J. (eds.) IPMI 2015. LNCS, vol. 9123, pp. 588–599. Springer, Cham (2015). doi:10.1007/978-3-319-19992-4_46
15. Shin, H.C., Roth, H., et al.: Deep convolutional neural networks for computer-aided detection: CNN architectures, dataset characteristics and transfer learning. IEEE Trans. Med. Imaging **35**, 1285–1298 (2016)
16. Siminski, K.: Clustering with missing values. Fundam. Inform. **123**(3), 331–350 (2013)
17. Starosolski, R.: New simple and efficient color space transformations for lossless image compression. J. Vis. Commun. Image Represent. **25**(5), 1056–1063 (2014)
18. Sternberg, S.: Biomedical image processing. IEEE Comput. **16**(1), 22–34 (1983)
19. Wang, J., Chan, K.L.: Active contour with a tangential component. J. Math. Imaging Vis. **51**(2), 229–247 (2014)
20. Wang, Q., et al.: HOSVD-based 3D active appearance model: segmentation of lung fields in CT images. J. Med. Syst. **40**(176), 1–11 (2016)
21. Zghidi, H., Walczak, M., et al.: Image processing and analysis of textile fibers by virtual random walk. In: Proceedings of the 2015 Federated Conference on Computer Science and Information Systems, vol. 5, pp. 717–720 (2013)
22. Zhang, T.Y., Suen, C.Y.: A fast parallel algorithm for thinning digital patterns. Commun. ACM **27**(3), 236–239 (1984)

Improved Automatic Face Segmentation and Recognition for Applications with Limited Training Data

Dane Brown[1,2(✉)] and Karen Bradshaw[1]

[1] Department of Computer Science, Rhodes University, Grahamstown, South Africa
dbrown@csir.co.za, k.bradshaw@ru.ac.za
[2] Council for Scientific and Industrial Research, Modelling and Digital Sciences,
Pretoria, South Africa

Abstract. This paper introduces varied pose angle, a new approach to improve face identification given large pose angles and limited training data. Face landmarks are extracted and used to normalize and segment the face. Our approach does not require face frontalization and achieves consistent results. Results are compared using frontal and non-frontal training images for Eigen and Fisher classification of various face pose angles. Fisher scales better with more training samples only with a high quality dataset. Our approach achieves promising results for three well-known face datasets.

Keywords: Biometrics · Face · Identification · Landmarks

1 Introduction

Face recognition is an important research area as it is one of the more visible and user-friendly biometrics [6]. In cases where the face is not used as an authentication biometric, it is still often bound to the primary identification device or system, such as an ID card or a criminal fingerprint database. As a primary means of identification, the face is challenging in uncontrolled applications due to varied pose angles and occlusions. Furthermore, real-world conditions often result in degradation of image quality.

Constructing a consistent face recognition system under these conditions is an ongoing research area and requires a new approach. Recent advancements include an automatic face alignment system requiring milliseconds per face [9] and the frontalization of all face images up to 60° pose angles [5,12]. Three-dimensional (3D) face modelling systems generally produce high recognition rates and versatility, but require greater computational power and more training data [7].

The contribution of this paper is a fast accurate approach, combining two recent advancements in the literature and improving feature extraction, without using 3D modelling. A particular improvement includes removal of frontal training data dependence completely by accepting varied angle face data. This

© Springer International Publishing AG 2017
S. Kozielski et al. (Eds.): BDAS 2017, CCIS 716, pp. 415–426, 2017.
DOI: 10.1007/978-3-319-58274-0_33

caters for applications that often have access to limited data but require a real-time response, such as tracking a felon using street camera feeds. Illumination changes and low quality data are also considered in the approach. The system is expected to produce a comparable accuracy when using frontal or varied pose angle training data. The system's limits with respect to facial pose angle are also investigated.

The rest of the paper is organized as follows: Sect. 2 presents related face verification and identification systems found in the literature. Section 3 discusses the construction and application of the face identification system. The experimental analyses and results are discussed in Sect. 4. Section 5 concludes the paper and discusses future work.

2   Related Studies

Cootes *et al.* [3] compared active shape models (ASMs) and active appearance models (AAMs). ASM was found to be faster and more accurate at feature point location, but AAM produced a better match to the image texture. ASM is susceptible to initialization failure when no face is detected. A simple and effective way to prevent this is pre-alignment using affine transformation steps, for which 3D deformable models are generally the most accurate. However, their applications are limited by slow 3D modelling and high training sample requirements [12].

Kazemi and Sullivan [9] produced an ASM using gradient boosting for learning an ensemble of regression trees. The face landmarks are automatically estimated directly from a sparse subset of pixel intensities, surpassing realtime performance at approximately 1 ms per image. Their approach optimizes the sum of square error loss and automatically handles missing or partially labelled data. The resulting coordinates for face landmarks are the first step toward reliably normalizing multiple face samples of individuals. The following face recognition systems all perform automatic alignment for normalized face segmentation.

Yi *et al.* [15] proposed a robust face recognition algorithm, geared towards large pose variations. Their approach consists of a 3D deformable model generated to estimate the pose of face images. A set of Gabor filters is transformed according to generated parameters of the model, resulting in extraction of the relevant Gabor features into a vector. Principal Component Analysis (PCA) is applied to capture the most relevant data from the feature vector. The "half-face" trick – using the least occluded side of the face under occlusion conditions – improved the system accuracy by 46%. The dot product is used to compute the similarity between resulting feature vectors of the training and test face images for matching. Their method offers a significant improvement only on large pose variation face data. They deduced that "traditional" approaches are only sufficient for near-frontal face data.

A well-known deep learning based study, known as DeepFace [13], uses a nine-layer neural network with over 120 million parameters. It achieves a high accuracy by training over 4 million labelled faces.

A recent approach by Haghighat *et al.* [5] automatically segments the face for robust recognition using only a single training sample, without 3D modelling or deep learning. Their approach is similar to that of Sagonas *et al.* but includes "half-face", which improves the accuracy for viewing angles that are greater than 45°, and larger angles than supported by Sagonas *et al.* The automatic segmentation method results in a single region of interest (ROI) consisting of the whole face and is vertically halved only when "half-face" is required. The AAMs used by Haghighat *et al.* aim to frontalize all faces using a base mesh per individual without any face detection preprocessing. The image features include a fusion of Histograms of Oriented Gradients (HOG) and a vast number of Gabor features using Canonical Correlation Analysis (CCA). The resulting feature vectors are matched using minimum distance classification of Discrete Cosine Transform (DCT) features. The disadvantage of their approach is the added time complexity of AAMs and 40 Gabor filters in five scales and eight orientations. Furthermore, near-frontal training samples are preferred for achieving high accuracy.

The half-face frontalization method improves the results for the systems by both Yi *et al.* [15] and Haghighat *et al.* [5], outperforming nine similar systems. In fact, Haghighat *et al.* conducted a comprehensive comparison and achieved very similar accuracies to Yi *et al.*'s 3D deformable model, on the FERET b-series [10] face dataset. On the other hand, a large number of relevant features are still lost with the half-face method.

This paper builds on Haghighat *et al.*'s system, using Kazemi and Sullivan's ASM method, and aims to further improve the accuracy at larger pose angles without negatively affecting narrow pose angles. In the process, time complexity is also reduced by combining a Láplacian of Gaussian (LoG) filter and extended local binary patterns (ELBP), instead of HOG and Gabor, for improved feature discrimination and an approximate 10-fold speed up – due to omitting the 40 Gabor filters. The addition of the LoG and ELBP combination, known as LLBP, produces significant improvement in Eigen and Fisher classification methods [2].

3   Proposed System

The following subsections detail the proposed face recognition solution. The system was coded in C++ using the OpenCV and Dlib image processing libraries. Henceforth, the proposed face recognition solution is referred to as varied reference angle (VRA). VRA focuses on increasing the versatility and accuracy of the face normalization and segmentation processes. Figure 1 provides an overview of the automatic face segmentation and recognition process.

3.1   Face Landmarks

An initial ROI is determined by first detecting the face using HOG combined with a linear support vector machine (SVM) classifier and a sliding window. While no face is found, face detection is repeated by rotating about the x-axis in 15° increments in both directions. A cascade of regressors is built by continually

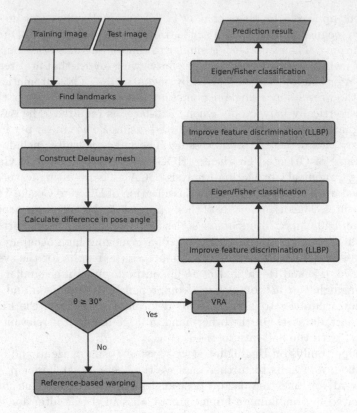

Fig. 1. Overview of automatic face segmentation and recognition using VRA.

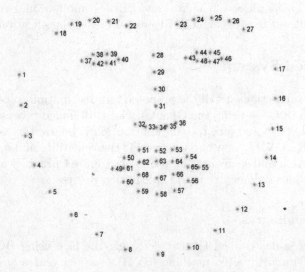

Fig. 2. Training data landmarks [11].

(a) Training image.

(b) Test image.

Fig. 3. Training and test images of the same person (PUT face dataset) [8].

updating a vector, within the initial ROI, consisting of 68 coordinates shown in Fig. 2 [9]. Training data samples were acquired from the iBUG 300-W face landmark dataset [11]. The learning stage is conducted by training coordinates as a set of triplets forming the input of the learning function for the next regressor in an iterative process. A decision is made based on the difference between the intensities of two pixels at each split node in the regression tree. The difference defines coordinates of the mean shape of the face. During the landmark prediction stage, the mean shape is warped according to the triplets of the learned model by calculating a similarity transform between the corresponding coordinates. The time complexity is reduced by calculating the transform and warping only once at each level of the cascade.

3.2   Face Normalization and Segmentation

A face mesh is constructed based on the detected landmarks. The face mesh is based on Delaunay triangulation, such that no landmark is inside the circumcircle of any triangle. The training image, used for the reference mesh, of a particular individual (class) is shown in Fig. 3a. The reference mesh is shown in Fig. 4a. In this paper, unseen test images are normalized and segmented based on a training image. An example test image, with an approximate 54° pose angle, of the correct class is shown in Fig. 3b with the corresponding mesh shown in Fig. 4b. The correlation between triangle vertices of the training and test images are based on the ordered landmarks in Fig. 2. The images are warped based on the corresponding triangle vertices.

An advantage of VRA is that the training image does not need to be frontal or near-frontal for that matter, increasing application versatility. The pose angle is determined based on six coordinates consisting of the nose, eyes, left mouth edge, right mouth edge and chin as shown in the figures. Furthermore, the pose angle of the test image is mirrored if the reference angle is in an opposite direction, reducing the difference in angle. The pose angle is determined using the OpenCV's implementation of the Perspective-n-Point [4] method, using the six coordinates. This requires camera calibration parameters such as the focal length, principal image point and skew, which are easily accessed through OpenCV's API.

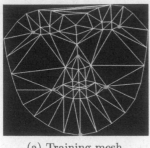

(a) Training mesh.

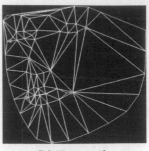

(b) Test mesh.

Fig. 4. Training and test image meshes of the same person.

The example training and test images have negligible vertical pose angles (less than 10°). The 10° buffer caters for less symmetrical faces, where the nose is not centred for frontal poses. Therefore, VRA only applies to the horizontal pose angle in this explanation. The accuracy is expected to increase by avoiding the half-face method and introducing VRA.

The resulting normalization of the test image, when using the non-VRA method (warp according to reference image), is shown in Fig. 5b. Only the right side of the face is used due to self-occlusion. While the training image is virtually distortion free as seen in Fig. 5a, the test image is severely distorted. The VRA method avoids the need for half-face during self-occlusion.

For VRA, the average of the reference (training) and test angle as the final pose angle θ, for both training and test images, was found to produce optimal results during preliminary testing. Thus it is an ideal replacement for the half-face method under self-occlusion conditions, especially prevalent at angles greater than or equal to 30°. VRA warps both the training and test image according to the average mesh when the pose angle is greater than or equal to 30°. The test image is warped according to the training mesh when the pose angle is less than 30°. VRA also increased the pose angle limit from 60° to 90°. Although, VRA is not practical in identification systems it is suited to verification systems. Classifying using multiple one-to-many training models (identification) for the various angles would require a substantial amount of processing power and storage. Therefore, VRA is limited to the angle closest to 0°, 30°, 45° and their respective mirrors for face identification in this study. The normalized and segmented training and test images at 30°, are shown in Fig. 6a, b. There are still differences between the training and test images in certain parts, but the eyes are well normalized due to their relatively simple and consistent shape in the mesh. Facial expression normalization is a non-trivial research area and is considered for future work.

Figures 5 and 6 compared non-VRA and VRA of a matching test image. The same test is thus conducted on an impostor test image, using the same reference mesh, since the test data in a real-time identification system is unseen – the class

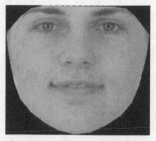

(a) Training image.

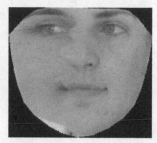

(b) Frontalized test image.

Fig. 5. Normalized and segmented training and test images not using VRA.

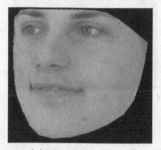

(a) Training image.

(b) Test image.

Fig. 6. Normalized and segmented training and test images warped to 30° using VRA.

is unknown and requires prediction at the classification step. The normalized and segmented training and test images, at 30°, are shown in Fig. 6a, b.

3.3 Feature Discrimination

In Subsect. 3.2, intra-class variation was minimized. In this subsection inter-class variation is maximized using a combination of LoG and ELBP (LLBP) [2]. The LoG filter removes unwanted features on the low and high frequency spectrum before enhancing the remaining features, effectively increasing the DC component. The Gaussian and Laplacian kernels were 15×15 and 7×7, respectively. This also further lowers intra-class variation by reducing subtle differences in images of the same person often caused by facial expressions and distortions due to warping. The ELBP operator parameters – one pixel radius and eight neighbour pixels – were multiplied by four and averaged with the normalized original image to enable its use as a standalone feature selector. This was applied to the normalized and segmented training and test images. When the ELBP operator is used this way, it reduces lighting differences without the typical noise side effect. The resulting images are resized to 75×75 before classification (Fig. 7).

(a) Impostor test image.

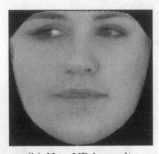

(b) Non-VRA result.   (c) VRA result.

Fig. 7. Normalized and segmented impostor images with/without VRA [8].

3.4 Classification

The Eigen classifier maximizes the total variance in data based on linear combination of features. The largest variance in data is contained within the first few principal components which are modelled into classes. The trained and test models are compared based on the distances between eigenvalues during matching.

Given N number of sample images x_k the total scatter matrix is defined as [1]:

$$S_t = \sum_{k=1}^{N} (x_k - \mu)(x_k - \mu)^T,$$

where $m \in \mathbb{R}^n$ is the mean image obtained from the samples.

The Fisher classifier takes the Eigen result and performs extra class-specific dimensionality reduction by respecting between-class and within-class scatter matrices. Fisher learns a class-specific transformation matrix and is expected to outperform Eigen on a high quality dataset when using more than one training sample. Fisher's training and testing times are lower than Eigen due to the reduced dimensionality.

Given C number of classes, the between-class scatter matrix is defined as [1]:

$$S_b = \sum_{i=1}^{C} N_k(\mu_k - \mu)(x_k - \mu)^T$$

and the within-class scatter matrix is defined as:

$$S_w = \sum_{i=1}^{C} \sum_{x_k \in \mathbb{X}_i} (x_k - \mu)(x_k - \mu)^T,$$

where $C - 1$ is the maximum number of non-zero generalized eigenvalues, which leads to extra dimensionality reduction.

The false acceptance rate (FAR) and false rejection rate (FRR) are calculated for both Eigen and Fisher as follows:

$$FAR = \frac{\text{false matches}}{C \times N}$$

and

$$FRR = \frac{\text{false non-matches}}{C \times N}$$

The system accuracy metric is used throughout the results as the related studies in Sect. 2 use only this metric. Finally, the system accuracy is calculated as $100 - \frac{FAR + FRR}{2}$. The Eigen and Fisher results are discussed in the next section.

4 Experimental Analysis and Results

This section documents of three experiments using different datasets. Training samples are removed from the available samples and the rest are used for testing, thereby ensuring that the test data is unseen.

4.1 Experiment 1

Table 1 compares the proposed system to the best system accuracies in the related studies on the FERET b-series [10] dataset consisting of 200 individuals and eight pose angles, obtained from Haghighat *et al.*'s experimental results. The images were captured at a resolution of 512×768 fine quality. Frontal images were used for training.

The similar results obtained for VRA, compared with Haghighat *et al.*'s system, are promising as VRA's full potential is realized when using non-frontal face images for training. The next set of experiments test VRA with frontal and non-frontal training images.

4.2 Experiment 2

The PUT face dataset [8], consisting of 88 samples per 100 individuals, was used in this experiment. Each sample was at a different pose angle in all directions permissible by the neck, ranging up to $60°$. The images were captured at a resolution of 2048×1536 fine quality.

Table 1. Face recognition rates of different approaches at eight viewing angles on the FERET b-series dataset (in %).

Method	+60°	+45°	+25°	+15°	−15°	−25°	−45°	−60°
Sagonas *et al.* [12]	–	96	100	100	100	99	96.5	–
Yi *et al.* [15]	93.75	98	98.5	99.25	99.25	98.5	98	93.75
Haghighat *et al.* [5]	91.5	96	100	100	100	100	99	93.5
VRA	93.75	98	100	100	100	100	98.5	94.25

The results using frontal as a reference are henceforth referred to as frontref, while results using left pose at $60°, 56°$ and $52°$ as reference are henceforth, referred to as leftref.

Table 2 summarizes the identification accuracies for the PUT dataset. A very minor accuracy decrease is observed when comparing frontref and leftref using a single training sample. This is encouraging as VRA is robust to an uncontrolled environment with limited training data. The best performer is Fisher frontref with a 99.69% accuracy when using three training samples. Face verification achieved 100% accuracy for all Eigen and Fisher versions.

Table 2. Eigen and Fisher results on the PUT dataset when using one and three training samples (in %).

	Eigen frontref (%)	Eigen leftref (%)	Fisher frontref (%)	Fisher leftref(%)
1	99.07	98.99	99.10	98.99
3	99.29	98.98	**99.69**	99.01

4.3   Experiment 3

For this experiment, the results using frontal as a reference are referred to as frontref, while results using left pose at $60°$, $45°$ and $15°$ as reference are referred to as leftref.

The FEI face dataset [14], consisting of 14 samples per 200 individuals, was used in this experiment. Each sample was at a different pose angle in all directions permissible by the neck ranging up to $90°$ angles. The images were captured at a resolution of 640×480, but at a noticeably lower focus quality than the previous two datasets.

Table 3 summarizes the identification accuracies for the FEI dataset. Fisher leftref achieved the best accuracy at 91.19% using 3 training samples. This is expected due to the limited number of total samples and the aggregation of features from three pose angles. Comparing that to Fisher frontref, shows how Fisher is affected when training data does not cover all intra-classes [1,2]. On the other hand, accurately classifying poses above $75°$ with a frontal training sample continues to be a problem for both Eigen and Fisher. Furthermore, the results on this dataset are the worst out of the three. This is attributed to face pose angles over 75° only containing half or less than half of the face, and also to the lower quality data. After removing the 75° and 90° test samples the average accuracy improved by 15%. Furthermore, the confusion matrix shows that the system achieved almost 55% accuracy for angles between 75° and 90°. This confirms that VRA, like other methods, has limitations when using extra-large pose angles to the reduced number of correlating features. However, VRA lowers distortions, effectively improving accuracies between 45° and 60° and enables a limited degree of identification for extra-large pose angles without 3D modelling, when compared with the related studies.

Table 3. Eigen and Fisher results on the FEI dataset when using one and three training samples (in %).

	Eigen frontref (%)	Eigen leftref (%)	Fisher frontref (%)	Fisher leftref(%)
1	72.48	69.21	72.68	69.26
3	85.54	88.29	85.36	**91.19**

5 Conclusion and Future Work

VRA loses less features than the half-face method and improves application versatility by requiring only a single sample at an angle up to 60°, effectively mitigating typical self-occlusions. The significant reduction in computational requirements allows for real-time use with less powerful equipment. The half-face method is still useful for other occlusions. A comparison was performed on VRA and the best performing non-VRA methods in the first experiment. The results show that normalizing pose angles between 45° and 60° was improved. The second experiment demonstrated its robustness by correctly identifying up to 87 pose angles per individual. Only a minor accuracy reduction was recorded when using non-frontal training samples. The third experiment showed that VRA is also capable of classifying angles up to 90°, however, at a significantly lower accuracy. This provides future research with a foundation for increasing the accuracy when nearing the pose angle limit.

Future work includes investigating facial expression normalization and further improvements on lowering distortion at large pose angles.

References

1. Belhumeur, P.N., Hespanha, J.P., Kriegman, D.: Eigenfaces vs. fisherfaces: recognition using class specific linear projection. IEEE Trans. Pattern Anal. Mach. Intell. **19**(7), 711–720 (1997)

2. Brown, D., Bradshaw, K.: An investigation of face and fingerprint feature-fusion guidelines. In: Kozielski, S., Mrozek, D., Kasprowski, P., Małysiak-Mrozek, B., Kostrzewa, D. (eds.) BDAS 2015–2016. CCIS, vol. 613, pp. 585–599. Springer, Cham (2016). doi:10.1007/978-3-319-34099-9_45

3. Cootes, T.F., Edwards, G., Taylor, C.: Comparing active shape models with active appearance models. In: Proceedings of British Machine Vision Conference, pp. 173–182. BMVA Press (1999)

4. Gao, X.S., Hou, X.R., Tang, J., Cheng, H.F.: Complete solution classification for the perspective-three-point problem. IEEE Trans. Pattern Anal. Mach. Intell. **25**(8), 930–943 (2003)

5. Haghighat, M., Abdel-Mottaleb, M., Alhalabi, W.: Fully automatic face normalization and single sample face recognition in unconstrained environments. Expert Syst. Appl. **47**, 23–34 (2016)

6. Jain, A., Hong, L., Pankanti, S.: Biometric identification. Commun. ACM **43**(2), 90–98 (2000)

7. Kafai, M., Eshghi, K., An, L., Bhanu, B.: Reference-based pose-robust face recognition. In: Kawulok, M., Celebi, M.E., Smolka, B. (eds.) Advances in Face Detection and Facial Image Analysis, pp. 249–278. Springer, Cham (2016). doi:10.1007/978-3-319-25958-1_9

8. Kasinski, A., Florek, A., Schmidt, A.: The put face database. Image Process. Commun. **13**(3–4), 59–64 (2008)

9. Kazemi, V., Sullivan, J.: One millisecond face alignment with an ensemble of regression trees. In: Proceedings of the IEEE Conference on Computer Vision and Pattern Recognition, pp. 1867–1874 (2014)

10. Phillips, P.J., Wechsler, H., Huang, J., Rauss, P.J.: The FERET database and evaluation procedure for face-recognition algorithms. Image Vis. Comput. **16**(5), 295–306 (1998)

11. Sagonas, C., Antonakos, E., Tzimiropoulos, G., Zafeiriou, S., Pantic, M.: 300 faces in-the-wild challenge: database and results. Image Vis. Comput. **47**, 3–18 (2016)

12. Sagonas, C., Panagakis, Y., Zafeiriou, S., Pantic, M.: Face frontalization for alignment and recognition. arXiv preprint arXiv:1502.00852 (2015)

13. Taigman, Y., Yang, M., Ranzato, M., Wolf, L.: DeepFace: closing the gap to human-level performance in face verification. In: Proceedings of the IEEE Conference on Computer Vision and Pattern Recognition, pp. 1701–1708 (2014)

14. Thomaz, C.E., Giraldi, G.A.: A new ranking method for principal components analysis and its application to face image analysis. Image Vis. Comput. **28**(6), 902–913 (2010)

15. Yi, D., Lei, Z., Li, S.Z.: Towards pose robust face recognition. In: Proceedings of the IEEE Conference on Computer Vision and Pattern Recognition, pp. 3539–3545 (2013)

Emotion Recognition: The Influence of Texture's Descriptors on Classification Accuracy

Karolina Nurzynska[✉]

Institute of Informatics, Silesian University of Technology,
Akademicka St 16, 44-100 Gliwice, Poland
karolina.nurzynska@polsl.pl

Abstract. This work describes experiments dedicated to analysis of the descriptive properties of several, most widely applied, texture operators in emotion recognition domain. Many researchers apply Gabor filters, histogram of oriented gradients, or local binary patterns in complex set-ups with different classification approaches and image processing methodologies, but nowhere it was verified, how each part of the system influences the resulting performance. Therefore, several experiments with Cohn-Kanade AU-Coded Facial Expression and Karolinska Directed Emotional Faces Databases were performed. These experiments reviled, that exploiting the histogram of oriented gradients overcomes other texture operators in most cases.

Keywords: Emotion recognition · Texture operators · Histogram of oriented gradients · Local binary patterns · Gabor filters · Classification

1 Introduction

The creation of digital camera allowed to record every event of daily life and view the collected data many times later on a personal computer. However, in the beginning the size of hard drive as well as the necessity to remember to carry the equipment were some limitation to the amount of gathered data and its possible applications. Nowadays, everyone has at least one mobile device, which enables acquisition, on-line analysis, and transfer of data between people. Such an easy access to visual information, makes the demands for its understanding bigger and broader.

One of the possible applications of visual data is the analysis of emotions depicted in the people's faces and it is the problem addressed in this research. There are several emotions, which strength of expression varies between cultures and personal differences. Yet, from the works of Darwin [7] which were later continued by Ekman [9,10], we know that there are six basic expression easily recognizable all around the world. This set of emotions consists of: anger, disgust, fear, happiness, sadness, and surprise, which examples are presented in Figs. 1 and 2.

© Springer International Publishing AG 2017
S. Kozielski et al. (Eds.): BDAS 2017, CCIS 716, pp. 427–438, 2017.
DOI: 10.1007/978-3-319-58274-0_34

The motivation for development of a system of automatic emotion recognition using images of facial gestures is huge [15]. It could find many applications in the medicine: for instance the amount of happiness displayed by the smiling expression is correlated to the severity of depression, hence automatic analysis of films recorded during sessions with physician may give information about the progress or regress of this disease; proper description of patient condition is also based on pain measurement, especially after a surgery, here automatic analysis of the pain emotion could support the physician as well. System which can detect emotions could support people suffering from autism, or blind ones, with information which emotional state is depicted in other's face and thus allow for better communication. Finally, one can imagine a game plot created basing on the information derived from user interest and mood [28] or a computer system exploring information about the user temper and presenting its functionality accordingly. But also more generally an understanding of human actions can be investigated [12].

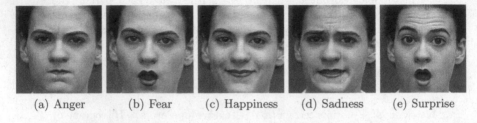

(a) Anger   (b) Fear   (c) Happiness   (d) Sadness   (e) Surprise

Fig. 1. Exemplary images presenting various emotions in CK database.

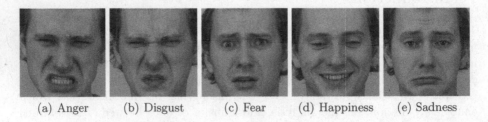

(a) Anger   (b) Disgust   (c) Fear   (d) Happiness   (e) Sadness

Fig. 2. Exemplary images presenting various emotions in KDEF database.

Automatic classification of emotions have been already discussed in the image processing domain. Moreover, it was stated that it is difficult to say how people see the face for its recognition [14]. Most of the existing approaches explored several classification techniques, which were applied for feature vectors extracted from the facial image data. In the case of Gabor filters (GF) and Local Binary Patterns (LBP), the image was processed as a whole [16,29], or was divided into sub-images, for which the feature was calculated [11,25]. Additionally, when LBP was exploited some weighting functions were used to derive the most discriminative information [13,27]. Similar approaches were presented when Histogram of Oriented Gradients (HoG) was applied. There were also solutions based on topological distances between characteristic points spotted on the face [1,18,26]. The calculated feature vectors may be very long and need preprocessing before the classification takes place. Here several dimension reduction techniques may be applied. Most frequently it is a principal component analysis [19,20]. For classification, the support vector machine with various kernels is applied usually, but there exists solution using k-nearest neighbours, deep learning, neural networks, and many others.

The broad literature concerning emotion recognition problem, addresses many techniques for data description for classification needs. Some solutions seem better than others, yet according to the author's knowledge the influence of texture analysis of feature vector descriptive properties has not been evaluated. In the literature one can find many combinations of the input image description ideas with chosen classification methodologies, yet it is difficult to claim clearly, which part of the system contributed significantly to the achieved result. Therefore, this work concentrates on comparison of several texture operators willingly used in this domain.

The paper is structured as follows. Section 2 presents chosen texture operators applied for emotion description. Then, Sect. 3 discusses details of the experiment preparation and gives brief information about used image databases. Next, the results are discussed in Sect. 3. Finally, Sect. 5 draws the conclusions.

2   Image Description Methods

Since it is difficult to explain how the emotion is recognized, the data derived from whole face is analysed. One of the approaches, elaborated in emotion recognition domain, suggest to exploit the information from texture by application of one from texture operators described in this section.

2.1   Local Binary Patterns

LBP is a texture operator which was developed to work with a monochromatic image I [21–24]. For each pixel $I(x_c, y_c)$, an $LBP_{R,P}$ code is calculated which describes the local neighbourhood defined by a radius R. The illumination changes, between the pixel of interest $g_c = I(x_c, y_c)$ and P points sampled at equal intervals on the circumference, portray characteristic features of an image patch as:

$$LBP_{P,R}(x_c, y_c) = \sum_{p=0}^{P-1} s(g_p - g_c) \cdot 2^p, \qquad (1)$$

where g_p is a grey scale value recorded for the pth-point; $p = 1, \cdots, P$ is the order of points; and s is a tresholding function given with a formula:

$$s(z) = \begin{cases} 1, z \geq 0, \\ 0, z < 0. \end{cases} \qquad (2)$$

The histogram of computed codes is used as an image content description. Since the length of histogram depends on the number of sampled points according to the relation 2^P, usually $P = 8$ what results in 256 element feature vector. In many cases, such length is still too long for further processing (especially, when one needs to concatenate several LBP histograms), hence uniform pattern version of this texture operator is exploited. It was noticed [24], that some codes convey more information that the others. Considering binary representation of the codes, those which have up to two transitions between 0 and 1 are named uniform and seem to be the most informative ones, thus each of them has a separate bin in the histogram, all other codes are collected in one additional bin. This assumption allows to shorten the length of obtained feature vector to 59 elements without loosing accuracy in data description.

2.2 Histogram of Oriented Gradients

HoG [2,5,6] is another method, which derives descriptive information from mono-chromatic image I. This time, the edges computed for each pixel using horizontal $[1, 0, -1]$ and vertical $[1, 0, -1]^T$ gradient operators provide an information, which is collected in a form of histogram describing an image patch called a cell. The histogram bins evenly split detected angles from 0 to 180 degrees range for unsigned directions. The cell is a rectangular or circular region, which size is a parameter. Each pixel belonging to the cell votes for chosen edge direction with computed magnitude. Since changes of illumination and contrast might influence locally collected data, several spatially connected cells determine a bigger area for which the histograms are normalized. Finally, the concatenated histograms constitute a feature vector.

2.3 Gabor Filter

Application of GF [8,30] to images enables an edge detection and therefore a description of its content. The filter is claimed to analyse the images in the same manner as human visual system works, thus seems appropriate for texture analysis and recognition. For a pixel (x,y), it is formulated as a Gaussian function multiplied by a sinusoidal wave following the formula:

$$\text{GF}(x,y) = \exp(-\frac{\hat{x}^2 + \gamma^2 \hat{y}^2}{2 \cdot \sigma^2}) \cdot \exp(i \cdot \frac{2\Pi}{\lambda}\hat{x} + \varphi), \tag{3}$$

and

$$\hat{x} = x \cdot \cos(\theta) + y \cdot \sin(\theta), \tag{4}$$
$$\hat{y} = -x \cdot \sin(\theta) + y \cdot \cos(\theta),$$

where other parameters are responsible for output function generation: θ – orientation (in presented research 8 orientation sampled every $45°$ were applied); γ – aspect ratio (set to 1); σ – effective width (responsible for five scale variation); φ – phase (set to 0); and λ – wavelength (set to 1). The filters obtained for each setting were convolved with the original image to obtain a feature vector part, which was later concatenated for all of them.

3 Experiment Set-Up

The comparison of presented texture operators as a method for facial gestures descriptors was performed using two different emotion databases: Cohn-Kanade AU-Coded Facial Expression Database and Karolinska Direct Emotional Faces Database. In the supplied data presenting emotions a face was detected using Viola-Jones method [31], the cropped face was scaled to constant size of 100×100 pixels and converted to a monochromatic scale, if necessary.

3.1 Parameter Settings

In order to better describe the image content when LBP operator was applied, the image was divided into 10×10 blocks of pixels for which the LBP histogram was calculated. The final feature vector was a concatenation of all computed characteristics. When HoG approach was executed, the cell side was 8 pixels, while the region consisted of 2×2 cells and the histogram contained 9 entries. The GF was calculated for 8 orientations and 5 scales.

The classification was performed with support vector machine SVM (using the LIBSVM implementation [3]). For linear kernel the c parameter was adjusted in range from 10^{-5} to 10^5. For radial basis function (RBF) kernel additionally γ parameter was searched in the same range. Moreover, the k-nearest neighbours (kNN) classifier was applied with several distance metrics: citiblock, correlation, Euclidean, Hamming, Jaccard, etc., and the k parameter was explored for values in range $1, \cdots, 15$. The experimental set-up assumed ten-fold cross-validation of the results, where one testing fold was examined over nine training ones.

3.2 Databases

Cohn-Kanade AU-Coded Facial Expression Database. Cohn-Kanade AU-Coded Facial Expression Database CK [4] consists of image sequences, where subjects present the change of facial gesture from the neutral one to the most expressive representation of one of the basic emotions. There are 29 examples displaying anger, 53 individuals expressing disgust, 40 image sequences for fear, 97 samples of happiness, 29 facial gestures of sadness, and 70 expression of surprise. Figure 1 depicts selected expressions. In order to prepare a class containing a neutral expression the first frame from each image sequences was selected. The images are monochromatic with 640×480 pixel resolution. The lighting condition varies in the images, and the subjects do not wear any covering elements, such as glasses or beards.

Karolinska Directed Emotional Faces Database. Karolinska Directed Emotional Faces Database KDEF [17] collects images presenting six basic emotions and the neutral facial expression. The database consists of images taken from 70 subjects. There were taken 2 images from 5 different angles for expression of each emotion. However, in presented experiments only frontal faces are considered, thus 140 images belongs to each group. The images are recorded in 32 bit RGB color and with 562×762 pixel resolution, as presented in Fig. 2. The subjects are from 20 to 30 years of age wearing any covering elements and make-up.

4 Results

In order to verify the descriptive capabilities of each texture operator three experiments were prepared. In all of them the radius R was set to 1, and the number of points P to 8 for LBP operator. The HoG was computed for regions consisting of 2×2 cells, where each cell was a square with side equal to 8 pixels. The GFs were convolved with whole image.

4.1 Texture Descriptor Performance

The aim of the first experiment was to verify the accuracy of image content description using one of several proposed texture processing techniques. Additionally, the influence of chosen classifier on overall performance was verified. The settings of classifiers parameters are summarised in Table 1.

Figures 3, 4, and 5 gather results achieved for all considered data description techniques performed on both datasets with SVM where the linear and RBF kernels were used, and kNN classifier was executed respectively. In most cases the best performance was obtained for the HoG data description technique, slightly overcoming the LBP texture operator and was far ahead of GF method. This approach loosed to LBP only once when RBF SVM was applied for CK dataset. Considering these results one could assume, that the HoG features are

Table 1. Settings of classifier's parameters.

Database	Classifier	Linear SVM	RBF SVM		kNN	
	Texture operator	c	c	γ	distance	neighbours
CK	LBP	10^0	10^0	10^{-1}	Correlation	15
	HoG	10^{-1}	10^0	10^{-1}	Cityblock	15
	GF	10^{-2}	10^1	10^{-4}	Spearman	15
KDEF	LBP	10^0	10^5	10^{-3}	Cityblock	15
	HoG	10^{-1}	10^4	10^{-5}	Correlation	11
	GF	10^{-1}	10^3	10^{-3}	Cityblock	15

the strongest to describe the differences in the images. On the other hand, it is not prone to errors in classification. Tables 2 and 3 present the confusion matrices for linear SVM classification with HoG operator for both datasets. From here, one can notice, that the worse performance of methods on CK dataset results

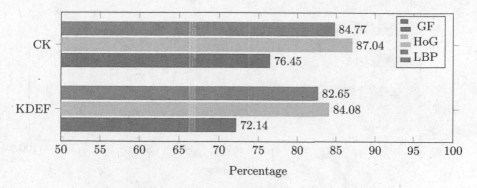

Fig. 3. Emotion classification accuracy for a linear SVM.

from huge difficulties in distinguishing between anger, fear, sadness and neutral expression as well as sadness and happiness (what is depicted in the rows of Table 2). These problems also existed in KDEF database, but with minor effect. Another reason, which may be responsible for the performance differences, is the amount of collected images within a dataset. The first set is much smaller than the other one, hence it is much probable that there are not enough examples to properly train the classifier.

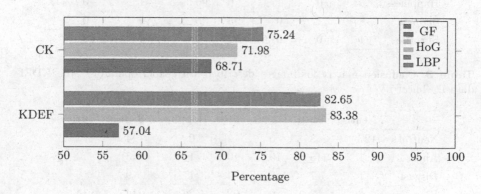

Fig. 4. Emotion classification accuracy for an RBF SVM.

4.2 Cross-Examination of Datasets

The aim of the second experiment was to verify whether the data collected in one dataset describes well the differences in emotion displays. It uses one data collection in training phase, while images from the other are used for testing. It enables to check how well the method describes the emotion, not the lighting condition and variation, which are present between various data sets. Moreover, this experiment is performed only for the classifier, which showed the highest

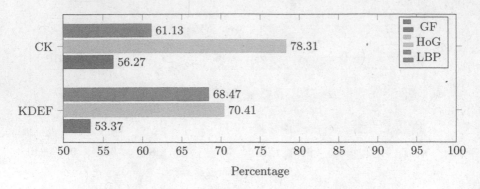

Fig. 5. Emotion classification accuracy for a kNN.

Table 2. Confusion matrix for images described with HoG operator – the CK dataset (linear SVM).

	Neutral	Anger	Disgust	Fear	Happiness	Sadness	Surprise
Neutral	314	0	0	0	2	2	0
Anger	16	10	1	0	1	1	0
Disgust	3	3	42	0	2	3	0
Fear	16	1	0	16	0	5	2
Happiness	5	0	1	0	90	1	0
Sadness	8	1	0	5	6	9	0
Surprise	1	0	0	2	0	0	67

Table 3. Confusion matrix for images described with HoG operator – the KDEF dataset (linear SVM).

	Neutral	Anger	Disgust	Fear	Happiness	Sadness	Surprise
Neutral	127	3	0	4	0	4	2
Anger	1	115	10	10	1	3	0
Disgust	1	7	118	4	1	9	0
Fear	6	3	9	92	2	14	14
Happiness	1	0	0	1	137	0	1
Sadness	7	5	2	13	3	106	4
Surprise	1	0	0	10	0	0	129

accuracy (linear SVM), as the goal is to compare the texture operators accuracies, not the classifiers power.

Figure 6 presents the obtained results for the linear SVM with all texture operators exploited for data description. Once again the HoG texture operator overcame LBP, and the GF shows very weak descriptive properties.

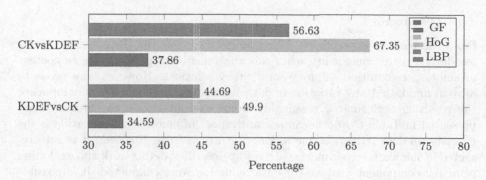

Fig. 6. Emotion classification accuracy when one database is used for training and the other for testing.

Moreover, the general performance was much worse, than in the previous case. That suggested the necessity of creating bigger data collection to these ones for better verification of emotions. Additionally, it is interesting to observe that when the smallest CK dataset was used for training better results were obtained in classification of KDEF images (CKvsKDEF). This probably results from more expressive facial gestures collected in CK.

4.3   Common Test for Datasets

Finally, this experiment collects all data from CK and KDEF datasets, to use them as a one bigger collection of images. In this case, the division of data into folds was similar as in the first experiment. Figure 7 presents the results obtained for linear SVM. Increasing the number of images diminished the performance from 84–87% to 82 in case of HoG feature, and from 82–84% to 82 when the LBP operator was used. Results for GF are not presented due to insufficient RAM (the experiments were conducted on PC with 16 GB RAM) for SVM result calculation. Moreover, the overall classification accuracy stayed on the same level for both examined techniques. It allows to draw a conclusion that the descriptive power of LBP and HoG features is comparable, yet there are cases when HoG overcomes the LBP feature descriptor.

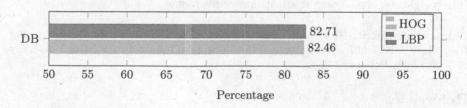

Fig. 7. Emotion classification accuracy for both datasets used together.

4.4   Discussion

From the realised experiments, it can be claimed that the HoG texture operator describes the images in such a way that classifiers performance in context
of emotion recognition improves over other solutions. However, one needs to
have in mind, that any variation in data description methodology may influence
the resulting performance reasonably. Some examples from the literature are
presented in Table 4. Another important aspect of emotion classification is the
necessity of data reduction due to long features vectors. However, it is not presented in this work, experiments with set-up described in this work and exploiting
principal component analysis for data reduction were conducted. It is possible
to claim, that this technique diminishes the data dimensions significantly, but is
correlated with slight deterioration of classification performance – similarly as it
was presented in case of smile recognition [20].

Table 4. Other methodologies for data description.

Authors	Feature description	Classification	Dataset	Results [%]
[27]	GF	Linear SVM	CK+	86.6
[27]	Weights for LBP blocks	Linear SVM	CK+	88.1
[11]	Radial encoding for GF	SVM	CK	91.51
[25]	LBP for chosen regions	SVM	CK	94.48

5   Conclusions

This work discusses the influence of a texture operator choice on emotion recognition accuracy. For the comparison three most commonly exploited texture
operators (GF, HoG, and LBP) were applied. Then the performance was verified exploiting various classification techniques based on SVM and kNN. The
experiments were conveyed on two databases frequently used in emotion recognition research. The experiments reviled that the most powerful texture operator
is HoG. Yet, one need to have in mind, that its superiority over LBP approach
is minimal, and when more complex image processing methodologies are applied
(like weighting, region selection), it maybe overcome.

Acknowledgement. This work was supported by statutory funds for young
researchers (BKM/507 /RAU2/2016) of the Institute of Informatics, Silesian University
of Technology, Poland.

References

1. Abdat, F., Maaoui, C., Pruski, A.: Human-computer interaction using emotion
   recognition from facial expression. In: 2011 Fifth UKSim European Symposium on
   Computer Modeling and Simulation (EMS), pp. 196–201, November 2011

2. Carcagni, P., del Coco, M., Leo, M., Distante, C.: Facial expression recognition and histograms of oriented gradients: a comprehensive study. SpringerPlus **4**, 645 (2015)

3. Chang, C.C., Lin, C.J.: LIBSVM: a library for support vector machines. ACM Trans. Intell. Syst. Technol. **2**, 27:1–27:27 (2011). http://www.csie.ntu.edu.tw/cjlin/libsvm

4. Cohn, J., Zlochower, A., Lien, J., Kanade, T.: Automated face analysis by feature point tracking has high concurrent validity with manual facs coding. Psychophysiology **36**(2), 35–43 (1999)

5. Dahmane, M., Meunier, J.: Emotion recognition using dynamic grid-based HoG features. In: 2011 Face and Gesture, pp. 884–888, March 2011

6. Dalal, N., Triggs, B.: Histograms of oriented gradients for human detection. In: CVPR 2005 Proceedings of the 2005 IEEE Computer Society Conference on Computer Vision and Pattern Recognition (CVPR 2005) - Volume 1 - Volume 01, pp. 886–893. IEEE Computer Society, Washington, DC, USA (2005). http://dx.doi.org/10.1109/CVPR.2005.177

7. Darwin, C.: The Expression of the Emotions in Man and Animals. John Murray, London (1872). freeman #1141

8. Daugman, J.G.: Complete discrete 2D Gabor transform by neural networks for image analysis and compression. IEEE Trans. Acoust. Speech Signal Process. **36**(7), 1169–1179 (1988). http://citeseer.nj.nec.com/context/16741/0

9. Ekman, P., Friesen, W.V.: Constants across cultures in the face and emotion. J. Pers. Soc. Psychol. **17**(2), 124–129 (1971)

10. Ekman, P., Friesen, W.V.: Pictures of Facial Affect. Consulting Psychologists Press, Palo Alto (1976)

11. Gu, W., Xiang, C., Venkatesh, Y.V., Huang, D., Lin, H.: Facial expression recognition using radial encoding of local gabor features and classifier synthesis. Pattern Recogn. **45**(1), 80–91 (2012)

12. Hachaj, T., Ogiela, M.R.: Human actions recognition on multimedia hardware using angle-based and coordinate-based features and multivariate continuous hidden markov model classifier. Multimed. Tools Appl. **7523**, 16265–16285 (2016). http://dx.doi.org/10.1007/s11042-015-2928-3

13. Huang, D., Shan, C., Ardabilian, M., Wang, Y., Chen, L.: Local binary patterns and its application to facial image analysis: a survey. IEEE Trans. Syst. Man Cybern. Part C: Appl. Rev. **41**(4), 1–17 (2011). http://liris.cnrs.fr/publis/?id=5004

14. Kasprowski, P.: Mining of eye movement data to discover people intentions. In: Kozielski, S., Mrozek, D., Kasprowski, P., Małysiak-Mrozek, B., Kostrzewa, D. (eds.) BDAS 2014. CCIS, vol. 424, pp. 355–363. Springer, Cham (2014). doi:10.1007/978-3-319-06932-6_34

15. Kołakowska, A., Landowska, A., Szwoch, M., Szwoch, W., Wróbel, M.R.: Emotion recognition and its applications. In: Hippe, Z.S., Kulikowski, J.L., Mroczek, T., Wtorek, J. (eds.) Human-Computer Systems Interaction: Backgrounds and Applications 3. AISC, vol. 300, pp. 51–62. Springer, Cham (2014). doi:10.1007/978-3-319-08491-6_5

16. Li, J., Lam, E.Y.: Facial expression recognition using deep neural networks. In: IEEE International Conference on Imaging Systems and Techniques (IST), pp. 1–6, September 2015

17. Lundqvist, D., Flykt, A., Öhman, A.: The Karolinska Directed Emotional Faces - KDEF. CD ROM from Department of Clinical Neuroscience, Psychology section, Karolinska Institutet (1998)

18. Luoh, L., Huang, C.C., Liu, H.Y.: Image processing based emotion recognition. In: 2010 International Conference on System Science and Engineering, pp. 491–494, July 2010

19. Nurzynska, K., Smolka, B.: Smiling and neutral facial display recognition with the local binary patterns operator. J. Med. Imaging Health Inf. **5**(6), 1374–1382 (2015)

20. Nurzynska, K., Smolka, B.: PCA application in classification of smiling and neutral facial displays. In: Kozielski, S., Mrozek, D., Kasprowski, P., Małysiak-Mrozek, B., Kostrzewa, D. (eds.) BDAS 2015. CCIS, vol. 521, pp. 398–407. Springer, Cham (2015). doi:10.1007/978-3-319-18422-7_35

21. Ojala, T., Pietikäinen, M., Harwood, D.: A comparative study of texture measures with classification based on featured distributions. Pattern Recognit. **29**(1), 51–59 (1996)

22. Ojala, T., Pietikäinen, M., Mäenpää, T.: A generalized local binary pattern operator for multiresolution gray scale and rotation invariant texture classification. In: Singh, S., Murshed, N., Kropatsch, W. (eds.) ICAPR 2001. LNCS, vol. 2013, pp. 399–408. Springer, Heidelberg (2001). doi:10.1007/3-540-44732-6_41

23. Ojala, T., Pietikäinen, M., Mäenpää, T.: Multiresolution gray-scale and rotation invariant texture classification with local binary patterns. IEEE Trans. Pattern Anal. Mach. Intell. **24**(7), 971–987 (2002)

24. Pietikäinen, M., Hadid, A., Zhao, G., Ahonen, T.: Computer Vision Using Local Binary Patterns. Computational Imaging and Vision, vol. 40. Springer, Heidelberg (2011)

25. Sadeghi, H., Raie, A.A., Mohammadi, M.R.: Facial expression recognition using geometric normalization and appearance representation. In: 8th Iranian Conference on Machine Vision and Image Processing (MVIP), pp. 159–163, September 2013

26. Salmam, F.Z., Madani, A., Kissi, M.: Facial expression recognition using decision trees. In: 2016 13th International Conference on Computer Graphics, Imaging and Visualization (CGiV), pp. 125–130, March 2016

27. Shan, C., Gong, S., McOwan, P.W.: Facial expression recognition based on local binary patterns: a comprehensive study. Image Vis. Comput. **27**(6), 803–816 (2009)

28. Szwoch, M.: On facial expressions and emotions RGB-D database. In: Kozielski, S., Mrozek, D., Kasprowski, P., Małysiak-Mrozek, B., Kostrzewa, D. (eds.) BDAS 2014. CCIS, vol. 424, pp. 384–394. Springer, Cham (2014). doi:10.1007/978-3-319-06932-6_37

29. Tsai, H.H., Lai, Y.S., Zhang, A.Y.C.: Using SVM to design facial expression recognition for shape and texture features. In: 2010 International Conference on Machine Learning and Cybernetics, vol. 5, pp. 2697–2704, July 2010

30. Turner, M.R.: Texture discrimination by Gabor functions. Biol. Cybern. **55**(2–3), 71–82 (1986). http://dl.acm.org/citation.cfm?id=11682.11683

31. Viola, P., Jones, M.J.: Robust real-time face detection. Int. J. Comput. Vis. **57**(2), 137–154 (2004)

Industrial Applications

The Use of the TGŚP Module as a Database to Identify Breaks in the Work of Mining Machinery

Jarosław Brodny[1], Magdalena Tutak[2], and Marcin Michalak[3(✉)]

[1] Institute of Production Engineering, Faculty of Organization and Management,
Silesian University of Technology, ul. Roosevelta 26, 41-800 Zabrze, Poland
Jaroslaw.Brodny@polsl.pl
[2] Institute of Mining, Faculty of Mining and Geology,
Silesian University of Technology, ul. Akademicka 2, 44-100 Gliwice, Poland
Magdalena.Tutak@polsl.pl
[3] Institute of Informatics, Silesian University of Technology,
ul. Akademicka 16, 44-100 Gliwice, Poland
Marcin.Michalak@polsl.pl

Abstract. The article presents the results of the causes of breaks in selected mining machines work. The studied machines belong to the mechanized longwall system. Identification of the causes of these breaks was carried out using author's database that was created on the basis of the information coming from the application, which is an integral part of the Means of Production Management Module (TGŚP). This module is one of the basic parts of the integrated enterprise management system SZYK2. The results clearly show that the developed solution enables more efficiently to identify the causes of breaks in examined machines work than previously used systems. It is the result of acquiring the tacit knowledge from dispatchers thanks to developed solutions.

Keywords: Database · Mining machinery · Management · Breaks in the work

1 Introduction

In the industry, also in coal mining, more and more widely, the advanced informatics systems for monitoring the work of any type of equipment and especially machines are used [7]. Large amounts of data registered in this process require proper processing and systematization, because in the original form they are very difficult to be adopted and practically used [6]. It is necessary to properly select data and make it compression in order to change them into source of information that can be used for example in management activities [2,9].

In the coal mining industry in the area of the machines use, especially machines working underground, large problems in the process of obtaining operating data about machine's work are observed. Moreover, the already obtained

© Springer International Publishing AG 2017
S. Kozielski et al. (Eds.): BDAS 2017, CCIS 716, pp. 441–452, 2017.
DOI: 10.1007/978-3-319-58274-0_35

data is not fully used. It refers primarily to data, which enables to determine the availability of machines and identify the causes of breaks occurring during their normative work.

Lack of reliable and credible information in the area of mining machines use makes it impossible to take proper and effective decisions and that can cause their underutilization and then, problems with the production continuity in mining enterprises. Therefore it is necessary to take actions aimed at improving this situation by increasing the level of use of information come derived from objective sources, which are undoubtedly industrial automation systems.

Currently data collected by these systems and showing the monitoring process is used almost only for current visualization and particular parameters reporting.

In practice, more and more often the need for broader use of the obtained data is observed. The purpose of these actions is the more complete identification and defining a more advanced diagnostic model of monitored machines and devices. This model should take into account the full cycles of machines work and make it possible to identify all kinds of breaks.

To improve this situation, actions, aimed at providing access to diagnostic data about mining machines belonging to the mechanized longwall system, were taken. A system of obtaining, archiving and analyzing data based on functioning industrial automation system elements was developed. Designed and built data warehouse enabled the centralization, archiving and analytical processing of data acquired from different machines.

These actions were taken as part of the project, realized by the Silesian University of Technology and consortium members, entitled "Aplication of the Overall Equipment Effectiveness method to improve the effectiveness of the mechanized longwall systems work in the coal exploitation process". The project is focused on the high level of the use of data, obtained from industrial automation system, about longwall system machines, the use of which in the current practice of coal exploitation was low.

The data obtained from industrial automation systems will be used to identify the machines work time intervals, which can be qualified as losses in the production process.

Losses are described in several categories. The first category is a losses layer that is equal to the loss of machine or entire mechanized longwall system availability. Usually, random events causing failures and, in consequence, downtime are the origin of these losses. The second layer consists of performance losses that occur when mechanized longwall system works. These losses are included in the next layer in which also the effects of decline in quality of products are observed. For each of the layers, appropriate percentile effectiveness indicator is calculated and the product of partial indicators is the aggregate effectiveness indicator.

To draw meaningful conclusions from the values of effectiveness indicators, it is necessary to identify the events, according to the quantitative and based on values qualification, which refers to the duration of event that is described by the state of the industrial automation parameters. In this case, quantitative qualification of these events is also necessary. This kind of qualifications should describe the specific reason of failure, downtime or performance loss. In the end, this reason can only be diagnosed by worker occupying the production process or an employee of maintenance service. This ascertainment leads directly to necessity of referring to and using of failure and downtime registration, and credible determination of their reasons.

This type of registration is carried out applying informatics solutions that are mostly dedicated to the maintenance service or managers responsible for means of production [2–4, 8–10].

In order to believably identify breaks in work and their reasons, it was necessary to develop a new method of their recording.

For this purpose, the integrated management system SZYK2, which is used in mines, was used. For this system, an additional application for the Means of Production Management Module (TGŚP) was made.

This application allows dispatchers to precisely indentify the reasons of downtime occurring in machines work. Machines, which were examined, were included in the mechanized longwall system (the longwall shearer, the armoured face conveyor, the beam stage loader and the crusher).

The article discusses the management system in the company, called SZYK2, with extensive TGŚP module used to the registration of the reasons of examined machines downtime. Results of failure cause registration for one of the exploited longwalls in the new TGŚP module were presented. The results unequivocally confirm that the use of the new solution has enabled the identification of about 67% of breaks in examined machines work. The result is efficiency improvement of breaks reasons identification and it is approximately by 200% better than using the previous system.

To sum up, it can be claimed that thanks to the application of the new TGŚP module, the following fundamental problems have been solved:

– obtaining the reliable data on the causes of downtime during the studied machine' work
– explanation of the causes of the breaks recorded by other industrial automation systems [8],
– the possibility of conducting the effective analysis of great amounts of obtained data on the breaks in the machine' work and the causes of these breaks.

Additionally, the development of the new methods of obtaining data on the causes of downtime, and the new database structure that enables a simple and clear registration of this data can be considered a scientific achievement. It all plays a significant role due to the environment in which this system works.

The paper does not present the detailed structure of the TGŚP module (as a database) and the used algorithms because of their commercial character.

The authors hope that the presented solution will be widely and practically used also in others than mining industries.

2   Characteristics of the SZYK2 System

Most of the national mining companies use integrated management system type of SZYK2 to support the management process [1]. This system was developed and distributed by COIG company. It is a consistent set of integrated applications based on the concepts and standards of ERP solutions [4,5].

It includes, among others, complexes dedicated to finance, logistics, production, sale and HR support. All solutions are available through a Web browser. Architecturally, all is based on the relational Oracle database. Each complex has its own database, and logically separated extensive database of central catalogues and files, being common resource of SZYK2 system, makes it all coherent. As a result, each of these complexes "sees" its own database and central resources. Application of this approach makes it possible to obtain high quality level of performance parameters of the system. For multi–plant mining companies it is very important that there is a possibility of using a single database in this system commonly.

Figure 1 shows a scheme of the SZYK2 system [1].

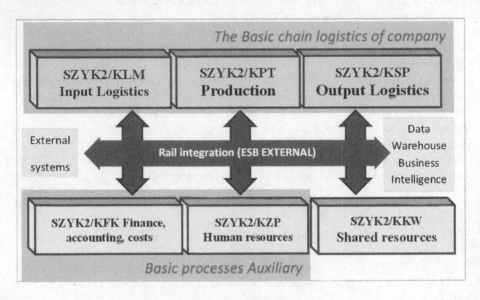

Fig. 1. A scheme of the SZYK2 system [1].

One of the most important complexes of SZYK2 system is the Production–Technical Complex (KPT). This complex is responsible for supporting business processes associated with the production, its preparation and maintenance. It allows operating the central part of the logistics chain in the company.

The Production–Technical Complex (KPT) of SZYK2 system includes modules for planning, scheduling and production monitoring, portfolio managing of projects and tasks, and a wide range of application that is included in the Means of Production Management module (TGŚP). Each complex and so KPT must provide the possibility of resources using from the local level — the resources dedicated to the specific mine, as well as from the central level — all the resources needed to mean of production management (including machines parks) from the company level.

Therefore, appropriate relations as parts of one KPT complex base were used for achieving access mechanisms.

The functionality of the modules included in the KPT complex is dedicated to processes in mining companies. As part of KPT, failure, downtime and other events affecting the production process are registered. The context of these events can be different depending on groups of functionality that are intended to production monitoring or means of production management.

From the studied problem need's point of view, which refers to machine resources and their functioning, it is justified to use the TGŚP module that is intended to means of production management.

Other complexes included in the integrated enterprise management system type of SZYK2 are as following:

- material logistics complex (supply, warehouse management, materials management, inventory consumption and inventory management),
- sale complex (distribution and products sale),
- shared files complex (it shares common files and dictionaries with modules in other complexes included in the system),
- financial and salary complex,
- employment and salary complex.

The SZYK2 system is used in holding and multi-branch structures. It can also work in mono-branch companies as well as in territorially dispersed structures, where lots of users work [1]. Universal structure of the system is its advantage and it influence on its broad use in various industries.

3   Characteristics of the TGŚP Module

In order to accurately identify the causes of work downtime of machines belonging to the mechanized longwall system, the application for the breaks reasons registration was made. This application has become an integral part of Means of Production Management Module (TGŚP). This module is part of the Production–Technical Complex of the Integrated Enterprise Management System. As an application EAM class (Enterprise Asset Management), it supports means of production (technical objects) and closely related to them overhaul management [5,11]. It helps to conduct the failure analysis and extraction analysis and manage mechanized roof supports. Functionality of the system is adapted to work, using mobile tools that enable storage management of machines and devices elements and to conduct underground logistics work on machine components.

Fig. 2. A view of the main application window of the TGŚP module [1]

Figure 2 shows a view of the main application window of the TGŚP module [1].

The standard TGŚP includes the following functional classes:

- managing a common file with information about types of machines and devices used in all mines branches of the mining company containing, in particular, all kind of technical data, documentation, technical drawings, diagrams, etc.
- machines and devices operation service, as well as income, taking on record, lease, transfer of means of production in one plant (between work places) and between plants, and liquidation,
- operations service of related to the usability of the machine (assembly, disassembly, exchange of particular elements)
- service of failure evidence of particular machines,
- the process of machines and devices maintenance service as well as scheduling, carrying out inspections, conservations, periodic and after–failure repairs of machines,
- service of operating work parameters monitoring for the history of using of the machine according to exploitation criteria (eg. progress etc.).

To provide information completeness in the TGŚP module and opportunities to use information, whose source is the TGŚP module, in the SZYK2 system, thanks to use of central resources, the integration of many components is made. This integration includes:

- resources of structural and organizational file (eg. the longwall face, regions, branches of the location, movement, determining the ownership time),
- resource of central file of contractors (eg. producers, tenants, service providers of commissioned works),
- resources of fixed assets management (eg. displacement, value change, liquidations),

- resource of production scheduling (machine resources assigning, parameters of completed work, schedules of longwalls and longwall faces work),
- resource of production process monitoring (evidences of completed works and events, and incidents, eg. failures),
- resource of tasks and projects management module (eg. orders related to the means of production),
- resource of material logistics (eg. the area of spare parts, exploitation means, material needs for repair works).

Presented functional range and integration of the system should make it possible to get a broad image of areas that are necessary for the completing particular tasks, especially, for qualitative identification of losses times in the level of industrial automation system.

Mining companies use the functionality of failure and incidents monitoring in the TGŚP module as well as in production process monitoring module. The way and scope of losses evidence and downtime causes identification were insufficient for the needs related to applied systems of the breaks registration.

The main reasons for this state were resulted from insufficient accuracy of attributes (features) describing failures (stoppages), and assumed too high value of lower threshold level of failure (downtime) time qualifying it for being placed into failure record.

Therefore, one of the mines used developed and dedicated functionality of the TGŚP module related to failure evidence, which was characterized by significantly increased level of failures registration accuracy and of their attributes. At the same time, the lower time limit deciding whether incident is qualified to be placed into registration was canceled.

A few months of this solution exploitation made it possible to gain knowledge in the area of use and the failure frequency of machines belonging to the mechanized longwall system. This knowledge was previously explicit. Revealing of this knowledge will allow achieving a new quality in the area of downtime causes identification of examined machines and obtaining of additional information about machines exploitation.

A module, dedicated to mining companies, aimed at supporting dispatcher's work is Production Processes Reporting Module (TMRPP2) [1]. This module allows registering and documenting the daily work in the mine based on files and dictionaries. Originally, it aimed also at failures and downtime data processing and making, on their basis, Daily Reports and other daily and periodic reports. This module gathering failures and downtime data should have also conducted advanced analysis of their causes, as well as registered the events and accidents in the mine.

In practice, it turned out that the amount of stored events in the module significantly differs from those registered by industrial automation system.

Table 1 shows the statement of the number of breaks in studied set of machines work for four weeks of working registered by the industrial automation system (IAS) and TMRPP2 module. The registration referred to the time before activating the modified TGŚP system. The results unambiguously show that the number of breaks registered in the TMRPP2 module significantly differs from the real number of breaks in these machines work. In practice, this module recorded only breaks, whose duration was more than 30 min.

Table 1. Statement of the number of breaks in studied set of machines

System	TPMRPP2	IAS	(TMRPP2/IAC)×100%
Week 1	10	76	13.2%
Week 2	12	73	16.4%
Week 3	14	69	20.3%
Week 4	12	106	11.4%

The large discrepancy in downtime evidence made it impossible to identify the real causes. This led the authors of the article to develop the new application that effectively works in the TGŚP module.

4   The Results of Registration and Analysis

Development and implementation of a dedicated application for registering of the breaks causes for studied set of machines made it possible to identify most of the reasons.

On this basis, database containing chronological report of the downtime causes was established. It also had division into particular machines and their subgroups. A potential cause of a downtime, a place of its occurrence and its duration were also registered. In addition, a preliminary type classification of the failure causing downtime and description of the reason were also made.

The data included in the database in the range of reported breaks was related to their amount recorded by the industrial automation system. It was assumed that industrial automation systems, due to the 1 Hz frequency, register all breaks, in machinery work time, whose duration is longer than one second. However, for technical reasons evidence of breaks in the TGŚP module in the analyzed period (assumed as preliminary) included only those downtime whose time was more than or equal to one minute.

Figure 3 presents quantitative breaks report about surveyed machines. These breaks were registered by the industrial automation systems and a modified version of the TGŚP module for six weeks of longwall work. These data are summarized in weekly blocks. This report includes all the breaks whose time is more than or equal to one minute. During this period, industrial automation system registered 987 breaks but a modified version of the TGŚP module included 661 breaks and that is 67% of breaks registered by industrial automation system (IAS).

Figure 4 shows the quantitative division of recorded, by both systems, breaks in machines work based on their duration. These breaks were divided into seven

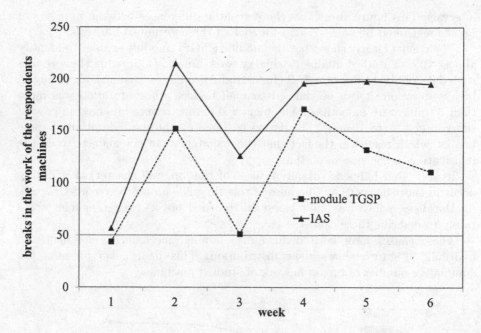

Fig. 3. Quantitative breaks report about surveyed machines

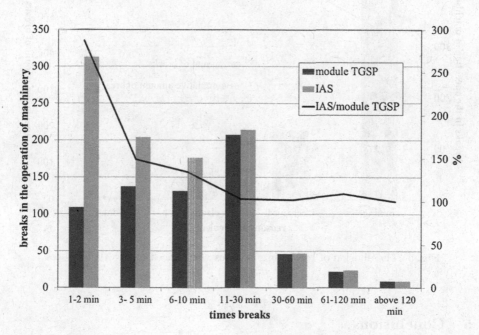

Fig. 4. The quantitative division of breaks in machines work recorded by both systems due to duration of breaks

intervals. This figure also shows the percentile differences between numbers of breaks registered by both systems for each of the determined intervals.

The results clearly show that the modified TGŚP module enables to identify almost 70% of studied mining machinery work breaks. Comparing these results with data obtained from the IAS (Industrial Automation System), according to breaks duration, it can be claimed that all breaks, whose duration was more than 30 min, were indentified. The biggest differences were observed in case of short intervals (to 10 min). The reason is probably the short duration of these breaks, which results in the fact that the dispatcher can not manage to do the registration in the system on time.

It also often happens (usually in case of the longwall shearer) that breaks occur in short intervals and because of this they all can not be recorded.

Database, which was built based on recorded breaks causes, enables accurately to describe these causes.

These causes have been divided into mining, mechanical, electrical and hydraulic. Figure 5 shows their distribution. This figure also presents the cumulative number of breaks in work of studied machines.

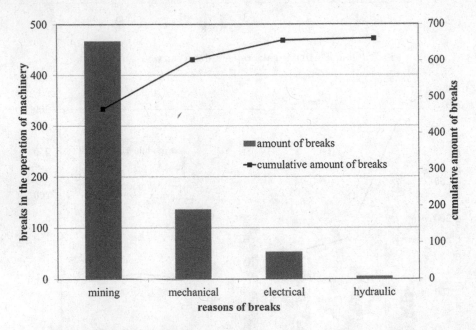

Fig. 5. The division of breaks in machines work based due to their causes

5   Conclusions

In the mining industry, like in other sectors of the economy, the effective use of technical means is very important. Especially, it refers to mining machines, which are directly involved in the coal exploitation process. Due to the specific nature of the machine work environment, machines have to be highly reliable and productive. Additionally, high costs of these machines make mining companies strive to maximize the use. Therefore, it is necessary to monitor the state of their work continuously and reduce all types of breaks.

The way of breaks causes identification, presented in this paper, using the TGŚP module refers to this process. The application, developed for this module, enabled to build a database of the causes of occurring breaks in studied machines work. Thanks to this application, a lot of information, which is so–called tacit knowledge and which has crucial meaning for improving the effectiveness of use of studied machines, was obtained.

Built base enables to analyze the recorded data in terms of obtaining credible and reliable knowledge about the real causes of recorded breaks in machines work. Afterwards, this knowledge should be used to reduce these breaks and improve the availability of these machines.

The results proved that the chosen direction of research in this area is right and correct. Causes of almost 70% of all breaks in machines work were indentified. Referring to breaks lasting longer than 10 min the rate is approximately 96%. In previous studies it rarely succeeded to identify 25% of occurring breaks.

Additionally, the built database allowed to conduct many analyzes to determine different kind of indicators that could be used for instance by maintenance services.

Therefore, it is justified that developed solution should be widely applied in practice. It is also necessary to conduct further research on its improvement and develop user–friendly interface, especially in the field of short breaks (lasting up to 5 min) registration in the machines work.

Acknowledgments. This article is the result of the research project No. PBS3/B6/25/2015 "Aplication of the Overall Equipment Effectiveness method to improve the effectiveness of the mechanized longwall systems work in the coal exploitation process" realized in 2015–2017, financed by NCBiR.

References

1. Integrated system for management support. www.coig.pl/dokumenty/katalog_informacyjny_szyk2.pdf. Accessed on 06 Dec 2016
2. Brodny, J., Stecuła, K., Tutak, M.: Application of the TPM strategy to analyze the effectiveness of using a set of mining machines. In: 16th International Multidisciplinary Scientific GeoConference SGEM 2016, pp. 65–72 (2016)
3. Devenport, H.T., Prusak, L.: Working Knowledge: How Organizations Manage What That Know. Harvard Business School Press, Boston (1998)
4. Drucker, P.F.: The emerging theory of manufacturing. Harvard Business Review, pp. 94–102 (1990)
5. Kelly, A.: Strategic Maintenance Planning. Butterworth-Heinemann, Oxford (2006)

6. Kozielski, M., Sikora, M., Wróbel, Ł.: Decision support and maintenance system for natural hazards, processes and equipment monitoring. Eksploatacja i Niezawodność - Maint. Reliab. 18(2), 218–228 (2016)
7. Michalak, M., Sikora, B., Sobczyk, J.: Diagnostic model for longwall conveyor engines. In: Gruca, A., Brachman, A., Kozielski, S., Czachórski, T. (eds.) Man–Machine Interactions 4. AISC, vol. 391, pp. 437–448. Springer, Cham (2016). doi:10.1007/978-3-319-23437-3_37
8. Stecuła, K., Brodny, J.: Application of the OEE model to analyze the availability of the mining armored face conveyor. In: 16th International Multidisciplinary Scientific GeoConference SGEM 2016, pp. 57–64 (2016)
9. Wrycza, W.: Analysis and Project of Management Informatics Systems: Methods, Techniques, Tools (in Polish). PWN, Warsaw (1999)
10. Yeates, D., Cadle, J.: Project Management for Information Systems. Pearson Education, Edinburgh Gate (1996)
11. Żółtowski, M.: Computer Aided Exploatation System Management in a Manufacturing Company (in Polish). Polish Society of Production Management Press, Opole (2011)

A Data Warehouse as an Indispensable Tool to Determine the Effectiveness of the Use of the Longwall Shearer

Jarosław Brodny[1], Magdalena Tutak[2], and Marcin Michalak[3(✉)]

[1] Faculty of Organization and Management, Institute of Production Engineering,
Silesian University of Technology, ul. Roosevelta 26, 41-800 Zabrze, Poland
Jaroslaw.Brodny@polsl.pl
[2] Faculty of Mining and Geology, Institute of Mining,
Silesian University of Technology, ul. Akademicka 2, 44-100 Gliwice, Poland
Magdalena.Tutak@polsl.pl
[3] Institute of Informatics, Silesian University of Technology,
ul. Akademicka 16, 44-100 Gliwice, Poland
Marcin.Michalak@polsl.pl

Abstract. The effective use of machines and devices in the mining industry is significant. In a competitive energy market, such effectiveness can decide about the further functioning of the company in many cases. The article subject refers to these issues, in particular, to the way of determining the availability of a longwall shearer used in underground coal mining. The article presents the proposal of using the data warehouse to determine the level of load of a longwall shearer during its work on the basis of shearer motor power consumption time series.

Keywords: Overall Equipment Effectiveness · OEE · Machine monitoring · Data warehouse

1 Introduction

From the decades the increase of production of electric energy from renewable resources can be observed. Simultaneously, hands the efforts to limit the negative influence of traditional energy production technologies are bore. Though in many European countries the coal—hard or brown—remains the main source of energy.

For example in Poland in 2012 83% of produced energy was generated from a coal (50% from a hard and 33% from a brown coal) while in Germany the total amount of a coal based energy production was at a level of 45%.

One of possible aspects of limitation the negative influence of hard coal mining into an environment is a proper operation of a mining machines. The proper operation covers the accessibility of the whole longwall system and also a assuring the optimal (nominal) conditions of machines work (current consumption etc.).

© Springer International Publishing AG 2017
S. Kozielski et al. (Eds.): BDAS 2017, CCIS 716, pp. 453–465, 2017.
DOI: 10.1007/978-3-319-58274-0_36

In the essential part, the paper focuses on the analysis of the longwall shearer work. The aim of this analysis was to determine the load of the shearer main engines which are responsible for cutting and moving. Registered values of current consumed by these engines were assumed as a measure of the load. Then, these values recorded as a time function were referred in addition to the shearer position in the longwall. The position was determined in reference to the section of the roof support, at which the shearer was at that time.

The combination of the shearer position and the current consumption values enabled to determine the load of these engines, depending on the shearer work phase. Shearer work during exploitation has been divided into three basic phases which include slotting, main work and final work.

The usage of the data warehouse enabled to carry out the analysis whose results clearly showed that the load of the shearer engines depends on its position in the longwall. Moreover, the breaks during work were determined. The obtained data is an important source of information for staff supervising the work of mining machinery and it should be used to make decisions in order to optimize the use of the machines.

The paper is organized as follows: it starts with the short description of efficiency analysis, then some introduction into machine monitoring is presented. Afterwards the detailed specification of the case analyzed in the paper is shown. Then the short description of data warehouse used is presented, followed by the explanation of the data taken into consideration in this approach and an overview of developed reports. The paper ends with some final conclusions.

2   Background

One of the main problems of national mining companies extracting coal is ineffective use of available technical means. Especially, it refers to the mining machinery used in the coal–cutting process, horizontal and vertical transport and crew area protection.

Taking into account the specifics of the mining industry, it can be assumed that a set of machines responsible for the coal and rock cutting out from the rock mass and transporting them from the zone of a longwall face is the most important in the exploitation process. These machines form a set called mechanized longwall system. This set of machines includes the longwall shearer, the armoured face conveyor, the beam stage loader, the crusher and the power roof support.

From the reliability point of view, this set is characterized by the serial structure. The most important machine of this set is the longwall shearer cutting and loading excavated material onto the armoured face conveyor. The effectiveness of the whole exploitation process of the longwall depends on the efficiency and effectiveness of longwall shearer's work.

To assess the level of the longwall shearer use in the underground coal mining process the Overall Equipment Effectiveness model (OEE) has been used [13] which was successfully applied in industry [14,16] and is still being developed [12]. This model is a quantitative tool for assessing the efficiency of the

Total Productive Maintenance (TPM) strategy implementation. This strategy includes a set of actions and activities aimed at the keeping machines failure–free and faultless by reducing failures, unplanned downtime, lacks and unscheduled service. In the consequence, it will directly result in economic efficiency improvement [1,11,17]. The purpose of the strategy application for longwall shearer work analysis (and then for the whole mechanized longwall system) is to identify all kinds of losses and eliminate or limit their sources and causes.

The application of the OEE model for assessing the level of the longwall shearer use in the coal exploitation is associated to the necessity of identification of the shearer state in the areas of availability, performance and quality of the product (in this case—coal). For each of these areas the partial indicators are determined and their product designates the Overall Equipment Effectiveness indicator of examined machine. Figure 1 shows the scheme of the Overall Equipment Effectiveness indicator determination [1,8,11].

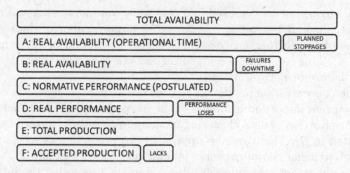

Fig. 1. The scheme of determining the Overall Equipment Effectiveness (OEE) indicator

According to Fig. 1 the final formula of its calculation becomes as follows:

$$OEE = \frac{B}{A} \cdot \frac{D}{C} \cdot \frac{F}{E}$$

Taking into account the specifics of the mining industry the most important meaning, during designating the OEE value, has the value of availability indicator, but basically, the value of the real availability. Essential for the credibility of the designated availability value is data selection, obtaining, processing and collecting. In the process of collecting information about the longwall shearer work parameters the industrial automation system, which inspects the exploitation process in the type of the SCADA (*Supervisory Control and Data Acquisition*), was used.

Based on the recorded data, analyzes have been conducted and their goal was to determine the longwall shearer availability. It should be noticed that the determination of the availability in the range like in this study without the support of a data warehouse would be practically impossible.

The methodology of availability indicator determination for mining machines (especially for longwall shearer) applied in study, which uses innovative for coal mining informatics tools, should improve the effectiveness of use of mining machines, and as a consequence should also have an influence on the increase of the overall effectiveness of the exploitation process.

The application the OEE model in underground mining exploitation conditions aimed ultimately at helping to establish the appropriate normative rules, good practices and improvement programs of production process, depending on the level of the indicators as the main content of the OEE model. As a consequence, this will lead to increased effectiveness of the entire production process, thereby to rationalization of production costs in the areas of the production and labor means.

3   Machine Monitoring

The typical environments deployed in a coal mine are monitoring and dispatching systems. Their main role is collecting the large number of data which for the purpose of the further analysis e.g., on–line prediction of the sensor measurements [6]. This can address different aspects of coal mine operation such as equipment failure or natural hazards.

Methane concentration prediction and seismic hazard analysis are one of the most popular and important examples of the research in the field of natural hazards [15]. Application of data clustering techniques to seismic hazard assessment was presented in [7]. There are also approaches to prediction of seismic tremors by means of artificial neural networks [4] and rule–based systems [3].

Another important role of monitoring systems is presenting the information about the diagnostic state of machines. Usually, machine learning methods are used to diagnostics of mining equipment and machinery, like in [2,6,9,10]. The issue of mining industry devices diagnostics was raised among others in the works [5].

4   Problem Specification

In the presented study the longwall shearer was analysed. It is the most important machine in the mechanized longwall system. The purpose of the analysis is a credible determination of the real availability of the longwall shearer that is the basis for availability indicator calculation in the OEE model. This indicator is the ratio of the real work time of the machine to the normative (planned) time, set in the technical and economic plan in the company.

The research problem is to develop efficient method of diagnostic data obtaining, processing and archiving for the studied longwall shearer. Such solution will allow credible determining of the indicator value.

Previous methods of determining the real work time of the longwall shearer were based on documents made by dispatchers or the registration of the shearer state in the "0–1" system. Definitely, the data registered in this "0–1" system had more credibility. The disadvantage of this method of longwall shearer work recording was the fact that the indication of "1" (work) occurred in case of working at least one of the five shearer's engines. In practice, it turned out that very often assumed work time was a period in which the longwall shearer did not work but just one of the engines was turned on.

The least credible registration system of shearer real work time was the registration system made by dispatchers. Therefore, it is justified to claim that the fundamental meaning for determining the partial indicators as well as the Overall Equipment Effectiveness indicator has the way of obtaining input data.

In case of the studied longwall shearer, its parameter registration system was used and it was based on indication of the industrial automation system in the range of the diagnostic parameters registration.

The parameters, which were used to determine the real work time of the longwall shearer, were current consumption waveforms consumed by the engines of the cutting heads, feed drive, hydraulic pump drive, the feed speed of the shearer and its position in the longwall.

These parameters were registered with a frequency of 1 Hz and this resulted in obtaining a huge database of these parameters due to 8-month period of study. Due to the fact that the other parameters of the shearer work such as oil temperature in the gear, set and actual frequency of the variable frequency drives were also registered, databases became very extensive.

For the realization of the research goals, it became necessary to build a data warehouse, which allowed efficient and effective archiving and analytical processing of obtained data. This data warehouse was also used to analyze the work of other machines belonging to the mechanized longwall system and the whole system treated as a single mega–machine.

5   Data Warehouse Description

The data repository used in our research is a data warehouse Vertica, the product of the HP company. In the measurement–based part it contains a snowflake architecture which centre is the WOEEADM.WOEE_NORM table (presented in the Fig. 2) with additional partially dictionary based information about the moment of measurement and its location in the longwall complex structure: machine id (WNO_WMA_ID), longwall id (WNO_WSC_ID), parameter id (WNO_WPA_ID) and time stamp (WNO_DATACZAS).

The warehouse contains also additional tables describing the structure of all equipment, additional parameters describing the mentioned elements as well as information about the first step of incoming data preprocessing. The total number of tables is 44.

One of the most important task of the ETL procedures applied in the warehouse data feeding is the analysis of the missing data. It is assumed that if the time of the data missing is less than 30 s it is filled with the last observed value.

The another step performed at the stage of loading the data is a normalization. Due to the fact that data comes from system of different producers (or

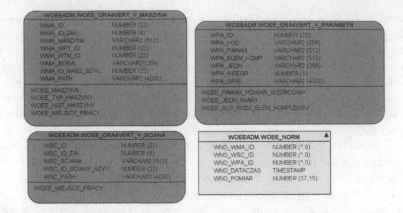

Fig. 2. The main table for measurements storage and three data views as a dictionary tables.

sometimes there are different formats of the data coming from the same type of machine of the same producer) all coming data should be represented in the same common way. The normalization translates the incoming format of the data into the structure of the mentioned snowflake hierarchy.

The warehouse stored procedures supplies also methods of cleaning the outstanding data. For every measured signal its maximal and minimal values are stored in the data warehouse and in the case when a new observation exceeds those limitations there are removed (and the procedure of missing data introduction is responsible for solving the problem).

For the purpose of the analysis presented in the paper the data warehouse has the possibility to generate the text reports of selected variables ant their values from the selected period of time. These text files become the important input for the further analysis, presented in the next section.

6   Analysis and Results

6.1   Data Description

In the analysis the two sets of monthly data describing the "typical operation" of a heading machine were used. The considered months were the following: February and March 2016. As the "typical operation" only the operation whose goal was the coal exploitation were considered. The observations from time of shut–down, inspection and maintenance works were excluded from the analysis. The remaining data contained also missing values which percentages for selected months are presented in the (Table 1).

From the wide range of observed variables only five of them were taken into consideration:

– the left tractor engine current (marked as KMB_pcl, due to the original Polish definition "KOMBAJN — prąd silnika ciągnika lewego),

Table 1. Description of the data from the two month of typical operation of a heading machine

	Feb	Mar
Total record count	1 881 500	1 546 799
Percentage of records with missing values	67.07	55.46

- the right tractor engine current (KMB_pcp — "prąd silnika ciągnika prawego"),
- the left organ engine current (KMB_pol — "prąd silnika organu lewego"),
- the right organ engine current (KMB_pop — "prąd silnika organu prawego"),
- the position of the machine, in a reference to the number of the section from the powered roof support (marked as KMB_poz).

Analyses were performed in the R environment, which required text files as an input. The sample file format is presented in Table 2.

Table 2. Several records from the analysed data file

Time	KMB_pcl	KMB_pcp	KMB_pol	KMB_pop	KMB_poz
2016-03-13 12:38:46	152.940	157.3500	71.30000	31.8	37.4
2016-03-13 12:38:47	143.880	162.5500	70.65000	31.8	37.4
2016-03-13 12:38:48	140.450	149.9200	94.65000	28.4	37.4

6.2   Diagnostic Report

The description of the proposed diagnostic report is presented on the measurements from the second of two mentioned months (March 2016).

To reflect the quality of the data it is important to point the distribution of missing data lengths. The distribution for the given period of time is presented in the Table 3. The ranges of bins are presented in seconds (first two bins), minutes (four bins), hours (four bins) and days (last two bins). For each bin the total number of data missing durations from the specified range is given.

The most general descriptive statistics for the considered currents are deciles that may help to compare the nonequality of the operation of corresponding left and right organ and engines.

The longwall consists of 106 powered roof support sections. In the Fig. 3 histograms of current consumption of left and right engines of a heading machine are presented. The Table 4 presents the selected percentiles of current values of four engines. These data corresponds to the whole range of machine position. It must be mentioned that all histograms are presented with the Y axis logarithmic.

It is implied by the specific operation condition on the both ends of the longwall that the current consumption differs from the typical coal exploitation. In the Fig. 4 the same histograms are presented and the difference is easy to be observed (lacks of the data correspond to the value 1 while bars going down correspond to the value 0). It is not presented on figures but histograms for the currents at the beginning of the longwall (section from 0th to 7th) looks similar.

Table 3. Distribution of missing values durations.

	[s]		[min]			
Bin	(0, 30)	[30, 60)	[1, 5)	[5, 15)	[15, 30)	[30, 60)
Count	0	102	171	268	40	10
	[h]				[day]	
Bin	[1, 3)	[3, 6)]	[6, 12)	[12, 24)	[1, 2)	[2,...]
Count	17	23	2	1	0	0

Table 4. Percentiles of a currents values.

	0%	10%	20%	30%	40%	50%	60%	70%	80%	90%	100%
KMB_pcl	0	0.7	1.25	2.2	6.27	26.86	35.8	41.8	47.6	55.35	162.3
KMB_pcp	0	0.75	1.3	2.33	7.17	30.3	39.87	46.14	52.23	60.52	164.9
KMB_pol	0	21.9	26.9	27.6	28.1	28.6	29.1	29.8	30.8	32.6	244.7
KMB_pop	0	23.23	25.6	26.2	26.7	27.45	28.3	29.2	31.05	35.2	243.7

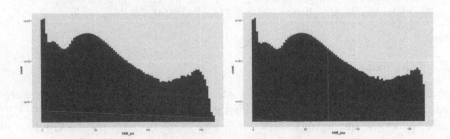

Fig. 3. Histograms of a current consumption for left and right engines

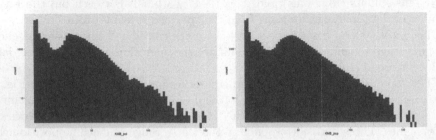

Fig. 4. Histograms of a current consumption for left and right engines from the 99th to the 106th section

That makes interesting to observe how do the histograms for the power consumption in the middle sections look like. They are presented in the Figs. 5 and 6.

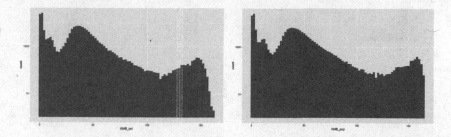

Fig. 5. Histograms of a current consumption for left and right engines from the 8th to the 98th section

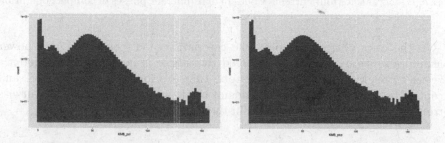

Fig. 6. Histograms of a current consumption for left and right engines from the 98th to the 8th section

There exists a simple explanation why there is a different characteristic of the power consumption during moving the shearer from the 8th to 98th section and back. It is due to the fact that in this case shearer tears off the coal only while moving from the first sections to the last. In the opposite direction it just comes back to the beginning of the longwall.

To stress the differences between following phases of longwall complex operation the charts presenting histograms of four currents in four different aspects of longwall complex operation are presented in the Fig. 7. Rows represent the same aspect for four engines and the following rows reflect the characteristic for the sections at the beginning of the longwall, for the sections at the end of the longwall, for the middle sections while the tearing off the coal and finally for the middle sections while the complex was going back to the beginning of the longwall.

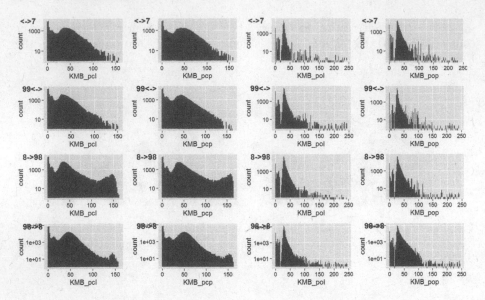

Fig. 7. Histograms for four currents values in four different phases of complex operation

On the basis of available data the correlations between all engines power consumptions were checked. It occurred that there is no significant difference between correlations measured in the total time of a longwall complex work and except of bound operations on all ends of the longwall. The charts of pairwise correlations comparison is presented in the Fig. 8.

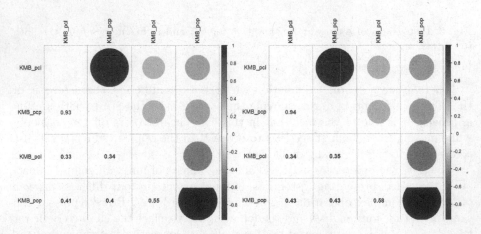

Fig. 8. Correlations between all currents (left) and only from currents of the middle sections (right).

7   Conclusions and Further Works

The necessity of the modern mining is the implementation of innovative diagnostic systems, which are based on approved methodologies of machine parks management and combine innovative technical solutions in the field of industrial automation system and information technology. The system, which was developed as a part of the project, will have such features and it will be based on the adaptation of methodology aimed at improving the overall effectiveness of coal exploitation. The system application will enable to control the rational use and machines operation. It will lead to improved exploitation efficiency, which in practice should translate into extend time of their use, reduced failures and downtime, and the proper organization and realization of work related to their operation and maintenance.

The article presents the issue which is a part of this area and which is an example of the practical application of the advanced informatics tools in the economy. The purpose of the analysis conducted and presented in the paper was to show that it is almost impossible to assess reliably the effectiveness of the use of the machines without using the modern tools, such as data warehouse. The presented example confirmed it.

The material shown in this paper represents a small part of the research conducted in the area of effectiveness of the mining machines use. The results clearly show that the diagnostic data obtained from the studied machinery in order to assess the level of their use, is the most credible and reliable source of information about their work. This data, combined with great analytical opportunities referring to the developed data warehouse, gives a chance to effective analyze and use the results practically.

Presented example shows how the recorded values of current consumed by the longwall shearer motors can be used to analyze the level of its use. Linking these values with the longwall shearer position enables to determine its load depending on the phase of work.

The results confirmed the typical pattern of the longwall shearer work, which shows that there is much higher loading of the engines in slotting phase. The determined correlations of load between the shearer engines confirm these patterns as well.

Histograms, presented in the paper, show summary durations of specific values of current consumed by the individual longwall shearer engines in the studied time. On the basis of these results it is possible to precisely determine the load of the longwall shearer and evaluate the places and the areas where there is its overload. However, the area of interpretation of the results is much broader and dependent on the expectations of the potential users of this information.

To sum up, it is justified to claim that the results confirm the validity of the methodology of research and they can be a solid basis for conducting the process of optimization of the use of the machines in the mining industry.

Acknowledgments. This article is the result of the research project No. PBS3/B6/25/2015: "Application of the Overall Equipment Effectiveness method to improve the effectiveness of the mechanized longwall systems work in the coal exploitation process", realized in 2015–2017, financed by NCBiR.

References

1. Brodny, J., Stecuła, K., Tutak, M.: Application of the TPM strategy to analyze the effectiveness of using a set of mining machines. In: 16th International Multi-disciplinary Scientific GeoConference SGEM 2016, pp. 65–72 (2016)
2. Głowacz, A.: Diagnostics of synchronous motor based on analysis of acoustic signals with the use of line spectral frequencies and K-nearest neighbor classifier. Arch. Acoust. **39**(2), 189–194 (2015)
3. Kabiesz, J., Sikora, B., Sikora, M., Wróbel, Ł.: Application of rule-based models for seismic hazard prediction in coal mines. Acta Montanist. Slovaca **18**(3), 262–277 (2013)
4. Kabiesz, J.: Effect of the form of data on the quality of mine tremors hazard forecasting using neural networks. Geotech. Geol. Eng. **24**(5), 1131–1147 (2006). http://dx.doi.org/10.1007/s10706-005-1136-8
5. Kacprzak, M., Kulinowski, P., Wędrychowicz, D.: Computerized information system used for management of mining belt conveyors operation. Eksploatacja i Nieza-wodność – Maint. Reliab. **50**(2), 81–93 (2011)
6. Kalisch, M., Przystałka, P., Timofiejczuk, A.: Application of selected classification schemes for fault diagnosis of actuator systems. In: 2014 Federated Conference on Computer Science and Information Systems, pp. 1381–1390 (2014)
7. Leśniak, A., Isakow, Z.: Space-time clustering of seismic events and hazard assessment in the Zabrze–Bielszowice coal mine, Poland. Int. J. Rock Mech. Min. Sci. **46**(5), 918–928 (2009)
8. Mazurek, W.: OEE indicator – theory and practice (in Polish) (2014). http://www.neuron.com.pl/pliki/oee.pdf. Accessed 06 Dec 2016
9. Michalak, M., Sikora, M., Sobczyk, J.: Analysis of the longwall conveyor chain based on a harmonic analysis. Eksploatacja i Niezawodność – Maint. Reliab. **15**(4), 332–336 (2013)
10. Michalak, M., Sikora, B., Sobczyk, J.: Diagnostic model for longwall conveyor engines. Adv. Intell. Syst. Comput. **391**, 437–448 (2016). http://dx.doi.org/10.1007/978-3-319-23437-3_37
11. Nakajima, S.: Introduction to TPM: Total Productive Maintenance. Productivity Press, Portland (1988)
12. Ng, K., Chong, K., Goh, G.: Improving overall equipment effectiveness (OEE) through the six sigma methodology in a semiconductor firm: a case study. In: IEEE International Conference on Industrial Engineering and Engineering Management, pp. 833–837 (2014)
13. Pomorski, T.: Managing overall equipment effectiveness OEE to optimize factory performance. In: 1997 IEEE International Symposium on Semiconductor Manu-facturing Conference Proceedings (Cat. No. 97CH36023), pp. A33–A36 (1997)
14. Ramlan, R., Ngadiman, Y., Omar, S., Yassin, A.: Quantification of machine performance through overall equipment effectiveness. In: 2015 International Symposium on Technology Management and Emerging Technologies, pp. 407–411 (2015)

15. Sikora, M., Sikora, B.: Improving prediction models applied in systems monitoring natural hazards and machinery. Appl. Math. Comput. Sci. **22**(2), 477–491 (2012). http://dx.doi.org/10.2478/v10006-012-0036-3
16. Singh, R., Gohil, A.M., Shah, D.B., Desai, S.: Total productive maintenance (TPM) implementation in a machine shop: a case study. Procedia Eng. **51**, 592–599 (2013). http://www.sciencedirect.com/science/article/pii/S1877705813000854
17. Stecuła, K., Brodny, J.: Application of the OEE model to analyze the availability of the mining armored face conveyor. In: 16th International Multidisciplinary Scientific GeoConference SGEM 2016, pp. 57–64 (2016)

Computer Software Supporting Rock Stress State Assessment for Deep Coal Mines

Sebastian Iwaszenko[✉] and Janusz Makówka

Central Mining Institute, Pl. Gwarków 1, 40-166 Katowice, Poland
siwaszenko@gig.eu

Abstract. Underground deep mining always influences the surrounding rock mass. The excavated voids cause changes in stress fields in their neighbourhood, which can lead to rock bursts. This can cause very serious threat to working people and equipment. Thus properly determined or predicted rock mass state is a crucial information for many purposes in mining activity. Not only does it allow to identify potential rock burst hazard in advance, but it also allows to design the parameters of mining technologies and excavation schedule to minimize the risk. One of the most important parameters describing the rock conditions is the seismic wave propagation anomaly, related to the rock mass stress state.

Methods for its prediction have been researched in GIG for over 40 years. The article presents the computer software for calculating a seismic waves propagation anomaly and stress state anomaly. The calculation algorithm based on the method developed in GIG was designed. The algorithm was implemented using C++ language and became a crucial component for calculation supporting computer software. The software was developed as a Windows application and uses Microsoft SQL Server as the database management system. It also allows importing input data from several file formats. The special attention was paid to appropriate handling of spatial and temporal information. The system is capable of visualizing the calculation area as well as exporting the results in Surfer format for further analysis. The developed software is a valuable tool supporting prediction of rock burst threats in deep coal mines.

Keywords: Database · Software design · Relational data model

1 Introduction

Despite the development of novel, green energy sources the fossil fuels will still fulfill the energy demands for several years. The hard coal is one of most commonly used fuels for the power industry. However, special attention has to be paid to both its excavation and utilization. The coal is widely known for its adverse effect to the environment. Recently several research projects have been done around the development of so called Clean Coal Technologies (CCT). Among them, the coal gasification is one of most interesting. The process was targeted not only by the experimental work [11], but was also widely investigated by

© Springer International Publishing AG 2017
S. Kozielski et al. (Eds.): BDAS 2017, CCIS 716, pp. 466–479, 2017.
DOI: 10.1007/978-3-319-58274-0_37

numerical modeling. Significant work was also dedicated to the appropriate visualisation of the process [12]. In spite of the aforementioned attempts to develop a new technology for coal processing and utilization, excavation of the underground deposits in deep mines will still be necessary.

Working in deep coal mines is continuously connected with exposition to several hazards. Methane, water, explosive dusts or rock burst are not only common but also very dangerous. One of methods of natural hazards risk limitation is application of ICT solutions. Information and database systems have already been used to monitor and visualize safety parameters [8]. The information systems can analyze and visualize data gathered from a network of sensors spread through the mine galleries. It is possible to identify the potential threat by analyzing the stream of data. However, once the threat is identified, there is very little time for human and equipment rescue. It is also difficult to guess if any particular mining technology application can increase or decrease the probability of an observed effect. Therefore wide attempts are made for using modeling and numerical methods for the prediction of threats.

The paper presentes a computer application and database backend dedicated for seismic waves and stress anomaly calculation. Developed software solutions help with prediction of rock burst hazard and is dedicated as a supporting tool for experts dealing with safety in deep mines.

2   Determination Stress State Anomaly in the Rock Mass

Stress state in rock mass is one of basic parameters in the point of view of mining openings' stability, a proper design of roof support and safety of miners. In the scale of a mine it mainly comes from disturbances generated by the mining of minerals. Mining is simply taking out the mineral and creating an empty space which is rarely filled out by other material, because it is troublesome and thus expensive. The voids sooner or later are tightened or filled up spontaneously by the surrounding rocks because of the pressure of the adjacent ground. This phenomenon generates successive deformations and coupled stress state alteration, because - imaging this evocatively - the rock parts held up by the volume of the removed minerals have to be carried out by the volumes left.

How to analyze the new state of rock mass with its deformations and stresses? It is possible to apply some of universal numerical methods as finite element method (FEM), boundary element method (BEM) implemented in modeling tools as COSMOS, ANSYS, NASTRAN and many others. Some specialized codes were developed, mainly based on the boundary element method as MUL-SIM NL, developed in 1980's by US Bureau of Mines [15], continued by LaModel in 2000's [5,6]. Another approach, also using BEM, but allowing an easy core code supplementation by additional libraries or routines, is presented in software developed in INERIS as Suit3D [13] and Fault3D [14]. Recently, the codes by Itasca CG as FLAC3D [10] and 3DEC, implementing distinct element method [3], are most commonly used in mining and civil engineering.

Although universal and giving precise results (if well implemented), all the codes are expensive in terms of the model construction and calculations (usually in multiple stages). It is understandable if one realises what large volumes of rock mass become influenced by a single exploitation panel - usually hundreds meters in each direction.

This complexity could be simplified in some situations. For the exploitation of hard coal underground in multi seam coal basins (as in the Upper Silesian Coal Basin which spreads from southern Poland to the northern Czech) it is possible to simplify the numerical model to multiple two dimensional planes with some exploitation panels surrounded by a rock mass with relatively common geomechanical parameters. It opens the possibility to generalise an influence of a single exploitation in one panel and to elaborate a set of characteristics of influence dependent on a few parameters. This is the basic concept for a fast and efficient stress calculation method called SigmaZ.

3   System Structure and Implementation

The computer software was developed based on a set of single edge of exploitation characteristics expressed in the seismic wave propagation anomaly developed by Dubiński [4], reflecting stress state anomaly. It has been the foundation for the calculation engine developed by Kabiesz and Makówka [7].

Let us consider a single excavation area, where coal (or other mineral) is excavated using the long wall exploitation system. As the coal is being removed, the roof falls down, forming the gob area. The exploited part of the coal seam (and probably surrounding strata) forms a polygonal shape (Fig. 1). The polygonal area of exploitation is called a panel. Being a disturbance in the rock mass, the panel is a source of observed anomalies and is a principal object of interest. The seismic waves anomaly along with the rock stress anomaly are distributed outside and inside of the panel. However, there is usually more than one panel formed during the mining operation. The multiple reciprocal interactions in the exploitation areas have to be taken into account. The principle of the calculations is to determine a characteristic of seismic wave anomaly for the given type, size and period of excavation for each panel. After that, their mutual geometrical placement and interactions are considered. The characteristic itself is a function defined on a line which is normal to the panel's edge. It is assumed that a coordinate system is bound to the line, with zero point at the panel's edge and positive values directed outside the panel.

The proposed solutions consists of three main parts:

- database of measured seismic waves anomalies,
- calculations formulas and procedure,
- cases and results database.

Each of the mentioned parts will be described in details in the next sections.

3.1   Measured Seismic Waves Anomalies Database

The measured seismic waves anomalies database is of the crucial importance for calculations. The data describing the anomaly characteristics are of multidimensional nature. They present the deviation from normal state of seismic wave

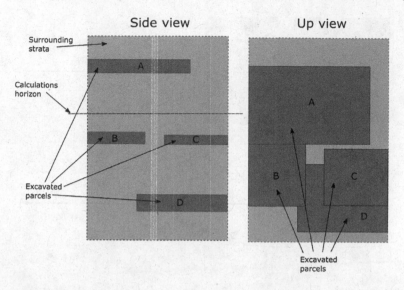

Fig. 1. The excavation panels and relationships between them. The interactions between panel B and C have to be considered: they are close to each other. The top of panel B and panel C is also close to calculation horizon, which have to be taken into account. Panels A and D are far from other panels and calculation horizon - no special treatment is necessary.

along the axis, which is normal to the panel's edge. The shape and amplitude of deviations are dependent on:

- distance from the edge measured along the axis,
- location - the points outside the panel have different characteristic from the ones placed inside,
- the vertical distance between plane containing considered panel and the plane at which anomaly is determined,
- time span from the end of panel exploitation,
- roof control method.

For each combination of quantities listed above, a value of a seismic wave can be determined. The basic characteristics have been elaborated by Dubiński [4] and are stored in a database. The data are organized as a kind of multidimensional cube. The first selector differentiates the data sets according to the roof control methods. The sets for two main used methods: backfill and caving, have been considered. Among each roof control type, 4 timespan values are considered: 1, 3, 5 and 10 years. Within the single year, the data are organized into two dimensional array with an x coordinate denoting the distance from the panels edge and a z coordinate denoting the vertical distance from the panel's layer and the layer containing the point in which the anomaly is calculated. The characteristics approximate to the shape of a given roof control type, timespan and vertical distance are shown in the Fig. 2. Distribution of anomaly value of

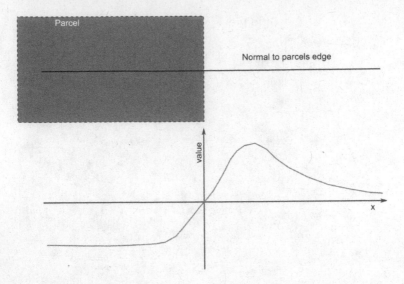

Fig. 2. The seismic wave anomaly characteristics definition.

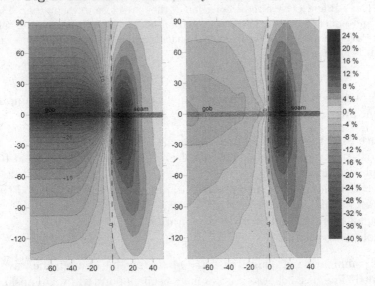

Fig. 3. Distribution of anomaly value of velocity of longitudinal seismic wave A_n in the surroundings of mining by caving edge: (a) for 1 year since mining operation completed, (b) for 10 years since mining operation completed.

velocity of longitudinal seismic wave A_n in the surrounding strata caused by caving edge was presented in Fig. 3.

The database for storing the characteristics is organized as follows. It has to be stressed, that 'read' operations on the database of characteristics are much more frequent than 'write'. In fact, the writing only occurs, when the data of

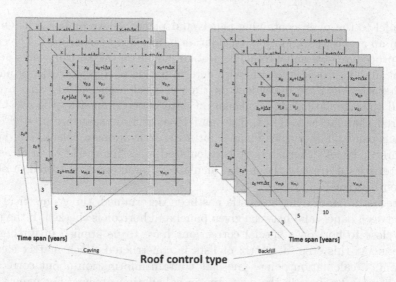

Fig. 4. Data model for storing the characteristics of the seismic wave anomaly.

a new measurement are available. It is hardly possible that such event takes place more often than once in a few years. Moreover, such data should be carefully prepared as they form the foundation for all calculations. Everyday use of the database utilizes 'read' operation. Therefore, the database structure should be optimized for 'read' operations. Furthermore, the integrity of the database requires, that there is only one set of characteristics for the given roof control type and timespan. Therefore the type and timespan of the roof control were chosen as index values for addressing an appropriate subset of characteristics data. As far as characteristic data are concerned, the x coordinate as well as the z coordinate change from the given start value with constant interval (usually different for x and z axis). Therefore, they are organized as a 2D grid. The rows of grid contain sets of values pinned to a chosen z axis coordinate. The columns of the grid are annotated with a x coordinate. Each grid structure is supplemented with a tag describing the x and z coordinate starting value and interval. The idea of the data model is presented in the Fig. 4. The roof control type has been represented as an enumerated type. As far as a time span is concerned, it was chosen to use a integer type for its representation according to the POSIX standard.

3.2 Calculations Formulas and Procedure

The calculations are based on computing the influence each panel has on seismic wave and stress anomaly in given point. It is assumed that all panels can be treated as two dimensional and lays in few horizontal plains, at different depths. Above or below the panels' plain another plain is considered: the layer, called 'horizon', where the seismic wave or stress anomaly has to be determined. It is parallel to the plain containing panels and placed above or below it. On the horizon an evenly spaced grid of calculation points has been defined. The calculation procedure determine total effect of all panels on the horizon grid's points. The calculation procedure starts with determination of characteristic for panels. The characteristics are dependent on time elapsed from end of panel excavation till the calculation date (calculation date is determined by the user). The measured characteristics for 1, 3, 5, and 10 years time span are used (Fig. 4). For each panel a time which elapsed form its excavation to calculation date is determined. Panel's characteristic is calculated as an interpolation of measured characteristics nearest to the date. Calculated characteristic is stored in object representing panel.

After each panel's characteristic has been determined, the geometrical relation between panels and between given panel and horizon is checked. If the panel is too close to horizon a special corrections have to be applied in further calculations. At this step, a matrix of lists is constructed. Each matrix element represents a calculation point. The lists contain information about corrections which have to be done in the next step in calculation process. For panels close to horizon, a new geometry is calculated - they are deflated in proportion to distance to the horizon. The other type of corrections to characteristic is necessary when panels are close each other. Then an internal and external part of characteristic has to be rescaled, proportionally to the distance between panels. Once every panel is checked the corrections, the matrix is prepared for further calculations. Each element pints to the list containing description on operations to be performed by calculating code. The main computation loop iterates for each point in horizon representing grid. For each panel, its influence is calculated. First, its characteristics is interpolated to depict panels distance from horizon. Then the distance from grid's point to panels nearest edge is determined. Using it, the value of seismic wave anomaly is calculated. Then, a correction matrix entry for given point is looked up. Any correcting operations described in the list arc applied, and resulting value is added to the results grid. The panels' loop iterates through all panels, then next horizon point is taken for calculations. The calculation algorithm was described in following pseudo-code.

```
Read Characteristics from Database;
Read Case definition from Database;
For every panel pk in panels' set:
    Calculate characteristics for pk's time span and distance form horizon.;
End
Prepare corrections matrix, corrMatrix;
Prepare results grid RG;
For every point P(x, y) in RG:
    For every panel pk in case panels' set:
    Calculate d as distance from P(x, y) to nearest edg;
    Calc r, the characteristic value for given d value;
    If corrMatrix(P(x, y)) not empty
        Applay corrections described in the list for P(x, y);
    End
    Add r to value stored in P(x, y)
    End;
End;
```

3.3   Data Representation

Calculations of seismic wave and stress anomaly requires handling data representing geometry and objects parameters. The panels were represented as a set of consecutive points forming polygon. The direction of point sequence define interior and exterior of the panel. All points used for panel's definition are stored in an array. Every panel holds a pointer to that array. The indexes of points which form a panel are stored in internal structure in each panel. The panel keeps track of its depth, thickness, rock properties and date of mining excavation. There is also internal array for storing current characteristics. The panel makes its internal structures available using setters and getters. Each panel has its own, unique identifier. The panels are gathered in an array in workset object. The workset represents a calculation case. Apart from storing a collection of panels, it also holds results grid and parameters for calculation case (e.g. calculation date, horizon depth, points' buffer). The workset determine each panel distance from horizon. It is responsible for panel's objects management, points buffer management and organizing result set. Neither panel nor workset perform calculations on its own. There is a dedicated calculation component responsible for computation. It operates on workset object and panels objects. Internally it holds a mentioned correction matrix as well as objects implementing numeric subroutines. The numeric subroutines are accessible through interfaces. The calculation component is configured during startup with dynamically created objects implementing calculation subroutines (c.g. for interpolation). Therefore it is relatively easy to change the behavior of calculation component. Visualization of panels and results is dedicated to graphic engine component. The component use set of classes wrapping data structures such as panel or results grid and visualize them on giver canvas object.

3.4   Cases and Results Database

The measured seismic waves anomalies do not exhaust the necessity for an information storage. There is also a set of data defining each calculation work set. The information can be divided into two parts: the definition of the calculation case and set of results (each calculation case can be associated with many sets of results). Calculation case data includes:

- Panels definition
- Calculation horizon definition
- Calculation case parameters

The set of results contains the value of a seismic wave and stress anomaly calculated in each point of the defined calculation grid. The set of results also contains the information about the panels which were selected for calculations (it is sometimes useful to exclude some of the panels from calculations) and the parameters used for calculations (e.g. depth).

For the storing of the calculation cases and the set of results, a database and an object model of the domain were developed. The object model is presented in the Fig. 5. The interpretation and responsibilities for presented classes are as follows.

CWorkSet: class implements the workset functionality. It holds a collection of panels (implemented as **CPanel** class), collection of results (**CResultSet** class) and points buffer (**CVertexBuffer** class). Objects of **CWorkSet** class represent a calculation case.

CPanel: Class representing a panel. It contains structures for holding panel's geometry (a pointer to **CVertexBuffer** class and indexes array) as well as panels parameters necessary for calculations (depth, roof control type, excavation date, characteristics).

CResultSet: class implements data structures and functionality necessary for managing the calculation's results. Apart from **CGrid2D** object, used for results grid representation, it remembers parameters values used for calculations.

CVertexBuffer: class holds coordinates of all points used in panels definition. The points are stored in the array. The panels use indexes to chose appropriate points. The point is represented by **CVertex** class.

CEdgeCharacteristic: class used for storing and accessing seismic wave anomaly characteristic information. The characteristics itself are stored as **CGrid2D** objects. **CEdgeCharacteristics** functionality include calculating characteristic interpolation for given timespan and xz coordinate.

CGrid2D: class represents a function sampled over regular grid of 2D points. Used for storing results and characteristics.

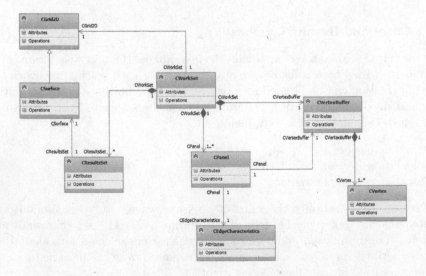

Fig. 5. Object model for calculations sets and results representation.

For the data management of the calculation set a simple relational model was developed (Fig. 6). The database allows a convenient storage and management of the calculations cases. Calculation data is persisted using binary serialization. After serialization, they are stored in binary field 'data'; in database structure. The database structure does not directly allow for manipulating the calculation cases data nor the set of results. The database links the additional, descriptive information with the data of binary serialized calculation cases. It helps users with organizing and managing the simulations and research work performed with the software.

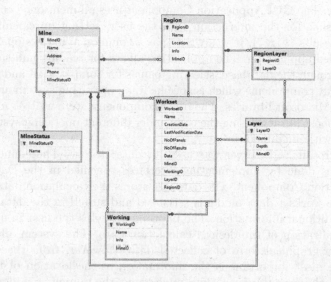

Fig. 6. Relational data model for calculations sets and results representation.

4   Software Design and Implementation

The software was developed using C++ and Mirosoft Visual Studio 2015 IDE. The software consists of two parts:

- End user application
- Database.

The end user application is composed of a few parts. It consists of the Data Structures Component (DSC), the Calculation Engine Component (CEC), the Data Repository abstraction Component (DRC) and the GUI Application Component (AC). All components were designed using the object oriented analysis and design approach [1,2]. The Data Structures Component implements the data model described in the previous sections. Besides storing all the data needed for calculations, the objects are capable of performing a set of basic operations (e.g. recalculating distance from the panel layer to the calculations horizon). They are also responsible for ensuring the consistency in the data. The calculation engine component operates on the data structures defined in DSC. The component is responsible for performing all data transformations and calculations necessary for the calculations of the seismic wave and stress anomaly. The Data Repository abstraction Component is responsible for the serialisation of the data structures and their storing in the repository. The relational database is considered a main and basic repository, but the component also provides services allowing for the storage of the data in the file system. It is also possible to save a part of data (e.g. results) in different format for further processing and visualization. The GUI Application Component uses all mentioned components. It is responsible for the interaction with the user, setting up calculations and presenting results (Fig. 7). It was written as a common Windows operating system controlled application. The application allows for reading and saving data, browsing through the database, selecting panels for computation and so on. The AC is the only component, which is specific to an operating system and uses the proprietary Microsoft libraries. The other components were written using C++, STL and a boost library. Using the code under different operating system should be possible without difficulties.

The Microsoft SQL server was used for data storage. The connection to a database was done by implementing interfaces specified in the Data Repository abstraction Component. The database stores the calculation data as binary objects. The workset data model is serialized and stored in the database along with the additional information. The information helps the user in a quick and efficient localization of the desired calculation data. The system allows coarse preview of a graphical form of collected data. However, from the user's perspective the most important is the information of the location of the case of calculations, the formal information (mine name, the mining area identification) and a few most important parameters.

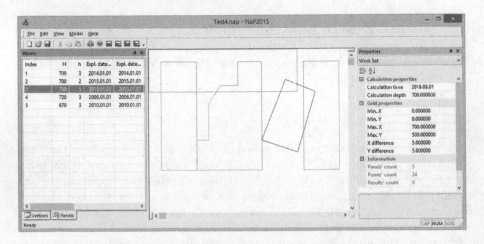

Fig. 7. The main window of developed application.

5   Results and Discussion

The developed methodology and software were tested on several sets of prepared data. Though the data were prepared specially for testing purpose, they depict the panels locations and geometry, which can be found in many real coal mines. There were 24 points and 5 panels defined. The projection of panels are presented in the Fig. 7. The panels share some of their edges. They are located at different depths, though some of them lay in the same plane. The calculation horizon was placed at 700 m below ground level, while panels depths vary form 670 to 720 m. A results grid size was 700×500 m, while the distance between grid nodes were 5 m both in x and y direction. The results set was exported to Golden Software Surfer for visualization. The Fig. 8 presents the obtained results.

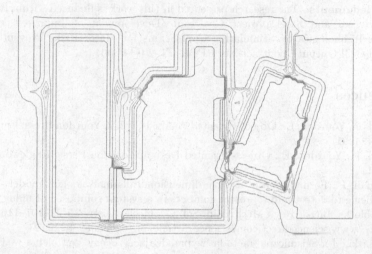

Fig. 8. Example of obtained calculations results - seismic stress anomaly.

The results are in a good agreement to the expectation. The interactions between panels were correctly calculated in most cases. However, some artifacts can be observed (e.g. elevated values in the direction of the common edge of panels 1 and 2). They reveal the limitations of the used methodology. As the method of calculations is still an active research area, it can be expected that artifacts will be removed in the near future. The developed software proved its usefulness in the prediction of the potentially hazardous conditions in the deep mine operations. It is also a convenient tool for the development of the calculations methodology. Due to the application of the object oriented principles, and especially Liskov's substitution principle [9], it is relatively easy to exchange one calculation algorithm with another.

6   Conclusions

The new software for the calculation of seismic waves and stress anomalies caused by an exploitation of mineral deposits (coal) has been developed. The software uses an original method based on measured characteristics. The software is composed of a few cooperating components. A database is used for characteristics and calculation cases data storage. Several test calculations were performed. Obtained calculations results proved correctness of application. However, a few minor artifacts were observed, which shows that further development of the method is necessary. The application along with the database are a useful tool for the prediction of potentially hazardous conditions and helps in improving safety in coal mining.

Acknowledgments. The research presented in this work is financed within Research Fund for Coal and Steel project GasDrain (Development of Improved Methane Drainage Technologies by Stimulating Coal Seams for Major Risks Prevention and Increased Coal Output), contract No. RFCR-CT-2014-00004.

References

1. Coad, P., Yourdon, E.: Object-Oriented Analysis, vol. 2. Yourdon Press, Englewood Cliffs (1991)
2. Coad, P., Yourdon, E.: Object-Oriented Design. Yourdon Press, Englewood Cliffs (1991)
3. Cundall, P.: Formulation of a three-dimensional distinct element model - Part I. A scheme to detect and represent contacts in a system composed of many polyhedral blocks. Int. J. Rock Mech. Min. Sci. Geomech. Abstr. **25**(3), 107–116 (1988). http://www.sciencedirect.com/science/article/pii/0148906288922930
4. Dubiński, J.: Sejsmiczna metoda wyprzedzajacej oceny zagroienia wstrzqsami gérniczymi w kopalniach wegla kamiennego. Prace GIG, Seria dodatkowa, Katowice (1989)
5. Hardy, R., Heasley, K.: Enhancements to the LaModel stress analysis program. In: 2006 SME Annual Meeting and Exhibit, Denver, Colorado, pp. 06–67 (2006)
6. Heasley, K.A.: Numerical modeling of coal mines with a laminated displacement-discontinuity code. Ph.D. thesis, advisor: Miklos D.G. Salamon, Department of Mining Engineering, Colorado School of Mines (1998)
7. Kabiesz, J., Makówka, J.: Empirical-analytical method for evaluating the pressure distribution in the hard coal seams. Min. Sci. Technol. (China) **19**(5), 556–562 (2009)
8. Kabiesz, J., Michalak, M., Iwaszenko, S.: Tworzenie interfejsu uzytkownika w oparciu o szablon MiSS Framework, vol. 30 (2009)
9. Liskov, B., Wing, J.M.: Family values: a behavioral notion of subtyping. Technical report, DTIC Document (1993)
10. Manual, FLAC Users: Itasca Consulting Group Inc., Minnesota, USA (1998)
11. Mocek, P., Pieszczek, M., Swiadrowski, J., Kapusta, K., Wiatowski, M., Stanczyk, K.: Pilot-scale underground coal gasification (UCG) experiment in an operating Mine "Wieczorek" in Poland. Energy **111**, 313–321 (2016). http://www.sciencedirect.com/science/article/pii/S0360544216307046

12. Nurzyńska, K., Iwaszenko, S., Choroba, T.: Database application in visualization of process data. In: Kozielski, S., Mrozek, D., Kasprowski, P., Małysiak-Mrozek, B., Kostrzewa, D. (eds.) BDAS 2014. CCIS, vol. 424, pp. 537–546. Springer, Cham (2014). doi:10.1007/978-3-319-06932-6_52
13. Sylla, M., Al Heib, M., Piguet, J.P.: 3D numerical modelling of underground excavations in a faulted rock mass using the Boundary Elements Method (BEM). In: 53rd Canadian Geotechnical Conference (2000)
14. Sylla, M., Senfaute, G., Al Heib, M., Derrien, Y., Josien, J.P.: Prévision du comportement des terrains sous l'influence des failles par méthodes numériques et microsismiques. In: 28th International Conference on Safety in Mines Research Institutes (1999)
15. Zipf, R.K.: MULSIM/NL theoretical and programmer's manual (1992)

An Ontology Model for Communicating with an Autonomous Mobile Platform

Rafal Cupek[1](✉), Adam Ziebinski[1], and Marcin Fojcik[2]

[1] Silesian University of Technology, Gliwice, Poland
{rcupek,aziebinski}@polsl.pl
[2] Western Norway University of Applied Sciences, Førde, Norway
marcin.fojcik@hvl.no

Abstract. This document presents an ontology-based communication interface dedicated for an autonomous mobile platform (AMP). All data between the Platform and other controllers such as PCs or AMPs are exchanged using standardized services. This solution not only allows the required measurement information and its states to be received from an AMP but also control of the Platform. The first advantage is that all of the information is available through an XML file. The second advantage is better possibility for controlling the AMP using external machines that can monitor the route of the AMP. In the case of avoiding obstacles, an external machine can, with the existing sensors and services, help the AMP come back via the correct route. The structure for the data for the Platform is described as set of standardized services such as information about the existing configuration and the status of any installed sensors. The XML format helps to structure information by adding metadata. To create a fully functioning system, it is necessary to add a semantic model (relations between the elements and services) of the AMP services. This paper describes one possible solution for creating ontology model, using the current configuration, services for monitor and services for control of the AMP.

Keywords: Autonomous systems · Cyber Physical Systems · Ontology · XML · Semantic model · OPC UA

1 Introduction

Cyber Physical Systems are places in which the embedded world meets the Internet world [19]. They deploy embedded cyber capabilities and join them with the physical world, including humans, infrastructure and platforms, that transform interactions with the physical world. The automation pyramid is no longer a canon in industrial IT systems. CPS are formed by networks of distributed operating entities that supply the production process and makes them capable of operating in a smart factory environment through the application of the Internet of Services [25]. Such services can be supported by agent-based

© Springer International Publishing AG 2017
S. Kozielski et al. (Eds.): BDAS 2017, CCIS 716, pp. 480–493, 2017.
DOI: 10.1007/978-3-319-58274-0_38

systems that support flexible and transparent IT services in order to make collaborative manufacturing compliant with CPS [8]. One of the important aspects
of contemporary manufacturing systems are the flexible internal logistics systems that support agile manufacturing [6] by providing support to the flexible
production paths. The recent progress in advanced ADAS (Advanced Driver
Assistance Systems) [28] make the idea of the Autonomous Mobile Platforms
for internal logistics in manufacturing systems more feasible. This paper focuses
on one of the crucial problems related to the application of autonomous vehicles
the problem of communication between autonomous vehicles, human users and
IT systems through flexible and self-describing services that make an AMP an
integral part of CPS.

An ontology is a semantic tool that is understandable by humans and computers that consists of a formalized representation of the knowledge about a
field of discourse, which is defined as "a formal, explicit specification of a shared
conceptualization" [23].

In the field of ADAS, an ontology-based Intelligent Speed Adaptation system
that can detect speeding situations by accessing an ontology-based Knowledge
Base was described in [26]. Ontology models about autonomy levels and situation assessment for Intelligent Transportation Systems have been introduced
for co-driving [18]. An ontology-based traffic model that can represent typical
traffic scenarios such as intersections, opposing traffic, multi-lane roads and bidirectional lanes was introduced in [20].

The ontology proposed in [3] to provide a description of all road entities
along with their interaction allows inferences about the knowledge about the
situation of a vehicle with respect to the environment in which it is navigating.
In paper [27], an ontology-based Knowledge Base and a decision making system
that can assist autonomous vehicles at narrow two-way roads and uncontrolled
intersections were described. Lean ontology close to real-time was described in
[12]. It contains the concepts of vehicles, crossings, crossing traffic lights and
traffic signs and lane and road connections. Vehicles collected, measured and
processed the data. The XML format allows open access to that data. XML is
a widely used markup language [14, 22] that defines a set of rules for encoding
documents in a format that is readable by both machines and humans. The data
in the XML format can be migrated [11] to a relational or NoSQL Database and
then processed.

In order to enable the physical implementation of the proposed ontology and
to convert it to a type of communication services that are easy to implement
in manufacturing system, the authors propose the application of the OPC UA
(IEC 62541) standard. OPC UA (Open Production Connectivity Unified Architecture) is a service-based architecture that relies on Web Services for communicating with enterprise management systems and TCP-based communication
for communication with control and HMI (Human Machine Interfaces). OPC is
maintained by the OPC Foundation [1] and is recommended by the Industry
4.0 guidelines. OPC UA supports data modelling and annotating raw input
data with useful semantic information that supports object-oriented vertical

communication in flexible manufacturing systems [5] and it can be implemented on different platforms including embedded systems that have very low computing resources [7]. All of the above-listed factors mean that although the proposed ontology for AMP can be implemented on a variety of service-oriented communication platforms, the OPC UA solution appears to be the most appropriate for industrial applications.

The rest of this paper is organised as follows: the concept of the multi-layer hardware architecture for the AMP is presented in Sect. 2. The proposed ontology that reflects the hardware model being used and the required logistics services is given in Sect. 3. Some of the implementation details including a representation of data model in XML and its implementation in the OPC UA address space are described in Sect. 4. The conclusions are presented in Sect. 5.

2   Architecture for an Autonomous Mobile Platform

Nowadays is possible to find many solutions of small mobile platforms. They were also created solutions for development control software and simulation for robots, on example Robot Operating System (ROS) [2]. ROS is implemented for Linux Ubuntu LTS. It provides standard services, hardware abstraction with low-level device control, message-passing between processes and others. ROS is not a real-time OS therefore was prepared own application for control and communication with AMP in C with functionality close to a real-time OS.

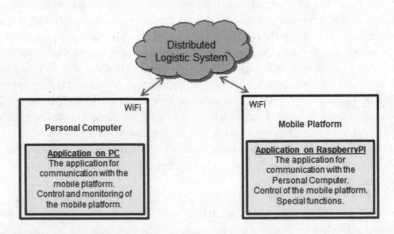

Fig. 1. The main idea of communication with an Autonomous Mobile Platform

The main idea of the proposed solution is to build an autonomous mobile platform that is equipped with a software system that is built on the RaspberryPi platform (Fig. 1). The AMP [28] is supervised, controlled and monitored through a Distributed Logistics System by the software application on a Personal Computer via the Internet. The application that is installed on the PC allows the AMP to be controlled by the services that specify the communication tasks for the AMP. Motion control of the AMP is executed through an application running on RaspberryPi that interprets the signals that are collected from the sensors that are installed on the platform and they also perform the services that are received from the PC. Communication between the PC and the RaspberryPi is performed via a Wi-Fi Internet connection.

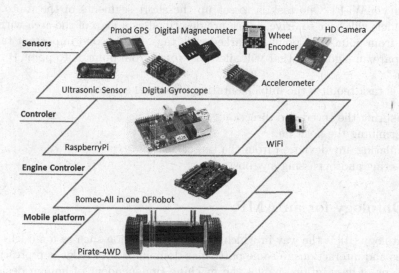

Fig. 2. The hardware architecture of an Autonomous Mobile Platform

The basic service allows data to travel from point A to point B by following a known route on a flat terrain.

An AMP is designed as a modular project that is based on a simple mobile platform, an engine control module and a movement control system. The main assumptions of its hardware architecture are presented in Fig. 2. The authors assume that the mobile platform and engine controls are one of the commercially available solutions and the control system is realized on a RaspberryPi system. Additionally, the system can be equipped with an FPGA circuit, which would allow special functions to be implemented [29].

The movement control system of an AMP allows it to drive a vehicle forward and backward, to turn left and right. In order to achieve this functionality, an AMP can be equipped (depending on the AMP configuration) with a set of sensors [9,21] that includes a:

- 3-axis Digital Gyroscope
- 3-axis Accelerometer
- Magnetometer
- GPS

- Encoder
- Ultrasound
- Camera

A software application for an AMP system is divided between a RaspberryPi and a Personal Computer. The RaspberryPi application is responsible for many individual tasks. The RaspberryPi allows the tasks to be received from the PC system via Wi-Fi. One task is to set up the short segments of the route that the vehicle will have to drive, which are described as a map of the area with the access from point A to point B marked with tags. The second important task is to prepare an algorithm that will allow the route from point A to point B to be driven.

This task requires the implementation of several solutions:

- measuring the speed and direction,
- determining the position,
- calculating any deviation from the target,
- detecting and bypassing any obstacles.

3   Ontology for an AMP

An ontology [10] is the way in which specific information such as a model of the entities and interactions in some particular domain of knowledge is represented. This type of description enables the machine (independent of human decision) interpretability of web content the parameters and the relations between them. There are many types of ontologies that can describe different models. Ontology Web Language is the language for defining Description Logics used in ontology creation [4].

Ontologies are often used in the area processing natural language or unstructured text documents [15]. In [13], an algorithm was presented for creating a mapping between selected semantic features and words. In [16], a classification method of potentially innovative domains acquired from the Internet to estimate whether they represent companies that are innovative or not was described. Another example is fuzzy ontology that is based on Fuzzy Semantic Petri Nets that was used for the automated recognition of complex video events [24].

An AMP is designated to work both indoors and outdoors. Previous experiments showed that it does not always work properly. Sometimes, an AMP cannot find the correct way to its destination. This can happen in a situation in which the way is blocked by obstacles or when the selected sensors are not precise to correct its actual position. Errors in the correct position accumulate, for example, inside buildings. In other situations such as a message from other AMP or service about traffic problems, there is a need to modify the route. A Platform should be ready to communicate with external systems in order to obtain current information or to ask for help. In order to use an AMP in today's Internet of Things [24] world, it is necessary to prepare a description of communication structures to cooperate with web services. An AMP can be equipped with different kinds and types of sensors depending on the needs. Sensors can use different techniques in different units to obtain the measurements.

A single measurement is usually insufficient to control a Platform. An acceleration measurement by itself gives no information about the position, GPS cannot work properly indoors, ultrasound can have problems in rain, etc.

The obvious method for realize communication between different devices is to use the Semantic Web, in which will be possible to use standard protocols and services to transfer, understand and process information from and to an AMP, regardless of its type, the sensors being used for the control methods. To create a Semantic Web, it is necessary to prepare information in a standardized form of ontology. In this situation, the ontology should contain a representation of the elements of the AMP such as its sensors, states, measurements, controls as well as the relations between them. In addition for ontology, we proposed three types of services to allow data exchange for different types of AMPs.

- Discovery – The presentation of the AMP. The Platform announces its Position, Movement, Task, Address and Sensor list. A position is defined by Longitude, Latitude and Altitude. Movement by Azimuth and Speed. Task as ID and Distance.
- Measurements – The AMP send, on demand, the values and information about the selected sensors. Different Sensors have different classes and properties.
- Control – The AMP receives the new target. This can be realized by a direct command such as Right, Left, Forward, Back, Step, Stop or by sending a list with new commands.

For sake of simplicity, this paper does not describe functionality [17] and rules of security such as authentication, rights to control or users priority.

By using the ontology description and web services, the AMP can exchange the necessary information with the control station. Is possible to choose and collect measurements (series of measurements) to control the AMP in the best possible way. The ontology proposed here is the structural framework for organizing the available information. It consists of the concepts in the AMP (sensors, states, commands) and the relations between them. It makes it possible to exchange data automatically regardless of the type of AMP and other equipment. All of the previous ontologies can be combined into one structure.

The AMP address space defines its own types, which are related to the proposed hardware model of an AMP and the software services required by the logistics system. The main idea of the proposed representation is shown in Fig. 3. The graphical description of the AMP ontology corresponds to the one used by the OPC UA standard. The graphic symbols used for representation are given by an Agenda. The Objects reflect the different data structures that are defined by the model. Some of the objects such as the ADAS1, Accelerometer, Magnetometer, Gyroscope, Encoder and Ultrasound are defined by their types, which provide additional information about the internal structure of the modelled devices. The Objects group actual information about the physical values that are available

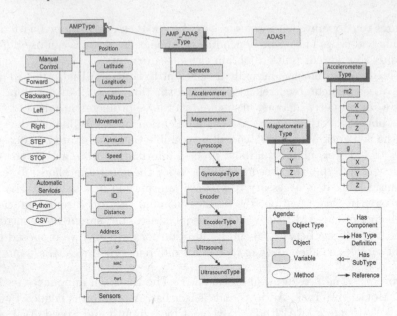

Fig. 3. The ontology for an Autonomous Mobile Platform

on an AMP that are collected and explored by variables. Variables are used to present the actual measurements or the AMP parameters that change during the execution of the required logistics tasks. The dependencies between the objects and variables are defined by a HasComponent reference. The object-oriented structure is also supported by events mechanisms, which means that there is no longer a need for cyclical data pulling and continuous object state checking. The information is automatically sent to the client that has subscribed to a given event and that receives the required message in the event of its evidence. A subscription mechanism is used to track any data changes in a consistent manner. These methods are used to represent the services that are available to the AMP. In the proposed model, services are divided between Manual Control and Automatic Services that are grouped by corresponding objects. The supported logistic services are visible on the left side of the picture.

To ensure the consistency of the information between the different implementations of the AMP, a type hierarchy was proposed. In our case, the ADAS1 object was created as an instantiation of Type AMP_ADAS_TYPE.

AMP_ADAS_TYPE is a child type of a generic AMP platform that is described by AMPType. The additional components that are added by the child type are sensors that can be specific for different implementations of the AMP. In the case of AMP_ADAS, the following sensors are used: Accelerometer, Magnetometer, Gyroscope, Encoder and Ultrasound. Moreover, in the case of the sensors, the relevant types expose their internal structure. The generic information that should be supported by any AMP are defined by the parent type AMPType, which includes information about the AMP location (Position), its speed and directional movement (grouped by Movement object) and details that are relevant to the logistics task (Task) being performed. The Address object contains the details that are necessary to establish communication with AMP.

4 Implementation

A WWW application was created on a PC to prepare an outdoor route for a mobile platform (Fig. 4) that could be exported in a .CSV file. The application in .NET on a PC allows the route data to be imported from a .CSV file and sent to the AMP. The file contains information about the following points along the route, e.g. the position of the AMP, the distance to be driven by the vehicle, the direction of the route and the specification of orders (turn around, ...)

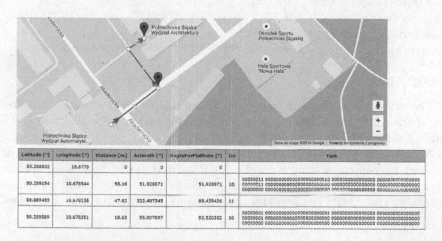

Fig. 4. The WWW application to prepare an outdoor route for the Autonomous Mobile Platform

The route in the .CSV file is described below (with short specification of tasks).

```
50.288832;18.6779;0;0;0;;
50.289154;18.678544;58.16;51.926971;51.926971;10;00000011
50.289493;18.678136;47.62;322.487545;89.439426;11;
50.289589;18.678351;18.63;55.007897;92.520352;10;00000001
```

Additionally, the route can be described in the Python language. The description in CSV and Python allows the Automation Services to be used. The XML file for the AutomaticServices is described below.

```
</AutomaticServices>
        <Python=\"0\"/>
        <CSV=\"0\"/>
</AutomaticServices>
```

Another main functionality is the continuous recording of the route and this information being sent from the AMP to the PC using an XML file. The information that is sent includes the measurement data from the sensors that are implemented on the AMP accelerometer, gyroscope, magnetometer, etc. The XML file for AMP_ADAS_Type, which included the data measured by the AMP, is described below.

```
<AMP_ADAS_Type>
   <timestamp ms=\"3787547561\"/>
   <accelerometer>
     <m2 x=\"0.650181\" y=\"0.0382459\" z=\"8.98779\"/>
     <g x=\"0.0663\" y=\"0.0039\" z=\"0.9165\"/>
   </accelerometer>
   <magnetometer x=\"-3.03\" y=\"178.77\" z=\"-254.52\"/>
   <gyroscope x=\"0.09625\" y=\"-1.07625\" z=\"0.4375\"/>
   <encoder distance=\"59\"/>
   <ultrasound distance=\"45\"/>
   < ... >
</AMP_ADAS_Type>
```

The measurement data are processed by the RaspberryPi and allow the Position of the AMP, the Movement with the actual Speed and the direction on the route (Azimuth) to be specified.

Additionally, the AMP can introduce an Address (IP and MAC address, port number) for communicating with the AMP as well as the task being realized (Task ID the point which it goes from the platform) and the distance from the last point. The AMPType provides these data as an XML file as shown below.

```
<AMPType>
        <Position>
                <Latitude=\"50.289154\"/>
                <Longitud=\"18.678544\"/>
                <Altitude=\"18.67854\"/>
        </Position>
        <Movement>
                <Azimuth=\"58.16\"/>
                <Speed=\"4\"/>
        </Movement>
        <Task>
                <ID=\"2\"/>
                <Distance=\"35\"/>
        </Task>
        <Address>
                <IP=\"192.168.1.0\"/>
                <MAC=\"0023546723C6\"/>
                <PortNumber=\"8000\"/>
        </Address>
</AMPType>
```

Some of the information above is presented on the AMP website that is available at the AMP IP address (Fig. 5).

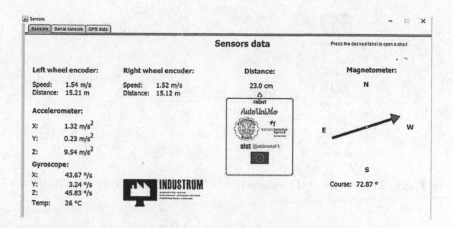

Fig. 5. The measurement data presented by WWW application on the AMP

Additionally, the AMP website contains the commands for the AMP that allow the AMP to be controlled manually in modes STOP or STEP (STEP - step or continuous movement), direction - forward, backward, left or right (Fig. 6). The XML file for ManualControl is described below.

```
<ManualControl>
        <Forward=\"0\"/>
        <Backward=\"1\"/>
        <Left=\"0\"/>
        <Right=\"0\"/>
        <STEP=\"0\"/>
        <STOP=\"0\"/>
</ManualControl>
```

All of this information can be exchanged between the AMP and several PCs, which allows the AMP to be monitored and controlled from several systems. All of these systems can help the AMP get to the destination. The use of the XML format allows for a fast connection and exchange of data between the AMP and other systems including new systems.

The ontology for the AMP presented in Fig. 3 was implemented using the OPC UA address space. The OPC UA address space is presented in Fig. 7. It is available for all OPC UA clients (in our case for UAExpert) by browsing services. The instance of an AMP_ADAS_Type is named ADAS1 and was created in the Objects folder on the server. The ADAS1 object exposes all of the information and services defined by type hierarchy including the variables defined for its parent type AMPType and additional information about the sensors being used

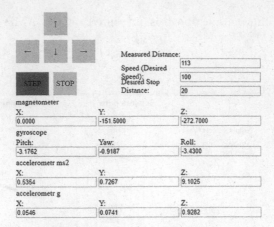

Fig. 6. The control of the vehicle by the WWW application on the AMP

for the specific implementation of the given AMP. The connections between ADAS1 and the other nodes defined by the OPC UA address space is specified by References. References show the type hierarchy and internal components of the object. The Attributes window shows the OPC UA node attributes (in this case, the attributes of the object node) that are used for the description and identification of the node on the server side.

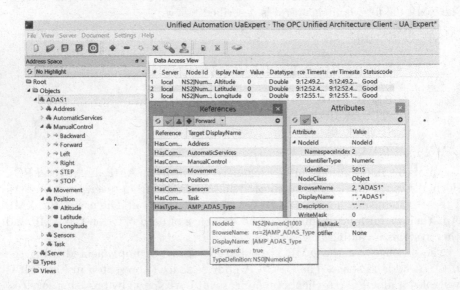

Fig. 7. Implementation of the OPC UA address space based on the AMP ontology

5   Conclusions

The authors propose an ontology-based communication interface that is dedicated for an autonomous mobile platform. The main advantage of the proposed communication interface is its self-descriptive nature. An AMP not only explores the data that is relevant to its state or the services that it is able to execute but also exposes the relations between different parts of the information model that are expressed by classes that are defined according to the proposed ontology. The presented model shows that ontology-based communication services are easier to maintain and to use by any client. Laboratory tests also proved that the presented concept can easily be implemented using well-accepted industrial communication standards. The OPC UA address space was defined according to the presented ontology. The practical tests of simple services (manual commands) were performed by a Human Machine Interface based on a web browser.

Acknowledgements. This work was supported by the European Union from the FP7-PEOPLE-2013-IAPP AutoUniMo project "Automotive Production Engineering Unified Perspective based on Data Mining Methods and Virtual Factory Model" (grant agreement no. 612207) and research work financed from funds for science in years 2016–2017 that are allocated to an international co-financed project (grant agreement no. 3491/7.PR/15/2016/2).

References

1. OPC Foundation: Unified Architecture. https://opcfoundation.org/about/opc-technologies/opc-ua/
2. Robot Operating System. http://www.ros.org/
3. Armand, A., Filliat, D., Ibañez-Guzman, J.: Ontology-based context awareness for driving assistance systems. In: 2014 IEEE Intelligent Vehicles Symposium Proceedings, pp. 227–233. IEEE (2014)
4. Bechhofer, S.: OWL: web ontology language. In: Liu, L., Özsu, M.T. (eds.) Encyclopedia of Database Systems, pp. 2008–2009. Springer, New York (2009)
5. Cupek, R., Fojcik, M., Sande, O.: Object oriented vertical communication in distributed industrial systems. In: Kwiecień, A., Gaj, P., Stera, P. (eds.) CN 2009. CCIS, vol. 39, pp. 72–78. Springer, Heidelberg (2009). doi:10.1007/978-3-642-02671-3_8
6. Cupek, R., Maka, A.: OPC UA for vertical communication in logistic informatics systems. In: 2010 IEEE Conference on Emerging Technologies and Factory Automation (ETFA), pp. 1–4. IEEE (2010)
7. Cupek, R., Ziebinski, A., Franek, M.: FPGA based OPC UA embedded industrial data server implementation. J. Circuits Syst. Comput. **22**(08), 1350070 (2013)
8. Cupek, R., Ziebinski, A., Huczala, L., Erdogan, H.: Agent-based manufacturing execution systems for short-series production scheduling. Comput. Ind. **82**, 245–258 (2016)
9. Czernek, W., Margas, W., Wyzgolik, R., Budzan, S., Ziebinski, A., Cupek, R.: GPS and ultrasonic distance sensors for autonomous mobile platform. Stud. Inform. **37**, 51–67 (2016)

10. Guarino, N.: Formal ontology and information systems. In: Proceedings of FOIS, vol. 98, pp. 81–97 (1998)
11. Harezlak, K., Skowron, R.: Performance aspects of migrating a web application from a relational to a NoSQL database. In: Kozielski, S., Mrozek, D., Kasprowski, P., Małysiak-Mrozek, B., Kostrzewa, D. (eds.) BDAS 2015. CCIS, vol. 521, pp. 107–115. Springer, Cham (2015). doi:10.1007/978-3-319-18422-7_9
12. Hülsen, M., Zöllner, J.M., Weiss, C.: Traffic intersection situation description ontology for advanced driver assistance. In: 2011 IEEE Intelligent Vehicles Symposium (IV), pp. 993–999. IEEE (2011)
13. Jastrząb, T., Kwiatkowski, G., Sadowski, P.: Mapping of selected synsets to semantic features. In: Kozielski, S., Mrozek, D., Kasprowski, P., Małysiak-Mrozek, B., Kostrzewa, D. (eds.) BDAS 2015-2016. CCIS, vol. 613, pp. 357–367. Springer, Cham (2016). doi:10.1007/978-3-319-34099-9_28
14. Kozielski, M.: Multilevel conditional fuzzy c-means clustering of XML documents. In: Kok, J.N., Koronacki, J., Lopez de Mantaras, R., Matwin, S., Mladenič, D., Skowron, A. (eds.) PKDD 2007. LNCS (LNAI), vol. 4702, pp. 532–539. Springer, Heidelberg (2007). doi:10.1007/978-3-540-74976-9_55
15. Maedche, A.: Ontology Learning for the Semantic Web, vol. 665. Springer Science & Business Media, New York (2012)
16. Mirończuk, M., Protasiewicz, J.: A diversified classification committee for recognition of innovative internet domains. In: Kozielski, S., Mrozek, D., Kasprowski, P., Małysiak-Mrozek, B., Kostrzewa, D. (eds.) BDAS 2015-2016. CCIS, vol. 613, pp. 368–383. Springer, Cham (2016). doi:10.1007/978-3-319-34099-9_29
17. Pamula, D., Ziebinski, A.: Hardware implementation of the MD5 algorithm. IFAC Proc. Vol. **42**(1), 45–50 (2009)
18. Pollard, E., Morignot, P., Nashashibi, F.: An ontology-based model to determine the automation level of an automated vehicle for co-driving. In: 2013 16th International Conference on Information Fusion (FUSION), pp. 596–603. IEEE (2013)
19. Poovendran, R.: Cyber-physical systems: close encounters between two parallel worlds [point of view]. Proc. IEEE **98**(8), 1363–1366 (2010)
20. Regele, R.: Using ontology-based traffic models for more efficient decision making of autonomous vehicles. In: 4th International Conference on Autonomic and Autonomous Systems (ICAS 2008), pp. 94–99. IEEE (2008)
21. Rybka, P., Wosik, K., Szczepanski, L., Ziebinski, A., Cupek, R., Wyzgolik, R., Budzan, S.: Power management and sensors handling on the autonomous mobile. Stud. Inform. **37**, 69–87 (2016)
22. Stanek, D., Mrozek, D., Małysiak-Mrozek, B.: MViewer: visualization of protein molecular structures stored in the PDB, mmCIF and PDBML data formats. In: Kwiecień, A., Gaj, P., Stera, P. (eds.) CN 2013. CCIS, vol. 370, pp. 323–333. Springer, Heidelberg (2013). doi:10.1007/978-3-642-38865-1_33
23. Studer, R., Benjamins, V.R., Fensel, D.: Knowledge engineering: principles and methods. Data Knowl. Eng. **25**(1), 161–197 (1998)
24. Szwed, P.: Video event recognition with fuzzy semantic Petri nets. In: Gruca, D.A., Czachórski, T., Kozielski, S. (eds.) Man-Machine Interactions 3. AISC, vol. 242, pp. 431–439. Springer, Cham (2014). doi:10.1007/978-3-319-02309-0_47
25. Tao, F., Cheng, Y., Xu, L., Zhang, L., Li, B.H.: CCIot-CMfg: cloud computing and internet of things-based cloud manufacturing service system. IEEE Trans. Ind. Inform. **10**(2), 1435–1442 (2014)

26. Zhao, L., Ichise, R., Mita, S., Sasaki, Y.: An ontology-based intelligent speed adaptation system for autonomous cars. In: Supnithi, T., Yamaguchi, T., Pan, J.Z., Wuwongse, V., Buranarach, M. (eds.) JIST 2014. LNCS, vol. 8943, pp. 397–413. Springer, Cham (2015). doi:10.1007/978-3-319-15615-6_30
27. Zhao, L., Ichise, R., Yoshikawa, T., Naito, T., Kakinami, T., Sasaki, Y.: Ontology-based decision making on uncontrolled intersections and narrow roads. In: 2015 IEEE Intelligent Vehicles Symposium (IV), pp. 83–88. IEEE (2015)
28. Ziebinski, A., Cupek, R., Erdogan, H., Waechter, S.: A survey of ADAS technologies for the future perspective of sensor fusion. In: Nguyen, N.-T., Manolopoulos, Y., Iliadis, L., Trawiński, B. (eds.) ICCCI 2016. LNCS (LNAI), vol. 9876, pp. 135–146. Springer, Cham (2016). doi:10.1007/978-3-319-45246-3_13
29. Ziebinski, A., Swierc, S.: Soft core processor generated based on the machine code of the application. J. Circuits Syst. Comput. **25**(04), 1650029 (2016)

Data Mining Tools, Optimization and Compression

Relational Transition System in Maude

Bartosz Zieliński[✉] and Paweł Maślanka

Faculty of Physics and Applied Informatics, Department of Computer Science,
University of Łódź, Pomorska 149/153, 90-236 Łódź, Poland
bzielinski@uni.lodz.pl

Abstract. Transition systems in which the state is described by a relational database found applications in artifact centric business process modeling, where the business artifacts are often modeled relationaly. We describe a framework implemented in term rewriting system Maude for specifying and model checking relational transition systems. The system was created to be a part of the future artifact centric business process modeling framework, but is of general interest on its own.

Keywords: Term rewriting · Databases · Transition systems · Business artifacts

1 Introduction

Interactive applications and business processes are commonly modeled as transition systems. To alleviate complexity, the state space of such systems is often structured, e.g., as algebra terms (in case of algebraic formalisms such as π-calculus [21]) or Petri net markings ([23], cf. [24]). In most cases, this state space is used to encode control flow information only. Data operated on, if any, is abstracted away and delegated to other formalisms.

For example, business process specifications in BPMN are concerned only with orchestration of activities within a workflow, and it almost completely ignores the business data modified or accessed by activities, even though this data does influence the execution of activities, e.g., in conditional splits.

This approach has obvious merits. Usually it makes the state space finite and provides a reasonable separation of concerns, e.g., when the structure of a business process is rigid and only losely connected with data. On the other hand, there are cases when those concerns are not so easily separated, particularly for data driven applications, frequently encountered in business practice.

This problem is well recognized (see e.g., [6, 14, 22]) and several solutions were proposed. Let us mention the two most relevant for this work:

Relational transducers [1] model processes as state automatons, where each state consist of instances of several relational databases. External actors interact with the process by inserting rows into tables in input and output databases.

Artifact centric business process modeling [5, 11, 14] is built around, unsurprisingly, *business artifacts*, that is

© Springer International Publishing AG 2017
S. Kozielski et al. (Eds.): BDAS 2017, CCIS 716, pp. 497–511, 2017.
DOI: 10.1007/978-3-319-58274-0_39

[...] business-relevant objects that are created, evolved, and (typically) archived as they pass through a business [11].

An artifact's data model describes components of artifact instances' state. Artifact centric business process modeling methodology is agnostic about data modeling formalism, but a significant amount of research is devoted to the use of relational model and its restricted variations (see e.g., [7,8,12,13]).

A common theme here is that at least part of the specification defines a (labeled) transition system in which states are given by relational databases. This shows a need for support for modeling and testing such systems.

In this paper we present a framework implemented in term rewriting system Maude for specifying and model checking relational transition systems. The system was created to be a part of the future artifact centric business process modeling framework but is of general interest on its own. The implementation of the system is in the "proof of concept" stage, and thus has many limitations. For example, the only queries supported so far are conjunctive queries (non-recursive datalog queries). While this might seem restrictive, some works on artifact-centric business process modeling (see e.g., [8,12]) actually assume this restriction by design, as conjunctive queries are sufficiently expressive for many practical applications, while at the same time enjoying good properties.

1.1 Prior Work

An alternative approach to artifact centred business process modeling was presented in [17], where the user specifies business process and data model in UML. In contrast, our work aims to support business process specification based on enriched process algebras (cf. [25]).

There exist various formalizations of relational databases such as formalization of relational model [4] and the certified implementation of the full relational database system [16]. Note that those formalizations deal with relational model as used in relational databases, whereas our purpose is to model transition systems in which states are described relationally.

In [2,3] an efficient Maude implementation of Datalog is presented. This implementation requires translating each datalog program (including base facts) into specific Maude modules, which is reasonable when evaluating queries against static knowleadge base but awkward when facts dynamically change.

2   Preliminaries

2.1   Term Rewriting and Maude

Maude [10] is a language and execution system based on term rewriting [19,20] and many sorted equational logic [18]. Rewriting systems that can be defined in Maude are of the form $(\mathcal{S}, \Sigma, \mathcal{M}, \mathcal{E}, \mathcal{A}, \mathcal{R})$, where

- $(\mathcal{S}, \Sigma)$ is a many sorted, ordered signature with $\mathcal{S}$ being the poset of sorts and Σ a collection of function signatures of the form $f : S_1 S_2 \ldots S_n \to S$. Connected components of $\mathcal{S}$, referred to as "kinds", correspond to types. We denote by $[S]$ the kind of S. Sorts of a kind are like predicates on the values in the kinds and can be interpreted as (a hierarchy of) subtypes. Variables are also sorted and only terms which are correct with respect to kinds can be formed. It is expected that a term either has the unique smallest sort with respect to subsort ordering or has only kind.
- $\mathcal{M}$ (resp. $\mathcal{E}$) is the collection of implicitly universally quantified conditional membership (resp. equality) axioms of the form "$T : S$ if C" (resp. "$T_1 = T_2$ if C") which declare all terms of the form T to have sort S (resp. terms of the form T_1 and T_2 to be equal). Conditions are conjunctions of equalities and (unconditional) membership declarations. In conditions, equalities of the form $T = \text{true}$ are shortened to T. Equality axioms "$T_1 = T_2$ if C" are interpreted internally as (conditional) rewritings rules of the form "$T_1 \Rightarrow T_2$ if C", and it is required that the rewriting system thus defined by equality axioms is terminating and confluent. Equations define a quotient $T_{(\mathcal{S},\Sigma)}/\mathcal{E}$ of ground terms in signature $(\mathcal{S}, \Sigma)$ by the congruence defined by equations, where every equivalence class can be represented by the (unique) normal form of the term under rewriting system defined by the equations.
- Properties such as commutativity or associativity are easily written as equalities but they do not define terminating and confluent rewriting system. Such properties ($\mathcal{A}$) can be defined in Maude by equational attributes instead.
- $\mathcal{R}$ is the collection of conditional rewriting rules.

Definitions of rewriting systems are collected in Maude modules, either functional (which cannot include rewriting rules and simply define quotient algebras) or system ones which can include rewriting rules. Modules can be parametrized by theories which define properties of classes of systems with which the module can be instantiated.

2.2   Conjunctive Queries and Dependencies

Let us recall some well known notions to fix the notation and terminology.

We assume a fixed set **Val** of atomic values and (at least countable) set of variables **Var**. A database schema $\mathcal{S} = (\mathcal{R}, ar)$ consists of a finite set of relation names $\mathcal{R}$ and arity assignement $ar : \mathcal{R} \to \mathbb{N}$. An instance I over the schema $\mathcal{S}$ is an assignement to each relation symbol $R \in \mathcal{R}$ a relation $R^I \subseteq \textbf{Val}^{ar(R)}$.

Terms are elements of **Val** $\cup$ **Var**. Given a schema $\mathcal{S} := (\mathcal{R}, ar)$, an atom is a an expression of the form $R(t_1, \ldots, t_{ar(R)})$ where $R \in \mathcal{R}$ and each t_i is a term. We often abbreviate list of terms $t_1, \ldots, t_n$ to t. Any partial map $\sigma :$ **Var** $\rightharpoonup$ **Val**, called substitution, can be extended to terms (where the extension is denoted by the same symbol) by setting $\sigma(v) = v$ for $v \in \textbf{Val} \cup (\textbf{Var} \setminus Dom(\sigma))$. A conjunctive query is an expression of the form

$$?(s) : -R_1(t^1) \wedge \cdots \wedge R_n(t^n), \tag{1}$$

where $R_1, \ldots, R_n$ are relation names, and s is a list of terms such that each variable in s also appears in one of the lists $t^1, \ldots, t^n$. Query arity is the size of s. Given an instance I over $\mathcal{S}$, the query Q from Eq. 1 defines a relation

$$Q^I := \{\sigma(s) \mid \sigma : \mathbf{Var} \rightharpoonup \mathbf{Val} \land \sigma(t^1) \in R_1^I \land \cdots \land \sigma(t^n) \in R_n^I\} \qquad (2)$$

which we consider to be the answer to Q.

Tuple generating dependencies (TGD's) are defined by formulas of the form

$$\forall \boldsymbol{x}\big(\Phi(\boldsymbol{x}) \Rightarrow \exists \boldsymbol{y}\Psi(\boldsymbol{x}, \boldsymbol{y})\big), \qquad (3)$$

where $\Phi(\boldsymbol{x})$ and $\Psi(\boldsymbol{x}, \boldsymbol{y})$ are conjunctions of atoms such that $\boldsymbol{x}$ (resp. $\boldsymbol{x}, \boldsymbol{y}$) is the list of distinct variables appearing in $\Phi(\boldsymbol{x})$ (resp. $\Psi(\boldsymbol{x}, \boldsymbol{y})$). Similarly, *equality generating dependencies* (EGD's) are defined by first order formulas of the form

$$\forall \boldsymbol{x}\big(\Phi(\boldsymbol{x}) \Rightarrow t_1 = t_2\big), \qquad (4)$$

where again $\Phi(\boldsymbol{x})$ is a conjunction of atoms such that $\boldsymbol{x}$ is the list of all the variables appearing in $\Phi(\boldsymbol{x})$ which also contains the variables in t_1 and t_2.

3   Relational Transition Systems

Our relational transition systems will be passive, responding only to external actions modeled as transition labels. Let us denote by $\mathcal{I}_{\mathcal{S}}$ the set of all instances of schema $\mathcal{S}$ and by **Rel** the set of all (finite) relations on **Val** of arbitrary arity.

Definition 1. *Relational transition system* $(\mathfrak{S}, \Lambda, \Gamma, \Theta, \rightarrow, \rightarrow_Q, \in)$ *consists of*

- *a database schema* $\mathcal{S} := (\mathcal{R}, ar)$,
- *set* Λ *(resp.* Γ*, resp.* Θ*) of action (resp. query, resp. predicate) labels of the form* $f(\boldsymbol{v}, \boldsymbol{R})$*, where* f *is the primary symbol of the label,* $\boldsymbol{v}$ *is a list of atomic values and* $\boldsymbol{R}$ *is a list of relations (with expected arities in a given position),*
- *a ternary (transition) relation* $\rightarrow \subseteq \mathcal{I}_{\mathcal{S}} \times \Lambda \times \mathcal{I}_{\mathcal{S}}$,
- *a ternary query relation* $\rightarrow_Q \subseteq \mathcal{I}_{\mathcal{S}} \times \Gamma \times \mathbf{Rel}$,
- *a binary predicate* $\in \subseteq \mathcal{I}_{\mathcal{S}} \times \Theta$*. We read* $I \in \theta$ *as "in the instance* I *the condition* θ *holds".*

Actions model changes to the database through the set of predefined operations identified by the primary symbol and parametrized with label arguments. Queries and predicates must be predefined too, and they also can accept parameters through labels. Queries and predicates do not modify the queried instance (provided as the source of $\rightarrow_Q$). The target of $\rightarrow_Q$ is interpreted as the answer. Queries and actions are not assumed to be total or deterministic. Thus:

- a given instance might admit only some of the transitions and queries. If for no instance I' (resp. relation R) $I \xrightarrow{\lambda} I'$ (resp. $I \xrightarrow{\Gamma}_Q R$) then we say that the action λ (resp. query γ) is inactive at the instance I.

– actions and queries need not be deterministic. In both cases non-determinism represents arbitrary choices by the user.

Non-determinism for queries requires some further explanation. Suppose one of the actors is a customer of e-commerce venue. There is an action which adds set of products (passed as the relational argument) to the basket. The products are not arbitrary but chosen from the set of available ones. This choice can be modeled as a non-deterministic query to the product information system. Alternatively, non-deterministic selection could be included in the (now argumentless) "add to basket" action. Which of the approaches is correct depends on the situation but making choice of products and adding them to basket separate events allows for modeling the situation in which when the customer finally makes the decision, the product is no longer available.

4   Relations, Queries and Conditions in Maude

In this section we describe implementation in Maude of relations, deterministic queries and dependencies. To keep the presentation self contained we explain some less obvious elements of the Maude language as they appear in the code. Otherwise, the reader is referred to the Maude Manual [9].

4.1   Relations and Tuples

We represent tuples (`Tuple`) as lists of atomic values. A single value is also a list. For queries and conditions we also need tuples of terms. A term (`ATerm`) is either an atomic value (`AValue`) or a variable (`AVar`). A tuple of values is in particular a tuple of terms (`VTuple`). Each of these syntactic elements has its corresponding sort and the subsort order serves as "is a" relation:

```
sorts AValue AVar ATerm Tuple VTuple .
subsorts AValue < Tuple < VTuple .
subsorts AVar AValue < ATerm < VTuple .
```

We do not implement any domain computations nor type control, but it helps readability to be able to construct atomic values from standard data types such as strings or natural numbers, e.g., with constructors

```
op {_} : Nat -> AValue [ctor]. op {_} : String -> AValue [ctor].
```

Here we use Maude's mixfix syntax, where the places for arguments are indicated by underscores. Thus, {''Bartek''} and {1} are terms of sort `AValue`. Variables are built from Maude's quoted identifiers (`Qid`'s) with `op $ : Qid -> AVar`. Finally, we define lists of values and terms:

```
op nilTuple : -> Tuple .
op __ : VTuple VTuple -> VTuple [assoc id: nilTuple ctor] .
```

We allow for tuples of arity 0 (constant `nilTuple`). A relation of arity 0 can be either empty or contain a single `nilTuple`. Lists are appended with the associative binary operator (with empty syntax) which has `nilTuple` as a neutral element. Those properties are not defined with equations. Instead, they are declared with attributes `assoc` and `id`. Thus, the following is a term of sort `VTuple`:

```
{1} {''aaa''} nilTuple $('A) {''bbb''} {'aaa} {2}
```

All sorts defined in this module belong to the same kind. Maude permits overloading of functions. However, in case of overloading for sorts in the same kind, Maude requires that definitions agree on common elements. Such a subsort overloading enables assigning finer sort for particular values, e.g.,

```
op __ : Tuple Tuple -> Tuple [ditto].
```

makes the result of appending two `Tuple`'s a `Tuple`.

We represent relations directly as finite sets of tuples, but, to simplify code and improve efficiency, we do not check that all the tuples in the relation have the same arity. Operations we define on relations do not depend on this assumption. The predefined `SET` module is parametrized with the `TRIV` theory. In order to instantiate this module we first need to declare the view which links the sort `Elt` in `TRIV` to the intended sort of set elements. In the case of tuples, this is achieved with the declaration

```
view Tuple from TRIV to TUPLE is sort Elt to Tuple. endv
```

where `TUPLE` is the module which defines the sort `Tuple`. Then `Set {Tuple}` is the sort of sets of tuples (the name in the bracket is the name of the view). Similarly we define for later use views `AValue`, `AVar`, and `Relation` (in the latter case `Elt` is linked with `Set {Tuple}`).

As relation names we use terms of sort `RelVar` which can be built with `op R : Qid -> RelVar` from quoted identifiers. One might also use constructors for names prefixed by a namespace. An instance (sort `Inst`) is represented as a map from `RelVar`'s to relations. As maps we utilize predefined `MAP` module, parametrized by `TRIV` theories of keys and values. The following command includes a copy of instantiation of `MAP` for `RelVar`'s as keys and relations as values, with the sort `Map{RelVar, Relation}` renamed to `Inst`:

```
pr MAP{RelVar, Relation} * (sort Map{RelVar, Relation} to Inst).
```

4.2   Syntax of Queries and Conditions

We implement Horn queries (Eq. (1)). Conditions implemented are either tgd's (Eq. (3)) or egd's (Eq. (4)). We need the sorts for atoms, lists of atoms (which represent conjunctions of atoms) equalities, tgd's, egd's, conditions (which are both tgd's and egd's) and queries:

```
sorts Atom AtomList Equality Tgd Egd Condition Query.
subsorts Tgd Egd < Condition.
```

Lists of atoms (including empty ones) are defined in a standard way:

```
subsort Atom < AtomList. op nil : -> AtomList [ctor].
op _,_ : AtomList AtomList -> AtomList [assoc comm id: nil].
```

Atoms are relation names (`RelVar`) followed by a tuple (`VTuple`) of terms:

```
op _(_) : RelVar VTuple -> Atom [ctor prec 3].
```

The queries (Eq. (1)), tgd's (Eq. (3) and egd's (Eq. (4)) are represented with

```
op ?(_):-_ : VTuple AtomList -> Query [ctor].
op [_=_] : ATerm ATerm -> Equality [ctor].
op _=>_ : AtomList Equality -> Egd [ctor].
op _=>_ : AtomList AtomList -> Tgd [ctor].
```

In the second argument of the tgd constructor variables which do not appear in the atoms from the first argument are considered to be existentially quantified.

4.3 Implementation of Query Answering and Condition Checking

We use a variant of the standard conjunctive query answering algorithm which creates, for each matching combination of tuples, a substitution for variables present in the body of the query:

```
pr MAP{AVar,AValue} * (sort Map{AVar,AValue} to VarToVal).
```

Every such substitution (of sort `VarToVal`) is applied to the tuple of terms in the query's head yielding a tuple of values in the answer using the function

```
op sub : VTuple VarToVal -> [Tuple].
```

Its codomain is in the kind and not sort because it can fail (e.g., when not all variables in the first argument appear as keys in the second) returning the constant `tset-error` in the kind `[Tuple]`. With correct arguments $sub((h_1 \ldots h_n), M) = (v_1 \ldots v_n)$, where $v_i = h_i$ if h_i is a value and $v_i = M[h_i]$ if h_i is a variable.

Another function updates the substitution by matching a tuple of values with a tuple of terms returning `match-err` in the kind when matching is impossible:

```
op match : VarToVal VTuple Tuple -> [VarToVal].
```

More precisely, $match(M, H, V) = $ match-err when H and V are of different length and $match(M, \mathrm{nilTuple}, \mathrm{nilTuple}) = M$. Otherwise,

$$
match(M, (h\ H), (v\ V)) = \begin{cases}
match(M, H, V) \text{ if } h : AVar \text{ and } M[h] = v, \\
match(M, H, V) \text{ if } h : AValue \text{ and } h = v, \\
match(M \cup \{h \mapsto v\}, H, V) \\
\quad \text{if } h : AVar \text{ and } M[h] = \text{undefined}, \\
\text{match-err if } h : AValue \text{ and } h \neq v
\end{cases}
$$

We now define the function which returns the set of tuples answering the query (or the constant `tset-error` in the kind if the query is ill formed):

```
op execQuery : Query Inst -> [Set{Tuple}].
```

The first step is to extract the relevant information from the arguments and reformat it in the form which is better suited for efficient rewriting:

```
op $execQuery : [VarToVal] VTuple [MatchList] -> [Set{Tuple}].
op toMatchList : AtomList Inst -> [MatchList].
eq execQuery(?(VT) :- AL, I)
        = $execQuery(empty, VT, toMatchList(AL, I)).
```

The first argument of $execQuery is a substitution to be be filled for each combination of source tuples by matching with VTuple's in the body AL of the query. The function toMatchList converts a list of atoms of the form $R_1(t^1), \ldots, R_n(t^n)$ to a list (of sort MatchList) of the form:

$$\{t^1 \mid R_1^I\}; \cdots ; \{t^n \mid R_n^I\}. \tag{5}$$

Here R_i^I is the set of tuples currently assigned by the instance (Inst) to the relation name R_i. This way we avoid costly searches for keys for every tuple.

The execQuery recurses on the elements of the MatchList:

```
eq $execQuery(M, VT, nilMatchList) = sub(VT, M).
eq $execQuery(M, VT, ({PVT | R} ; ML))
        = gatherTuples(M, VT, PVT, empty, R, ML).
eq $execQuery(MK, VT, MLK) = tset-error [owise].
```

In the base case, when the list is empty, the substitution M is applied to the tuple of terms VT from the head of the query. The step is given by the second equality which delegates iterating over the tuples in R to the function:

```
op gatherTuples : VarToVal VTuple VTuple
            Set{Tuple} Set{Tuple} MatchList -> [Set{Tuple}].
```

The fourth and fifth arguments of this function are, respectively, the "tuples in the query answer so far" (Ans) and the "tuples in the currently considered relation yet to process". The base case is "nothing more to process":

```
eq gatherTuples(M, VT, PVT, Ans, empty, ML) = Ans.
```

The step, in which we process the next tuple T in R, is more complicated:

```
eq gatherTuples(M, VT, PVT, Ans, (T, R), ML)
 = gatherTuples(M, VT, PVT, unionOrError(Ans,
      $execQuery(match(M, PVT, T), VT, ML)), delete(T,R), ML).
```

unionOrError returns the error term if its second argument is an error term. Otherwise, it reduces to the set union of both arguments. Thus, the step adds to ans the tuples returned by the recursive call of $execQuery with the MatchList ML corresponding to as yet unconsidered atoms and the VarToVal substitution enriched by matching with the current tuple T.

Note that iteration over tuples of the current relation is implemented by tail recursion. Maude performs the tail recursion optimization. Recursion over MatchList element is not tail but its depth will usually be very small.

Checking of Tgd's and Egd's for the given Inst is implemented with

```
op verifyTgd : Tgd Inst -> [Bool].
op verifyEgd : Egd Inst -> [Bool].
```

Their codomain is the kind of Bool instead of the sort as they return the error term in the kind if the checked dependency was ill formed.

Terms $\mathbf{verifyTgd}(\mathcal{A} \Rightarrow \mathcal{A}_1, I)$ and $\mathbf{verifyEgd}(\mathcal{A} \Rightarrow E, I)$ are reduced to boolean constants as follows. Each combination of tuples matching the atoms in $\mathcal{A}$ yields a substitution σ. The algorithm returns true if for each such substitution all atoms in $\mathcal{A}_1$ (in case of tgd) or the equality E (in case of egd) is satisfied. Equality $t_1 = t_2$ is satisfied if all the variables in this equality are defined by σ and $\sigma(t_1) = \sigma(t_2)$. List of atoms $\mathcal{A}_1$ satisfies σ if all the atoms in $\mathcal{A}_1$ satisfy σ. The atom $R(t)$ satisfies σ if there exists a tuple $v \in R^I$ which matches with t in the presence of σ, that is iff $match(\sigma, t, v)$ (defined in the previous subsection) does not return an error term. We optimize the algorithm by replacing (using the function toMatchList) the lists of atoms with MatchList's (cf. Eq. (5)).

5 Defining Transitions

5.1 DML Operations and Generic Transition Syntax

Transitions are defined in terms of insert and delete operations:

```
sort Dml.
op insert_<=_ : Atom AtomList -> Dml [ctor].
op delete_<=_ : Atom AtomList -> Dml [ctor].
```

They are evaluated against an instance using the function

```
op exec : Inst Dml -> [Inst].
```

which returns a new instance. Let $\Psi(x)$ be a conjunction of atoms with variables x (represented as an AtomList), and let t be a list of terms. Then

- $\mathbf{exec}(I, \mathbf{insert}\ R(t) \Leftarrow \Psi(x)) = I'$, where I' is identical to I, except that it maps R to $R^{I'} := R^I \cup \{\sigma(t) \mid \sigma : \mathbf{Var} \rightharpoonup \mathbf{Val} \land \Psi(\sigma(x))\}$.
- $\mathbf{exec}(I, \mathbf{delete}\ R(t) \Leftarrow \Psi(x)) = I'$ where I' is identical to I, except that it maps R to $R^{I'} := R^I \setminus \{\sigma(t) \mid \sigma : \mathbf{Var} \rightharpoonup \mathbf{Val} \land \Psi(\sigma(x))\}$.

In case of error, $\mathbf{exec}()$ returns an error term in kind but not sort.

The basic operators and sorts supporting transitions are as follows:

```
sorts Label QueryLabel ActionLabel Comm.
subsorts QueryLabel ActionLabel < Label.
op act : Inst Label -> Comm.
op result : Inst -> Comm [ctor].
op result : Inst Set{Tuple} -> Comm [ctor].
op result : Inst Bool -> Comm [ctor].
```

For each transition system those declarations have to be supplemented with the module defining constructors for action labels (of sort Label) as well as equations and rewriting rules which allow to reduce/rewrite terms of the form $\mathbf{act}(I, L)$ to the terms of the form $\mathbf{result}(I')$ or $\mathbf{result}(I', A)$ where I' is a new instance after the action labeled by L and, in the case of query actions, A is the relation or a boolean value which answers the query. We describe building such a concrete module later in this section.

5.2 Choosing Rows Probabilistically

A common task for the user is to choose values (or rows) from some set, e.g., a student may choose courses. As indicated earlier, we simulate such choices inside non-deterministic actions and queries with probabilistic choice. We use the standard sampling algorithm (see e.g., [15]): Visit every element of an M-element set. If K elements have been chosen after first t visits then select $(t + 1)$'st element with probability $\frac{N-K}{M-t}$. We implement sampling in the module parametrized by the TRIV theory X of the elements of sampled set.

First we need random numbers. The predefined random() function maps natural numbers to the respective elements of a pseudo-random sequence (generated by Mersenne Twister Random Number Generator with the seed chosen at the start of Maude interpreter). The following function maps this sequence into the real interval $[0, 1]$:

```
op rand : Nat -> [Float]. vars N M CN : Nat.
eq rand(CN) = float(random(CN)/4294967295).
```

We need to keep track of which element of the pseudo-random sequence is to be used next, hence the additional (last) argument of the sampling operator sample_of_with_ defined below. The term sample N of S with CN reduces to the pair (of sort CPair) consisting of a random N-element sample of the set S and the first unused index of the pseudo-random sequence. The auxilliary function sampleRec makes the definition tail-recursive:

```
sort CPair . op [_|_] : [Set{X}] Nat -> CPair [ctor].
op sample_of_with_ : Nat Set{X} Nat -> CPair.
op sampleRec : Nat Set{X} Nat Set{X} Nat -> CPair.
vars S C : Set{X} . var E : X$Elt.
eq sample N of S with CN = sampleRec(N, empty, | S |, S, CN).
eq sampleRec(0, C, M, S, CN) = [ C | CN ].
eq sampleRec(s N, C, s M, (E, S), CN) =
```

```
if float((s N)/(s M)) >= rand(CN)
  then sampleRec(N, (E, C), M, delete(E, S), s CN)
  else  sampleRec(s N, C, M, delete(E,S), s CN) fi.
```

In `sampleRec(N, C, M, S, NC)`, M (resp. S) is the number (resp. the set) of elements not yet considered, and N (resp. C) is the number (resp. the set) of elements already selected. Deleting the considered element E is necessary as S may contain duplicates.

5.3   Example Specification

Here we specify a tiny fragment of a paper submission system. The schema contains a single relation symbol **Paper**(*auth, doc*) (in parentheses we indicate the intended meaning of each positional attribute). A given paper (*doc*) may have more than one author (*auth*). The only operation we consider is the submission, labeled submit(*doc, A*), of the paper *doc* written by authors enumerated in the set A. The paper can be submitted iff it has not been previously submitted. The following egd expresses this precondition of submit(*doc, A*) action:

$$\forall auth(\textbf{Paper}(auth, doc) \Rightarrow 1 = 2). \tag{6}$$

The action, when applicable, inserts rows (*auth, doc*) into the **Paper** relation for all $auth \in A$. Using our framework we encode this transition as follows:

```
op submit : AValue Set{Tuple} -> ActionLabel [ctor].
op author : -> RelVar.
var I : Inst . var A : Set{Tuple} . var D : AValue.
crl act(I, submit(D, A)) => result(remove(author, exec(
     insert(author, A, I),
     insert R('Paper) ($('auth) D) <= author ($('auth)))))
  if verifyEgd(R('Paper) ($('auth) D) => [{1} = {2}], I).
```

Note that the instance passed to the $act function is extended with the relation A (containing authors) named by a constant `author` guaranteed not to clash with other names. Now we can test the submission process by rewriting terms like

```
act(i, submit({10}, ( {''Bartosz''}, {''Pawel''}))).
```

where `i` is some initial instance, and we represent the new paper by the atomic value {10}. In order to demonstrate how can we discover errors in specifications using Maude, assume for the moment that we forgotten to add to the rewriting rule for `act` the condition preventing submission of the paper already submitted. First we need to define a rewriting system which repeatedly submits random papers. Then we check if the state in which a duplicate paper is submitted and the submission completed is reachable from the initial state.

```
var N CN : Nat. var A : Set{Tuple}. sort CnComm.
op <_|_> : Comm Nat -> CnComm [ctor].
op nextAct : Inst Nat CPair -> CnComm.
rl < result(I) | CN > => nextAct(I, random(CN) rem 10,
                        (sample 2 of people with (s CN))).
eq nextAct(I, N, [ A | CN ]) = < act(I, submit({N}, A)) | CN >.
```

Here people is a constant equal to a finite set (of 6 names, not shown) from which authors are randomly selected. Document id's are selected randomly from the set $\{0, \ldots, 9\}$. The sort CnComm and its constructor are needed to keep track of the last unused position in the pseudo-random number sequence generated by the random function. Now we can check if it is possible to submit a paper already submitted by executing a command

```
search [1] < result(i) | 0 > =>*
  < act(I:Inst, submit(D:AValue, A:Set{Tuple})) | CN:Nat >
such that not verifyEgd(R('Paper) ($('auth) D:AValue)
    => [{1} = {2}], I:Inst) /\ act(I:Inst,
      submit(D:AValue, A:Set{Tuple})) => result(I1:Inst).
```

In the case of incorrect implementation it finds an example of such double submission. In case of the incorrect one no such solution exists, but because here the system is finite state the command also terminates returning "no solution".

6   Evaluation of Query Answering

We expect that relational states in tested systems will typically be relatively small (even though in principle they can grow unboundedly). Thus, we do not need our framework to be able to efficiently execute queries against instances with thousands or milions of tuples. On the other hand, we still expect the queries to be reasonably efficiently evaluated on small instances. Here we test our framework on a family of instances I_n, indexed by natural numbers, of a schema with a single 2-ary relation symbol E. The instance I_n encodes a binary tree with the set of nodes $\{0, \ldots, n-1\}$ and edges encoded by the tuples of E^{I_n}, where $E^{I_n} = \{(m, \lfloor m/2 \rfloor) \mid 0 < m < n\}$. We executed the query

$$?(x,t) : -E(x,y), E(y,z), E(z,t)$$

against instances I_n for several values of n using a computer with 2.8 GHz Intel i7 processor and 16 GB of memory. The number of rewrites and the time duration of query computation are presented in the table below.

n	Rewrites	Time [ms]
10	2013	0
50	54213	3 ms
100	218463	19 ms
200	876963	113 ms
500	5492463	779 ms

7 Conclusion

We described a framework implemented in term rewriting system Maude for specifying relational transition systems. To the best of our knowleadge it is the first such framework. The framework allows only for conjunctive queries which simplifies the implementation, and is sufficient in most cases. The significance and usefulness of our approach lies in the fact that relational transition systems are in general infinite state and thus the only available automated way to verify such systems is to partially check the correctness through simulation, which our framework (and Maude) is designed for.

The system was created to be a part of the future artifact centric business process modeling framework (and not the replacement for relational database). In particular, we plan to couple it with some implementation of the π-calculus in Maude, modifying the process algebra in such a way that it can perform actions on the relational database in addition to the standard π process transitions.

Our framework is also interesting on its own, as it shows the viability of simulating the relational database within term rewriting system. It also supports probabilistic exploration of reachable states which in case of infinite state systems may provide more effective way of finding errors. Another original aspect of our system is that the actions and queries are parametrised not only by atomic values (as in [8]) but also by relations. This allows for similar expressivity as in the case of relational transducers [1] without introducing a global input database.

References

1. Abiteboul, S., Vianu, V., Fordham, B., Yesha, Y.: Relational transducers for electronic commerce. In: Proceedings of the Seventeenth ACM SIGACT-SIGMOD-SIGART Symposium on Principles of Database Systems, pp. 179–187. ACM (1998)
2. Alpuente, M., Feliú, M.A., Joubert, C., Villanueva, A.: Defining datalog in rewriting logic. In: De Schreye, D. (ed.) LOPSTR 2009. LNCS, vol. 6037, pp. 188–204. Springer, Heidelberg (2010). doi:10.1007/978-3-642-12592-8_14
3. Alpuente, M., Feliú, M.A., Joubert, C., Villanueva, A.: Implementing datalog in maude. In: Peña, R. (ed.) Proceedings of the IX Jornadas sobre Programación y Lenguajes (PROLE 2009) and I Taller de Programación Funcional (TPF 2009), pp. 15–22, September 2009

4. Benzaken, V., Contejean, É., Dumbrava, S.: A Coq formalization of the relational data model. In: Shao, Z. (ed.) ESOP 2014. LNCS, vol. 8410, pp. 189–208. Springer, Heidelberg (2014). doi:10.1007/978-3-642-54833-8_11

5. Bhattacharya, K., Caswell, N.S., Kumaran, S., Nigam, A., Wu, F.Y.: Artifact-centered operational modeling: lessons from customer engagements. IBM Syst. J. **46**(4), 703–721 (2007)

6. Calvanese, D., De Giacomo, G., Montali, M.: Foundations of data-aware process analysis: a database theory perspective. In: Proceedings of the 32nd ACM SIGMOD-SIGACT-SIGAI Symposium on Principles of Database Systems, pp. 1–12. ACM (2013)

7. Calvanese, D., Montali, M., Patrizi, F., De Giacomo, G.: Description logic based dynamic systems: modeling, verification, and synthesis. In: Proceedings of the 24th International Conference on Artificial Intelligence, pp. 4247–4253. AAAI Press (2015)

8. Cangialosi, P., De Giacomo, G., De Masellis, R., Rosati, R.: Conjunctive artifact-centric services. In: Maglio, P.P., Weske, M., Yang, J., Fantinato, M. (eds.) ICSOC 2010. LNCS, vol. 6470, pp. 318–333. Springer, Heidelberg (2010). doi:10.1007/978-3-642-17358-5_22

9. Clavel, M., Duran, F., Eker, S., Lincoln, P., Marti-Oliet, N., Meseguer, J., Talcott, C.: Maude Manual (Version 2.6) (2011)

10. Clavel, M., Durán, F., Eker, S., Lincoln, P., Martí-Oliet, N., Meseguer, J., Talcott, C.: The maude 2.0 system. In: Nieuwenhuis, R. (ed.) RTA 2003. LNCS, vol. 2706, pp. 76–87. Springer, Heidelberg (2003). doi:10.1007/3-540-44881-0_7

11. Cohn, D., Hull, R.: Business artifacts: a data-centric approach to modeling business operations and processes. Bull. IEEE Comput. Soc. Tech. Comm. Data Eng. **32**(3), 3–9 (2009)

12. De Giacomo, G., De Masellis, R., Rosati, R.: Verification of conjunctive artifact-centric services. Int. J. Coop. Inf. Syst. **21**(02), 111–139 (2012)

13. Hariri, B.B., Calvanese, D., De Giacomo, G., De Masellis, R., Felli, P.: Foundations of relational artifacts verification. In: Rinderle-Ma, S., Toumani, F., Wolf, K. (eds.) BPM 2011. LNCS, vol. 6896, pp. 379–395. Springer, Heidelberg (2011). doi:10.1007/978-3-642-23059-2_28

14. Hull, R.: Artifact-centric business process models: brief survey of research results and challenges. In: Meersman, R., Tari, Z. (eds.) OTM 2008. LNCS, vol. 5332, pp. 1152–1163. Springer, Heidelberg (2008). doi:10.1007/978-3-540-88873-4_17

15. Knuth, D.: The Art of Computer Programming. Seminumerical Algorithms, vol. 2, 3rd edn. Addison-Wesley, Reading (1997). xiv+762 pp., ISBN 0-201-89684-2

16. Malecha, G., Morrisett, G., Shinnar, A., Wisnesky, R.: Toward a verified relational database management system. In: ACM Sigplan Notices, vol. 45, pp. 237–248. ACM (2010)

17. Merouani, H., Mokhati, F., Seridi-Bouchelaghem, H.: Formalizing artifact-centric business processes-towards a conformance testing approach. In: ICEIS (2), pp. 368–374 (2014)

18. Meseguer, J.: Membership algebra as a logical framework for equational specification. In: Presicce, F.P. (ed.) WADT 1997. LNCS, vol. 1376, pp. 18–61. Springer, Heidelberg (1998). doi:10.1007/3-540-64299-4_26

19. Meseguer, J.: Conditional rewriting logic as a unified model of concurrency. Theor. Comput. Sci. **96**(1), 73–155 (1992)

20. Meseguer, J., Rosu, G.: The rewriting logic semantics project. Theor. Comput. Sci. **373**(3), 213–237 (2007). http://www.sciencedirect.com/science/article/pii/S0304397506009042

21. Milner, R.: Communicating and Mobile Systems: the pi Calculus. Cambridge University Press, Cambridge (1999)
22. Reichert, M.: Process and data: two sides of the same coin? In: Meersman, R., Panetto, H., Dillon, T., Rinderle-Ma, S., Dadam, P., Zhou, X., Pearson, S., Ferscha, A., Bergamaschi, S., Cruz, I.F. (eds.) OTM 2012. LNCS, vol. 7565, pp. 2–19. Springer, Heidelberg (2012). doi:10.1007/978-3-642-33606-5_2
23. Van Der Aalst, W.M.P., Ter Hofstede, A.H.M.: YAWL: yet another workflow language. Inf. syst. **30**(4), 245–275 (2005)
24. Wibig, M.: Optimisation of business processes using petri nets and dynamic programming. Stud. Inform. **31**(2B), 173–180 (2010)
25. Zieliński, B., Sobieski, Ś., Kruszyński, P., Sysak, M., Maślanka, P.: Object π-calculus and document workflows. In: Bellatreche, L., Manolopoulos, Y. (eds.) MEDI 2015. LNCS, vol. 9344, pp. 227–238. Springer, Cham (2015). doi:10.1007/978-3-319-23781-7_18

A Performance Study of Two Inference Algorithms for a Distributed Expert System Shell

Tomasz Xięski[(✉)] and Roman Simiński

Institute of Computer Science, University of Silesia, Katowice, Poland
{tomasz.xieski,roman.siminski}@us.edu.pl

Abstract. The rule knowledge-based systems are still popular in the real-world applications and the rules are considered as a standard form of knowledge representation in intelligent information systems. While the number of knowledge-based applications grows, the number of tools for building such systems grows much more slowly. This work is the part of research focused on the development of new methods and tools for building rule-based expert systems. The software components mentioned in this work are the main parts of the distributed expert system shell. The realized implementation assumes, that the inference is performed on the preloaded knowledge base stored in the memory. But such a way of using rule bases may be unrealisable or ineffective for large ones, especially when a weak hardware configuration (mobile applications, embedded systems) is used. In this work the utilization of a database stored procedures is considered. This approach minimizes the network traffic and is independent from the used programming tools—only a connection to the database server is required. The main goal of the experiments was to describe an experimental implementation of the forward chaining inference algorithm (as the stored procedure) and to evaluate this approach in comparison to performing inference on preloaded (real-world) knowledge bases.

Keywords: Knowledge base · Expert system shell · Web application

1 Introduction

The rule representation and forward/backward chaining inference are still popular in real-world applications [2,23]. The rules are considered as a standard output form of data mining methods, and are again an important and useful material for constructing knowledge bases for different types of decision support systems [21]. The number of expert systems' applications grows, but unfortunately the number of tools for building knowledge based systems grows much more slowly. The set of ready to use tools is still restricted to the well-known systems (e.g. JESS [14], CLIPS [4], DROOLS [5], EXSYS [8]), and interesting new systems are not so popular yet (XTT [17], SPHINX [31]). The authors of

© Springer International Publishing AG 2017
S. Kozielski et al. (Eds.): BDAS 2017, CCIS 716, pp. 512–526, 2017.
DOI: 10.1007/978-3-319-58274-0_40

this publication argue that there is still a need for expert system building tools, especially for modern, easy to use systems which are able to work over the WWW network.

This work is the part of research focused on the development of new methods and tools for building knowledge-based decision support systems. The software components presented in this work are the main parts of the distributed expert system shell. The KBExplorer is a WWW application which allows the user to create, edit and share the rule-based knowledge bases [28]. The KBExpertLib is a software library, which allows programmers to implement domain knowledge based systems using the Java programming language [25]. This library makes it possible to run different kinds of inference (classical and modified forward and backward chaining algorithms [19,20,29] on rule-based knowledge bases stored in the KBExplorer database or saved locally in the form of XML files.

The KBExploratorDesktop [20] is a desktop application which allows to analyse knowledge bases created by the KBExplorer. This system utilizes the KBExpertLib library and is implemented as an standard JavaFX GUI program. The KBExploratorDesktop allows the user to perform different kinds of forward and backward chaining inference, rules clustering, and also provides tools for viewing the whole rule base structure and methods of visualization of rules groups. The prototype version of KBExplorer and the demo version of KBExploratorDesktop are available on-line: http://kbexplorer.ii.us.edu.pl. The Fig. 1 presents the main software modules of the proposed distributed expert system shell. The system is still under development, currently the new, enhanced versions of the software are in the test phase.

This article presents selected implementation issues, directly connected with database topic. We assume that the knowledge bases are stored in the database of the KBExplorer system and may be exported to an XML file for offline analysis. For the purpose of this work we are going to perform inference from a client side software. The primary scenario implemented in the KBExpertLib assumes, that the content of a particular knowledge base is retrieved from the XML file or a relational database and is stored in the RAM. Furthermore, the inference is performed on these data structures and the effectiveness of this process depends on the hardware configuration of the local machine and will vary due to the size of the knowledge base.

When we consider software working on a weak hardware configuration (mobile applications, embedded systems) such a way of using rule bases may be unrealisable for large ones. That is why, in this work we also analyse a different approach. We assume that the inference process is being realized fully on the server side. Worth to note is that we considered different ways of the server side implementation—as a PHP module, by the usage of Rest API services and as a utilization of stored procedures (within the database server). The last approach will be described in this work (the two other approaches are still in consideration haven't been completed yet). The utilization of stored procedures minimizes the network traffic, as only a single request is necessary. The main advantage of this approach is independence from the used programming tools—only a connection to the database server is required. The main goal of the experiments was to describe an experimental implementation of the forward chaining inference algorithm (as the stored procedure) and to evaluate this approach in comparison to performing inference on preloaded knowledge bases. Experiments were conducted on knowledge bases, storing respectively 416, 1119, 4438 and 22190 rules.

The remainder of this paper is organized as follows. The related works section provides an overview of currently available web-based expert systems building tools. Section 3 presents background information and the problem description considered in this work. Section 4 outlines the proposed methods and tools—by a description of the proposed software and two implementations of forward chaining inference algorithms. Section 5 presents the experiments and discussion of obtained results. Finally, the conclusions section summarizes the paper.

2   Related Works

Several tools and languages are available for developing decision support systems [26]. Traditional rule based systems were developed as desktop applications and a number of development tools are available for such systems. Meanwhile, web applications have grown rapidly and have had significant impact on the application of a traditional expert system [22]. The detailed discussion and comparison of modern tools goes beyond the scope of this study. Some aspects of such review can be found in [18,22] and also in [6,10,12]—in this work only basic information is presented.

The Acquire system [1] provides an ability to develop web-based user interfaces through a client–server development kit that supports Java and ActiveX controls. The ExSys system [8] provides the Corvid Servlet Runtime and implements the Exsys Corvid Inference Engine as a Java Servlet. Developed at NASA, the C Language Integrated Production System (CLIPS) is a rule-based programming language useful for creating expert systems [4]. Jess is a popular rule engine for the Java platform. JESS or Java Expert System Shell is the skeleton of expert systems developed by Sandia National Laboratories. Jess is written in Java and it is possible to run code in this language using Jess. It uses a syntax similar to Lisp [14]. It is compatible with both the Windows and Unix systems. Rules written using Jess are saved in the form of an XML file which must contain a *rule-execution-set* element [3]. JESS is a rule engine and scripting language, which provides a console for programming and enables basic input/output operations—it cannot be used directly in a web-based application but it is possible to use JESS within the JSP platform.

Another commercial expert system building tool is XpertRule [33], which offers a Knowledge Builder Rules Authoring Studio. The XpertRule KBS interfaces over the Web with a thin client using Microsofts Active Server Page technology. Applications developed using the Knowledge Builder Rules Authoring Studio can be generated as Java Script/HTML files for deployment as Web applications. The Web Deployment Engine is a JavaScript rules runtime engine which runs within a browser. Similar concepts share the eXpertise2Go's Rule-Based Expert System, which provides free building and delivery tools that implement expert systems as Java applets, Java applications and Android apps [7].

Next interesting system is Drools, a Business Rules Management System solution. It provides a core Business Rules Engine, a web authoring and rules management application (Drools Workbench) and an Eclipse IDE plugin for core development [5]. In the literature we can find some attempts to build web-based domain independent expert systems [9,32] as well as specialized domain expert systems, e.g. [15,34]. This work introduces an another decision support system building tool—the KBExplorator system. It is a web application and it allows the user to create, edit and share rule knowledge bases. It is also connected with the KBExpertLib—a software library, which allows programmers to use different kinds of inference within any software projects implemented in Java programming language.

3   Background Information and Problem Description

In this work, the following formal description of a knowledge base is assumed: a knowledge base is a pair $\mathcal{KB} = (\mathcal{R}, \mathcal{F})$ where $\mathcal{R}$ is a non-empty finite set of rules and $\mathcal{F}$ is a finite set of facts. Furthermore, $\mathcal{R} = \{r_1, r_2, \ldots, r_n\}$ and each rule $r \in \mathcal{R}$ will be represented in the form of Horn's clause: $r : p_1 \wedge p_2 \wedge \cdots \wedge p_m \to c$, where m is the number of literals in the conditional part of rule r $(m \geq 0)$, p_i is the i-th literal in the conditional part of rule r $(i = 1 \ldots m)$ and c denotes the literal of the decisional part of rule r.

For each rule $r \in \mathcal{R}$ we define the following functions: $concl(r)$—returns the conclusion literal of rule r: $concl(r) = c$; $cond(r)$—the value of this function is the set of conditional literals of rule r: $cond(r) = \{p_1, p_2, \ldots p_m\}$, $literals(r)$—the value of this function is the set of all literals of rule r: $literals(r) = cond(r) \cup \{concl(r)\}$, $csizeof(r)$—conditional size of rule r, equal to the number of conditional literals of rule r $(csizeof(r) = m)$: $csizeof(r) = |cond(r)|$, $sizeof(r)$—total size of rule r, equal to the number of conditional literals of rule r increased by 1 for a single conclusion literal: $sizeof(r) = csizeof(r) + 1$. We will also consider *facts* as clauses without any conditional literals. The set of all such clauses f will be called *set of facts* and will be denoted by $\mathcal{F}$: $\mathcal{F} = \{f : \forall_{f \in \mathcal{F}} \ cond(f) = \emptyset \wedge f = concl(f)\}$.

For the purpose of this work, rule's literals will be denoted as pairs of attributes and their values. Let A be a non-empty finite set of conditional and decision attributes. For every symbolic attribute $a \in A$ the set V_a will be defined as the set of possible values of attribute a. Attribute $a \in A$ may be simultaneously a conditional and decision attribute. Also a conclusion of a particular rule r_i can be a condition in an other rule r_j. It means that rules r_i and r_j are connected and it is possible that inference chains may occur. The literals of the rules from $\mathcal{R}$ are considered as attribute-value pair (a, v), where $a \in A$ and $v \in V_a$. Furthermore, the notation (a, v) and $a = v$ for symbolic attributes is equivalent. We also consider numeric attributes and literals containing such attributes are represented in the form of an attribute-relation-value triple.

3.1   Inference Algorithm

The software implementation considered in this work is able to perform forward and backward chaining inference. The *KBExpertLib* provides classical versions of the algorithms as well as their modified versions [19,20,24]. In this paper two different implementations of a classical forward chaining inference will be considered. This type of inference is also called *data* or *facts driven*. Each rule is analysed whether all its' premises are satisfied. If that is the case, the rule is activated and its' conclusions are added to the set of facts. The algorithm stops either there are no more rules which can be activated or when the starting hypothesis is added to the set of facts [11,16]. The inference process can be divided into three stages: matching, choosing and execution. At first, the premises of every rule are analysed to match them to the current set of facts – the conflict rules set is selected. Then, if such rules exists, the inference engine selects single rule to activate and at the execution stage, the conclusion of the selected rule is appended to the facts set. The rule for activation is selected according to the conflict resolution strategy (the selection strategies used in proposed inference implementation are briefly described in the Sect. 4.1).

Require: $\mathcal{R}$: the set of rules, $\mathcal{F}$: the set of facts, g: the goal
Ensure: $\mathcal{F}$
    procedure forwardInference($\mathcal{R}$, $\mathcal{F}$, *selStrategy*)
    var R, A : RuleSet
    var r : Rule
    begin
    $A \leftarrow \emptyset$
    select $R \subseteq \mathcal{R} : \forall r \in R, cond(r) \subseteq \mathcal{F}$
    while $R \neq \emptyset \wedge \neg g \in \mathcal{F}$ **do**
      select $r \in R : r$ **where** r **fulfills** *selStrategy*
      $\mathcal{F} \leftarrow \mathcal{F} \cup \{concl(r)\}$
      $\mathcal{A} \leftarrow \mathcal{A} \cup \{r\}$
      select $R \subseteq \mathcal{R} - A : \forall r \in R, cond(r) \subseteq \mathcal{F}$
    end while
    end procedure

Forward chaining inference leads to a massive growth of new facts and a time-consuming rules' matching process, therefore a modification of the forward and backward chaining algorithms was proposed and described in the previous works [13,19,20,24]. The proposed modification of the forward chaining algorithm regards the reduction of the time necessary to realize the matching phase of inference by choosing only the rules from a particular rules' group—detailed description is presented in [19]. The proposed modification of the backward algorithm consists of the reduction of the search space by choosing only the rules from a particular rules' group, according to a generated (decision oriented) rules partitioning and the estimation of the usefulness of each recursive call for sub-goals. Therefore, only the necessary rules are processed and only promising recursive calls of the classical backward chaining algorithm are made.

In this work two custom versions of a classical forward chaining inference implemented in the KBExplorer system will be considered. In the experiments the worst possible case of inference is considered (in which all the rules should be eventually activated) – this allows the authors to evaluate the boundaries of a practical implementation of the proposed solutions. These experiments will be in the future repeated with use of the aforementioned, modified inference algorithms.

3.2 Software Description

The software components discussed in this section are the main parts of the distributed expert system shell. The KBExplorer is a WWW application which allows the user to create, edit and share the rule-based knowledge bases. It works on the client side and requires only the usage of a typical modern web browser. Unlike desktop applications, the KBExplorer system do not have to be installed on a local machine. Each user can register his own account—knowledge bases created by the user are stored in a relational database, and may be shared between the registered system's users. Moreover, it is also possible to download the knowledge base as an XML file (for further analysis). The KBExpertLib is a software library, which allows programmers to implement domain knowledge based systems using the Java programming language. This library makes it possible to run different kinds of inference (classical and modified forward and backward chaining algorithms) on rule-based knowledge bases stored in the KBExplorer database or saved locally in the form of XML files.

The KBExpertLib may be used on the server side, but the main scope of its application are client side desktop programs (developed in Java). The KBExpertLib is an object-oriented library, which classes are divided into the following packages: kbcore—the main, essential classes for representing rules, kbinfer—classes providing the classical and modified inference algorithms, kbpartition—classes allowing the decomposition of rule bases, and kbtools—additional helper classes. The KBExploratorDesktop should be considered as an offline version of the KBExplorator. This system utilizes the KBExpertLib library and is implemented as a JavaFX program. The Fig. 1 presents the main software modules.

The rule-based knowledge bases are usually physically stored in a relational database. Any registered user of the KBExplorer system can create and manage multiple knowledge bases, each one containing a custom set of attributes, their values, facts and rules. Rule bases could be also shared between users—it is possible to share particular rule bases for editing or only in the read-only mode.

3.3 Two Approaches to Inference Implementation

In the recent papers [27,30] the connections with other modern expert systems building tools were discussed. For this reason, in this work the detailed analysis and comparison between existing tools has been omitted. In the paper [30] we also presented a relational database model for the rule-based knowledge base

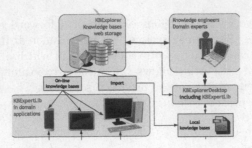

Fig. 1. Main components of the distributed expert systems shell

(presented in the `KBExplorer` system). Three main issues were discussed – a logical model of the rule base for the proposed web expert system shell, the architecture of such a system and the transformation of the proposed model into a relational database. These works also presented the evaluation of the rules retrieval effectiveness (in the proposed model)—experiments were conducted to determine the duration of retrieving of a single rule or group of rules in large rules sets, counting up to 20000 rules. The paper [27] presented experiments concerning the evaluation of time efficiency of loading rules from local XML files into the internal, object oriented data structures (defined in the package `kbcore` of `KBExpertLib`) as well as the estimation of memory usage for such data structures. What is more the effectiveness evaluation of inference algorithms implemented in the `KBExpertLib` library was presented.

In this work an another inference framework is considered. We assume that the knowledge bases are stored in the database of the `KBExplorer` system and we are going to perform inference from a client side software connected with the `KBExplorer` database via Internet. The primary scenario implemented in the `KBExpertLib` assumes, that the content of a particular knowledge base is retrieved from the relational database and is stored in the object-oriented data structures placed in the RAM. The inference is performed on these data structures.

This method is similar to the approach described in the [27], where a rule base is loaded from a local XML file. Unfortunately, the time needed to perform the knowledge base loading from XML files is satisfactory only for small bases. For example, for a knowledge base aggregating 22190 rules the loading time exceeds 8 min [27] and is not acceptable in real-world usages. In this work we try to evaluate the time efficiency of a direct on-line access to the database containing a rule knowledge base. We are going to check whether loading data directly from the relational database (on the fast internet connection) may be an alternative for using local XML files.

The second approach for the inference realization assumes, that the inference process is being realized fully on the server side. It is worth to mention that the authors have considered different ways of the server side implementation—as a PHP module, by the usage of Rest API services and as a utilization of stored procedures (within the database server). Only the last approach will be described in this work as the two other approaches are not completed yet. The utilization of stored procedures minimizes the network traffic, as only a single request is necessary. The main advantage of this approach is an independence from the used programming tools—only a connection to the database server is required.

4   Methods and Tools

This section presents in detail the first of the previously mentioned approaches of inference implementation. The KBExpertLib is a software library dedicated for the Java programming language. It allows the programmers to use knowledge bases and inference algorithms within any software projects implemented in the Java programming language. The inference algorithm works on data stored in computer memory, which must be retrieved from Java beforehand by a series of SQL statements.

4.1   Forward Chaining Inference as the Java Code

The KBForwardInferer is the main class providing forward chaining inference algorithms (previously described). These algorithms use the interface, which gathers simple statistics about the inference process (KBForwardInferTracer). The following example presents the real forward chaining inference code with the usage of the tracer interface. We decided to show the original code, because it clearly represents the whole structure of the inference process:

```
public boolean classicInference( KBInferer.RuleSelStrategy strategy )
{
    if( base == null ) return false;
    if( getGoal() != null && base.isFact( goal ) ) // Exit if the goal is a fact
        return setGoalConfirmed( true );
    tracer.startTracing();          // Starts inference tracing
    tracer.traceInitialFacts();     // Dump the initial facts to the tracer object
    setNewFactsInferred( false );   // We assume inference failure
    setGoalConfirmed( false );
    tracer.traceBegin();            // Starts time measurement
    KBRules matchingRules = new KBRules( base );  // A container for fact matching rules
    base.selectFactsMatchingRules( matchingRules ); // Select fact matching rules
    while( matchingRules.numOfRules() > 0 ) // While matching fact rules set is not empty
    {
        if( goal != null && base.facts.isFact( goal ) ) // Check goal if defined
            break;
        tracer.traceNewIteration(); // Update the iterations counter
        // Appends matching rules to the tracer file
        tracer.traceMatchingRules( matchingRules );
        // Select a rule to be processed
        KBRule rule = selectRule( matchingRules, strategy );
        base.facts.addNewFact( rule.conclusion ); // Appends conclusion to the facts set
        // Appends information about the activated rule to the report
        tracer.traceActivatedRule( rule );
        setNewFactsInferred( true ); // Store information about new fact
        rule.setActivated(); // Set the rule's activation flag
        // Select rules matching to the facts set
        base.selectFactsMatchingRules( matchingRules );
    }
    if( setGoalConfirmed( base.facts.isFact( goal ) ) )
        tracer.traceConfirmedGoal( goal );
    if( newFactsInferred() )
```

```
        tracer.traceNewFacts();
    tracer.traceEnd();
    tracer.traceStatistic();
    tracer.stopTracing();
    return newFactsInferred();
}
```

The detailed descriptions of the algorithm's steeps are included in presented code as the comments. The `KBForwardInferer` actually provides only four simple conflict resolution strategies. The inference algorithm may choose first or last rule from the conflict rules set, alternatively rule with shortest or longest conditional part can be selected (`KBInferer.RuleSelStrategy` strategy allows to select applied strategy).

4.2   Forward Chaining Inference as the Stored Procedure

This section presents the second of the earlier mentioned approaches of inference realization. This approach works in the database server layer as a stored procedure, which uses native properties of the MySQL engine.

Algorithm 1. Forward chaining inference as the stored procedure

---

Data: $R = \{r_1, \ldots, r_n\}$ - rules from knowledge base, which have not been activated yet; $g := (a_j, v_j)$ - inference goal as a descriptor
Result: F - the set of facts given as descriptors;
begin
  Read rules from knowledge base;
  /* For every not-activated rule   */
  foreach r_i *in* R **do**
    /* Check if the inference goal is a fact. If so, exit
       the procedure   */
    if $g \in F$ **then**
    | **return** *Success*
    else
      /* Select the ID of the first rule from the
        knowledge base, which hasn't been activated yet
        */
      $rId \leftarrow \text{selectFirst}(r_i)$;
      if $rId = NULL$ **then**
      | **return** *Failure*
      end
      /* Get the rule data and insert its conclusion to
        the set of facts   */
      $F \leftarrow F \cup \{concl(r_i \text{ with } rId)\}$;
      /* Mark the rule as activated   */
      $R = R \backslash \{r_i \text{ with } rId\}$;
    end
  end
end

Please note, that a simplified version of the inference algorithm is presented, which assumes that any selected rule from the knowledge base may be fired, because we are interested only in the worst-case scenario. The pseudocode of the described approach is presented in Algorithm 1 and the exemplary implementation of stored procedure is as follows:

```
main_infer:BEGIN
  DECLARE rId INT;
  read_loop: LOOP
  #Check if the inference goal is a fact. If so, exit the procedure.
  IF (SELECT COUNT(*) FROM infer WHERE FK_attributeID = 212 AND operator = '==' AND
  FK_valueID = 17) THEN
    LEAVE main_infer;
  END IF;
  #Select the ID of the first rule from the knowledge base, which hasn't been activated yet.
  #If there isn't any, exit the procedure.
  SET rId = (SELECT ruleId FROM rule WHERE wasChecked = 0 AND FK_knowledgeBase = kbID LIMIT 1);
  IF rId IS NULL THEN
    LEAVE read_loop;
  END  IF;
  #Get the rule data and insert its conclusion to the set of facts.
  INSERT INTO infer SELECT FK_attributeID, operator, FK_valueID, continousValue, FK_knowledgeBase
  FROM 'attributeValue', 'attribute' WHERE FK_ruleID = rId AND attributeID = FK_attributeID
  AND isConclusion = 1;
  #Mark the rule as activated.
  UPDATE rule SET wasChecked = 1 WHERE ruleId = rId;
  END LOOP;
END
```

The above-metioned implementation assumes that the initial set of facts is empty and the user has set the inference goal to a descriptor expressed internally as a pair of $attributeID = 212$ and $valueID = 17$ (which is also the conclusion of the last rule in the knowledge base). If the goal of the inference already belongs to the set of facts, it is confirmed and the procedure ends. This is not the case, so the procedure select the ID of the first (not analysed) rule from the knowledge base (as only such a simple strategy was implemented in this version). If such a rule exists, its conclusion is added to the set of facts and its marked as checked. If all rules were checked the procedure ends.

5   Experiments and Discussion

This section presents the experiments performed on four real-world knowledge bases. The first used knowledge base (bud4438) is used by a builder company in Poland and currently consists of 4438 rules. There are 2802 symbolic attributes and 51 numeric. The formed (by domain experts) rules were generally very short, because the conditional part of a given rule was between one and eighteen literals long. The average length of the stored rules (defined as the number of conditional literals) was only 2,66. This is because the creators of the knowledge base were experts from the building or architecture fields and their understanding of the decission support systems field and related concepts was minimal. This also means that the structure of the knowledge base was flat (as generally such bases should be hierarchical). The second base (bud22190) was generated by duplicating the first one five times with random modifications (because gaining access to large, real-world knowledge bases is very difficult). The third (eval416) and fourth (eval1199) knowledge bases regarded the topic of effectiveness evaluation of sales representatives and consisted of 416 and 1119 rules. More information about the structure and experimental evaluation of rules partitioning concerning these knowledge bases can be found in [27].

The aim of the first experiment was to evaluate the loading times and memory usage of the knowledge base exported to an XML file for offline analysis. Furthermore data loading time form a relational database was also considered. The XML file format was chosen by the authors because of its clear advantages such as being technology agnostic, human readable, extensible and allows quick validation. Being plain text, XML is technology independent and can be used by any technology for data storage or transmission. The simplicity of the XML file format ensures that it is easily human readable and understandable. What is more, custom tags can be created very easily and by using XSD as well as proper namespaces the generated structure can be validated in an efficient manner. Unfortunately, this format has also one big disadvantage which is high redundancy. Normally XML file contains a lot of repetitive terms, which implies high storage requirements (and potentially increased transmission costs). This fact is also confirmed by the results shown in Table 1.

Table 1. Knowledge base mass storage usage and average loading times

Knowledge base	Loading time from XML [s]	Loading time from database [s]	XML file size [B]	XML file line count	RAM usage [B]
eval416	0,810	2,092	480	12537	95964
eval1199	3,643	5,185	1153	39976	285316
bud4438	45,570	29,238	6797	167118	1197148
bud22190	516,551	188,195	27241	667344	4646348

Results presented in Table 1 clearly indicate, that because of the chosen structure[1] the file size and line count can be (and usually is) much bigger than the stored rules count. It is especially visible for the bud4438 and bud22190 knowledge bases, because they have a lot of attributes which definitions (and a list of possible values) are also stored besides the rule set. An increased file size implies also higher loading times. A loading time of nearly nine minutes (from the XML file) for the biggest database can be regarded as significant. Loading times from the database are better for bigger knowledge bases, but still can be

---

[1] The chosen XML structure was discussed in more detail in our previous publications such as [27,28].

treated as high. That is why the authors wanted to check if performing operations directly on the database server can be an alternative to having to wait for the data loading phase to finish in order to perform e.g. inference in the client's application.

The goal of the second experiment was to compared two (previously described in this work) methods of forward chaining inference—a classical version implemented in the Java programming language and one that operates directly on the database server. The results of this experiment are presented in Table 2.

Table 2. Analysis of rules' retrieval time (in seconds) based on their identifiers

Knowledge base	Rule count	Data source	Minimum	Maximum	Mean	Median	σ
eval416	416	Database	16,1710	16,7098	16,5007	16,5365	0,2252
eval416	416	RAM	0,0042	0,0056	0,0050	0,0050	0,0005
eval1119	1119	Database	45,3419	46,9881	46,3444	46,5064	0,6963
eval1119	1119	RAM	0,0206	0,0709	0,0300	0,0217	0,0201
bud4438	4438	Database	7,1753	10,1816	8,5748	8,5324	1,1248
bud4438	4438	RAM	0,2334	0,2828	0,2487	0,2376	0,0205
bud22190	22190	Database	94,0995	96,0619	95,2647	95,5558	0,8120
bud22190	22190	RAM	10,3705	10,8686	10,7426	10,6849	0,1806

It is obvious that inference realized on objects placed in the RAM of a client's computer will be faster than performing the same process directly in the database layer (even when using proper column indexes). But the results in Table 2 should be interpreted in the context of knowledge base loading times (see Table 1). Then one can observe an advantage of the stored procedure method (in case of knowledge bases bud4438 and bud22190) compared to running inference on the desktop application. In case of the Java program the user has to wait a time period of about 29–45 (in case of bud4438) or 188–516 (in case of bud22190) seconds for the database to load and additionally 0,2 or 10 s for the inference process to finish. When performing inference directly on the database this time is reduced to a bit over 7 or 94 s respectively. This means that the user will be able to perform inference at least two to five times (directly on the database) whereas the data would still be loading in the desktop application. Of course when the user plans to perform inference multiple times it is still better to load the data into the desktop application, because it's a one-time operation only. As far as the eval416 and eval1119 knowledge bases are concerned, the direct database approach performs worse than the traditional one. This is caused by the structure of these knowledge bases. Although they store less rules, they form a hierarchical structure, and a chain of rules is activated and analysed instead of only a single rule like in the case of the bud4438 and bud22190 bases which have a flat structure.

6   Conclusions

In this work two approaches for inference implementation, which uses knowledge bases stored in the form of an XML file or relational database, were introduced.

The first approach requires a priori loading of the knowledge base contents to the RAM of the client's computer. The inference process is performed later on using only data stored in the RAM. The second approach involves implementing inference as a stored procedure, run in the environment of the database server.

Experiments were conducted on real-world knowledge bases with a relatively large number of rules. Experiments prepared so that one could evaluate the pessimistic complexity of the inference algorithm. The results confirmed, that the inference implemented in object-oriented data structures loaded into memory is effective. The times of inference in the worst-case scenario did not exceed 11 s. However, loading the contents of the knowledge base proved to be time consuming. For small bases these times were about a couple of seconds, but for the largest one they reached nearly 9 min (when loading from an XML file). Such a case is acceptable in systems where the knowledge base is reloaded rarely, and the waiting time (for data loading) is tolerable from the user's point of view. For applications where there is need for frequent reloading of the knowledge base, this solution is inconvenient and may be cumbersome for the user. This may be the case for knowledge bases which are frequently updated, e.g. by programs that use specific algorithms to automatically generate rules.

The inference implemented in the form of a stored procedure runs significantly slower than the solution described previously, which is not a surprising result. However, when comparing the inference times and adding the time of loading data from the knowledge base, the solution using a stored procedure turns out to be faster (in specific conditions). This solution is especially convenient when the knowledge base is updated frequently.

Implementation of the inference in the form of a stored procedure is an interesting solution and will eventually be permanently included into the described KBExplorer system and the KBExpertLib library. This will allow the programmer implementing a domain expert system to select the desired mode of inference. Alternatively, we consider the usage of the KBExpertLib library in a server side implementation of inference – as a service available via REST API. This solution will be analysed as the next stage of research.

References

1. Acquired Intelligence: Acquired Intelligence Home Page. http://aiinc.ca. Accessed Oct 2015
2. Akerkar, R., Sajja, P.: Knowledge-Based Systems. Jones and Bartlett Publishers, Burlington (2010)
3. Canadas, J., Palma, J., Túnez, S.: A tool for MDD of rule-based web applications based on OWL and SWRL. In: Knowledge Engineering and Software Engineering (KESE6), p. 1 (2010)
4. CLIPS: CLIPS NASA Home Page. http://www.siliconvalleyone.com/founder/clips/index.htm. Accessed Nov 2016
5. DROOLS: DROOLS Home Page. https://www.drools.org. Accessed Nov 2016
6. Duan, Y., Edwards, J.S., Xu, M.: Web-based expert systems: benefits and challenges. Inf. Manag. **42**(6), 799–811 (2005)

7. eXpertise2Go: eXpertise2Go Home Page. http://expertise2go.com. Accessed Nov 2016
8. Exsys: Exsys Home Page. http://www.exsys.com. Accessed Nov 2016
9. Gensym Corporation: Gensym Corporation Announces Gensym G2 8.4R2 Platform. http://www.marketwired.com. Accessed Jan 2017
10. Grove, R.: Internet-based expert systems. Expert Syst. **17**(3), 129–135 (2000)
11. Grzymala-Busse, J.W.: Managing Uncertainty in Expert Systems, vol. 143. Springer Science & Business Media, New York (2012)
12. Huntington, D.: Web-based expert systems are on the way: Java-based web delivery. PC AI **14**(6), 34–36 (2000)
13. Jach, T., Xięski, T.: Inference in expert systems using natural language processing. In: Kozielski, S., Mrozek, D., Kasprowski, P., Małysiak-Mrozek, B., Kostrzewa, D. (eds.) BDAS 2015. CCIS, vol. 521, pp. 288–298. Springer, Cham (2015). doi:10.1007/978-3-319-18422-7_26
14. JESS: JESS Information. http://herzberg.ca.sandia.gov. Accessed Nov 2016
15. Li, D., Fu, Z., Duan, Y.: Fish-expert: a web-based expert system for fish disease diagnosis. Expert Syst. Appl. **23**(3), 311–320 (2002)
16. Ligeza, A.: Logical Foundations for Rule-based Systems, vol. 11. Springer, Heidelberg (2006)
17. Ligeza, A., Nalepa, G.J.: Knowledge representation with granular attributive logic for XTT-based expert systems. In: FLAIRS Conference, pp. 530–535 (2007)
18. Mathkour, H., Al-Turaiki, I., Touir, A.: The development of a bilingual fuzzy expert system shell. J. King Saud Univ.-Comput. Inf. Sci. **21**, 27–44 (2009)
19. Nowak-Brzezinska, A., Siminski, R.: New inference algorithms based on rules partition. In: Proceedings of the 23th International Workshop on Concurrency, Specification and Programming, Chemnitz, Germany, 29 September–1 October, 2014, pp. 164–175 (2014). http://ceur-ws.org/Vol-1269/paper164.pdf
20. Simiński, R., Nowak-Brzezińska, A.: Goal-driven inference for web knowledge based system. In: Wilimowska, Z., Borzemski, L., Grzech, A., Świątek, J. (eds.) Information Systems Architecture and Technology: Proceedings of 36th International Conference on Information Systems Architecture and Technology – ISAT 2015 – Part IV. AISC, vol. 432, pp. 99–109. Springer, Cham (2016). doi:10.1007/978-3-319-28567-2_9
21. Polkowski, L.: Rough Sets in Knowledge Discovery 2: Applications, Case Studies and Software Systems, vol. 19. Physica, Heidelberg (2013)
22. Ruiz-Mezcua, B., Garcia-Crespo, A., Lopez-Cuadrado, J., Gonzalez-Carrasco, I.: An expert system development tool for non AI experts. Expert Syst. Appl. **38**(1), 597–609 (2011)
23. Sajja, P.S., Akerkar, R.: Knowledge-based systems for development. Adv. Knowl. Based Syst.: Model Appl. Res. **1**, 1–11 (2010)
24. Simiński, R.: Extraction of rules dependencies for optimization of backward inference algorithm. In: Kozielski, S., Mrozek, D., Kasprowski, P., Małysiak-Mrozek, B., Kostrzewa, D. (eds.) BDAS 2014. CCIS, vol. 424, pp. 191–200. Springer, Cham (2014). doi:10.1007/978-3-319-06932-6_19
25. Simiński, R.: The kbexpertlib software library for java-functionality properties and performance study. Studia Inform. **37**(1), 125–134 (2016)
26. Simiński, R.: Multivariate approach to modularization of the rule knowledge bases. In: Gruca, A., Brachman, A., Kozielski, S., Czachórski, T. (eds.) Man–Machine Interactions 4. AISC, vol. 391, pp. 473–483. Springer, Cham (2016). doi:10.1007/978-3-319-23437-3_40

27. Simiński, R.: The experimental evaluation of rules partitioning conception for knowledge base systems. In: Borzemski, L., Grzech, A., Świątek, J., Wilimowska, Z. (eds.) Information Systems Architecture and Technology: Proceedings of 37th International Conference on Information Systems Architecture and Technology – ISAT 2016 – Part I. AISC, vol. 521, pp. 79–89. Springer, Cham (2017). doi:10. 1007/978-3-319-46583-8_7

28. Simiński, R., Nowak-Brzezińska, A.: KBExplorator and KBExpertLib as the tools for building medical decision support systems. In: Nguyen, N.-T., Manolopoulos, Y., Iliadis, L., Trawiński, B. (eds.) ICCCI 2016. LNCS (LNAI), vol. 9876, pp. 494–503. Springer, Cham (2016). doi:10.1007/978-3-319-45246-3_47

29. Siminski, R., Wakulicz-Deja, A.: Rough sets inspired extension of forward inference algorithm. In: Proceedings of the 24th International Workshop on Concurrency, Specification and Programming, Rzeszow, Poland, 28–30 September 2015, vol. 2, pp. 161–172 (2015)

30. Simiński, R., Xięski, T.: Physical knowledge base representation for web expert system shell. In: Kozielski, S., Mrozek, D., Kasprowski, P., Małysiak-Mrozek, B., Kostrzewa, D. (eds.) BDAS 2015-2016. CCIS, vol. 613, pp. 558–570. Springer, Cham (2016). doi:10.1007/978-3-319-34099-9_43

31. SPHINX: SPHINX Home Page. https://aitech.pl. Accessed Nov 2016

32. Wang, W., Yang, M., Seong, P.H.: Development of a rule-based diagnostic platform on an object-oriented expert system shell. Ann. Nucl. Energy **88**, 252–264 (2016)

33. Xpert Rule: Xpert Rule Home Page. http://www.xpertrule.com. Accessed Nov 2016

34. Zetian, F., Feng, X., Yun, Z., XiaoShuan, Z.: Pig-vet: a web-based expert system for pig disease diagnosis. Expert Syst. Appl. **29**(1), 93–103 (2005)

DUABI - Business Intelligence Architecture for Dual Perspective Analytics

Bartosz Czajkowski[✉] and Teresa Zawadzka

Department of Software Engineering, Gdańsk University of Technology,
ul. Gabriela Narutowicza 11/12, 80-233 Gdańsk, Poland
czajkowski92@gmail.com, tegra@eti.pg.gda.pl

Abstract. A significant expansion of Big Data and NoSQL databases made it necessary to develop new architectures for Business Intelligence systems based on data organized in a non-relational way. There are many novel solutions combining Big Data technologies with Data Warehousing. However, the proposed solutions are often not sufficient enough to meet the increasing business demands, such as low data latency while still maintaining high functionality, efficiency and reliability of Data Warehouses. In this paper we propose DUABI - the BI architecture that enables both traditional analytics over OLAP Cube as well as near real-time analytics over the data stored in the NoSQL database. The presented architecture leverages features of NoSQL databases for scalability and fault-tolerance with the use of mechanisms like sharding and replication.

Keywords: Business Intelligence · NoSQL · Big Data · Real-time · Analytics · Scalability · Fault-tolerance

1 Introduction

At the beginning of the second decade of the 21st century there was discussed the problem of relation between Big Data and Data Warehouses. Do the Big Data solutions replace Data Warehouses? Today, we know that Big Data is a type of data and the set of technologies allowing processing them, while Data Warehouse is an architecture which allows managing clean, high-quality integrated data. Thus the question arises how to integrate Big Data with Data Warehouses [14, 15]. This integration is the process that happens for the last few years. Let us have a closer look at available solutions:

Firstly we have got the traditional technologies used in Data Warehouses. There are two standard architectures: introduced by Inmon [13] and Kimball [16]. In the business applications any combination of these architectures are used. All of them have some common features:

– data are integrated into Data Warehouse and stored in relational structures (in Kimball architecture these structures, star schemas, are prepared in advance with the focus on analytics [16]),

© Springer International Publishing AG 2017
S. Kozielski et al. (Eds.): BDAS 2017, CCIS 716, pp. 527–538, 2017.
DOI: 10.1007/978-3-319-58274-0_41

- data structures are populated with batch loading (often once a day),
- effective analysis requires rewriting relational structures to analytical ones (often column oriented databases),
- analytical structures are difficult to maintain (altering relational structures is cumbersome),
- they are not suitable for Big Data.

Next, we have Big Data and technologies that allow processing them. Let us mention the 3Vs model of Big Data defined by Doug Laney: Volume - the size of generated and stored data, Velocity - the speed at which new data is generated and Variety - the different types of data, also semi-structured or unstructured [17]. Among the Big Data technologies the most widely used is Hadoop [1], an open source framework allowing distributed processing of large datasets across clusters of computers using simple programming model - MapReduce [8].

Finally, during the last years there were developed NoSQL databases which have the following features: effective operation in distributed model, structures other than relational, lack of rigid schemas and application of BASE paradigm rather than ACID one [19].

The paper, proposing a DUABI (Business Intelligence Architecture for Dual Perspective Analytics), in some way, concerns all these three fields of computer science. While developing DUABI the following assumptions in each of these fields were made:

1. The standard architecture for Data Warehouses is extended by introducing new components, which should not have an influence on traditional Business Intelligence analysis. However, the new components can provide new high-quality data.
2. The architecture should allow processing Big Data, characterized with Volume, Velocity and Variety. Having these assumptions in mind the new architecture should enrich the standard analysis with new information provided by Big Data, as well as near real-time analysis of Big Data. This duality justifies the name of our proposed architecture.
3. Big Data are stored in NoSQL databases.

2 Rationale for New BI Solutions Architectures Using NoSQL Databases

NoSQL databases have been mainly created for storing and processing large volumes of different types of data and for ensuring their high availability. They have not been designed to use them for analytics and OLAP purposes. Therefore, there is a lack of mechanisms that allow using these databases in an efficient way for decision making processes and, generally speaking, Business Intelligence. What is more, making the *ad hoc* queries using NoSQL databases requires more programming skills and knowledge of the API in comparison to relational ones, as NoSQL databases intentionally skip some of the advanced features for performance reasons and thus such advanced features must be programmed manually in application [12].

The increasing need to develop new methods and solutions for described problems has resulted in several concepts that can be used in order to gain business value from NoSQL databases. These concepts can be divided into three groups:

1. Dedicated BI applications,
2. Integration with existing data sources,
3. Virtualization [9].

Let us briefly consider each group of concepts. The first one consists of architectures that require creating specialized software in order to use a NoSQL database for analytics. The first solution in this group is presented in Fig. 1.

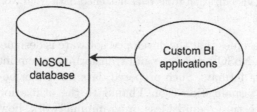

Fig. 1. Dedicated BI application approach (based on [9])

This architecture ensures that the data latency will be zero due to the direct connection to the NoSQL database. However, the lack of the preprocessed aggregates results in the need of on-the-fly calculations and, therefore, the analysis latency increases. To avoid this drawback, the solution can be upgraded as it is shown in Fig. 2.

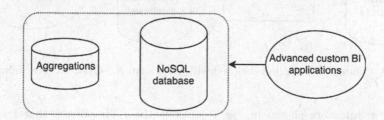

Fig. 2. Dedicated BI application approach extended with preprocessed aggregates (based on [9])

In this case, the developed software has to be even more specialized and able to both create and use aggregates that may be stored either in a NoSQL database or outside of it. Unfortunately, none of these two architectures provides the possibility to issue the true *ad hoc* queries - depending on the level of advancement, created software may allow the user to make only the specific queries that can be handled. What is more, because of the NoSQL interface, none of these architectures supports the use of common BI tools. Such tools can be used in the third architecture, shown in Fig. 3.

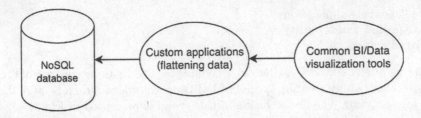

Fig. 3. Common BI visualization tools over flattened data from NoSQL (based on [9])

This solution assumes that the developed software is designed only for pulling the data out of the NoSQL database and transforming them into one of the typical structured data formats. Such prepared data may then be used as an input for a common BI/visualization tool. Thanks to the utilization of the common BI tool, this architecture requires less programming work, however, it still does not provide the true *ad hoc* queries.

In the following paragraphs we briefly mention group of concepts consisting of solutions that integrate existing data sources and allow users to make the true *ad hoc* queries.

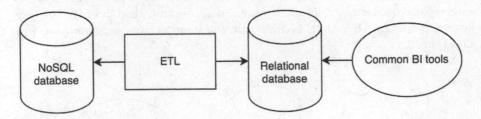

Fig. 4. Common BI solution enriched with data from NoSQL data (based on [9])

The architecture shown in Fig. 4 contains the ETL process that transfers the data from NoSQL database into relational one. Due to the large volume of data, this step requires making some pre-aggregates before loading into SQL database. After the successfully performed ETL process, data may be then used in a common BI tool and the true *ad hoc* queries can be made. This solution may also be upgraded - the relational database may be replaced with a Data Warehouse. This is shown in Fig. 5.

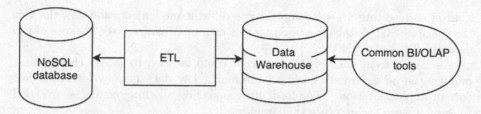

Fig. 5. Common BI solution enriched with Data Warehouse and data from NoSQL database (based on [9])

As in the previous solution, this architecture also requires the ETL process with pre-aggregating the data. However, in this case the data are loaded directly into a Data Warehouse and, therefore, can be analyzed with the information from another data sources using common BI/OLAP tools. Despite the possibility to issue the *ad hoc* queries, these two solutions are not sufficient enough to meet current business demands simply because of the high data latency - data freshness in the relational database or Data Warehouse depends on the frequency of the ETL process. In most cases it is around 24 h.

The last group of concepts concerns the virtualization. Solutions based on this mechanism are often used in the EII *(Enterprise Information Integration)* systems. The goal of such systems is to *"provide uniform access to multiple data sources without having to first loading them into a data warehouse"* [11]. The example of this kind of concept using NoSQL database is shown in Fig. 6.

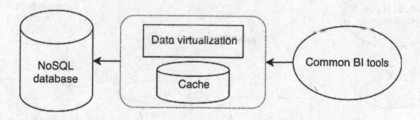

Fig. 6. Architecture utilizing virtualization approach (based on [9])

In this architecture, the common BI tools make the SQL queries that are being translated in real-time into proper NoSQL commands. The process of translation is possible by using a middleware layer and *Data virtualization* module that behaves as a wrapper - the NoSQL database is being seen as a relational one. Besides, the layer also consists of the *Cache* module which may store the aggregates and support the efficient *ad hoc* queries. The main drawbacks of this kind of architecture are its complexity, a difficult implementation and the fact that each NoSQL database will require its own middleware layer because of the different types of APIs.

As can be seen, none of these concept could be used to satisfy the need for making the *ad hoc* queries while maintaining low data latency level and low cost of implementation. Therefore, after analyzing all these concepts, we tried to develop such architecture that would:

– be relatively easy to implement,
– support the *ad hoc* queries and the aggregates,
– offer low data latency and could be used for the near real-time analytics.

3   Hybrid Architecture - DUABI

As stated in previous point, none of described approaches fulfills requirements of enabling both traditional analytics over OLAP Cubes as well as near real-time analytics over raw data. Therefore, the combination of the approaches must take place. The analysis showed, that having in mind features and implementation effort needed, the mixture of *Dedicated BI application approach extended with preprocessed aggregates* and *Common BI solution enriched with Data Warehouse and data from NoSQL database* has the highest features to cost ratio. Such approach is presented in Fig. 7.

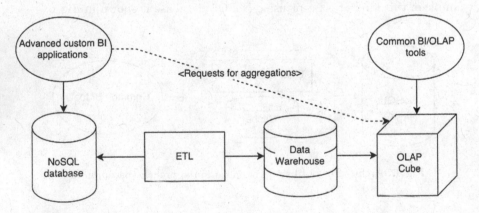

Fig. 7. Concept of mixture of *Dedicated BI application approach extended with pre-processed aggregates* and *Common BI solution enriched with Data Warehouse and data from NoSQL database*

Besides the ETL process that transfers pre-aggregated data into the Data Warehouse, the architecture designed in this way requires also an OLAP Cube and two different BI tools:

– the advanced application that would connect to the NoSQL database and use the aggregates stored in the OLAP Cube,
– the common BI application that would connect only to the OLAP Cube.

Such architecture allows not only for analyzing the data in real-time by reading them directly from the NoSQL database, but also would enable the traditional analytics and would provide the possibility to issue *ad hoc* queries against the OLAP Cube.

What is more, to maximize the advantages of that architecture, we considered also two main features of NoSQL databases, namely sharding and replication.

The use of the sharding mechanism would make the architecture horizontally scalable and would provide a higher bandwidth capacity due to the database fragmentation and geographical distribution - with proper selection of shard key the incoming data could be written to nodes that are less loaded or are located closer to the user.

The replication mechanism, in turn, would reduce the NoSQL database load - all kind of analytics could be performed on the data stored not only in primary node, but also its replicas. Using mentioned *primary-secondary* (also known as *master-slave*) replication model would let for efficient analysis of huge amounts of data. In the same time, the use of the replication would make this architecture fault-tolerant. If there was a failure of the primary node, replication mechanism would automatically switch to one of its replicas.

Eventually, the final form of DUABI is presented in Fig. 8 (for the sake of clarity, the arrows in the diagram indicate the data flow).

In DUABI, the operational data that seamlessly arrive in the system are loaded to the NoSQL database shards (*PRIMARY nodes*) and after that, they are asynchronously replicated to the replicas (*SECONDARY nodes*). When replicated, the data may then be used by the *Advanced custom BI application* for near to real-time analytics. However, because of the time required to synchronize replicas (so called a *replication lag*), it is not a true real-time. In addition, the application may use the aggregates that are stored in the *OLAP Cube*. In order to fill that multidimensional structure with the data, the proper *ETL* processes need to be performed. Depending on the frequency, the operational data stored in the *NoSQL database replica* are being pre-aggregated and loaded into the *Staging area*. Despite these data, the *Staging area* may also contain the data from several other data sources (eg. *Flat files* or *Source systems*). Then, all of the data are being integrated together and in that form are being loaded to the *Data Warehouse*. Afterwards, at the specified time, the *OLAP Cube* is being processed and may be used both for the traditional analytics by the *Common BI/OLAP tool* and for supporting the near real-time analytics by the *Advanced custom BI application*. Such designed architecture enables, therefore, dual perspective analytics.

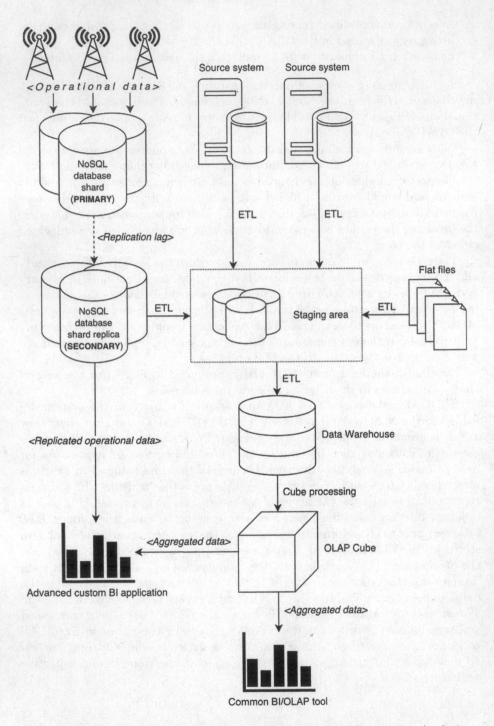

Fig. 8. DUABI - Business Intelligence Architecture for Dual Perspective Analytics

4   MeteoAnalytics System as a Validation of DUABI

In order to confirm the assumed characteristics and features of DUABI, we designed and implemented a prototype that is based on our architecture - Meteo-Analytics System. The prototype simulates an environmental monitoring system for weather observations and analysis. Such system should be able to handle a lot of data incoming from various sensors and transmitters. What is more, these data may come from many different and specialized systems, and thus may not have a common, rigid structure [10]. To implement the MeteoAnalytics System we used the following tools:

- MongoDB 3.2 [2] as a NoSQL database,
- csv files as an external data source,
- Microsoft SQL Server 2012 [3] as an RDBMS for the *Staging area* and the Data Warehouse,
- SQL Server Data Tools [4] as a tool for creating the ETL processes and the OLAP Cube,
- Microsoft SQL Server Analysis Services as a tool for manage the OLAP Cube,
- RealTime Meteo Dashboard as our custom BI application,
- Bilander Platform [5] as an external BI tool.

To reliably reproduce the actual environmental monitoring system, we created a data generator that was regularly (every ten seconds) loading the weather data into the MongoDB database. These data, stored in the BSON documents, consisted of different weather parameters, such as: surface air temperature, atmospheric pressure, humidity. Next, the generated data were replicated to the MongoDB replica and then used in the way that was described in DUABI.

MeteoAnalytics System differs a bit from general architecture of DUABI in the terms of replication mechanism. To ensure that our system would be true fault-tolerant, we used the arbiter node. Thereby, any failure of the primary node results in an immediate voting and switching between the *PRIMARY* and the *SECONDARY* node so that the system would still load the new weather data.

As an advanced custom Business Intelligence application we used RealTime Meteo Dashboard - homemade software that connects directly to the replica of the MongoDB database and visualize the incoming data in almost real-time. The actual values of each weather parameter are shown in the separate plots and are regularly updated as new data arrive. The examples of the plots are shown in Fig. 9.

These plots present not only the actual weather conditions, but also the corresponding last weeks values. This kind of visualization was possible through the use of the aggregated data stored in the OLAP Cube.

To show off the main feature of DUABI - the possibility of dual perspective analytics, we also created several typical Business Intelligence reports and dashboards using the standard Business Intelligence software - Bilander Platform and its module for analytics. As an example, Fig. 10 presents a visualization of aggregated weather data that was created with Bilander Platform.

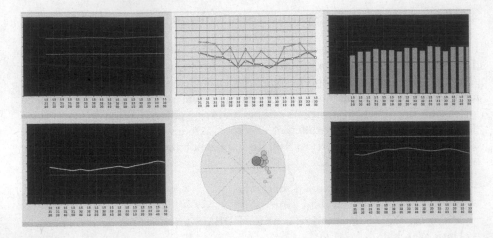

Fig. 9. Near real-time data presented in RealTime Meteo Dashboard

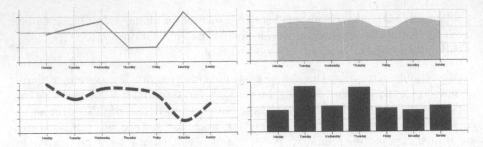

Fig. 10. Aggregated data presented in Bilander Platform

5   Related Work

The concepts that are currently being designed to support analytics over NoSQL databases are mainly based on the *Dedicated BI application approach extended with preprocessed aggregates* or the *Architecture utilizing virtualization approach* and often use Hadoop framework. However, the solutions from the first group (such as Kyvos [6]) are not mature enough to support the true *ad hoc* queries, while the solutions from the second group (such as Apache Kylin [7]) does not support the full richness of SQL when translating to NoSQL or MapReduce commands.

There are some other concepts of using both Hadoop and relational databases in conjunction for dual perspective analytics (such as one shown in Fig. 11 - an approach to implement Big Data Warehouse soultion [18]). However, we cannot indicate the existing examples based on this kind of architectures.

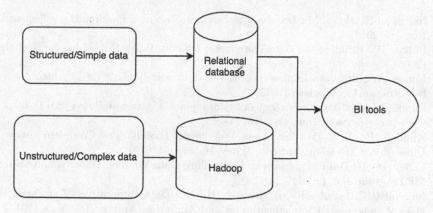

Fig. 11. Architecture using both Hadoop and RDBMS simultaneously (based on [18])

6   Conclusions

The paper presents the architecture for BI solution that enables both traditional analytical processing, as well as near real-time analytical processing of Big Data. The case study of presented architecture was based on processing of meteorological data. There were also performed theoretical analyzes of scalability of the proposed solution. As a part of further work the scalability of the proposed architecture should be examined for large sets of data incoming with different frequencies. The application in real conditions will allow its further development and adjustments to fulfill all defined requirements.

References

1. http://hadoop.apache.org
2. https://www.mongodb.com/mongodb-3.2
3. https://www.microsoft.com/en-us/download/details.aspx?id=29062
4. https://www.microsoft.com/en-us/download/details.aspx?id=42313
5. http://www.bilandergroup.com
6. http://www.kyvosinsights.com
7. http://kylin.apache.org
8. Dean, J., Ghemawat, S.: MapReduce: simplified data processing on large clusters (2004)
9. Goodman, N.: BI/Analytics on NoSQL: Review of Architectures. NoSQL Now! (2011)
10. Günther, O., Radermacher, F., Riekert, W.: Environmental monitoring models, methods, and systems. In: Avouris, N.M., Page, B. (eds.) Environmental Informatics - Methodology and Applications of Environmental Information Processing, pp. 13–38. Springer, Heidelberg (1995)
11. Halevy, A., Ashish, N., Bitton, D., Carey, M., Draper, D., Pollock, J., Rosenthal, A., Sikka, V.: Enterprise information integration: successes, challenges and controversies. In: Proceedings of the 2005 ACM SIGMOD International Conference on Management of Data (2005)

12. Indrawan-Santiago, M.: Database research: are we at a crossroad? - reflection on NoSQL (2012)
13. Inmon, W.: Building the Data Warehouse, 4th edn. Wiley Publishing Inc, Hoboken (2005)
14. Inmon, W.: Big data implementation vs. data warehousing (2013). http://www.b-eye-network.com/view/17017
15. Inmon, W.: Big data technology does not replace a data warehouse (2013). http://www.b-eye-network.com/view/16714
16. Kimball, R., Ross, M.: The Data Warehouse Toolkit: The Complete Guide to Dimensional Modeling, 3rd edn. Wiley, Hoboken (2013)
17. Laney, D.: 3D Data Management: Controlling Data Volume, Velocity, and Variety. META group Inc. (2001)
18. Mohanty, S., Jagadeesh, M., Srivatsa, H.: Big Data Imperatives: Enterprise Big Data Warehouse. BI Implementations and Analytics. Apress, New York (2013)
19. Sadalage, P., Fowler, M.: NoSQL Distilled: A Brief Guide to the Emerging World of Polyglot Persistence. AddisonWesley, Boston (2009)

Comparative Analysis of JavaScript and Its Extensions for Web Application Optimization

Adam Mlynarski and Karolina Nurzynska[✉]

Institute of Informatics, Silesian University of Technology,
Akademicka Street 16, 44-100 Gliwice, Poland
Karolina.Nurzynska@polsl.pl

Abstract. This work is dedicated to analysis and comparison of the efficiency of several extensions of JavaScript. Analysis concentrates on the quality of delivered application performance in terms of web page update, database display refreshing, etc. The comparison is performed using three scenarios of: array data display, filling a form, and switching the views between application pages. The research addresses functionality of frameworks and libraries taken under consideration on the personal computer as well as on the mobile device. The results of comparison show, that it is difficult to find one solution, which works well in all circumstances. React, as a view of application, can be recommended for server side flow control, near the database, while Angular should be considered when a clear division into server and client side is sought.

Keywords: JavaScript · Extensions of Javascript

1 Introduction

A database alone, without a functionality enclosed within a software designed for its data presentation, is just a container of data. Yet, the ability of web surfers to explore a database by convenient web interface makes it very usable. There are many constraints connected with such software development. One of them is the quality and timing of data presentation on the web pages. This problem on the side of a database has been already addressed in [10]. But the ability of data presentation when timing constraints are considered by the Internet applications is also important.

It is hard to think about present world without dynamic web pages. Most of desktop applications is constantly replaced by Single Page Applications. There is no doubt that this approach has a lot of advantages, such as independence of device and operating system or the ability to access data from anywhere in the world. One of the major components which had an impact on the popularization of Internet applications was JavaScript (JS). Nowadays there are many extensions based on JS and it is hard to chose the best one. In this article four of them were compared: Angular, Angular 2, Backbone, and React.

Angular is one of the most popular JS frameworks and Angular 2 is its very promising successor. React is a library created and used by Facebook.

© Springer International Publishing AG 2017
S. Kozielski et al. (Eds.): BDAS 2017, CCIS 716, pp. 539–550, 2017.
DOI: 10.1007/978-3-319-58274-0_42

With React, developer creates components which can be used in many places. Backbone, in the contrary, is a library that can be adapt to many requirements - it is an extension that makes JavaScript more organized, but it does not force programmers to use only one solution (language, pattern etc.).

2   Experiments Preparations

Way of creating basic elements on the page is different for all of the compared JavaScript extensions. However, performance is one of the very important issues when creating a production application. Therefore, there were prepared three scenarios under which performance was measured:

Array scenario presents an array of elements on the page. Any object inside the array was placed in a row of `<table>` attribute. The refresh time for a random element on the page was 100 ms, in this time the operation seems smooth for a user [9]. This scenario not only displayed a list of items, but also performed the task of scrolling down the screen. This test could have a big impact on the comparison results, because a side scrolling is associated with increase of many complex operations in the browser, so it can freeze the page momentary. An exemplary problem is illustrated in Fig. 1. Implementation details:

- React: `map` function which return list item components,
- Angular: framework directive `ng-repeat`,
- Angular 2: framework directive `ngFor`,
- Backbone: jQuery `append` method.

Form scenario was designed to fill the `<input>` elements on the page with addition of new objects to existing array. Each new object was appended at the beginning of the array to check how framework and libraries can handle a shift of the entire list of objects. In contrast to the previous scenario, any random refreshing of elements on the page was applied. Implementation details:

- Angular: HTML `input` element with framework directive `ng-model`,
- Angular 2: HTML `input` element with framework directive `[(ngModel)]`,
- Backbone: HTML `input` element with jQuery `val()` method,
- React: HTML `input` element with `onChange` React method. 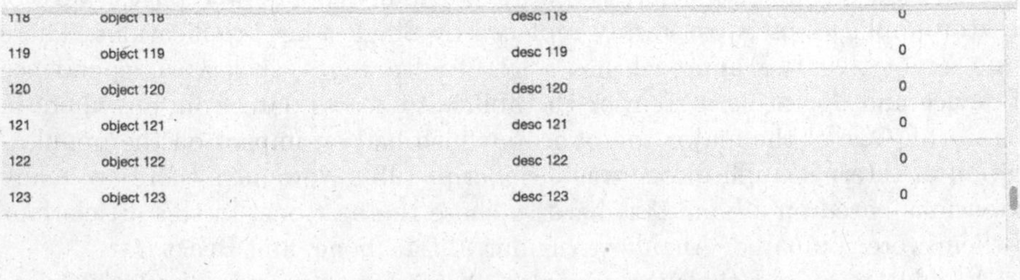

118	object 118	desc 118	0
119	object 119	desc 119	0
120	object 120	desc 120	0
121	object 121	desc 121	0
122	object 122	desc 122	0
123	object 123	desc 123	0

Fig. 1. Problem with refreshing page while scrolling.

Router scenario switches between three application views which were set inside main page. On each view there were elements which were assigned to them. This scenario was designed to check refreshing the view with loading new elements on the page. Implementation details:

- Angular: `ui-router` module extension,
- Angular 2: `RouterModule` extension,
- Backbone: `Backbone.Router`,
- React: `react-router` library.

Nowadays it is important to run web application on desktop browser and mobile device, too. Therefore, each scenario was run on a personal computer and also on a mobile device. All tests had to investigate two important application properties: building page from layers and checking frames management. These properties were split to verification of following parameters [2]:

Composite layers. It is the time which browser spent on composing rendered layers into a screen image. Every element of the site is located in a separate layer, which can overlap, penetrate, and may be transparent [11].

Update layer. It is the time spent on updating of single layer in a browser.

First paint. It is the time used immediately before rasterizing the page into bitmaps.

Frames per second (fps). It is a measure of a website's responsiveness. A frame rate of 60 fps is the target for smooth performance [4].

Dropped frame count. The number of frames produced by the GPU were over 16.6 ms apart (1000 ms/60).

Mean frame time. Average time taken to render each frame.

Each parameter was measured by browser-perf [2]. This tool is based on Chrome developer tools and requires chrome-driver [3] to work with. All applications were written with the same HTML attributes and IDs in order to enable automation of testing: scrolling page, filling elements, and switching between pages. To enable testing on mobile device it was also required to switch mobile device into developer mode. Debugging and collecting data was made with Android Debug Bridge [1]. Each test was repeated ten times in order to decrease risk of random results and the final results present mean average and standard deviation of obtained outcomes. Collected data were then grouped for analysis.

After each test, there were assigned points for each JavaScript extension in order to compare and show which performance is the best on mobile device, which on personal computer and which on both. Following point scale was adapted:

- 1st place (the best result) – 4 points,
- 2nd place – 3 points,
- 3rd place – 2 points,
- 4th place – 1 point.

Research were carried out on devices which parameters are listed in Table 1.

Table 1. Parameters of the devices.

	Personal computer	Mobile device
Name	MacBook Pro	Huawei P9
CPU	Intel Core i7 2.5 GHz	HiSilicon Kirin 955 2.5 GHz
Cores	4	8
RAM	16 GB	3 GB
OS	OS X El Capitan	Android 6.0 Marshmallow
Browser	Chrome 53.0.2785.143	Chrome 53.0.2785.124

3   Results

Final results are presented separately for each scenario. Tested parameters are shown in the graphs, with plotted values obtained for the mobile device and personal computer. Additionally, a summary of achieved performances is gathered in tables. In this article all scenarios tested on computer and mobile device are regarded as equally valid, hence their weights are equal.

Table 2. The array scenario: points for personal computer.

	React	Angular	Angular 2	Backbone
Composite layers	3	1	2	4
Update layer	3	2	1	4
First paint	4	3	1	2
Frames per second	3	3	3	1
Dropped frame count	2	3	4	1
Mean frame time	2	3.5	3.5	1
Summary	17	15.5	14.5	13

Table 3. The array scenario: points for mobile device.

	React	Angular	Angular 2	Backbone
Composite layers	4	2	1	3
Update layer	3.5	2	1	3.5
First paint	4	3	2	1
Frames per second	2	3.5	3.5	1
Dropped frame count	2	4	3	1
Mean frame time	2	4	3	1
Summary	17.5	18.5	13.5	10.5

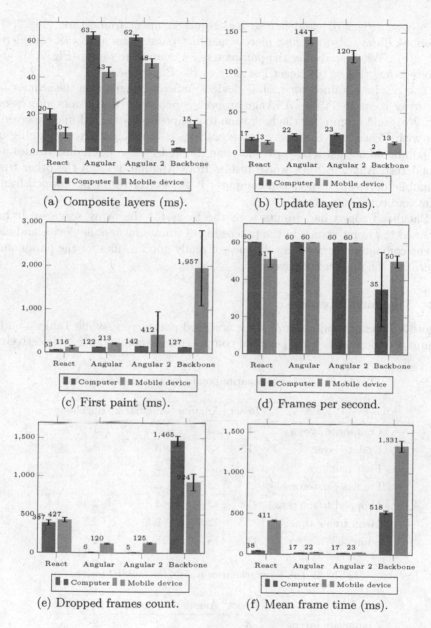

Fig. 2. Results for the array scenario.

3.1 The Array Scenario

Figure 2 presents results recorded for verified parameters, while Tables 2 and 3 summarise the results for a personal computer and mobile device, respectively.

The best results for the array scenario which has been taken on personal computer were obtained by React (Table 2). It got very good value of frames per second (Fig. 2(d)) and not so bad result in mean frame time (Fig. 2(f)). Facebook library also got first place when first paint parameter was considered (Fig. 2(c)). When analysing the parameters on composite layers (Fig. 2(a)) and update layers (Fig. 2(b)) React lost only to the Backbone.

The results obtained for mobile devices indicate Angular as the winner for the array scenario (Table 3). Angular best coped with the frames processing (Fig. 2(e)) and despite for the long time to composite layers and update layer it won with React, which was on the second place. Angular 2 after summing up the points did not get a better result than its predecessor. Excluding frames per second test, Angular 2 got worse results from Angular 1 in all the other tests on mobile devices. For personal computer it got better result in dropped frame count and composite layers.

Backbone coped bad results with the tasks for the array scenario. It has obtained the worst results in the processing of frames for mobile device and also for the computer. However, Backbone got really good results for the parameters describing the application layers.

3.2   The Form Scenario

Figure 3 presents results recorded for analysed parameters, while Tables 4 and 5 summarise the results for a personal computer and mobile device respectively.

Table 4. The form scenario: points for personal computer.

	React	Angular	Angular 2	Backbone
Composite layers	1	3	3	3
Update layer	1	3	3	3
First paint	4	3	2	1
Frames per second	2	3.5	3.5	1
Dropped frame count	1	3	4	2
Mean frame time	2.5	2.5	2.5	2.5
Summary	11.5	18	18	12.5

Table 5. The form scenario: points for mobile device.

	React	Angular	Angular 2	Backbone
Composite layers	3	4	1	2
Update layer	1	2	3	4
First paint	4	3	2	1
Frames per second	1	4	2.5	2.5
Dropped frame count	1	4	3	2
Mean frame time	1	4	3	2
Summary	11	21	14.5	13.5

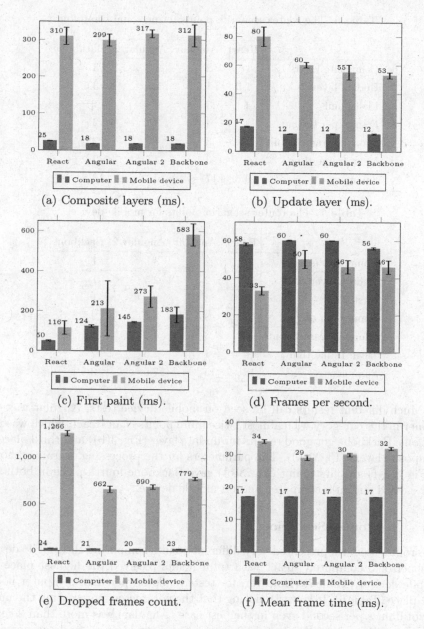

(a) Composite layers (ms).

(b) Update layer (ms).

(c) First paint (ms).

(d) Frames per second.

(e) Dropped frames count.

(f) Mean frame time (ms).

Fig. 3. The form scenario.

Analysing the graphs of the form scenario, it can be seen that each of the applications got very similar results on the personal computer. However, taking such a criterion, Google products had achieved the highest places (Tables 4 and 5). Although the results were very similar to each other, the last place in this scenario was taken by React.

Table 6. The router scenario: points for personal computer.

	React	Angular	Angular 2	Backbone
Composite layers	1	2	4	3
Update layer	1.5	1.5	3.5	3.5
First paint	4	3	2	1
Frames per second	1	4	3	2
Dropped frame count	1	3	4	2
Mean frame time	1	3.5	3.5	2
Summary	9.5	17	20	13.5

Table 7. The router scenario: points for mobile device.

	React	Angular	Angular 2	Backbone
Composite layers	4	2	3	1
Update layer	4	2.5	2.5	1
First paint	4	3	2	1
Frames per second	1	3.5	3.5	2
Dropped frame count	1	3	4	2
Mean frame time	4	1	2.5	2.5
Summary	18	15	17.5	9.5

Much different results can be seen on mobile device tests. Angular was the winner in this category with almost twice more points than React, which was the last one. Backbone got good results for update layer (Fig. 3(b)) and third place in composite layers (Fig. 3(a)). The parameters for the processing of frames shown in Fig. 3(d–f) and first paint (Fig. 3(c)) gave Backbone fourth place in both the test carried out on a computer and on mobile device.

3.3   The Router Scenario

The router scenario presented very different results comparing the mobile device and personal computer. React with total points received the highest place for mobile devices (Table 7). However, for tests made on personal computer it got last place (Table 6). It is worth noting that the parameter representing the number of frames per second even in the best case (Angular) was more than 2 times lower than the expected value of 60 frames per second in test made on personal computer (Fig. 4(d)). Even worse results achieved all test applications for the mobile device not exceeding the limit of 10 frames per second.

Figure 4(a) presenting composite layers and Fig. 4(b) presenting update layer show that results obtained on personal computer are much more better than mobile device. In this test Backbone got the last place. Also analysing first paint parameter (Fig. 4(c)), it indicates Backbone as the looser, especially on the mobile device.

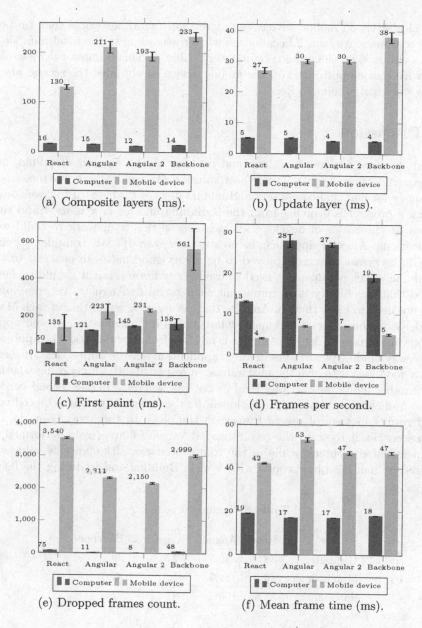

(a) Composite layers (ms).

(b) Update layer (ms).

(c) First paint (ms).

(d) Frames per second.

(e) Dropped frames count.

(f) Mean frame time (ms).

Fig. 4. The router scenario.

The router scenario proved to be very demanding on mobile devices. Figure 4(e) shows that the number of dropped frames for each JavaScript extension amounted to more than 2000 (mobile device). However, on the personal computer this number is always under 100 dropped frames. Also mean frame time (Fig. 4(f)) on mobile device was more than two times worse than on personal computer. Angular 2 became the winner according to the total number of points obtained on the personal computer (Table 6). This framework coped also well with routing performed on the mobile device. It obtained the second place, losing only half a point to React.

4 Discussion

Based on extensive investigations and ranking of the number of points, one can conclude that Angular is the best solution for applications written simultaneously for personal computers and mobile devices (Table 8). First version of Google's product is evolving from the beginning and for now is used and supported by thousands of developers which choose this framework to build web applications. Angular approach to create their own HTML compiler [8] and allowing to create directives proved to be a very good move. In contrast to the Backbone, developer does not need to use jQuery library, but it should be mentioned that this library is an important element inside Angular. Yet, developer does not need to know that. In Angular there is a visible separation of each MVC (model-view-controller) component, what makes implementation of web applications much more easier [7]. Angular is not free from any defects. The approach of two way binding of variables in a real application may be impractical. It seems that a good idea is to give the programmer a responsibility for choosing whether the variable should be two way bind or not (as was done in the second version of the framework). As already mentioned, first version of Angular achieved very good results in tests of use of frames on the page. In the array scenario and form scenario, it received the recommended value of 60fps (tests performed on the personal computer), while in the router scenario, although this value was lowered by half, also best coped with a task. Building and updating the layers

Table 8. Summary: points.

	React	Angular	Angular 2	Backbone
Computer	38	50.5	52.5	39
Mobile device	46.5	54.5	45.5	33.5
Summary	84.5	105	98	72.5

Table 9. Summary points for personal computer.

	React	Angular	Angular 2	Backbone
Array	17	15.5	14.5	13
Form	11.5	18	18	12.5
Router	9.5	17	20	13.5
Summary	38	50.5	52.5	39

Table 10. Summary points for mobile device.

	React	Angular	Angular 2	Backbone
Array	17.5	18.5	13.5	10.5
Form	11	21	14.5	13.5
Router	18	15	17.5	9.5
Summary	46.5	54.5	45.5	33.5

were weaker side of Angular where in most of tests (excluding the form scenario on mobile device) the better results were obtained by other extensions.

Based on the number of summary points the second place took Angular 2. This framework only in the router scenario made on personal computer got better results than its predecessor. Yet, this was sufficient to take first place in the ranking of the total number of points obtained on the personal computer (Table 9). In tests performed on the mobile device, Angular 2 was better than Angular 1 in router scenario (Table 10). Comparing the results of Google products obtained in testing, it seems that developers should not migrate from the first version to the other. Nowadays when the number of mobile devices grows very fast, it is expected from newer versions of frameworks to provide better support and higher performance. Taking also into consideration the fact that Angular 2 is a new product, support is also lower in comparison to the first version.

React in the overall ranking took the third place. In the array and form scenario performed on the personal computer, React has achieved the best results associated with the processing of frames. However, tests performed on the mobile device showed not so good results, especially of frames per second and dropped frames count. React in each of the tests achieved very good results for the parameter examining first paint. Facebook's product got better summary results than Angular 2 on mobile device and took the second place.

The lowest number of points in both tests performed on the personal computer and tests performed on the mobile device obtained Backbone. This library did not win in any of the conducted scenarios. One of the biggest Backbone problem is frame processing. Also it is really surprising that the number of dropped frames count performed on personal computer in the array scenario turned out to be much more higher than on the mobile device, while in another JS extensions better results were obtained on a personal computer. The first paint on mobile device also differs from the other. The cause of poor Backbone performance, may be the fact that any change in the element on the page will refresh all objects. The advantage of Backbone can be composite layers and update layer. Except the router scenario for mobile device this library got there good results.

Based on the above conclusions it can be inferred that Backbone should not be the part of new web applications. Analysing the use of library, it is impossible not to get the impression that this JS extension must be connected with jQuery. On the other hand, Backbone is really light but developers should keep in mind that all actions also need Underscore.js and jQuery.

5 Conclusions

React is the perfect choice in applications where there is no clear division between the client and server side [5]. The simplicity of implementation and the small size makes the Facebook product wide-ranging. Also when creating applications which contain many repeated components, developers should think about React. Facebook's support and very high popularity makes React also a good choice for near future.

Angular got the highest number of points, but considering the release of a new product from Google, it would be worth to think about the new version of Angular. Angular 2 combines the proses of Angular 1 and React. The native language of Angular 2 is TypeScript what should contribute to a more stable version of the application written by developers [6]. But it can not be forgotten that first version of Angular has a very strong support, which today is much better than Angular 2. The choice between these two frameworks can be difficult and will certainly depend on the size and time required for implementation of the project.

All of the presented JavaScript extensions are designed to support REST applications. Nowadays it is the most popular and operative way to get data from databases. One of the biggest advantage is simplicity and stateless. It should be also noticed, that the data exchanged is mostly performed using JSON format, which had the effect of reducing information sent to the browser.

Acknowledgements. Karolina Nurzynska work was supported by statutory funds for young researchers (BKM/507/RAU2/2016) of the Institute of Informatics, Silesian University of Technology, Poland.

References

1. Android debug bridge. https://developer.android.com/studio/command-line/adb.html. Accessed 16 Oct 2016
2. Browser-perf. https://github.com/axemclion/browser-perf. Accessed 16 Oct 2016
3. Chrome driver. https://sites.google.com/a/chromium.org/chromedriver/downloads. Accessed 16 Oct 2016
4. Frame rate. https://developer.mozilla.org/en-US/docs/Tools/Performance/Frame_rate. Accessed 16 Oct 2016
5. React. https://facebook.github.io/react/. Accessed 16 Oct 2016
6. Fenton, S.: Pro TypeScript: Application-Scale JavaScript Development. Apress, New York (2014)
7. Freeman, A.: Pro AngularJS. Apress, New York (2014)
8. Green, B.: AngularJS. O'Reilly Media, Sebastopol (2013)
9. Nielsen, J.: Usability Engineering. Morgan Kaufmann, Burlington (1993)
10. Szczyrbowski, M., Myszor, D.: Comparison of the behaviour of local databases and databases located in the cloud. In: Kozielski, S., Mrozek, D., Kasprowski, P., Małysiak-Mrozek, B., Kostrzewa, D. (eds.) BDAS 2015-2016. CCIS, vol. 613, pp. 253–261. Springer, Cham (2016). doi:10.1007/978-3-319-34099-9_19
11. Williams, R.: The Non Designer's Web Book. Peachpit Press, Berkeley (2005)

ALMM Solver - Database Structure and Data Access Layer Architecture

Krzysztof Rączka[✉] and Edyta Kucharska

Department of Automatics and Biomedical Engineering,
AGH University of Science and Technology,
30 Mickiewicza Av, 30-059 Krakow, Poland
{kjr,edyta}@agh.edu.pl

Abstract. The paper presents form of data storage in ALMM Solver and propose the structure of database. The solver is built on Algebraic-Logical Meta-Model of Multistage Decision Process (ALMM of MDP) methodology. Functional and non-functional requirements for data source are described. The detailed structure of database areas (Problem instance, Experiment data, Experiment parameters) and the architecture of the solver's database communication layer and specific architecture of the SimOpt module from the perspective of communication are presented. Proposed database structure takes into account that not only numeric variables, but also data sets and sequences defined system state can be stored. The paper presents an overview of selected solvers from the data source perspective too.

Keywords: Database structure · Design patterns · Layered architecture · Simulation · Optimization · Algebraic-logical meta-model

1 Introduction

The real issues faced on a daily basis by managers of production and logistics enterprises are, to a large extent, complex discrete optimization problems. Academic and commercial solvers can be used to fix such problems. They differ from each other in terms of the manner and scope of problems solved, as well as the methods used to find a solution. Among the available solvers, we can find software tools implemented on top of well-known spreadsheet applications (such as Frontline Solvers - Solver.com [1], Lindo Whats Best [2], TK. Solver 5.0 [3]) with the use of general methods. There are also applications based on various specialized problem models, which also enable us to use AI methods. An example of such a solver is JABAT, designed in the agent technology [14], based on the A-Team architecture and JADE Framework. Other available solvers are based on the Constraint Integer Programming (CIP) approach, that integrates constraint programming (CP), mixed integer programming (MIP) and satisfiability modeling and solving techniques (SAT) [4,12] and the Logical Constraint Programming (CLP) approach [5]. According to that approach, problems are

© Springer International Publishing AG 2017
S. Kozielski et al. (Eds.): BDAS 2017, CCIS 716, pp. 551–563, 2017.
DOI: 10.1007/978-3-319-58274-0_43

modeled by a finite, predetermined number of variables, their domains, and relations between them. Another example is LiSA - a software package for solving deterministic scheduling problems, where the main focus is shop-scheduling and one-machine problems. It allows us to identify the type of the problem using Graham's number in order to determine its complexity; and to solve selected problems with the use of an exact algorithm or an heuristic algorithm. Moreover, it enables us to develop algorithms for specific problems [6,15]. The article focuses on the ALMM Solver, which falls within the second group of software tools (i.e. it is an application based on specialized problem model) and allows us to solve complex combinatorial optimization problems. It is built on the basis of Algebraic-Logical Meta-Model of Multistage Decision Process (ALMM of MDP) methodology [18]. With this solver, the problem is modeled by the formal model based on the algebraic-logical meta-model methodology and can be solved with the use of different methods (exact and heuristic) as well as new meta-heuristic algorithms developed specifically for algebraic-logical meta-model. It also facilitates solving of non-deterministic scheduling problems.

In case of solvers designed to solve complex problems, especially NP-hard problems, data on the problem, its parameters and solutions obtained must be stored correctly. One of the key aspects of the ALMM paradigm is the analysis of the data structure of states from the perspective of data storage. It is variable and can store not only numeric variables, but also data sets and sequences.

The aim of the article is to discuss the data source and the data access layer in the ALMM Solver, in particular, to present requirements regarding data sources. Moreover, the paper proposes a database structure for the solver, which would facilitate not only storage of basic data required to obtain a solution, but also analysis and design of new methods and problem-solving algorithms, as well as a data access layer architecture. This paper is an extension of the work presented in [26], where an initial proposition for functionality and modular structure of the ALMM Solver and detailed description its core module "SimOpt" were given, and in [46], where the more detailed architecture of the ALMM Solver taking into account the functional and non-functional requirements as well as the practices, design patterns and principles, that was used to ensure the best quality of the solver software, were proposed.

2   Data Source in Selected Solvers

Various available solvers can be used to try to solve discrete optimization problems. The solvers differ in terms of the scope of issues/problems covered and the solving methods used. For the purpose of this paper, several solvers were analyzed from the perspective of data storage. The analysis focused on the following solvers: IBM ILOG CPLEX Optimization Studio, Lindo Whats Best, GoldSim, AIMMS, Frontline Solver (Solver.com), SuperDecisions, TK. Solver 5.0 and LISA. In terms of data storage, the solvers can be summarized as follows:

- IBM ILOG CPLEX Optimization Studio - data can be stored in memory, saved in the database, in Excel or exported to external files. The solver has advanced simulation progress monitoring tools (output data, computing time and memory usage analysis) [7].
- Lindo Whats Best - Lindo Whats Best - the solver was built as an Excel plug-in. During the simulation, data are stored in RAM memory, which creates issues in case of larger amounts of data (Insufficient Memory (Help Reference: MEMORY)). Once the simulation is complete, output data are saved in Excel [2].
- GoldSim - during the simulation, data are saved in RAM memory. Whereas output data are saved in *.gsm files, which store data in the internal GoldSim format. The solver can export data to Excel and text files [8].
- Frontline Solvers (Solver.com) - FrontlineSolvers (Solver.com) - the solver has been implemented as an Excel plug-in. Output data are saved directly in Excel. Integration with database systems is possible with the use of standard Excel functions. The implementation of this solver has opened multiple opportunities for integration with other systems. It can be used as an Excel plug-in or as an SDK for various programming languages. The solver's great flexibility makes it possible to store data in any way desired [1].
- SuperDecisions - the solver saves output and input data in files. Data are stored in the solver's internal format. The solver cannot be integrated with a database [9].
- TK. Solver 5.0 integrated directly with Excel. Output data can be saved in *.tkx files or in the database [3].
- LISA (A Library of Scheduling Algorithms) - the solver does not facilitate storage of data in the database and does not save the results in real time in files or any other data storage means. After the simulation, output data are saved in text files [6].
- JABAT Solver missing technical details on the solver. Impossible to analyze the solver's data storage mechanism [14].

A tool different from those listed above is AIMMS [10], which facilitates, among others, problem modeling. Modeled problems are solved with the use of any external solver available in the AIMMS system, integrated with the platform. The AIMMS platform uses the PostgreSQL database.

On the basis of the analysis performed, it can be concluded that most solvers allow export of output data to external systems, such as Excel, databases and file systems. Some solvers can save data in external systems in real time. Whereas others only save data once the simulation has been completed. This means that if the simulation is unexpectedly interrupted, all output data are lost.

3   Characteristics of ALMM Solver

The purpose of the paper is to present form of data storage in ALMM Solver and propose the structure of database. The concept and structure of this software tool was proposed in [26]. Paper [46] presents requirements regarding the solver, functional as well as non-functional and the best practices, design patterns and principles, that were applied to ensure the best quality of the solver software. A basic layout of the Problem Model Library is proposed in [24].

The solver is built on Algebraic-Logical Meta-Model of Multistage Decision Process (ALMM of MDP) methodology. The ALMM paradigm was proposed and developed by Dudek-Dyduch [17–21,23]. This approach has been evolved by research team and following approaches were presented: method uses a specially designed local optimization task and the idea of semi-metric [25], method based on learning process connected with pruning non-perspective solutions [39], substitution tasks method [22,27], switching of algebraic-logical models in flow-shop scheduling problem with defects [31,32], failure modes modeling for scheduling [47] and hybrid algorithm for scheduling manufacturing problem with defects removal [41,42]. According to ALMM theory, a problem is modeled as a multistage decision process (DMP) together with optimization criterion. Multistage decision process is defined as a sextuple: a set of decisions, a set of generalized states, an initial generalized state, a partial function called a transition function, a set of not admissible generalized states, a set of goal generalized states. The optimization task is to find an admissible decision sequence that optimizes quality criterion. Decision sequence is constructed from the initial state to goal state, a non-admissible state, or state with an empty set of possible decisions, in such a way that new state is calculated using a transition function and depends on the previous state and the decision made at this state. The solver has been developed to solve discrete optimization problems, in particular for NP-hard problems and contains different methods (exact and heuristic, especially new meta-heuristic algorithms developed specifically for algebraic-logical meta-model).

Solver Structure. Let us recall the structure of the solver [26,46]. The solver is composed of four basic functional modules [49]. First is ALM Modeler, which allows one to define the model of the problem in ALMM methodology. Second is Algorithm Module, which provides a collection of already implemented methods and algorithms (Algorithm Repository) and allows one to design new method (Algorithm Designer). The main module is SimOpt, which solves given problems on the basis of the ALMM methodology and consists of five components (State Chooser, Decision Generator, Decision Chooser, State Generator, Quality Criteria) and Coordinator, which coordinates the whole procedure of building a process trajectory (a sequence of states or sequence of decisions). The last module is the Results Interpreter, in which obtained solution (decision sequence and final value of global quality criterion) is identified.

Solution Generation. Solution Generation within the ALMM Solver is as follows. A selection of an appropriate model of the problem, methods and algorithms along with its parameters and uploading the instance data should be done initially. These settings are stored in the Controller component. Controller decides what actions State Chooser, Decision Generator, Decision Chooser, State Generator and Quality Criteria take. The solution finding process is done by executing in a cycle appropriate components of the SimOpt module. It is managed by Coordinator. In each cycle the next node of the state tree is generated. Firstly, according to chosen search strategy, the state for which the next state will be generated is selected (by StateChooser). Secondly, a whole or a part of a set of possible decisions for this state is generated by DecisionGenerator from this set. Next, in accordance to a particular procedure specified in Controller which decides about using in computations algorithm from the repository, one decision is selected by DecisionChooser. The decision (selected by DecisionChooser) and the current state (elected by StateChooser) are sent to StateGenerator to calculate a new state and to the QualityCriteria component to calculate the criterion value for this state. The state is verified whether it belongs to the set of non-admissible or goal states. All the information connected with new state generation is stored. Finally, Analyzer performs the analysis of the new state and, due to information gained during generation of the trajectory, can change the parameters stored in Controller to new cycle of state generation.

4 Functional and Non-functional Requirements for Data Source

The data structure within the solver is based on its functionality, as the solver should, among others, allow us to search for solutions to problems which concern prioritizing tasks of various classes. Moreover, the solver cannot be limited in terms of the number of machines used or the number of tasks. Consequently, data on status coordinates and decision coordinates depend on the given instance of the problem being solved. Another important aspect is the ability to use various methods and algorithms for the same problem instance. Therefore, it is necessary to store data on the construction of the solution graph (status graph), which is a method of determining the next status (graph node) by selecting a decision from a given status (for example, in the form of a local optimization criterion) and status selection strategy, on the basis of which the final section of the solution is constructed (process trajectory). The methods and algorithms used can be saved in the database and later read during the simulation experiment. The solver can also read and save predefined algorithms in the database with set parameter values from files of the right structure, compatible with the solver's database. Remembering the simulation results (intermediate and final), including predefined simulation experiment parameters, is another advantage. Such data make it possible not only to extract the problem solution found for the given input data, including the quality criterion and the sequence of decisions made, but also to visualize the result of the simulation as a status graph or a graphic representation of the course of the simulation (for example, visualization of the sequence of tasks performed by individual machines in the production process, or movement of vehicles in transport-related problems). Moreover, saving intermediate and final results makes it possible to later analyze the solution or historic solutions obtained to be able to select better simulation parameters with the use of, for example, learning algorithms, and to develop other heuristic algorithms.

Data generated by the solver may be needed for integration with other solvers. This means that they should be saved permanently, not only once the simulation is complete, but also during the simulation. It is possible to integrate the solver with external systems using the results generated by it in real time, for example, when the solver is used as an APS application integrated with ERP and MES systems [40]. The functional aspects of data storage are discussed above. We should also consider non-functional requirements for data source [13]:

– Reliability of the solver understood as fault-free operation over a specified period of time. To achieve this, data must be stored in a manner which reduces problems related to memory limitations.
– Data backup the ability to restore data in case of losing the main copy.

5   Data Storage in ALMM Solver

As the solver is a system which processes very large amounts of data in real time, it is necessary to provide a reliable repository. The solver makes it possible to perform simulations lasting up to several weeks (if using exact methods for large problems), which creates a need to store very large amounts of data. The solver's architecture was designed to allow storage of data in various repositories. Several data storage means were taken into consideration, such as RAM memory, relational database, file system, object database, NoSQL database. This paper focuses on two of those means, i.e. RAM memory and relational database. Use of the remaining data storage tools and their testing will be the subject of future research.

RAM Memory. Data storage in RAM is very easy to implement and does not require the solver to be integrated with external applications. It is a very efficient solution, because it is not necessary to download data from external systems as they are always available in RAM memory. This solution has significant disadvantages in case of long simulations. One of the issues is insufficient memory and unexpected interruptions at any random moment of the simulation. If the simulation is interrupted, all results generated will be lost and the simulation will need to be restarted. Despite the significant drawbacks, this solution can be used for simple simulations lasting no more than several minutes.

Relational Database. Relational databases allow for permanent storage of large amounts of data. To use them, the solver needs to be integrated with external software. This means that its implementation is slightly more complicated than in case of storing data in RAM. The use of databases facilitates fulfillment of most functional and non-functional requirements, and eliminates the shortcomings of storing data in RAM. The analysis performed shows that functional and non-functional requirements are satisfied to a higher extent by relational databases [34]. Easy implementation of RAM data storage solutions makes them very useful for short and small simulations.

6   Database Structure

In order to establish the database structure, the domain data model should be analyzed [44]. In case of ALMM solver one of the most difficult aspects proved to be the method of saving the states of the simulation in the database. It turned out that the state may comprise not only of numeric values, but also of set sequences. It was decided that state data should be stored in five tables: State, Machine_State, Machine_State_Item and Task_State, Task_State_Item. The two tables: Task_State, Task_State_Item are not provided for by the ALMM methodology. They allow the state to be represented in relational databases as a set or set of sequence. On the basis of the solver's data structure analysis, we propose the database structure presented in Fig. 1. The proposed structure is divided into four database areas:

– Problem instance area stores all data for the individual problem instance. The instance is defined by the number of the machines and tasks with their characteristics/properties, parameters, limitations and any possible relations. The data are stored in several tables. The list of tasks is stored in the Task table, whereas the characteristics are stored in the Task_Parameter table. The list of machines is stored in the Machine table. The Machine_Paremeter table contains the attributes of an individual machine. The Task_Parameter and Machine_Parameter tables can store any attributes. All problems have different attributes in terms of the tasks and the machine. The only limitation is that due to the structure of the database, it is recommended to store scalar values of the attributes. Storage of objects or lists of objects is rare, although practically possible, but not recommended.

– Experiment data area - saves data generated during the experiment, among others, the generalized status with quality criterion, and the set of possible decisions in the given state. The main table is the Experiment_Session. The records in this table represent each simulation launch. The actual status of the system is saved in the tables: State, Task_State, Task_State_Item, Machine_State and Machine_State_Item. The Task_State_Item and Machine_State_Item tables, due to their structure - similar to the respective tables in the instance area - can store any state of the machine and task respectively. This means that the data structure developed can store data on any models of algebraic and logistic problems. Moreover, each state has a reference to tables which store the set of possible decisions, i.e. the Decision and Decision_Item tables. One of the state characteristics is the decision on the basis of which the state was created.

– Experiment parameters area - stores input parameters of the simulation. These are parameters set by the user before starting the experiment. Moreover, the area stores simulation parameters necessary to carry out the experiment. Storage of these parameters makes it possible to analyze them and to use learning algorithms in order to select better parameters in the future and even to modify the parameters during the simulation. The parameters presented are stored in two tables: Session_Parameters and State_Parameters respectively. The tables created make it possible to store any parameters with scalar values.

- Algorithm repository area - stores all algorithms, methods and their input/default parameters found in the solver. The algorithm repository database will be researched separately in the future and is not described in detail in this paper.

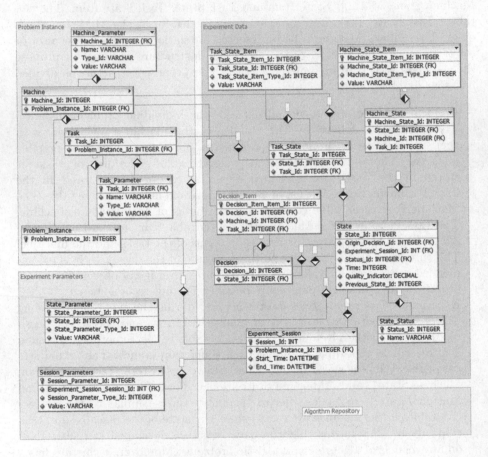

Fig. 1. ALMM solver database structure

7   Solver Communication Layer Architecture with Database

The solver's architecture (presented in Fig. 2) was designed in such a way as to keep the module which performs the simulation as independent as possible from all auxiliary modules/layers, such as communication between the solver and the database, simulation launch application, algorithm repository, integration bus and result interpreter [28]. Design patterns used reduce the direct relation between the simulation module and the database data entry layer [35, 36, 38, 45]. The relation is considered loosely coupled. It was decided that the simulation launch application should save all data in the database and launch the simulation. Whereas the SimOpt module, on the basis of data found in the database, should perform the simulation of the modeled problem. The Integration Bus and Result Interpreter should only download data from the database and there should be no connection with the SimOpt module [29, 30]. The main simulation module was designed in such a way as to ensure that the data access layer is not an integral part of this module. The data entry layer was implemented with the use of Aspect Oriented Programming (AOP) [33, 37]. This made it possible to ensure the correct division of the functions and to break down the SimOpt module into components related to the domain layer and components related to the data layers [48]. The key data storage components can be easily replaced. This makes it possible to change the data source and to use other tools and solutions for this purpose with ease [43].

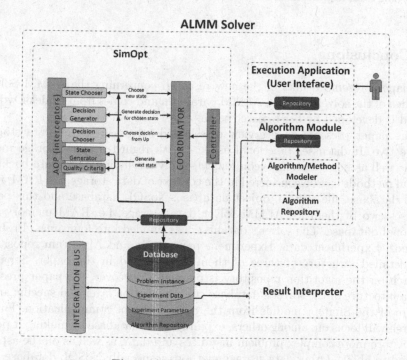

Fig. 2. ALMM solver architecture

After analyzing the operation of the SimOpt module, so-called AOP join points were identified (places where methods are triggered which return data that must be saved in the database). It was decided that they should be triggered by all the key components of the SimOpt module, i.e. State Chooser, Decision Generator, Decision Chooser, State Generator, Quality Criteria. Aspect were implemented with the use of the Castle Windsor framework, which makes it possible to create AOP implementing interceptors [11]. The interceptors implement logics, operating on the data layer. Data access was designed with the use of the repository design pattern and implemented with the use of ORM Nhibernate [16]. ORM facilitates object and relational mapping of the data model to produce a model understood by the database engine. The use of the framework significantly improves the architecture and makes system maintenance easier. It was considered whether to implement database communications through direct communication and SQL language implementation, which has a great advantage in the form of much better data access efficiency. However, implementation of the solver in the form of loosely coupled components facilitated use of multithreading. This made it possible to significantly improve the efficiency of communication with the database with the use of ORM. A downside of using the SQL language is the difficulty with maintaining the application developed (future modifications and development) and low transparency of the data access layer. This is an important aspect to consider in case of a dynamically developing solver.

8   Conclusions

The paper presents results of the new research concerning the ALMM Solver. The idea of the solver based on ALMM paradigm and comes from Dudek-Dyduch E. and is developed by her research team.

The paper presents an overview of selected solvers from the data storage perspective. It discusses functional aspects with regard to data storage methods, as well as non-functional requirements (solver reliability and data backup). Several methods can be considered in the context of data storage, i.e. RAM, relational database, file system, object database, NoSQL database and this paper analyses two of them for ALMM solver data storage, i.e. RAM memory and relational database. The solver's database was divided into four areas: Problem instance, Experiment data, Experiment parameters and Algorithm repository. The detailed structure of three of them is presented in the paper. Separate research on the algorithm repository is planned. Moreover, the paper presents the architecture of the solver's database communication layer and specific architecture of the SimOpt module from the perspective of communication. Future research will focus on, among others, expanding the database by making it possible to store more complex problem instances and data, as well as on the aspects referred to above. Other data storage methods, especially NoSQL database, will be researched further. Moreover, a comparison of relational database and NoSQL database performance will be analyzed. The application of parallel computing to the ALMM solver is also planned in future work.

References

1. www.solver.com
2. www.lindo.com
3. https://www.uts.com/
4. http://scip.zib.de/
5. http://www.jacop.eu/
6. www.math.ovgu.de/Lisa.html
7. https://www-947.ibm.com/
8. www.goldsim.com
9. www.superdecisions.com/
10. http://aimms.com/
11. Windsor castle. http://www.castleproject.org/
12. Achterberg, T., Berthold, T., Koch, T., Wolter, K.: Constraint integer programming: a new approach to integrate CP and MIP. In: Perron, L., Trick, M.A. (eds.) CPAIOR 2008. LNCS, vol. 5015, pp. 6–20. Springer, Heidelberg (2008). doi:10.1007/978-3-540-68155-7_4
13. Ameller, D., Franch, X.: How do software architects consider non-functional requirements: a survey. In: Wieringa, R., Persson, A. (eds.) REFSQ 2010. LNCS, vol. 6182, pp. 276–277. Springer, Heidelberg (2010). doi:10.1007/978-3-642-14192-8_25
14. Barbucha, D., Czarnowski, I., Jędrzejowicz, P., Ratajczak-Ropel, E., Wierzbowska, I.: JABAT middleware as a tool for solving optimization problems. In: Nguyen, N.T., Kowalczyk, R. (eds.) Transactions on CCI II. LNCS, vol. 6450, pp. 181–195. Springer, Heidelberg (2010). doi:10.1007/978-3-642-17155-0_10
15. Brasel, H., Herms, A., Mrig, M., Tautenhahn, T., Tusch, J., Werner, F.: Heuristic constructive algorithms for open shop scheduling to minimize mean flow time. Eur. J. Oper. Res. **189**, 856–870 (2008)
16. Chatekar, S.: Learning NHibernate 4. Packt Publishing, Birmingham (2015)
17. Dudek-Dyduch, E.: Simulation of some class of discrete manufacturing processes. In: Proceedings of European Congress on Simulation, Praha (1987)
18. Dudek-Dyduch, E.: Formalization and Analysis of Problems of Discrete Manufacturing Processes, vol. 54. Scientific Bulletin of AGH University, Automatyka (1990). (in Polish)
19. Dudek-Dyduch, E.: Learning based algorithm in scheduling. J. Intell. Manuf. **11**(2), 135–143 (2000)
20. Dudek-Dyduch, E.: Algebraic logical meta-model of decision processes - new metaheuristics. In: Rutkowski, L., Korytkowski, M., Scherer, R., Tadeusiewicz, R., Zadeh, L.A., Zurada, J.M. (eds.) ICAISC 2015. LNCS (LNAI), vol. 9119, pp. 541–554. Springer, Cham (2015). doi:10.1007/978-3-319-19324-3_48
21. Dudek-Dyduch, E.: Modeling manufacturing processes with disturbances - twostage AL model transformation method. In: The 20th International Conference on Methods and Models in Automation and Robotics, MMAR Proceedings, pp. 782–787 (2015)
22. Dudek-Dyduch, E., Dutkiewicz, L.: Substitution tasks method for discrete optimization. In: Rutkowski, L., Korytkowski, M., Scherer, R., Tadeusiewicz, R., Zadeh, L.A., Zurada, J.M. (eds.) ICAISC 2013. LNCS (LNAI), vol. 7895, pp. 419–430. Springer, Heidelberg (2013). doi:10.1007/978-3-642-38610-7_39

23. Dudek-Dyduch, E., Dyduch, T.: Learning algorithms for scheduling using knowledge based model. In: Rutkowski, L., Tadeusiewicz, R., Zadeh, L.A., Żurada, J.M. (eds.) ICAISC 2006. LNCS (LNAI), vol. 4029, pp. 1091–1100. Springer, Heidelberg (2006). doi:10.1007/11785231_114

24. Dudek-Dyduch, E., Korzonek, S.: ALMM solver for combinatorial and discrete optimization problems – idea of problem model library. In: Nguyen, N.T., Trawiński, B., Fujita, H., Hong, T.-P. (eds.) ACIIDS 2016. LNCS (LNAI), vol. 9621, pp. 459–469. Springer, Heidelberg (2016). doi:10.1007/978-3-662-49381-6_44

25. Dudek-Dyduch, E., Kucharska, E.: Learning method for co-operation. In: Jędrzejowicz, P., Nguyen, N.T., Hoang, K. (eds.) ICCCI 2011. LNCS (LNAI), vol. 6923, pp. 290–300. Springer, Heidelberg (2011). doi:10.1007/978-3-642-23938-0_30

26. Dudek-Dyduch, E., Kucharska, E., Dutkiewicz, L., Rączka, K.: ALMM solver - a tool for optimization problems. In: Rutkowski, L., Korytkowski, M., Scherer, R., Tadeusiewicz, R., Zadeh, L.A., Zurada, J.M. (eds.) ICAISC 2014. LNCS (LNAI), vol. 8468, pp. 328–338. Springer, Cham (2014). doi:10.1007/978-3-319-07176-3_29

27. Dutkiewicz, L., Kucharska, E., Rączka, K., Grobler-Dębska, K.: ST method-based algorithm for the supply routes for multilocation companies problem. In: Skulimowski, A.M.J., Kacprzyk, J. (eds.) Knowledge, Information and Creativity Support Systems: Recent Trends, Advances and Solutions. AISC, vol. 364, pp. 123–135. Springer, Cham (2016). doi:10.1007/978-3-319-19090-7_10

28. Evans, E.: Domain-Driven Design: Tackling Complexity in the Heart of Software, chaps. 2, 4, 5. Prentice Hall, Upper Saddle River (2003)

29. Fowler, M.: Patterns of Enterprise Application Architecture, chaps. 2, 3, 9. Addison-Wesley, Boston (2002)

30. Gamma, E., Helm, R., Johnson, R., Vlissides, J.: Design Patterns: Elements of Reusable Object-Oriented Software. Part Design Pattern Catalog. Addison-Wesley, Reading (1994)

31. Grobler-Dębska, K., Kucharska, E., Dudek-Dyduch, E.: The idea of switching algebraic-logical models in flow-shop scheduling problem with defects. In: 18th International Conference on Methods and Models in Automation and Robotics, MMAR Proceedings, pp. 532–537 (2013)

32. Grobler-Dębska, K., Kucharska, E., Jagodziński, M.: ALMM-based switching method for FSS problem with defects. In: 19th International Conference on Methods and Models in Automation and Robotics, MMAR Proceedings (2014)

33. Groves, M.: AOP in .NET: Practical Aspect-Oriented Programming. Manning Publications, New York (2013)

34. Harrison, N., Avgeriou, P.: Pattern-driven architectural partitioning: balancing functional and non-functional requirements. In: Second International Conference Digital Telecommunications, ICDT 2007 (2007)

35. Jurgens, D.: Survey on software engineering for scientific applications - reuseable software, grid computing and application. Informatikbericht 2009-02, Institut fur Wissenschaftliches Rechnen, Technische Universitat (2009)

36. Kaur, A., Mann, K.S.: Component based software engineering. Int. J. Comput. Appl. **2**(1), 105–108 (2010)

37. Kiczales, G., Lamping, J., Mendhekar, A., Maeda, C., Lopes, C., Loingtier, J.-M., Irwin, J.: Aspect-oriented programming. In: Akşit, M., Matsuoka, S. (eds.) ECOOP 1997. LNCS, vol. 1241, pp. 220–242. Springer, Heidelberg (1997). doi:10. 1007/BFb0053381

38. Krosche, M.: A generic component-based software architecture for the simulation of probabilistic models. Dissertation (2013)

39. Kucharska, E., Dudek-Dyduch, E.: Extended learning method for designation of co-operation. In: Nguyen, N.T. (ed.) TCCI XIV 2014. LNCS, vol. 8615, pp. 136–157. Springer, Heidelberg (2014). doi:10.1007/978-3-662-44509-9_7

40. Kucharska, E., Grobler-Dębska, K., Gracel, J., Jagodziński, M.: Idea of impact of ERP-APS-MES systems integration on the effectiveness of decision making process in manufacturing companies. In: Kozielski, S., Mrozek, D., Kasprowski, P., Małysiak-Mrozek, B., Kostrzewa, D. (eds.) BDAS 2015. CCIS, vol. 521, pp. 551–564. Springer, Cham (2015). doi:10.1007/978-3-319-18422-7_49

41. Kucharska, E., Grobler-Dębska, K., Rczka, K.: ALMM-based approach for scheduling manufacturing problem with defects removal. In: Advances in Mechanical Engineering (2017, to appear)

42. Kucharska, E., Grobler-Dębska, K., Rączka, K.: ALMM-based methods for optimization makespan flow-shop problem with defects. In: Borzemski, L., Grzech, A., Świątek, J., Wilimowska, Z. (eds.) Information Systems Architecture and Technology: Proceedings of 37th International Conference on Information Systems Architecture and Technology – ISAT 2016 – Part I. AISC, vol. 521, pp. 41–53. Springer, Cham (2017). doi:10.1007/978-3-319-46583-8_4

43. Lau, K., Wang, Z.: Software component models. IEEE Trans. Softw. Eng. **33**(10), 709–724 (2007)

44. Nurzyńska, K., Iwaszenko, S., Choroba, T.: Database application in visualization of process data. In: Kozielski, S., Mrozek, D., Kasprowski, P., Małysiak-Mrozek, B., Kostrzewa, D. (eds.) BDAS 2014. CCIS, vol. 424, pp. 537–546. Springer, Cham (2014). doi:10.1007/978-3-319-06932-6_52

45. Ravichandran, T., Rothenberger, M.: Software reuse strategies and component markets. Commun. ACM **46**(8), 109–114 (2003)

46. Rączka, K., Dudek-Dyduch, E., Kucharska, E., Dutkiewicz, L.: ALMM solver: the idea and the architecture. In: Rutkowski, L., Korytkowski, M., Scherer, R., Tadeusiewicz, R., Zadeh, L.A., Zurada, J.M. (eds.) ICAISC 2015. LNCS (LNAI), vol. 9120, pp. 504–514. Springer, Cham (2015). doi:10.1007/978-3-319-19369-4_45

47. Sękowski, H., Dudek-Dyduch, E.: Knowledge based model for scheduling in failure modes. In: Rutkowski, L., Korytkowski, M., Scherer, R., Tadeusiewicz, R., Zadeh, L.A., Zurada, J.M. (eds.) ICAISC 2012. LNCS (LNAI), vol. 7268, pp. 591–599. Springer, Heidelberg (2012). doi:10.1007/978-3-642-29350-4_70

48. Vernon, V.: Implementing Domain-Driven Design. Addison-Wesley Professional, Upper Saddle River (2013)

49. Wirfs-Brock, R., McKean, A.: Object Design: Roles, Responsibilities, and Collaborations. Addison Wesley, Upper Saddle River (2002)

Human Visual System Inspired Color Space Transform in Lossy JPEG 2000 and JPEG XR Compression

Roman Starosolski[✉]

Institute of Informatics, Silesian University of Technology,
Akademicka 16, 44-100 Gliwice, Poland
roman.starosolski@polsl.pl

Abstract. In this paper, we present a very simple color space transform HVSCT inspired by an actual analog transform performed by the human visual system. We evaluate the applicability of the transform to lossy image compression by comparing it, in the cases of JPEG 2000 and JPEG-XR coding, to the ICT/YCbCr and YCoCg transforms for 3 sets of test images. The presented transform is competitive, especially for high-quality or near-lossless compression. In general, while the HVSCT transform results in PSNR close to YCoCg and better than the most commonly used YCbCr transform, at the highest bitrates it is in many cases the best among the tested transforms. The HVSCT applicability reaches beyond the compressed image storage; as its components are closer to the components transmitted to the human brain via the optic nerve than the components of traditional transforms, it may be effective for algorithms aimed at mimicking the effects of processing done by the human visual system, e.g., for image recognition, retrieval, or image analysis for data mining.

Keywords: Image processing · Color space transform · Human visual system · Bio-inspired computations · Lossy image compression · ICT · YCbCr · YCoCg · LDgEb · Image compression standards · JPEG 2000 · JPEG XR

1 Introduction

For natural images, the correlation of the RGB color space primary color components red (R), green (G), and blue (B) is high [18]. Correlation results from the typical characteristic of RGB images and reflects that the same information is contained in two or all three components. For example, an image area which is bright in one component usually is also bright in others. Computer generated images also share such characteristic, since artificial images mostly are made to resemble natural ones. Recent image compression standards: JPEG 2000 [9,28] (as well as the DICOM incorporating JPEG 2000 [15]) and JPEG XR [4,10] compress independently the components obtained from an RGB image by using a

© Springer International Publishing AG 2017
S. Kozielski et al. (Eds.): BDAS 2017, CCIS 716, pp. 564–575, 2017.
DOI: 10.1007/978-3-319-58274-0_44

transform to a less correlated color space. Although the independent compression of transformed components is not the only method for color image compression, it is the most frequently used one. It allows to construct a color image compression algorithm based on a simpler grayscale image compression algorithm. As compared to compressing the untransformed components, by applying the color space transform we improve the image reconstruction quality or the lossless compression ratio (for lossy and lossless algorithms, respectively), since without the transform the same information would be independently encoded more than one time. However, alternative approaches are known that take advantage of inter-component correlations while encoding of untransformed or transformed components [1,6,7].

In this paper we present a human visual system inspired color space transform (HVSCT) for lossy image compression. We evaluate this transform by comparing it for 3 sets of test images and 2 image compression standards (JPEG 2000 and JPEG-XR) to transforms ICT/YCbCr and YCoCg.

The reminder of this paper is organized as follows. In Sect. 2 we discuss properties of irreversible color space transforms and present the ICT/YCbCr and YCoCg transforms used then for comparison with HVSCT. Section 3 introduces the new transform. Section 4 contains experimental procedure, results, and discussion; Sect. 5 summarizes the research.

2   Color Space Transforms

The Karhunen-Loève transform (KLT) is an image-dependent transform that for a specific image is constructed by using the Principal Component Analysis (PCA), it optimally decorrelates the image [16,18]. The computational time complexity of PCA/KLT is in practice too high to compute it each time an image gets compressed. Instead, fixed transforms are constructed based on PCA/KLT by performing PCA on a set of typical images. Then, assuming that the set is sufficiently representative also for other images, which were not included in the set, we use the obtained fixed KLT transform variant for all images. The frequently used color space transforms, for example, the YCbCr color space transform described below, are fixed transforms constructed based on PCA/KLT; however, there are algorithms constructing a color space transform for the specific image. An adaptive selection of the transform from a large family of 60 simple transforms was proposed by Strutz [26]; performing the selection slightly increases the overall cost of the lossless color image compression algorithm. In [27], an even larger family of 108 simple transforms is presented; adaptive transform selection is performed for the entire image or for separate image regions, however, the latter approach leads to only a small further ratio improvement. Singh and Kumar [19] presented an image adaptive method of constructing a color space transform based on the Singular Value Decomposition. Although this method is of significantly greater computational time complexity, than a method which directly selects a transform from a family of simple transforms, it is still simpler than computing PCA/KLT for a given image.

The probably most commonly used color space, but the RGB space, is YCbCr. It was constructed using PCA/KLT, but with an additional requirement: the transform should contain a component that approximates the luminance perception of the human visual system [14]. YCbCr contains the Y component that represents the luminance and two chrominance components: Cb and Cr. YCbCr was constructed decades ago for video data and nowadays is used both for video and for still image compression. There are many variants of the transform between RGB and YCbCr (resulting in respective variants of the YCbCr color space). Below we present one of them, ICT (Eq. 1), with inverse (Eq. 2):

$$\begin{bmatrix} Y \\ Cb \\ Cr \end{bmatrix} = \begin{bmatrix} 0.29900 & 0.58700 & 0.11400 \\ -0.16875 & -0.33126 & 0.50000 \\ 0.50000 & -0.41869 & -0.08131 \end{bmatrix} \begin{bmatrix} R \\ G \\ B \end{bmatrix}, \tag{1}$$

$$\begin{bmatrix} R \\ G \\ B \end{bmatrix} = \begin{bmatrix} 1.00000 & 0.00000 & 1.40200 \\ 1.00000 & -0.34413 & -0.71414 \\ 1.00000 & 1.77200 & 0.00000 \end{bmatrix} \begin{bmatrix} Y \\ Cb \\ Cr \end{bmatrix}. \tag{2}$$

ICT is defined in the JPEG 2000 standard for lossy compression [10].

Note, that if the transformed components are to be stored using integer numbers, then the transform is not exactly reversible—we say that it is irreversible or not integer-reversible. It is not a problem in a typical case of lossy coding, where distortions introduced by forward and inverse transform are much smaller than distortions caused by lossy compression and decompression. However, in the case of the very high quality coding, the color space transform may limit the obtainable reconstruction quality. The integer-reversible variants of ICT and of other transforms are constructed using the lifting scheme [2]. The reversibility is obtained at the cost of the dynamic range expansion of the transformed chrominance components by 1 bit (the dynamic range of a component is defined as a number of bits required to store pixel intensities of this component). The dynamic range expansion affects the transform applicability, since certain algorithms and implementations either do not allow or do not process efficiently images of depths greater than, e.g., 8 bits per component. Expansion may be avoided by the use of modular arithmetic (as in the RCT transform in the JPEG-LS extended standard [8]), however, such transform introduces sharp edges to transformed components, that worsen the lossy compression effects. In this research, we focus on typical transforms for lossy coding—not using the modular arithmetic and not expanding the dynamic range of transformed components.

A recent YCoCg transform is an another interesting transform (forward in Eq. 3 and inverse in Eq. 4):

$$\begin{bmatrix} Y \\ Co \\ Cg \end{bmatrix} = \begin{bmatrix} 1/4 & 1/2 & 1/4 \\ 1/2 & 0 & -1/2 \\ -1/4 & 1/2 & -1/4 \end{bmatrix} \begin{bmatrix} R \\ G \\ B \end{bmatrix}, \tag{3}$$

$$\begin{bmatrix} R \\ G \\ B \end{bmatrix} = \begin{bmatrix} 1 & 1 & -1 \\ 1 & 0 & 1 \\ 1 & -1 & -1 \end{bmatrix} \begin{bmatrix} Y \\ Co \\ Cg \end{bmatrix}. \tag{4}$$

It was obtained based on PCA/KLT constructed for a Kodak image-set (see Sect. 4.1 for the Kodak set description); YCoCg is an irreversible variant of a YCoCg-R transform included in the JPEG-XR standard [10,14]. The YCoCg transform is significantly simpler to compute, than ICT. The former requires 15 simple floating point operations (additions, subtractions, multiplications) for forward and 8 operations for inverse transform. The YCoCg forward transform may be computed in 6 integer operations (add, subtract, and bit-shift; the latter denoted by $\gg$):

$$t = (R + B) \gg 1;\ Y = (G + t) \gg 1;\ Co = R - t;\ Cg = Y - t;$$

inverse in 4 additions and subtractions only:

$$G = Y + Cg;\ t = Y - Cg;\ R = t + Co;\ B = t - Co;$$

3 New Transform Inspired by Human Visual System

We described in detail previously [21] the following interesting fact. A color space transform that results in a single luminance and 2 chrominance components is performed by our (i.e., human) visual system. There are three types of cone cells in our retinas that are most sensitive to three light wavelengths, these are S-cones (short wavelength with sensitivity peak in violet), M-cones (middle wavelength, sensitivity peak in green), and L-cones (long wavelength, peak in yellow). According to the common opinion, the cones simply respond to blue (S-cones), green (M-cones), and red (L-cones) light. However, this opinion is wrong not only because the cone sensitivity peaks are outside of red and blue wavelengths, but also since S-cones are sensitive to colors ranging from violet to green, whereas M-cones and L-cones are sensitive to the full visible spectrum. However, indeed the highest reaction to blue color, among all cone types, is shown by S-cones, to green by M-cones, and to red by L-cones. The response of cones is then transformed and to the brain, via the optic nerve, the below three components are transmitted:

- the luminance computed as a sum of responses of L-cones and M-cones,
- the red minus green color component (a difference between responses of L-cones and M-cones),
- and the blue minus yellow color component (a difference between response of S-cones and a sum of L-cones and M-cones responses, which may be also seen as a difference between the response of S-cones and the computed luminance).

For brevity, above mentioned were only certain aspects of human visual system reduced to essentials; for further details the Reader is referred to [5] and references therein.

In [21] we have investigated an integer-reversible color space transform for lossless image compression modeled on a transform performed by the human visual system (LDgEb); it resulted in very good average lossless compression ratios for several image sets and compression algorithms. The same transform has been presented by Strutz as a special case (denoted $A_{4,10}$) of a large family of simple integer-reversible transforms [27]. Here, evaluation was based on counting how many times given transform resulted in the best lossless ratio for an image from the test-set employed. Interestingly, while for a diverse set of photographic images it was the second one most often resulting in best ratio, for a more homogeneous set it has never been the best one. LDgEb is simple to compute, but as most of the transforms for lossless coding, it causes dynamic range expansion of chrominance components. In this research, based on LDgEb, we propose an irreversible color space transform for lossy image compression, named HVSCT. The new transform originates from the human visual system and is normalized to obtain the same dynamic range after the transform, as before it. In the HVSCT transform (Eq. 5, inverse presented in Eq. 6):

$$\begin{bmatrix} Y \\ Cd \\ Ce \end{bmatrix} = \begin{bmatrix} 1/2 & 1/2 & 0 \\ 1/2 & -1/2 & 0 \\ -1/4 & -1/4 & 1/2 \end{bmatrix} \begin{bmatrix} R \\ G \\ B \end{bmatrix}, \tag{5}$$

$$\begin{bmatrix} R \\ G \\ B \end{bmatrix} = \begin{bmatrix} 1 & 1 & 0 \\ 1 & -1 & 0 \\ 1 & 0 & 2 \end{bmatrix} \begin{bmatrix} Y \\ Cd \\ Ce \end{bmatrix} \tag{6}$$

the luminance is, as in humans, a sum of two longer wavelength components, but multiplied by 0.5: $Y = (R + G)/2$, chrominance components are normalized differences: between two longer wavelength components $Cd = (R - G)/2$ and between the shortest wavelength component and the luminance $Ce = (B - Y)/2$.

The HVSCT forward transform is even simpler to compute than YCoCg, since it may be computed in 5 integer operations (add, subtract, bit-shift):

$$Cd = (R - G) \gg 1;\ Y = R - Cd;\ Ce = (B - Y) \gg 1;$$

inverse HVSCT, as inverse YCoCg, requires 4 additions and subtractions:

$$R = Y + Cd;\ G = Y - Cd;\ B = Y + Ce + Ce;$$

4 Experimental Evaluation

4.1 Procedure

In the evaluation we included the ICT, YCoCg, and HVSCT transforms; we also report results for untransformed RGB images. Transforms were applied with the colortransf tool, version 1.1[1]; ICT was performed using real (double)

---

[1] http://sun.aei.polsl.pl/~rstaros/imgtransf/colortransf/.

numbers, whereas other transforms were implemented using integers. If a transformed pixel component exceeded its nominal range, then it was corrected to its nearest range limit. After the color space transform was applied to an image, the resulting components were compressed as a single image (all components together, with disabled color space transform of the compression algorithm). We measured resulting compression ratio in bits per pixel [bpp] (bitrate) and, after decompression and inverse color space transform, the PSNR image quality [dB] in the RGB domain.

We evaluated the effects of the color space transforms on image quality of JPEG 2000 and JPEG XR compression. JPEG 2000 is an ISO/IEC and ITU-T standard lossy and lossless image compression algorithm based on discrete wavelet transform (DWT) image decomposition and arithmetic coding [9,28]. JPEG XR is a recent ISO/IEC and ITU-T lossy and lossless standard algorithm designed primarily for high quality, high dynamic range photographic images; it is based on discrete cosine transform image decomposition and adaptive Huffman coding [4,10]. We used the JasPer implementation of JPEG 2000 by Adams, version 1.900[2] and a standard reference implementation of JPEG XR [11]. The former allows setting the desired bitrate, to which the actual one is close (not greater than). The JPEG XR implementation allows setting the quantization value for all components (we used this option) or for each component individually, but the resulting bitrate depends significantly on the image contents. We report average PSNR for bitrates from 0.25 bpp to 6 bpp. Since JPEG 2000 and JPEG XR for the below sets of images obtain average lossless ratios from 9.5 bpp to 13.3 bpp, the 6 bpp may be considered a high quality or nearly-lossless bitrate setting. The compression was performed with several (16) bitrate/quantization settings. Then, in the cases of both algorithms, to obtain average PSNR for a group of images at a desired bitrate, for each image the PSNR was interpolated (polynomial 3-point interpolation) based on results obtained for 3 bitrates nearest to the desired one.

The evaluation of transforms was performed for the following sets of 8-bit RGB test images:

- Waterloo[3] – a set ("Colour set") of color images from the University of Waterloo, Fractal Coding and Analysis Group repository, used for a long time in image processing research. The set contains 8 natural photographic and artificial images, among them the well-known "lena" and "peppers", image sizes vary from 512×512 to 1118×1105.
- Kodak[4] – a set of 23 photographic images released by the Kodak corporation, the set is frequently used in color image compression research. All images are of size 768×512.

---

[2] http://www.ece.uvic.ca/~mdadams/jasper/.

[3] http://links.uwaterloo.ca/Repository.html.

[4] http://www.cipr.rpi.edu/resource/stills/kodak.html.

– EPFL[5] – a recent set of 10 high resolution images used at the École polytechnique fédérale de Lausanne for evaluation of subjective quality of JPEG XR [3]. Image sizes from 1280×1506 to 1280×1600.

4.2   Results

In Tables 1 and 2 for JPEG 2000 and JPEG XR, respectively, we present average PSNR calculated for images from all groups; images were transformed with the ICT, YCoCg and HVSCT transforms and we also report the results of coding of untransformed RGB images. Due to limited space the data is in Tables presented for certain bitrate/quantization settings only, however, results for all the settings are presented in Fig. 1 (in panel A for JPEG 2000 and panel B for JPEG XR). Results are similar for both algorithms. Average PSNR image quality for low bitrates is for all 3 transforms close to each other and better than for untransformed RGB by about 2–3.5 dB (depending on the bitrate). From about 1bpp, the PSNR of YCoCg and HVSCT transformed images is still close to each other and better than for RGB, but the results of the ICT transform get closer to results obtained without a color space transform.

Table 1. JPEG 2000 average PSNR for all images from all groups.

Transform	Bitrate								
	0.25	0.50	1.00	1.50	2.00	3.00	4.00	5.00	6.00
RGB	25.798	28.008	30.847	32.872	34.516	37.151	39.242	41.083	42.655
ICT	27.790	30.559	33.881	36.001	37.562	39.737	41.388	42.550	43.630
YCoCg	27.846	30.667	34.087	36.342	38.063	40.611	42.735	44.512	46.037
HVSCT	27.639	30.402	33.781	36.046	37.784	40.351	42.487	44.223	46.048

Table 2. JPEG XR average PSNR for all images from all groups.

Transform	Bitrate								
	0.25	0.50	1.00	1.50	2.00	3.00	4.00	5.00	6.00
RGB	25.708	28.019	30.953	33.028	34.680	37.299	39.448	41.332	43.088
ICT	28.339	30.752	34.088	36.176	37.726	39.983	41.503	42.561	43.331
YCoCg	28.421	30.908	34.430	36.723	38.511	41.249	43.287	44.816	46.016
HVSCT	28.250	30.785	34.295	36.590	38.360	41.127	43.209	44.805	46.087

---

[5] http://documents.epfl.ch/groups/g/gr/gr-eb-unit/www/IQA/Original.zip.

A) JPEG 2000   **B)** JPEG XR

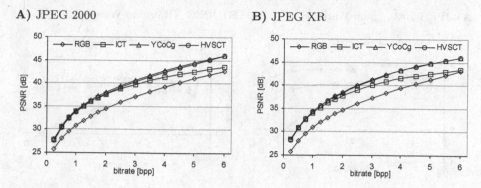

Fig. 1. JPEG 2000 (A) and JPEG XR (B) average PSNR for all images from all groups.

Good performance of the new transform may be surprising considering that, as opposed to others, it has not been designed based on analysis of digital images, but simply adopted from the analog transform of the human visual system. On the other hand, digital images are acquired or constructed for the human visual system, therefore digital image representation as well as algorithms processing images may be indirectly influenced by our visual system. The close relation of HVSCT to the human visual system may be beneficial for other image processing algorithms that exploit color space transforms and are aimed at mimicking the effects of image processing and analysis done by the human visual system, e.g., face or skin detection algorithms [12,13].

In order to compare effects of color space transforms on JPEG 2000 and JPEG XR more thoroughly, in Fig. 2 for each compression algorithm and for each image group we present PSNR relative to PSNR obtained without a color space transform. For a given transform its relative PSNR was calculated by subtracting from its absolute PSNR the PSNR obtained in the case of RGB. Dissimilarities in relative performance of transforms are noticeable rather between image groups, than between compression algorithms. Generally, results of HVSCT are close to results of YCoCg also for individual image groups. While for the lowest bitrates HVSCT is close, but worse than others, for high bitrates in many cases it obtains the best PSNR. Let's look at the results for the Waterloo group of images (Fig. 2, panels A and B). This is the oldest group and only this group contains both natural and computer generated images; the latter differ significantly from natural continuous-tone images, as some of them are dithered, have sparse histograms of intensity levels [17,20], or are composed from images and text. For the Waterloo images, YCoCg is better than HVSCT at all bitrates by from 0.17 to 0.79 dB. For more recent Kodak and EPFL continuous tone natural images (Fig. 2, panels C to F), the difference between HVSCT and YCoCg in average PSNR is at most 0.44 dB (0.20 for JPEG XR). Here, YCoCg is better up to a certain threshold (depending on algorithm and set from 1 bpp to 5.5 bpp), above which HVSCT obtains better PSNR.

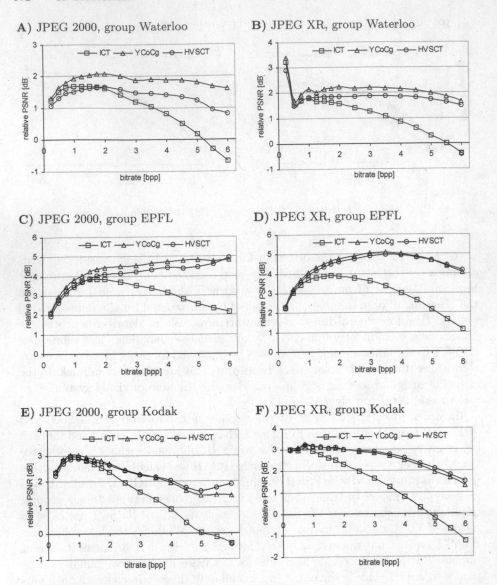

Fig. 2. JPEG 2000 and JPEG XR average PSNR image quality for individual groups of images, relative to quality obtained in the case of untransformed RGB.

We did not perform any systematic subjective assessment of transform effects on reconstructed image quality, however, in the cases of several images, which were viewed on a regular LCD IPS monitor, for both algorithms there was no noticeable quality difference between ICT, YCoCg, and HVSCT, while images compressed in RGB domain at lower bitrates contained more apparent artifacts and less details.

5   Conclusions

We presented the bio-inspired HVSCT transform for lossy image compression, which originates from the analog transform performed by the human visual system. The transform does not expand the dynamic range of transformed components and is of very low computational time complexity (5 simple integer operations per pixel for forward transform, 4 for inverse).

We evaluated effects of employing the new transform in JPEG 2000 and JPEG XR compression for 3 popular sets of test images and compared it with the ICT/YCbCr and YCoCg transforms. Although, as opposed to others, HVSCT was not designed based on analysis of digital color images, with respect to PSNR it obtains results close to the recent YCoCg transform. In most cases, for the greatest part of the tested bitrate range (0.25 to 6 bpp) YCoCg results in little better PSNR than HVSCT, whereas ICT in significantly worse. HVSCT appears the most useful in high quality lossy image compression, or in near-lossless compression, since at the highest bitrates it usually is the best among the tested transforms.

In order to improve the effects of HVSCT, we plan to apply to it the RDLS method. RDLS is a modification of the lifting scheme [2] that was found effective for improving the lossless compression ratios in the cases of several integer-reversible lifting-based color space transforms [23,24] and the discrete wavelet transform [22,25]. HVSCT may be implemented using the lifting scheme; the good results of RDLS obtained for lossless coding indicate that it may be also effective for the near-lossless compression. As compared to color spaces traditionally used in various algorithms in the digital image processing domain, like retrieval or recognition, that are aimed at mimicking the effects of processing happening in the human visual system, the components of HVSCT are closer to the components transmitted to the human brain via the optic nerve. This opens a promising field of future research. Checking whether by using HVSCT instead of traditional color space transforms the results of such algorithms will get closer to results we expect from experience with our own visual system is certainly worthwhile. Other potential fields of further research are subjective evaluation of image quality after the HVSCT followed by the lossy compression as well as application of HVSCT to video data.

Acknowledgment. This work was supported by BK-219/RAU2/2016 Grant from the Institute of Informatics, Silesian University of Technology.

References

1. Chen, X., Canagarajah, N., Nunez-Yanez, J.L.: Lossless multi-mode interband image compression and its hardware architecture. In: Gogniat, G., Milojevic, D., Morawiec, A., Erdogan, A. (eds.) Algorithm-Architecture Matching for Signal and Image Processing. LNEE, vol. 73, pp. 3–26. Springer, Dordrecht (2011). doi:10. 1007/978-90-481-9965-5_1

2. Daubechies, I., Sweldens, W.: Factoring wavelet transforms into lifting steps. J. Fourier Anal. Appl. **4**(3), 247–269 (1998). doi:10.1007/BF02476026

3. De Simone, F., Goldmann, L., Baroncini, V., Ebrahimi, T.: Subjective evaluation of JPEG XR image compression. In: Applications of Digital Image Processing XXXII, Proceedings of SPIE, vol. 7443, p. 66960A (2009). doi:10.1117/12.830714

4. Dufaux, F., Sullivan, G.J., Ebrahimi, T.: The JPEG XR image coding standard. IEEE Sig. Process. Mag. **26**(6), 195–199, 204 (2009). doi:10.1109/MSP.2009.934187

5. Gegenfurtner, K.R., Kiper, D.C.: Color vision. Annu. Rev. Neurosci. **26**, 181–206 (2003). doi:10.1146/annurev.neuro.26.041002.131116

6. Gershikov, E., Porat, M.: Correlation vs. decorrelation of color components in image compression-which is preferred? In: Proceedings of the 15th European Signal Processing Conference (EUSIPCO) 2007, pp. 985–989 (2007)

7. Gershikov, E., Porat, M.: Optimal color image compression using localized color component transforms. In: Proceedings of the 16th European Signal Processing Conference (EUSIPCO) 2008, pp. 133–139 (2008)

8. ISO, IEC, ITU-T: Information technology - lossless and near-lossless compression of continuous-tone still images: extensions. ISO/IEC International Standard 14495-2 and ITU-T Recommendation T.870 (2003)

9. ISO, IEC, ITU-T: Information technology - JPEG 2000 image coding system: core coding system. ISO/IEC International Standard 15444-1 and ITU-T Recommendation T.800 (2004)

10. ISO, IEC, ITU-T: Information technology - JPEG XR image coding system - image coding specification. ISO/IEC International Standard 29199-2 and ITU-T Recommendation T.832 (2012)

11. ISO, IEC, ITU-T: Information technology - JPEG XR image coding system - reference software. ISO/IEC International Standard 29199-5 and ITU-T Recommendation T.835 (2012)

12. Kawulok, M.: Skin detection using color and distance transform. In: Bolc, L., Tadeusiewicz, R., Chmielewski, L.J., Wojciechowski, K. (eds.) ICCVG 2012. LNCS, vol. 7594, pp. 449–456. Springer, Heidelberg (2012). doi:10.1007/978-3-642-33564-8_54

13. Kawulok, M., Kawulok, J., Nalepa, J.: Spatial-based skin detection using discriminative skin-presence features. Pattern Recognit. Lett. **41**, 3–13 (2014). doi:10.1016/j.patrec.2013.08.028

14. Malvar, H.S., Sullivan, G.J., Srinivasan, S.: Lifting-based reversible color transformations for image compression. In: Proceedings of SPIE, Applications of Digital Image Processing XXXI, vol. 7073, p. 707307 (2008). doi:10.1117/12.797091

15. National Electrical Manufacturers Association: Digital Imaging and Communications in Medicine (DICOM) Part 5: Data Structures and Encoding. NEMA Standard PS 3.5-2014a (2014)

16. Pearson, K.: On lines and planes of closest fit to systems of points in space. Philos. Mag. **2**(11), 559–572 (1901). doi:10.1080/14786440109462720

17. Pinho, A.J.: Preprocessing techniques for improving the lossless compression of images with quasi-sparse and locally sparse histograms. In: Proceedings of ICME 2002, vol. 1, pp. 633–636 (2002). doi:10.1109/ICME.2002.1035861

18. Pratt, W.: Digital Image Processing. Wiley, Hoboken (2001). doi:10.1002/0471221325

19. Singh, S., Kumar, S.: Novel adaptive color space transform and application to image compression. Signal Process. Image Commun. **26**(10), 662–672 (2011). doi:10.1016/j.image.2011.08.001

20. Starosolski, R.: Compressing high bit depth images of sparse histograms. In: Simos, T.E., Psihoyios, G. (eds.) International Electronic Conference on Computer Science, AIP Conference Proceedings, vol. 1060, pp. 269–272. American Institute of Physics (2008). doi:10.1063/1.3037069
21. Starosolski, R.: New simple and efficient color space transformations for lossless image compression. J. Vis. Commun. Image Represent. 25(5), 1056–1063 (2014). doi:10.1016/j.jvcir.2014.03.003
22. Starosolski, R.: Application of reversible denoising and lifting steps to DWT in lossless JPEG 2000 for improved bitrates. Sig. Process. Image Commun. 39(A), 249–263 (2015). doi:10.1016/j.image.2015.09.013
23. Starosolski, R.: Application of reversible denoising and lifting steps to LDgEb and RCT color space transforms for improved lossless compression. In: Kozielski, S., Mrozek, D., Kasprowski, P., Małysiak-Mrozek, B., Kostrzewa, D. (eds.) BDAS 2016. CCIS, vol. 613, pp. 623–632. Springer, Cham (2016). doi:10.1007/978-3-319-34099-9_48
24. Starosolski, R.: Application of reversible denoising and lifting steps with step skipping to color space transforms for improved lossless compression. J. Electron. Imaging 25(4), 043025 (2016). doi:10.1117/1.JEI.25.4.043025
25. Starosolski, R.: Skipping selected steps of DWT computation in lossless JPEG 2000 for improved bitrates. PLOS ONE 11(12), e0168704 (2016). doi:10.1371/journal.pone.0168704
26. Strutz, T.: Adaptive selection of colour transformations for reversible image compression. In: Proceedings of the 20th European Signal Processing Conference (EUSIPCO) 2012, pp. 1204–1208 (2012)
27. Strutz, T.: Multiplierless reversible colour transforms and their automatic selection for image data compression. IEEE Trans. Circ. Syst. Video Technol. 23(7), 1249–1259 (2013). doi:10.1109/TCSVT.2013.2242612
28. Taubman, D.S., Marcellin, M.W.: JPEG2000 Image Compression Fundamentals, Standards and Practice. Springer, New York (2004). doi:10.1007/978-1-4615-0799-4

Author Index